SEVENTH EDITION

RESEARCH METHODS IN PHYSICAL ACTIVITY

Jerry R. Thomas, EdD

Jack K. Nelson, EdD

Stephen J. Silverman, EdD

Human Kinetics

Library of Congress Cataloging-in-Publication Data

Thomas, Jerry R., author.
  Research methods in physical activity / Jerry R. Thomas, Jack K. Nelson, Stephen J. Silverman. -- Seventh edition.
     p. ; cm.
  Includes bibliographical references and indexes.
  I. Nelson, Jack K., author. II. Silverman, Stephen J., author. III. Title.
  [DNLM: 1. Motor Activity. 2. Kinesiology, Applied. 3. Physical Education and Training. 4. Research Design. 5. Sports
Medicine--methods. WE 103]
     GV361
     613.7'1072--dc23

                                        2014028466

ISBN: 978-1-4504-7044-5

The web addresses cited in this text were current as of February 2015, unless otherwise noted.

Acquisitions Editor: Amy N. Tocco
Developmental Editor: Melissa J. Zavala
Managing Editor: Carly S. O'Connor
Copyeditor: Patsy Fortney
Indexer: Nancy Ball
Permissions Manager: Dalene Reeder
Senior Graphic Designer: Keri Evans
Cover Designer: Keith Blomberg
Photo Asset Manager: Laura Fitch
Visual Production Assistant: Joyce Brumfield
Photo Production Manager: Jason Allen
Art Manager: Kelly Hendren
Associate Art Manager: Alan L. Wilborn
Illustrations: © Human Kinetics, unless otherwise noted
Printer: Walsworth Print

Printed in the United States of America   10 9 8 7 6 5 4 3 2

The paper in this book was manufactured using responsible forestry methods.

Human Kinetics
Website: www.HumanKinetics.com

United States: Human Kinetics
P.O. Box 5076
Champaign, IL 61825-5076
800-747-4457
e-mail: humank@hkusa.com

Canada: Human Kinetics
475 Devonshire Road Unit 100
Windsor, ON N8Y 2L5
800-465-7301 (in Canada only)
e-mail: info@hkcanada.com

Europe: Human Kinetics
107 Bradford Road
Stanningley
Leeds LS28 6AT, United Kingdom
+44 (0) 113 255 5665
e-mail: hk@hkeurope.com

Australia: Human Kinetics
57A Price Avenue
Lower Mitcham, South Australia 5062
08 8372 0999
e-mail: info@hkaustralia.com

New Zealand: Human Kinetics
P.O. Box 80
Mitcham Shopping Centre, South Australia 5062
0800 222 062
e-mail: info@hknewzealand.com

E6201

CONTENTS

With the publication of this seventh edition, 30 years have passed since the first edition was published in 1985. We take this opportunity to thank all the people who have used this book over the years. We hope that you have learned a lot about research methods in the study of physical activity. Maybe you have even enjoyed the humorous stories, jokes, and pictures that we have included to enliven the reading. We were particularly pleased to note the paper by Silverman and Keating (2002) in *Research Quarterly for Exercise and Sport* that indicated that 71% of the people responding to their survey about the research methods course used our book. Their survey results clearly show that the content of this book is well aligned with the topics that teachers of research methods believe are important. We are also delighted that many of you in other English-speaking countries have also used this book—*Good on ya, mates!* In addition, we appreciate that earlier editions have been translated into Chinese (twice), Greek, Korean, Italian, Japanese, Spanish, and Portuguese.

Dr. Jack Nelson has retired but continues to provide assistance and advice. Dr. Stephen Silverman joined us as a coauthor on the fifth edition and, in spite of our sense of humor, agreed to continue on the sixth and seventh editions. Dr. Silverman is a well-known scholar and methodologist in physical education pedagogy and is a former editor in chief of *Research Quarterly for Exercise and Sport.*

The main use of this text still appears to be in the first graduate-level research methods courses, although it is also being used in undergraduate research methods courses and as a resource for those engaged in research planning and analysis. Our use of the term *physical activity* in the book title is meant to convey the broadly conceived field of study often labeled kinesiology, exercise science, exercise and sport science, human movement, sport studies, or physical education, as well as related fields such as physical therapy, rehabilitation, and occupational therapy. We hope that everyone who reads, understands, plans, carries out, writes, or presents research will find the book a useful tool to enhance his or her efforts.

This seventh edition retains the basic organization of the sixth edition, as follows:

■ *Part I* is an overview of the research process, including developing the problem, using the literature, preparing a research plan, and understanding ethical issues in research and writing.

■ *Part II* introduces statistical and measurement issues in research, including statistical descriptions, power, relationships among variables, differences among groups, nonparametric procedures, and measurement issues in research.

■ *Part III* presents the types of, or approaches to, research, including historical, philosophical, research synthesis, survey, descriptive, epidemiological, experimental, qualitative, and mixed methods.

■ *Part IV* will help you complete the research process, which includes writing the results and discussion, organizing the research paper, developing good figures and tables, and presenting research in written and oral forms.

■ The appendix includes statistical tables.

Instructors using this text in their courses will find an instructor guide, test package, presentation package, and image bank at www.HumanKinetics.com/ResearchMethodsInPhysicalActivity.

The image bank includes most of the art, tables, and example elements from the text, which can be used to create custom presentations. The instructor guide includes chapter overviews, sample course syllabi, supplemental class activities, and student handouts. The test bank includes 302 questions.

Although the format of the book remains similar to that of the sixth edition, we have made a number of changes that we hope improve and update the text. Various people have read and reviewed the previous edition and provided helpful comments, including the many students we have taught and faculty who use the book to teach research methods. We truly do pay attention to the things you tell us. But sometimes, when we read reviews, we feel as Day (1983, p. xi) did when he read that a reviewer described his book as both good and original, but then went on to say that "the part that is good is not original and the part that is original is not good." Following is a short review of the changes in this seventh edition:

eBook
available at
your campus bookstore
or HumanKinetics.com

■ *Part I*: *Overview of the Research Process*. Each chapter includes minor revisions that reflect updated information and more recent reports. We have made a significant revision to chapter 2 about using library techniques by adding much more on electronic searches. In addition, chapter 5 on ethical issues has been updated with particular attention to procedures for the use of human and animal subjects.

■ *Part II*: *Statistical and Measurement Concepts in Research*. We strive in each edition to increase the relevance of the examples and provide easy-to-understand calculations for basic statistics. We have reduced the examples of hand calculations and formulas and replaced them with sample output from the Statistical Package for the Social Sciences. We have taken 2008 player performance data for outcome and skill variables from the Professional Golfers Association website as examples for analysis in the statistical chapters. Along with the chapter examples, the learning activities in the instructor guide should help students grasp the fundamentals of statistical techniques. We continue to use a unified approach to parametric and nonparametric techniques.

■ *Part III*: *Types of Research*. We have continued our use of expert authors to present coherent views of sociohistorical research (a new chapter by David Wiggins and Daniel Mason), philosophical research (R. Scott Kretchmar and Tim Elcombe), and epidemiological research in physical activity (Barbara Ainsworth and Charles Matthews). These three types of research are outside our expertise, and we wanted them presented by expert scholars. Another significant revision in part III is the qualitative research chapter. The many changes in the field have prompted us to expand and update our discussion of qualitative research. Because of the increasing simultaneous use of both quantitative and qualitative approaches within a single study, we have expanded chapter 20 on mixed-methods research. In addition, we have made minor revisions and updates to all the other chapters in this part.

■ *Part IV*: *Writing the Research Report*. The two chapters in this section remain essentially the same, with changes and updates. Chapter 21 has a substantial new section that includes these heads: Thesis and Dissertation Proposal, Advisor and Dissertation Committee, The Good Scholar Must Research and Write, Scientific Writing, and First Things Are Sometimes Best Done Last.

As we have said in each edition, we are grateful for the help of our friends, both for help that we acknowledge in various places in the book and for help in other places where we have inadvertently taken an idea without giving credit.

After the passage of time, one can no longer remember who originated what idea. After the passage of even more time, it seems to me that all of the really good ideas originated with me, a proposition which I know is indefensible (Day, 1983, p. xv).

We believe that this book provides the necessary information for both the consumer and the producer of research. Although no amount of knowledge about the tools of research can replace expertise in the content area, good scholars of physical activity cannot function apart from the effective use of research tools. Researchers, teachers, clinicians, technicians, health workers, exercise leaders, sport managers, athletic counselors, and coaches need to understand the research process. If they do not, they are forced to accept information at face value or on the recommendation of others. Neither is necessarily bad, but the ability to evaluate and reach a valid conclusion based on data, method, and logic is the mark of a professional.

Inserted into some chapters are humorous stories, anecdotes, sketches, laws, and corollaries. These are intended to make a point and enliven the reading without distracting from the content. Research processes are not mysterious events that graduate students should fear. To the contrary, they are useful tools that every academic and professional should have access to; they are, in fact, the very basis on which we make competent decisions.

Jerry R. Thomas
Jack K. Nelson
Stephen J. Silverman

Dear Student of Research Methods:

We want you to learn the material here, and most of you are learning it in a classroom setting as well as by reading the textbook. From many years of teaching research methods, we have arrived at the following recommendations:

1. Attend and participate in class—90% of life is showing up!
2. Take notes in class—writing it down is an effective way to learn.
3. Read the assigned materials before class—duh!
4. Plan for and ask at least one question in every class.
5. Develop and work with a study group.
6. Prepare for exams and tests—do not cram; study over several days.
7. Use campus resources to improve learning—library, computer, the Internet.
8. Visit often with your professor—those of us teaching research methods are likable folks!

The following list will help you determine your readiness to be a student of research methods.

Score ONE point for each of the following statements that describes you:

Your library carrel is better decorated than your apartment is.

You have taken a scholarly article to a bar or coffee shop.

You rate coffee shops on the availability of outlets for your electronic devices.

You have discussed academic matters at a sporting event.

You actually have a preference between microfilm and microfiche.

You always read the reference lists in research articles.

You think that the sorority sweatshirt Greek letters are a statistical formula.

You need to explain to children why you are in the 20th grade.

You refer to stories as "Snow White et al."

You wonder how to cite talking to yourself in APA style.

Scoring scale

5 or 6—definitely ready to be a student in research methods

7 or 8—probably a master's student

9 or 10—probably a doctoral student

Humorously yours,
Professors of Research Methods

ACKNOWLEDGMENTS

As with any work, numerous people contributed to this book, and we want to recognize them. Many are former students and colleagues who have said or done things that better developed our ideas as expressed in these pages. Also, a number of faculty members who have used previous editions have either written reviews or made suggestions that have improved the book. Although we cannot list or even recall all these contributions, we do know that you made them, and we thank all of you.

In particular, we thank Karen French at the University of South Carolina, Dick Magill at Louisiana State University (emeritus), Brad Cardinal at Oregon State University, and Kathi Thomas at the University of North Texas for allowing us to use materials that were published jointly with them. Scott Kretchmar, Tim Elcombe, David Wiggins, Daniel Mason, Barb Ainsworth, and Chuck Matthews made invaluable contributions with their chapters on research methods in the areas of philosophy, history, and exercise epidemiology, which are areas we simply could not write about effectively.

If you adopt this book for your class, we hope that you will make use of the class teaching materials available on the Human Kinetics website at www.HumanKinetics.com/Research MethodsInPhysicalActivity. Included are over 300 Microsoft PowerPoint slides, learning activities, test questions, and other course materials. We thank Phil Martin from Iowa State University for his contributions to these materials.

Finally, we thank the staff at Human Kinetics, in particular Amy Tocco, our acquisitions editor, Melissa Zavala, our developmental editor, and Carly S. O'Connor, our managing editor for this edition, for their support and contributions. They have sharpened our thinking and improved our writing.

Jerry R. Thomas
Jack K. Nelson
Stephen J. Silverman

PART I

Overview of the Research Process

> The researches of many have thrown much darkness on the subject and if they continue, soon we shall know nothing at all about it.
>
> —attributed to Mark Twain

This part provides an overall perspective of the research process. The introductory chapter defines and reviews the types of research done in physical activity and provides some examples. We define science as systematic inquiry, and we discuss the steps in the scientific method. This logical method answers the following four questions (Day, 1983, p. 4), which constitute the parts of a typical thesis, dissertation, or research report:

1. What was the problem? Your answer is the introduction.
2. How did you study the problem? Your answer is the materials and methods.
3. What did you find? Your answer is the results.
4. What do these findings mean? Your answer is the discussion.

We also present alternative approaches for doing research relative to a more philosophical discussion of science and ways of knowing. In particular, we address qualitative research, the use of field studies, and methods of introspection as strategies for answering research questions instead of relying on the traditional scientific paradigm as the only approach to research problems.

Chapter 2 suggests ways of developing a problem and using the literature to clarify the research problem, specify hypotheses, and develop the methodology. In particular, we emphasize the use of new electronic technology for searching, reading, analyzing, synthesizing, organizing, and writing literature reviews.

The next two chapters in part I present the format of the research proposal with examples. This information is typically required of the master's or doctoral student before collecting data for the thesis or dissertation. Chapter 3 addresses defining and delimiting the research problem, including the introduction, statement of the problem, research hypotheses, operational definitions, assumptions and limitations, and significance. Information is provided for both quantitative and qualitative approaches. Chapter 4 covers methodology, or how to do the research using either quantitative or qualitative methods. Included are the topics of participant selection, instrumentation or apparatuses, procedures, and design and analysis. We emphasize the value of pilot work conducted before the research and how cause and effect may be established.

Chapter 5 discusses ethical issues in research and scholarship. We include information on misconduct in science; ethical considerations in research writing, working with advisors, and copyright; and the use of humans and animals in research.

When you have completed part I, you should better understand the research process. Then comes the tricky part: learning all the details. We consider these details in part II (Statistical and Measurement Concepts in Research), part III (Types of Research), and part IV (Writing the Research Report).

Introduction to Research in Physical Activity

To each person, the word *research* conjures up a different picture. One might think of searching the Internet or going to the library; another might visualize a lab filled with test tubes, vials, and perhaps little white rats. Therefore, as we begin a text on the subject, we must establish a common understanding of research. In this chapter, we introduce you to the nature of research. We do this by discussing methods of problem solving and types of research. We explain the research process and relate it to the parts of a thesis. By the time you reach the end of chapter 1, you should understand what research really involves.

THE NATURE OF RESEARCH

The object of research is to determine how things are as compared to how they might be. To achieve this, research implies a careful and systematic means of solving problems and involves the following five characteristics (Tuckman, 1978):

- **Systematic.** Problem solving is accomplished through the identification and labeling of variables and is followed by the design of research that tests the relationships among these variables. Data are then collected that, when related to the variables, allow the evaluation of the problem and hypotheses.

- **Logical.** Examination of the procedures used in the research process allows researchers to evaluate the conclusions drawn.

- **Empirical.** The researcher collects data on which to base decisions.

- **Reductive.** The researcher takes individual events (data) and uses them to establish general relationships.

- **Replicable.** The research process is recorded, enabling others to test the findings by repeating the research or to build future research on previous results.

Problems to be solved come from many sources and can entail resolving controversial issues, testing theories, and trying to improve present practice. For example, a popular topic of concern is obesity and methods of losing weight. Suppose we want to investigate this issue by comparing the effectiveness of two exercise programs in reducing fat. Of course, we know that caloric expenditure results in the loss of fat, so we will try to find out which program does this better under specified conditions. (*Note:* Our approach here is to give a simple, concise overview of a research study. We do not intend it to be a model of originality or sophistication.)

This study is an example of **applied research**. Rather than try to measure the calories expended and so on, we approach it strictly from a programmatic standpoint. Say that we are operating a health club and offer aerobic dance and jogging classes for people who want to lose weight. Our research question is this: Which program is more effective in reducing fat?

Suppose we have a pool of participants to draw from and can randomly assign two-thirds of them to the two exercise programs and one-third to a control group. We have their scout's honor that no one is on a drastic diet or is engaging in any other strenuous activities while the study is in progress. Both the aerobic dance class and the jogging class are one hour long and are held five times a week for 10 weeks. The same enthusiastic and immensely qualified instructor teaches both classes.

Our measure of fatness is the sum of skinfold measurements taken at eight body sites. Of course, we could use other measures, such as the percentage of fat estimated from **hydrostatic weighing** (or total body water or some other estimate of fatness). In any case, we can defend our measures as valid and reliable indicators of fatness, and skinfolds are functional field measures. We measure all the participants, including those in the control group, at the beginning and the end of the 10-week period. During the study we try to ensure that the two programs are similar in procedural aspects, such as motivational techniques and the aesthetics of the surroundings. In other words, we do not favor one group by cheering them on and not encouraging the other, nor do we have one group exercise in an air-conditioned, cheerful, and healthful facility while the other has to sweat it out in a dingy room or a parking lot. We try to make the programs as similar as possible in every respect except the experimental treatments. The control group does not engage in any regular exercise.

After we have measured all the participants on our criterion of fatness at the end of the 10-week program, we are ready to analyze our data. We want to see how much change in skinfold thickness has occurred and whether differences have occurred between the two types of exercise. Because we are dealing with samples of people (from a whole universe of similar people), we need to use some type of statistics to establish how confident we can be in our results. In other words, we need to determine the significance of our results. Suppose the mean (average) scores for the groups are as follows:

- Aerobic dance: –21 mm
- Jogging: –25 mm
- Control: +8 mm

These values (which we made up) represent the average change in the combined skinfold thicknesses of the eight body sites. The two experimental groups lost fat, but the control group actually showed increased skinfold thicknesses over the 10-week period.

We decide to use the statistical technique of analysis of variance with repeated measures. We find a significant *F* ratio, indicating that significant differences exist among the three groups. Using a follow-up test procedure, we discover that both exercise groups are significantly different from the control group. But we find no significant difference between the aerobic dance and the jogging groups. (Many of you may not have the foggiest idea what we are talking about with the statistical terms *F ratio* and *significance,* but do not worry about it. All that is explained later. This book is directly concerned with those kinds of things.)

applied research—A type of research that has direct value to practitioners but in which the researcher has limited control over the research setting.

hydrostatic weighing—A technique for measuring body composition in which body density is computed by the ratio of a person's weight in air and the loss of weight underwater.

Our conclusion from this study is that both aerobic dance and jogging are effective (apparently equally so) in bringing about a loss in fatness of overweight people (such as the ones in our study) over 10 weeks. Although these results are reasonable, remember that this scenario is only an example. We could also pretend that this study was published in a prestigious journal and that we won the Nobel Prize.

Research Continuum

Research in our field can be placed on a continuum that has applied research at one extreme and **basic research** at the opposite extreme. The research extremes are generally associated with certain characteristics. Applied research tends to address immediate problems, use so-called real-world settings, use human participants, and involve limited control over the research setting, but it gives results that are of direct value to practitioners. At the other extreme, basic research usually deals with theoretical problems. It uses the laboratory as the setting, frequently uses animals as subjects, carefully controls conditions, and produces results that have limited direct application. Christina (1989) suggested that basic and applied forms of research are useful in informing each other as to future research directions. Table 1.1 demonstrates how research problems in motor learning might vary along a continuum from basic to applied depending on their goals and approaches.

To some extent, the strengths of applied research are the weaknesses of basic research, and vice versa. Considerable controversy exists in the literature on social science (e.g., Creswell, 2009; Jewczyn, 2013) and physical activity (e.g., Christina, 1989) about whether research should be more basic or more applied. This issue, labeled **ecological validity**, deals with two concerns: Is the research setting perceived by the research participant in the way intended by the experimenter? Does the setting have enough of the real-world characteristics to allow generalizing to reality?

Of course, most research is neither purely applied nor purely basic; rather, it incorporates some degree of both. We believe that systematic efforts are needed in the study of physical activity to produce research that moves back and forth across Christina's (1989) levels of research (table 1.1). Excellent summaries of this type of research and the accumulated knowledge are provided in three edited volumes representing exercise physiology, sport psychology, and motor behavior: *Physical Activity, Fitness, and Health* (Bouchard, Shepard, & Stephens, 1994), *Handbook of Sport Psychology* (Singer, Hausenblas, & Janelle, 2001), and *Cognitive Issues in Motor Expertise* (Starkes & Allard, 1993). An expert prepared each

basic research—A type of research that may have limited direct application but in which the researcher has careful control of the conditions.

ecological validity—The extent to which research emulates the real world.

Basic research and applied research can be thought of as two ends of a continuum. Basic research addresses theoretical problems, often in laboratory settings, and may have limited direct application. Applied research addresses immediate problems, often in less controlled real-world settings, and is more closely linked to application.

TABLE 1.1

Levels of Relevance of Motor Learning Research for Finding Solutions to Practical Problems in Sport

	Level 1 Least direct relevance Basic research	Level 2 Moderate direct relevance Applied research	Level 3 Most direct relevance Applied research
Ultimate goal	Develop theory-based knowledge appropriate for understanding motor learning in general with no requirement to demonstrate its relevance for solving practical problems	Develop theory-based knowledge appropriate for understanding the learning of sport skills in sport settings with no requirement to find immediate solutions to learning problems in sport	Find immediate solutions to learning problems in sport with no requirement to demonstrate or develop theory-based knowledge at either level 1 or level 2
Main approach	Test hypotheses in a laboratory setting using experimenter-designed motor tasks	Test hypotheses in a sport setting or in an applied laboratory setting using sport skills or motor tasks that have properties of those skills	Test solutions to specific learning problems in sport in the settings described under the applied research at level 2

From R.W. Christina, 1989, "Whatever happened to applied research in motor learning?" In *Future directions in exercise and sport science research*, edited by J.S. Skinner et al. (Champaign, IL: Human Kinetics). By permission of Robert W. Christina.

chapter in these books to summarize theories as well as to present basic and applied research about areas related to exercise physiology, sport psychology, and motor expertise. The novice researcher would do well to read several of these chapters as examples of how knowledge is developed and accumulated in the study of physical activity. We need more efforts to produce a related body of knowledge in the study of physical activity. Although the research base has grown tremendously in our field over the past 40 years, much remains to be done.

There is a great need to prepare proficient consumers and producers of research in kinesiology. To be proficient, people must thoroughly understand the appropriate knowledge base (e.g., exercise physiology, motor behavior, pedagogy, and the social and biological sciences) as well as research methods (qualitative, quantitative, and mixed methods). In this book we attempt to explain the tools necessary for consuming and producing research. Many of the same methods are used in the various areas of kinesiology (as well as in the fields of psychology, sociology, education, and physiology). Quality research efforts always involve some or all of the following actions:

- Identifying and delimiting a problem
- Searching, reviewing, and effectively writing about relevant literature
- Specifying and defining testable hypotheses
- Designing the research to test the hypotheses
- Selecting, describing, testing, and treating the participants
- Analyzing and reporting the results
- Discussing the meaning and implications of the findings

Practicality and Accessibility

We recognize that not everyone is a researcher. Many people in kinesiology have little interest in research per se. In fact, some have a decided aversion to it. Researchers are sometimes viewed as strange people who deal with insignificant problems and are out of touch with the real world (although we know that none of you think that way). In an informative yet entertaining book on writing scientific papers, Day and Gastel (2006) related the story about two men riding in a hot-air balloon who encountered some cloud coverage and lost their way. When they finally descended, they did not recognize the terrain and had not the faintest idea where they were. It so happened that they were drifting over the grounds of one of our more famous scientific research institutes. When the balloonists saw a man walking alongside a road, one of them called out, "Hey, mister, where are we?" The man looked up, took in the situation, and after a few moments of reflection said, "You're in a hot-air balloon." One balloonist turned to the other and said, "I'll bet that man is a researcher." The other balloonist asked, "What makes you think so?" The first replied, "His answer is perfectly accurate and totally useless" (pp. 204-205).

All kidding aside, the need for research in any profession cannot be denied. After all, one of the primary distinctions between a discipline or profession and a trade is that the trade deals only with how to do something, whereas the discipline or profession concerns itself not only with how but also with why something should be done in a certain manner (and why it should even be done at all). But although most people in a discipline or profession recognize the need for research, most do not read research results. This situation is not unique to our field. It has been reported that only 1% of chemists read research publications, that fewer than 7% of psychologists read psychological research journals, and so on. The big question is why. We guess that most professionals who do not read research believe that doing so is not necessary. They believe that research is not practical enough or does not directly pertain to their work. Another reason given by practitioners for not reading research publications is that they cannot understand them. The language is too technical, and the terminology is unfamiliar and confusing. This complaint is valid, but we could argue that if the professional preparation programs were more scientifically oriented, the problem would diminish. Nevertheless, the research literature is extremely difficult for the nonresearcher to understand.

Reading Research

Someone once said (facetiously) that scientific papers are meant not to be read but to be published. Unfortunately, we find considerable truth in this observation. We writers are often guilty of trying to use language to dazzle the reader and perhaps to give the impression that our subject matter is more esoteric than it really is. We tend to write for the benefit of a rather small number of readers—that is, other researchers in our field.

We have the problem of jargon, of course. In any field, whether it is physics, football, or cake baking, jargon confounds the outsider. The use of jargon serves as a kind of shorthand. It provides meaning to the people within the field because everyone uses those words in the same context. Research literature is famous for using a three-dollar word when a nickel word would do. As Day and Gastel (2006, p. 200) asked, what self-respecting writer would use a three-letter word such as *now* when he or she could use the elegant expression *at this point in time*? Researchers never *do* anything, they perform it; they never *start*, they *initiate*; and they *terminate* instead of *end*. Day and Gastel further remarked that an occasional author slips and uses the word *drug,* but most salivate like Pavlov's dogs in anticipation of using *chemotherapeutic agent.*

The need to bridge the gap between the researcher and the practitioner has been recognized for years. For example, the *Journal of Physical Education, Recreation and Dance* has a feature called "Research Works," which disseminates applied research information to teachers, coaches, and fitness and recreation leaders. The website for the American Kinesiology Association (www.americankinesiology.org) regularly has a section on applied research. Yet despite these and other attempts to bridge the gap between researchers and practitioners, the gap is still imposing.

It goes without saying that if you are not knowledgeable about the subject matter, you cannot read the research literature. Conversely, if you know the subject matter, you can probably wade through the researcher's jargon more effectively. For example, if you know baseball and the researcher is recommending that by shortening the radius, the hitter can increase the angular velocity, you can figure out that the researcher means to choke up on the bat.

One of the big stumbling blocks is the statistical analysis part of research reports. Even the most ardent seeker of knowledge can be turned off by such descriptions as this: "The tetrachoric correlations among the test variables were subjected to a centroid factor analysis, and orthogonal rotations of the primary axes were accomplished by Zimmerman's graphical method until simple structure and positive manifold were closely approximated." Please note that we are not criticizing the authors for such descriptions, because reviewers and editors usually require them. We are just acknowledging that statistical analysis is frightening to someone who is trying to read a research article and does not know a factor analysis from a volleyball. The widespread use of computers and "computerese" probably compounds the mystery associated with statistics. Many people believe anything that comes out of a computer. Others are more old-fashioned and check the computer's accuracy with their calculators. A classic, yet fictional, case of a computer mistake occurred in a high school in which the computer printed the students' locker numbers in the column where their IQs were supposed to go. It is classic because no one noticed the error at the time, but at the end of the year the students with the highest locker numbers got the best grades.

How to Read Research

Despite all the hurdles that loom in the practitioner's path when reading research, we contend that you can read and profit from (usually not materially, but then consider the data on the new Speedo swimsuit and its influence on the 2008 swimming results in Beijing) the research literature even if you are not well grounded in research techniques and statistical analysis. We would like to contend that after you read this book, you will be able to read any journal in any field, but the publisher would not let us say that. We offer the following suggestions on reading the research literature:

- Become familiar with a few publications that contain pertinent research in your field. You might get some help on choosing the publications from a professor or librarian.

- Read only studies that are of interest to you. This point may sound too trite to mention, but some people feel obligated to wade through every article.

- Read as a practitioner would. Do not look for eternal truths. Look for ideas and indications. No study is proof of anything. Only when it has been verified repeatedly does it constitute knowledge.

- Read the abstract first. This saves time by helping you determine whether you wish to read the whole thing. If you are still interested, then you can read the study to gain a better understanding of the methodology and the interpretations, but do not get bogged down with details.

- Do not be too concerned about statistical significance. Understanding the concept of significance certainly helps, but a little common sense serves you about as well as knowing the difference between the .02 and the .01 levels, or a one-tailed test versus a two-tailed test. Think in terms of meaningfulness. For example, if two methods of teaching bowling result in an average difference of 0.5 pins, what does it matter whether the difference is significant? On the other hand, if a big difference is present but not significant, further investigation is warranted, especially if the study involved a small number of participants. Knowing the concepts of the types of statistical analysis is certainly helpful, but it is not crucial to being able to read a study. Just skip that part.

- Be critical but objective. You can usually assume that a national research journal selects studies for publication by the jury method. Two or three qualified people read and judge the relevance of the problem, the validity and reliability of the procedures, the efficacy of the experimental design, and the appropriateness of the statistical analysis. Certainly, some studies are published that should not be. Yet if you are not an expert in research, you do not need to be suspicious about the scientific worth of a study that appears in a recognized journal. If it is too far removed from any practical application to your situation, do not read it.

You will find that the more you read, the more you understand, simply because you become more familiar with the language and the methodology, like the man who was thrilled to learn he had been speaking prose all his life.

An Example of Practical Research

To illustrate our research consumer suggestions, consider the lighthearted account of a young physical education teacher and coach named Sonjia Roundball (J.K. Nelson, 1988). In a moment of weakness, Sonjia glanced through the table of contents of *Research Quarterly for Exercise and Sport,* which had been left in her car by a graduate student friend of hers (they had used it to keep their tacos from dripping on the upholstery). She experienced a spark of interest when she noticed an article titled "The Effects of a Season of Basketball on the Cardiorespiratory Responses of High School Girls." With some curiosity, she turned to the article and began to read. In its introductory passages, the article stated that relatively little specific information was available on the physiological changes in girls because of sport participation. A short review of literature cited a few studies on swimmers and other sport participants, and the upshot was that female athletes possess higher levels of cardiorespiratory fitness than nonathletes do. The author emphasized that no studies had tried to detect changes in girls' fitness during a season of basketball.

The next section of the study dealt with methods. It noted the length of the season, the number of games, the number of practices and their length, and the breakdown of time devoted to drills, scrimmages, and individual practice. The participants were 12 girls on the high school basketball team and 14 girls in the nonparticipant group who were in physical education classes and had academic and activity schedules similar to those of the participants. All participants in the study were tested at the beginning and end of the season on maximal oxygen consumption and various other physiological measurements dealing with ventilation, heart rate, and blood pressure. Sonjia remembered those things from her exercise physiology course a number of years earlier and was willing to accept these as good indicators of cardiorespiratory fitness.

The results were then presented in tables. Sonjia did not understand these things but was willing to trust the author as to their appropriateness. The author discovered no significant increases in any of the cardiorespiratory measures from the preseason test to the postseason test for either group. This finding jolted Sonjia to the quick! Surely, a strenuous sport such as basketball should produce improvements in fitness. Something must be wrong here, she thought. She further read (with small consolation) that the basketball players had higher values of maximal oxygen consumption than the nonparticipants did at both the beginning and the end of the season. Sonjia then read the **discussion**, which mentioned things such as that the values were higher than similar values in other studies. (So what? Sonjia thought.) She read with more interest the observations by the author that boys' basketball programs were more strenuous in terms of length and number of workouts. Sonjia began to think about this. The author admitted that the number of participants was small and that some changes might not have been detected, and the author offered other speculations. The author concluded, however, that the training program used in this study was not strenuous enough to induce significant improvement in cardiorespiratory fitness.

Sonjia was sophisticated enough to realize the limitations of the single study. Nevertheless, the practice schedule and general practice routines used in the study were similar to her own. She noticed in the references for the article three studies from a journal called *Medicine & Science in Sports & Exercise.* She had never read that journal, but she decided to drive over to the university the next weekend to look up the publication in the library. When she located the journal, the latest issue happened to have an article on the conditioning effects of swimming on college women. Although the study dealt with a different sport and a different age group, she reasoned that the review of literature might prove fruitful. She was right. It cited a recent study on aerobic capacity, heart rate, and energy cost during a season of girls' basketball. Sonjia quickly located the study and then read with the excitement that comes from the personal discovery of ideas. She also was pleasantly surprised to find that it was easier reading than the first study because she was now more familiar with the terminology and the general organization of the article.

This study also reported no improvement in aerobic capacity during the season. The study involved monitoring heart rates during games by telemetry, and researchers infrequently observed heart rates of over 170 beats per minute (bpm). They concluded that the practice sessions were apparently too moderate in intensity and that the training should be structured to meet both the skill and the fitness demands of the sport.

Sonjia returned to her school determined to take a more scientific approach to her basketball program. To start, she had one of her managers chart the number of minutes players were actually engaged in movement in the practice sessions. Sonjia also had the players take their pulses at various intervals during the sessions. She was surprised to find that the heart rates rarely surpassed 130 bpm. As an outgrowth of her recent literature search, she remembered that an intensity threshold was necessary to bring about improvement in cardiorespiratory fitness. She knew that for this age group a heart rate of about 160 bpm was needed to provide a significant training effect. Consequently, she initiated some changes in

discussion—The chapter or section of a research report that explains what the results mean.

her practice sessions (including more conditioning drills) and made the scrimmages more intensive and gamelike. To end this saga of Sonjia, you will be happy to know that Coach Roundball's team went on to win all its games, the district and the state championships, and the world games.

Summarizing the Nature of Research

science—A process of careful and systematic inquiry.

Thomas Huxley, a famous British scientist who promoted Darwin's theory of evolution, wrote that **science** is simply common sense at its best. However, the status science holds is based on findings being correct most of the time as well as finding the instances in which reported findings are not correct. Science is systematic; if you and I do the same experiment at the same time or two years apart, we should get the same answer. Unfortunately, several recently reported attempts to replicate earlier studies have not been successful. Of the 1.4 million scholarly papers published in journals each year, an important question might be How many reported findings are wrong?

Having said that about science, discovery can be rewarding, whether that discovery is research that applies to and can improve your situation or new knowledge obtained while researching your thesis or dissertation. Research should be viewed more as a method of problem solving than as some dark and mysterious realm inhabited by impractical people who speak and write in baffling terms. We believe that practitioners can read research literature, and we are dedicated in this text to facilitating the process of becoming a research consumer.

UNSCIENTIFIC VERSUS SCIENTIFIC METHODS OF PROBLEM SOLVING

Although there are many definitions of research, nearly all characterize research activity as some sort of structured problem solving. The word *structured* refers to the fact that a number of research techniques can be used as long as they are considered acceptable by scholars in the field. Thus, research is concerned with problem solving, which then may lead to new knowledge.

scientific method of problem solving—A method of solving problems that uses the following steps: defining and delimiting the problem, forming a hypothesis, gathering data, analyzing data, and interpreting the results.

The problem-solving process involves several steps whereby the problem is developed, defined, and delimited; hypotheses are formulated; data are gathered and analyzed; and the results are interpreted with regard to the acceptance or rejection of the hypotheses. These steps are often referred to as the **scientific method of problem solving**. The steps also constitute the chapters, or sections, of the research paper, thesis, or dissertation. Consequently, we devote much of this text to the specific ways these steps are accomplished.

Some Unscientific Methods of Problem Solving

Before we go into more detail concerning the scientific method of problem solving, we should recognize some other ways by which humankind has acquired knowledge. All of us have used these methods, so they are recognizable. Helmstadter (1970) labeled the methods tenacity, intuition, authority, the rationalistic method, and the empirical method.

Tenacity

tenacity—An unscientific method of problem solving in which people cling to certain beliefs regardless of a lack of supporting evidence.

People sometimes cling to certain beliefs despite a lack of supporting evidence. Our superstitions are good examples of the method called **tenacity**. Coaches and athletes are notoriously superstitious. A coach may wear a particular sport coat, hat, tie, or pair of shoes because the team won the last time he wore it. Athletes frequently have a set pattern that they consider lucky for dressing, warming up, or entering the stadium. Although they acknowledge no logical relationship between the game's outcome and their particular routine, they are afraid to break the pattern.

For example, take the man who believed that black cats bring bad luck. One night when he was returning to his ranch, a black cat started to cross the road. The man swerved off onto the prairie to keep the cat from crossing in front of him and hit a hard bump that caused the headlights to turn off. Unable to see the black cat in the dark night, he sped frantically over rocks, mounds, and holes until he came to a sudden stop in a ravine, wrecking his car and sustaining moderate injuries. Of course, this episode just confirmed his staunch belief that black cats do indeed bring bad luck.

Obviously, tenacity has no place in science. It is the least reliable source of knowledge.

Intuition

Intuitive knowledge is sometimes considered common sense or self-evident. Many self-evident truths, however, are found to be false. That the earth is flat is a classic example of the intuitively obvious; that the sun revolves around the earth was once self-evident; that no one could run a mile in less than 4 min once was self-evident. Furthermore, for anyone to shot-put more than 70 ft (21 m) or pole-vault more than 18 ft (5.5 m) or for a woman to run distances over 0.5 mi (0.8 km) was impossible. One fundamental tenet of science is that we must be ever cognizant of the importance of substantiating our convictions with factual evidence.

Authority

Reference to some authority has long been used as a source of knowledge. Although this approach is not necessarily invalid, it does depend on the authority and the rigidity of adherence. But the appeal to authority has been carried to absurd lengths. Even personal observation and experience have been deemed unacceptable when they dispute authority. Supposedly, people refused to look through Galileo's telescope when he disputed Ptolemy's explanation of the world and the heavens. Galileo was later jailed and forced to recant his beliefs. Bruno also rejected Ptolemy's theory and was burned at the stake. (Scholars read and believed Ptolemy's book on astrology and astronomy for 1,200 years after his death!) In 1543 Vesalius wrote a book on anatomy, much of which is still considered correct today. But because his work clashed with Galen's theories, he met with such ridicule that he gave up his study of anatomy.

Perhaps the most crucial aspect of the appeal to authority as a means of obtaining knowledge is the right to question and to accept or reject the information. Furthermore, the authority's qualifications and the methods by which the authority acquired the knowledge also determine the validity of this source of information.

Rationalistic Method

In the rationalistic method, we derive knowledge through reasoning. A good example is the following classic syllogism:

All men are mortal (major premise).

The emperor is a man (minor premise).

Therefore, the emperor is mortal (conclusion).

Although you probably would not argue with this reasoning, the key to this method is the truth of the premises and their relationship to each other. For example,

Basketball players are tall.

Tom Thumb is a basketball player.

Therefore, Tom Thumb is tall.

In this case, however, Tom is very short. The conclusion is trustworthy only if it is derived from premises (assumptions) that are true. Also, the premises may not in fact be premises but rather descriptions of events or statements of fact. The statements are not connected in a cause-and-effect manner. Consider the following example:

> There is a positive correlation between shoe size and mathematics performance among elementary school children (i.e., children with large shoe sizes do well in math).
>
> Herman is in elementary school and wears large shoes.
>
> Therefore, Herman is good in mathematics.

Of course, in the first statement the factor common to both mathematics achievement and shoe size is age. Older children tend to be bigger and thus have bigger feet than younger children. Older children also have higher achievement scores in mathematics, but there is no cause-and-effect relationship. You must always be aware of this when dealing with correlation. Reasoning is fundamental in the scientific method of problem solving but cannot be used by itself to arrive at knowledge.

Empirical Method

empirical—A description of data or a study that is based on objective observations.

The word **empirical** denotes experience and the gathering of data. Certainly, data gathering is part of the scientific method of solving problems. But relying too much on your own experience (or data) has drawbacks. First, your own experience is limited. Furthermore, your retention depends substantially on how the events agree with your experience and beliefs, on whether things "make sense," and on your state of motivation to remember. Nevertheless, the use of data (and the empirical method) is high on the continuum of methods of obtaining knowledge as long as you are aware of the limitations of relying too heavily on this method.

Scientific Method of Problem Solving

The methods of acquiring knowledge previously discussed lack the objectivity and control that characterize the scientific approach to problem solving. Several basic steps are involved in the scientific method. Some authors list seven or eight steps, and others condense these steps into three or four. Nevertheless, all the authors are in general agreement as to the sequence and processes involved. The steps are briefly described next. The basic processes are covered in detail in other chapters.

Step 1: Developing the Problem (Defining and Delimiting It)

This step may sound contradictory, because how could the development of the problem be part of solving it? Actually, the discussion here is not about finding a problem to study (ways of locating a problem are discussed in chapter 2); the assumption is that the researcher has already selected a topic. But to design and execute a sound investigation, the researcher must be specific about what is to be studied and to what extent it will be studied.

Many ramifications constitute this step, an important one being the identification of the independent and dependent variables. The **independent variable** is what the researcher is manipulating. If, for example, two methods of teaching a motor skill are being compared, then the teaching method is the independent variable; this item is sometimes called the experimental, or treatment, variable.

independent variable—The part of the experiment that the researcher is manipulating; also called the *experimental variable* or *treatment variable*.

dependent variable—The effect of the independent variable; also called the *yield*.

The **dependent variable** is the effect of the independent variable. In the comparison of teaching methods, the measure of skill is the dependent variable. If you think of an experiment as a cause-and-effect proposition, the cause is the independent variable and the effect is the dependent variable. The latter is sometimes referred to as the yield. Thus, the researcher must define exactly what will be studied and what will be the measured effect. When this question is resolved, the experimental design can be determined.

Step 2: Formulating the Hypothesis

The **hypothesis** is the expected result. A person setting out to conduct a study generally has an idea as to what the outcome will be. This anticipated solution to the problem may be based on some theoretical construct, on the results of previous studies, or perhaps on the experimenter's experience and observations. The last source is probably least likely or defensible because of the weaknesses of the unscientific methods of acquiring knowledge discussed previously. Regardless, the research should have some experimental hypothesis about each subproblem in the study.

One of the essential features about the hypothesis is that it be testable. The study must be designed in such a way that the hypothesis can be either supported or refuted. Obviously, then, the hypothesis cannot be a type of value judgment or an abstract phenomenon that cannot be observed.

For example, you might hypothesize that success in athletics depends solely on fate. In other words, if a team wins, it is because it was meant to be; similarly, if a team loses, a victory was just not meant to be. Refuting this hypothesis is impossible because no evidence could be obtained to test it.

Step 3: Gathering the Data

Of course, before step 2 can be accomplished, the researcher must decide on the proper methods of acquiring the necessary data to be used in testing the research hypothesis. The reliability of the measuring instruments, the controls that are employed, and the overall objectivity and precision of the data-gathering process are crucial to solving the problem.

In terms of difficulty, gathering data may be the easiest step because in many cases it is routine. Planning the method, however, is one of the most difficult steps. Good methods attempt to maximize both the **internal validity** and the **external validity** of the study.

Internal validity and external validity relate to the research design and controls that are used. Internal validity refers to the extent to which the results can be attributed to the treatments used in the study. In other words, the researcher must try to control all other variables that could influence the results. For example, Jim Nasium wants to assess the effectiveness of his exercise program in developing physical fitness in young boys. He tests his participants at the beginning and then at the end of a 9-month training program and concludes that the program brought about significant improvement in fitness. What is wrong with Jim's conclusion? His study contains several flaws. The first is that Jim did not consider maturity. Nine months of maturation produced significant changes in size and in accompanying strength and endurance. Also, what else were the participants doing during this time? How do we know that other activities were not responsible, or partly so, for the changes in their fitness levels? Chapter 18 deals with these threats to internal validity.

External validity pertains to the generalizability of the results. To what extent can the results apply to the real world? A paradox often occurs for research in the behavioral sciences because of the controls required for internal validity. In motor-learning studies, for example, the task is often something novel so that it provides a control for experience. Furthermore, being able to measure the performance objectively and reliably is desirable. Consequently, the learning task is frequently a maze, a rotary pursuit meter, or a linear position task, all of which may meet the demands for control with regard to internal validity. But then you face the question of external validity: How does performance in a laboratory setting with a novel, irrelevant task apply to learning gymnastics or basketball? These questions are important and sometimes vexing, but they are not insurmountable. (They are discussed later.)

---

hypothesis—The anticipated outcome of a study or experiment.

If a hypothesis is testable, a study will either support or refute it. Testability is a necessary feature of a hypothesis.

internal validity—The extent to which the results of a study can be attributed to the treatments used in the study.

external validity—The generalizability of the results of a study.

Possible Misinterpretations of Results

- We will never run out of math professors because they always multiply.
- When the body is fully immersed in water, the telephone rings.
- If there are only two people in a locker room, they will have adjacent lockers.
- The ocean would be much deeper without sponges.

Step 4: Analyzing and Interpreting Results

The novice researcher finds this step to be the most formidable for several reasons. First, this step usually involves some statistical analysis, and the novice researcher (particularly the master's student) often has a limited background in and a fear of statistics. Second, analysis and interpretation require considerable knowledge, experience, and insight, which the novice may lack.

That analyzing and interpreting results is the most challenging step goes without question. It is here that the researcher must provide evidence for the support or rejection of the research hypothesis. In doing this, the researcher also compares the results with those of others (the related literature) and perhaps attempts to relate and integrate the results into some theoretical model. Inductive reasoning is employed in this step (whereas deductive reasoning is primarily used in the statement of the problem). The researcher attempts to synthesize the data from his or her study along with the results of other studies to contribute to the development or substantiation of a theory.

ALTERNATIVE MODELS OF RESEARCH

In the preceding section we summarized the basic steps in the scientific method of problem solving. Science is a way of knowing, often defined as structured inquiry. One basic goal of science is to explain things or to generalize and build a theory. When a scientist develops a useful model to explain behavior, scholars often test predictions from this model using the steps of the scientific method. The model and the approaches used to test the model are called a paradigm.

Normal Science

normal science—An objective manner of study grounded in the natural sciences that is systematic, logical, empirical, reductive, and replicable.

For centuries, the scientific approaches used in studying problems in both the natural and the social sciences have been what Thomas Kuhn (1970), a noted science historian, termed **normal science**. This manner of study is characterized by the elements we listed at the beginning of this chapter (i.e., systematic, logical, empirical, reductive, and replicable). Its basic doctrine is objectivity. Normal science is grounded in the natural sciences, which have long adhered to the idea of the orderliness and reality of matter—that is, that nature's laws are absolute and discoverable by objective, systematic observations and investigations that are not influenced by (in other words, independent of) humans. The experiments are theory driven and have testable hypotheses.

Normal science received a terrific jolt with Einstein's theory of relativity and the quantum theory, which indicated that nature's laws could be influenced by humans (that is, that reality depends to a great extent on how one perceives it). Moreover, some things, such as the decay of a radioactive nucleus, happen for no reason at all. The fundamental laws that had been believed to be absolute were now considered statistical rather than deterministic. Phenomena could be predicted statistically but not explained deterministically (Jones, 1988).

Challenges to Normal Science

Relatively recently (since about 1960), serious challenges have arisen regarding normal science's concept of objectivity (i.e., that the researcher can be detached from the instruments and conduct of the experiment). Two of the most powerful challengers to the idea of objective knowledge were Thomas Kuhn (1970) and Michael Polanyi (1958). They contended that objectivity is a myth. From the first inception of the idea for the hypothesis through the selection of apparatus to the analysis of the results, the observer is involved. The conduct of the experiment and the results can be considered expressions of the researcher's point of view. Polanyi was especially opposed to the adoption of normal science for the study of human behavior.

Kuhn (1970) maintained that normal science does not really evolve in systematic steps the way that scientific writers describe it. Kuhn discussed the **paradigm crisis phenomenon**, in which researchers who have been following a particular paradigm begin to find discrepancies in it. The findings no longer agree with the predictions, and a new paradigm is advanced. Interestingly, the old paradigm does not die completely but only develops varicose veins and fades away. Many researchers with a great deal of time and effort invested in the old paradigm are reluctant to change, so it is usually a new group of researchers who propose the new paradigm. Thus, normal science progresses by revolution, with a new group of scientists breaking away and replacing the old. Kuhn and Polanyi concurred that the doctrine of objectivity is simply not a reality. Nevertheless, normal science has been and will continue to be successful in the natural sciences and in certain aspects of the study of humans. But Martens (1987) contended that it has failed miserably in the study of human behavior, especially in the more complex functions.

As a sport psychologist, Martens asserted that laboratory experiments have limited use in answering questions about complex human behavior in sport. He considered his role as a practicing sport psychologist to have been far more productive in gaining knowledge about athletes and coaches and the solutions to their problems. Other workers in the so-called helping professions have made similar observations about both the limitations of normal science and the importance of alternative sources of knowledge in forming and shaping professional beliefs. Schein (1987), a noted scholar of social psychology, related an interesting (some might call it shocking) revelation concerning the relative influence of published research results versus practical experience. At a conference, he and a number of his colleagues were discussing what they relied on most for their classroom teaching. These professors seemed to agree that the data they really believed in and used in the classroom came from personal experience and information learned in the field. Schein was making the point that different categories of knowledge can be obtained by different methods. In effect, some people are more influenced by sociological and anthropological research models than by the normal science approach.

For some time, many scholars in education, psychology, sociology, anthropology, sport psychology, physical education, and other disciplines have proposed methods of studying human behavior other than those of conventional normal science. Anthropologists, sociologists, and clinical psychologists have used in-depth observation, description, and analysis of human behavior for nearly three-quarters of a century. For over 40 years, researchers in education have used participant and nonparticipant observation to obtain comprehensive, firsthand accounts of teacher and student behaviors as they occur in real-world settings. More recently, physical educators, sport psychologists, and exercise specialists have been engaged in this type of field research. A number of names given to this general form of research are

paradigm crisis phenomenon—The development of discrepancies in a paradigm leading to proposals of a new paradigm that better explains the data.

qualitative research—
A research method that often involves intensive, long-term observation in a natural setting; precise and detailed recording of what happens in the setting; and the interpretation and analysis of the data using description, narratives, quotes, charts, and tables. Also called *ethnographic, naturalistic, interpretive, grounded, phenomenological, subjective,* and *participant observational research.*

ethnographic, qualitative, grounded, naturalistic, and participant observational research. Regardless of the names, the commitments, and the beliefs of the researchers, this type of research was not well received at first by the adherents of normal science and the scientific method. In fact, this form of research (we include all its forms under the name **qualitative research**) has often been labeled by normal scientists as superficial, lacking in rigor, and just plain unscientific. As qualitative research methods have evolved, so has the thinking of many of these people. As you will see in chapters 19 and 20, many of the research tenets listed by Kuhn (1970) are found in contemporary qualitative research.

Martens (1987) referred to such adherents of normal science as the gatekeepers of knowledge because they are the research journal editors and reviewers who decide who gets published, who serves on the editorial boards, and whose papers are presented at conferences. Studies without internal validity are not published, yet studies without external validity lack practical significance. Martens (1987) charged that normal science (in psychology) prefers publishability to practical significance.

The debates between qualitative and normal (often classified as quantitative) research have been heated and prolonged. The qualitative proponents have gained confidence and momentum in recent years, and this point of view is now recognized as a viable method of addressing problems in the behavioral sciences. Credibility is established by systematically categorizing and analyzing causal and consequential factors. The naturalistic setting of qualitative research both facilitates analysis and precludes precise control of so-called extraneous factors, as does much other research occurring in field settings. The holistic interrelationship among observations and the complexity and dynamic processes of human interaction make it impossible to limit the study of human behavior to the sterile, reductionistic approach of normal science. **Reductionism**, a characteristic of normal science, assumes that complex behavior can be reduced, analyzed, and explained as parts that can then be put back together as a whole and understood. Critics of the conventional approach to research believe that the central issue is the unjustified belief that normal science is the only source of true knowledge.

reductionism—A characteristic of normal science that assumes that complex behavior can be reduced, analyzed, and explained as parts that can then be put back together to understand the whole.

Implications of These Challenges

The challenges to normal science involve many implications. For example, when we study simple movements, such as linear positioning in a laboratory to reflect cognitive processing of information, do we learn anything about movements in real-world settings such as the performance of sport skills? When we evaluate EMG activity in specific muscle groups during a simple movement, does the result really tell us anything about the way the nervous system controls movements in natural settings such as athletics? Can we study the association of psychological processes related to movement in laboratory settings and expect the results to apply in sport and exercise situations? When we conduct these types of experiments, are we studying nature's phenomena or laboratory phenomena?

Do not misinterpret the intent of these questions. They do not mean that nothing important can be discovered about physical activity from laboratory research. What they suggest is that these findings do not necessarily accurately model the way humans plan, control, and execute movements in natural settings associated with exercise and sport.

Kuhn's (1970) explanations of how science advances and the limitations of applying normal science to natural settings demonstrate that scientists need to consider the various ways of knowing and that the strict application of the normal scientific method of problem solving may sometimes hinder rather than advance science. If the reductionistic approach of the scientific method has not well served the natural scientists who developed it, then certainly researchers in human behavior need to assess the relative strengths and weaknesses of conventional and alternative research paradigms for their particular research questions.

Alternative Forms of Scientific Inquiry

Martens (1987, p. 52) suggested that we view knowledge not as either scientific or unscientific or as either reliable or unreliable but rather as existing on a continuum, as illustrated in figure 1.1. This continuum, labeled DK for *degrees of knowledge,* ranges from "don't know" to "damn konfident." Considered in this way, varying approaches to disciplined inquiry are useful in accumulating knowledge. As examples, Martens (1979, 1987) urged sport psychologists to consider the idiographic approach, introspective methods, and field studies instead of relying on the paradigm of normal science as the only answer to research questions in sport psychology. Thomas, French, and Humphries (1986) detailed how to study children's sport knowledge and skills in games and sports. Costill (1985) discussed the study of physiological responses in practical exercise and sport settings. Locke (1989) presented a tutorial on the use of qualitative research in physical education and sport. In later chapters we give greater detail about some of these alternative strategies for research, particularly historical, philosophical, qualitative, and mixed methods.

What we hope you gain from this section is that science is disciplined inquiry, not a set of specific procedures. Although advocates of alternative methods of research are often persuasive, we do not want you to conclude that the study of physical activity should abandon the traditional methods of normal science. We have learned much from these techniques and will continue to do so. Furthermore, we certainly do not want you to toss away this book as being pointless. We have not even begun to tell you all the fascinating things that we have learned over the years (it is hard to tell whether some of these things should be classified as normal or abnormal science). In addition, we have many funny stories yet to tell (abnormal humor). Aside from these compelling reasons for continuing with the book, we want you to realize and appreciate that so-called normal science is not the solution to all questions raised in our field. Furthermore, none of the alternative methods of research denounces the scientific method of problem solving.

The bottom line is that different problems require different solutions. As we said before, science is disciplined inquiry, not a set of specific procedures. We need to embrace all systematic forms of inquiry. Rather than argue about the differences, we should capitalize on the strengths of all scholarly methods to provide useful knowledge about human movement. The nature of the research questions and setting should drive the selection of approaches to acquiring knowledge. In fact, just as Christina (1989) suggested that researchers might move among levels of research (basic to applied), so researchers might move among paradigms (quantitative to qualitative to mixed methods) to acquire knowledge. In addressing this issue, we have expanded the qualitative and mixed-method research sections in chapters 19 and 20, which focus on using varying types of research approaches. Of course, we do not want to be perceived like Danae in the comic strip *Non Sequitur* (by Wiley). Danae (a young girl) says to her horse that she wants to grow up to be a preconceptual scientist. Her horse asks, "What is that?" to which Danae responds that

Figure 1.1 The degrees of knowledge theory with examples of methods varying in degree of reliability.

Reprinted, by permission, from R. Martens, 1987, "Science, knowledge, and sport psychology," *The Sport Psychologist* 1(1): 46.

it is "the new science of reaching a conclusion before doing any research and then simply dismissing anything contrary to your preconceived notions."

TYPES OF RESEARCH

Research is a structured way of solving problems. Different kinds of problems attend the study of physical activity; thus, different types of research are used to solve these problems. This text concentrates on five types of research: analytical, descriptive, experimental, qualitative, and mixed methods. A brief description of each follows.

Analytical Research

analytical research— A type of research that involves in-depth study and the evaluation of available information in an attempt to explain complex phenomena; can be categorized in the following way: historical, philosophical, reviews, and research synthesis.

As the name implies, **analytical research** involves in-depth study and the evaluation of available information in an attempt to explain complex phenomena. The types of analytical research are historical, philosophical, reviews, and research synthesis.

Historical Research

Obviously, historical research deals with events that have already occurred. Historical research focuses on events, organizations, institutions, and people. In some studies, the researcher is interested mostly in preserving the record of events and accomplishments. In other investigations, the researcher attempts to discover facts that will provide more meaning and understanding of past events to explain the present state of affairs. Some historians have even attempted to use information from the past to predict the future. The research procedures associated with historical studies are addressed in considerable detail in chapter 12.

Philosophical Research

Critical inquiry characterizes philosophical research. The researcher establishes hypotheses, examines and analyzes facts, and synthesizes the evidence into a workable theoretical model. Many of the most important problem areas must be dealt with by the philosophical method.

My research methods teacher will love this idea.

© Creativa/fotolia

Problems dealing with objectives, curricula, course content, requirements, and methodology are but a few of the important issues that can be resolved only through the philosophical method of problem solving.

Although some authors emphasize the differences between science and philosophy, the philosophical method of research follows essentially the same steps as other methods of scientific problem solving. The philosophical approach uses scientific facts as the basis for formulating and testing research hypotheses.

An example of such philosophical research was Morland's 1958 study in which he analyzed the educational views held by leaders in American physical education and categorized them into educational philosophies of reconstructionism, progressivism, essentialism, and perennialism.

Having an opinion is not the same as having a philosophy. In philosophical research, beliefs must be subjected to rigorous criticism in light of the fundamental assumptions. Academic preparation in philosophy and a solid background in the fields from which the facts are derived are necessary. Other examples and a more detailed explanation of philosophical research are given in chapter 13.

Reviews

A **review** is a critical evaluation of recent research on a particular topic. The author must be extremely knowledgeable about the available literature as well as the research topic and procedures. A review involves an analysis, evaluation, and integration of the published literature, often leading to important conclusions concerning the research findings up to that time (for good examples of reviews, see Blair, 1993, and Silverman and Subramaniam, 1999).

Certain publications consist entirely of reviews, such as *Psychological Review, Annual Review of Physiology, Review of Educational Research,* and *Kinesiology Review.* A number of journals publish reviews periodically, and some occasionally devote entire issues to reviews. For example, the 75th anniversary issue of *Research Quarterly for Exercise and Sport* (Silverman, 2005) contains some excellent reviews on various topics.

review—A critical evaluation of research on a particular topic.

Research Synthesis

Reviews of literature are difficult to write because they require the synthesis of a large number of studies to determine common underlying findings, agreements, or disagreements. To some extent this is like trying to make sense of data collected on a large number of participants by simply looking at the data. Glass (1977) and Glass, McGaw, and Smith (1981) proposed a quantitative means of analyzing the findings from numerous studies; this method is called meta-analysis. Findings between studies are compared by changing results within studies to a common metric called effect size. Over the years, many meta-analyses have been reported in the physical activity literature (e.g., Payne & Morrow, 1993; Rawdon, Sharp, Shelley & Thomas, 2012; Schieffer & Thomas, 2012; Thomas & French, 1985). This technique is discussed in more detail in chapter 14.

Descriptive Research

Descriptive research is concerned with status. The most prevalent descriptive research technique is the survey, most notably the questionnaire. Other forms of surveys include the interview (personal and by telephone) and the normative survey. Chapter 15 provides detailed coverage of these techniques. The following sections briefly describe three types of survey research techniques.

descriptive research—A type of research that attempts to describe the status of the study's focus. Common techniques are questionnaires, interviews, normative surveys, case studies, job analyses, observational research, developmental studies, and correlational studies.

Questionnaire

The main justification for using a questionnaire is the need to obtain responses from people, often from a wide geographical area. The questionnaire usually strives to secure information about present practices, conditions, and demographic data. Occasionally, a questionnaire asks for opinions or knowledge.

Interview

The interview and the questionnaire are essentially the same technique insofar as planning and procedures are concerned. Obviously, the interview has certain advantages over the questionnaire. The researcher can rephrase questions and ask additional ones to clarify responses and secure results that are more valid. Becoming a skilled interviewer requires training and experience. Telephone interviewing has become increasingly common in recent years. It costs half as much as face-to-face interviews and can cover a wide geographical area, which is generally a limitation in personal interviews. We discuss some other advantages of the telephone interview technique in chapter 15.

Normative Survey

A number of notable normative surveys have been conducted in the fields of physical activity and health. The normative survey generally seeks to gather performance or knowledge data on a large sample from a population and to present the results in the form of comparative standards, or norms. The *AAHPER Youth Fitness Test Manual* (American Association for Health, Physical Education and Recreation, 1958) is an outstanding example of a normative survey. Thousands of boys and girls ages 10 to 18 throughout the United States were tested on a battery of motor fitness items. Percentiles were then established for comparative performances to provide information for students, teachers, administrators, and parents. The AAHPER youth fitness test was developed in response to another survey, the Kraus-Weber test (Kraus & Hirschland, 1954), which revealed that American children scored dramatically lower on a test battery of minimum muscular fitness when compared with European children.

Other Descriptive Research Techniques

Among the other forms of descriptive research are the case study, the job analysis, observational research, developmental studies, and correlational studies. Chapter 16 provides detailed coverage of these descriptive research procedures.

Case Study

The case study is used to provide detailed information about an individual (or institution, community, and so on). It aims to determine unique characteristics about the subject or condition. This descriptive research technique is used widely in such fields as medicine, psychology, counseling, and sociology. The case study is also a technique used in qualitative research.

Job Analysis

This type of research is a special form of case study. It is done to describe the nature of a particular job, including the duties, responsibilities, and preparation required for success in the job.

Observational Research

Observational research is a descriptive technique in which behaviors are observed in the participants' natural setting, such as the classroom or play environment. The observations are frequently coded, and then their frequency and duration are analyzed.

Developmental Studies

In developmental research, the investigator is usually concerned with the interaction of learning or performance with maturation. For example, a researcher may wish to assess the extent to which the ability to process information about movement can be attributed to maturation as opposed to strategy, or to determine the effects of growth on a physical parameter such as aerobic capacity.

Developmental research can be undertaken by what is called the longitudinal method, whereby the same participants are studied over a period of years. Obvious logistical problems are associated with longitudinal studies, so an alternative is to select samples of participants from different age groups to assess the effects of maturation. This is called the cross-sectional approach.

Correlational Studies

The purpose of correlational research is to examine the relationship between performance variables, such as heart rate and ratings of perceived exertion; the relationship between traits, such as anxiety and pain tolerance; or the correlation between attitudes and behavior, as in the attitude toward fitness and the amount of participation in fitness activities. Sometimes correlation is employed to predict performance. For example, a researcher may wish to predict the percentage of body fat from skinfold measurements. Correlational research is descriptive in that a cause-and-effect relationship cannot be presumed. All that can be established is that an association is (or is not) present between two or more traits or performances.

Epidemiological Research

Another form of descriptive research that has become a viable approach to studying problems dealing with health, fitness, and safety is the epidemiological research method. This type of research pertains to the frequencies and distributions of health and disease conditions among populations. Rate of occurrence is the basic concept in epidemiological studies. The size of the population being studied is an important consideration in examining the prevalence of such things as injuries, illnesses, or health conditions in a specified at-risk population.

Although cause and effect cannot be established by incidence and prevalence data, a strong inference of causation can often be made through association. Chapter 17 is devoted to epidemiological research.

Experimental Research

Experimental research has a major advantage over other types of research in that the researcher can manipulate treatments to cause things to happen (i.e., establish a cause-and-effect situation). This approach contrasts with other types of research in which already existing phenomena or data from the past are observed and analyzed. As an example of an experimental study, assume that Virginia Reel, a dance teacher, hypothesizes that students would learn more effectively through the use of video. First, she randomly assigns students to two sections. One section is taught by the so-called traditional method (explanation, demonstration, practice, and critique). The other section is taught in a similar manner except that the students are filmed while practicing and can thus observe themselves as the teacher critiques their performances. After 9 weeks, a panel of dance teachers evaluates both sections. In this study, the teaching method is the independent variable, and dance performance (skill) is the dependent variable. After the groups' scores are compared statistically, Virginia can conclude whether her hypothesis can be supported or not.

In experimental research, the researcher attempts to control all factors except the experimental (or treatment) variable. If the extraneous factors can be controlled, then the researcher can presume that the changes in the dependent variable are due to the independent variable. Chapter 18 is devoted to experimental and quasi-experimental research.

experimental research—A type of research that involves the manipulation of treatments in an attempt to establish cause-and-effect relationships.

Qualitative Research

In the study of physical activity, qualitative research is the relatively new kid on the block, although it has been used for many years in other fields, such as anthropology and sociology. Researchers in education have been engaged in qualitative methods longer than researchers in our field have. As previously mentioned, several names are given to this type of research (ethnographic, naturalistic, interpretive, grounded, phenomenological, subjective, and participant observational). Some are simply name differences, whereas some have different approaches and points of focus. We have lumped them all under the heading of qualitative research, because that term seems to be the most commonly used in our field.

Qualitative research is different from other research methods. It is a systematic method of inquiry, and it follows the scientific method of problem solving to a considerable degree, although it deviates in certain dimensions. Qualitative research rarely establishes hypotheses at the beginning of the study, but instead uses more general questions to guide the study. It proceeds in an inductive process in developing hypotheses and theory as the data unfold. The researcher is the primary instrument in data collection and analysis. Qualitative research is characterized by intensive firsthand presence. The tools of data collection are observation, interviews, and researcher-designed instruments (Creswell, 2009). Qualitative research is described in chapter 19 and 20.

Mixed Methods of Research

In mixed methods of research, both quantitative and qualitative approaches are included (or mixed) within a research effort. This approach, often viewed as a pragmatic one, suggests that both qualitative and quantitative techniques are useful when studying real-world phenomena. The notion is that the approach that works best to capture behavioral data is what should be used. For mixed methods, qualitative and quantitative techniques are integrated, or mixed, within a single study; it is like two small studies, one quantitative and one qualitative.

OVERVIEW OF THE RESEARCH PROCESS

A good overview of the research methods course, which serves well as an introduction to this book, is provided in figure 1.2. This flowchart provides a linear way to think about planning a research study. After the problem area is identified, reading and thinking about relevant theories and concepts, as well as a careful search of the literature for relevant findings, lead to the specification of hypotheses or questions. Operational definitions are needed in a research study so that the reader knows exactly what the researcher means by certain terms. Operational definitions describe observable phenomena that enable the researcher to examine empirically whether the predictions can be supported. The study is designed, and the methods are made operational. The data are then collected and analyzed, and the findings are identified. Finally, the results are related back to the original hypotheses or questions and discussed in relation to theories, concepts, and previous research findings.

PARTS OF A THESIS: A REFLECTION ON THE STEPS IN THE RESEARCH PROCESS

This chapter has introduced the research process. The theme has been the scientific method of problem solving. Generally, a thesis or research article has a standard format. This feature is for expediency, so that the reader knows where to find the pieces of information, such as purpose, methods, and results. The format also reflects the steps in the scientific method of

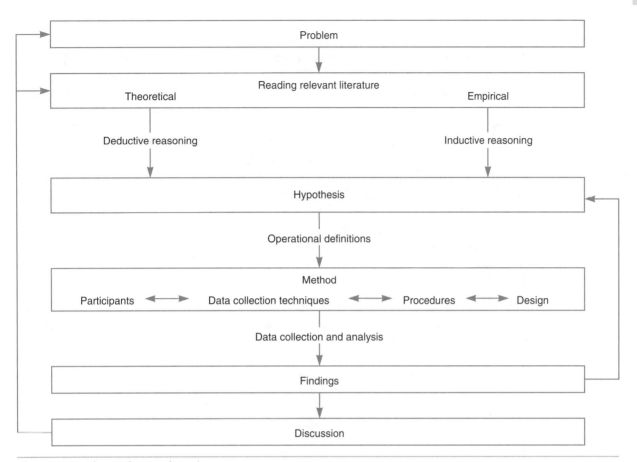

Figure 1.2   The total research setting.

problem solving. We now look at a typical thesis format and see how the parts correspond to the steps in the scientific method.

Sometimes, theses and dissertations are done in a chapter format in which each chapter represents a separate part of the research report (e.g., introduction). This model has been commonly used over the years. We believe that it is more appropriate for graduate students to prepare their theses or dissertations in a form suitable for journal publication because that is an important part of the research process. Writing the theses or dissertation in a journal format also prepares students for future writing. In chapter 22 we provide considerable detail about how to use a journal format for a thesis or dissertation and the value of doing so. Throughout this book, we refer to the typical parts of a research report. These can be considered either as parts of a journal paper or as chapters, depending on the format selected.

Introduction

In the introduction, the problem is defined and delimited. The researcher specifically identifies the problem and often states the research hypotheses. Certain terms critical to the study are operationally defined for the reader, and limitations and perhaps some basic assumptions are acknowledged.

The literature review may be in the introduction, or it may warrant a separate section. When it is in the first part, it more closely adheres to the steps in the scientific method of problem solving; that is, the literature review is instrumental in the formulation of hypotheses and the deductive reasoning leading to the problem statement.

Methods

The purpose of the methods section is to make the thesis format parallel to the data-gathering steps of the scientific method. First, the researcher explains how the data were gathered. The participants are identified, the measuring instruments are described, the measurement and treatment procedures are presented, the experimental design is explained, and the methods of analyzing the data are summarized. The major purpose of the methods section is to describe the study in such detail and with such clarity that a reader could duplicate it.

The introduction and methods section often comprise the **research proposal** and are presented to the student's thesis committee before the research is undertaken. For the proposal, these two parts are often written in future tense and then changed to past tense when the final version of the thesis is completed. The research proposal often contains some preliminary data demonstrating that the student has the required expertise to collect the data using the instrumentation needed.

research proposal—A formal preparation that includes the introduction, literature review, and proposed method for conducting a study.

Results

The results present the pertinent findings from the data analysis and represent the contribution to new knowledge. The results section corresponds to the step in the scientific method in which the meaningfulness and reliability of the results are scrutinized.

Discussion and Conclusions

In this last step in the scientific method, the researcher employs inductive reasoning to analyze the findings, compare the findings with those of previous studies, and integrate them into a theoretical model. In this part of a research paper, the acceptability of the research hypotheses is judged. Then, based on the analysis and discussion, conclusions are usually made. The conclusions should address the purpose and subpurposes that were specified in the introduction.

Qualitative and Mixed-Method Studies

When the qualitative or mixed-method approach is used, the format for the thesis or dissertation often varies from the preceding (i.e., introduction, method, results, and discussion). However, the general notion of explaining the problem, describing how data were collected, presenting the results, and providing a discussion remain the same (chapters 19, 20, 21, and 22 provide a more in-depth discussion of this).

SUMMARY

Research is simply a way of solving problems. Questions are raised, and methods are devised to try to answer them. There are various ways of approaching problems (research methods). Sometimes the nature of the problem dictates the method of research. For example, a researcher who wants to discover the origins of a sport would use the historical method of research. A researcher who wants to look at a problem from a particular angle may select a research method that can best answer the question from that angle.

Research on the topic of teaching effectiveness, for example, can be approached in several ways. An experimental study could be conducted in which the effectiveness of teaching methods in bringing about measurable achievement is compared. Or a study could be designed in which teachers' behaviors are coded and evaluated using some observational instrument. Or another form of descriptive research could be used that employs the questionnaire or the interview technique to examine teachers' responses to questions concerning their beliefs or practices. Or perhaps a qualitative study could be undertaken to observe and interview one teacher in one school systematically over an extended period to portray the teacher's experiences and perceptions in the natural setting. And of course, qualitative and quantitative approaches can be combined in a mixed-method format.

The point is that there is not just one way to do research. Some people do only one type of research. Some are critical of the methods used by others. However, anyone who believes that his or her type of research is the only "scientific" way to solve problems is narrow-minded and downright foolish. Science is disciplined inquiry, not a set of specific procedures.

Basic research deals primarily with theoretical problems, and the results are not intended to have immediate application. Applied research, on the other hand, strives to answer questions that have direct value to the practitioner. There is a need to prepare proficient consumers of research as well as researchers. Thus, one purpose of a book on research methods is to help the reader understand the tools necessary both to consume and to produce research.

We presented here an overview of the nature of research. The scientific method of problem solving was contrasted with "unscientific" methods by which people acquire information. Multiple research models were discussed to emphasize that there is not just one way to approach problems in our discipline and profession of kinesiology. We identified the five major types of research used in the study of physical activity: analytical, descriptive, experimental, qualitative, and mixed methods. These categories and the techniques they encompass are covered in detail in later chapters.

Check Your Understanding

1. Look through some recent issues of *Research Quarterly for Exercise and Sport*. Find and read a research article of interest that is quantitative in nature and another that you believe is qualitative. Which of these did you find easier to understand? Why?

2. Find an article that describes a study that you would classify as an applied research study. Also, find one that you believe describes a basic research study. Defend your choices.

3. Think of two problems that need research in your field. From the descriptions of the types of research in this chapter, suggest how each problem might be researched.

2

Developing the Problem and Using the Literature

Getting started is the hardest part of almost any new venture, and research is no exception. You cannot do any meaningful research until you have identified the area that you want to investigate, learned what has been published in that area, and figured out how you are going to conduct the investigation. In this chapter, we discuss ways to identify researchable problems, search for literature, and write the literature review.

IDENTIFYING THE RESEARCH PROBLEM

Of the many major issues facing the graduate student, a primary one is identifying a research problem. Problems may arise from real-world settings or be generated from theoretical frameworks. Regardless, a basic requirement for proposing a good research problem is in-depth knowledge about the area of interest. But sometimes, as students become more knowledgeable about a content area, everything seems to be already known. Thus, although you want to become an expert, do not focus too narrowly, because doing so can limit topics. Relating your knowledge base to other areas often provides insight into significant areas for research.

Ironically, we ask students to start thinking about possible research topics in their research methods course, typically a course they take in the first semester (or quarter) of graduate school before they have had the opportunity to acquire in-depth knowledge. As a result, many of their research problems are trivial, lack a theoretical base, and replicate earlier research. Although this shortcoming is considerable, the advantages of taking the research methods course early in the program are substantial in terms of success in other graduate courses. In this course, students learn the following:

- To approach and solve problems in a scientific way
- To search the literature
- To write in a clear, scientific fashion

- To understand basic measurement and statistical issue
- To use an appropriate writing style
- To be intelligent consumers of research
- To appreciate the wide variety of research strategies and techniques used in an area of study

How, then, does a student without much background select a problem? As you devote ever-greater effort to thinking of a topic, you may become increasingly inclined to think that all the problems in the field have already been solved. Adding to this frustration is the pressure of time. To assure you that important questions still wait to be addressed, we have provided the sidebar Top 10 Problems That Have Not Been Resolved by Humankind.

Guidelines for Finding a Topic

To help alleviate the problem of finding a topic, we offer the following suggestions. First, be aware of the research being done at your institution, because research spawns other research ideas. Often, a researcher has a series of studies planned. Second, be alert for any controversial issues in some area of interest. Lively controversy prompts research in efforts to resolve the issue. In any case, be sure to talk to professors and advanced graduate students in your area of interest and to use their suggestions to focus on a topic (the sidebar Look for Causes, Not Effects explains the importance of this advice when seeking a topic). Third, read a review paper (possibly in a review journal, research journal, or recent textbook). From there, read several research studies in the reference lists and locate other current research papers on the topic. Using all this information, make a list of research questions that appear unanswered or are logical extensions of the material you have read. Try to pick problems that are neither too hard nor too easy. The hard ones will take you forever, and you will never get your thesis done. No one cares about the easy ones.

Top 10 Problems That Have Not Been Resolved by Humankind

10. Is there ever a day when mattresses are not on sale?
9. If people evolved from apes, why are there still apes?
8. Why does someone believe you when you say there are 4 billion stars but checks when you say the paint is wet?
7. Why do you never hear father-in-law jokes?
6. If swimming is so good for your figure, how do you explain whales?
5. How do those dead bugs get into enclosed light fixtures?
4. Why does Superman stop bullets with his chest but duck when you throw a revolver at him?
3. Why do people constantly return to the refrigerator with the hope that something new to eat has materialized?
2. Why do banks charge a fee on "insufficient funds" when they know there is not enough money in the account?

    . . . drum roll . . .

1. Why doesn't Tarzan have a beard?

LOOK FOR CAUSES, NOT EFFECTS

In an interesting paper, Salzinger (2001) pointed out that scientists should not just look at the effects they observe but try to find the causes underlying them. Finding surprising results is always interesting, but the causes behind them are what we should seek. We often hear that the purpose of the scientific method is to disprove theories, or that we can never prove a theory. A single experiment, however, seldom disproves a theory. "Old theories, like old soldiers, just fade away when better theories come to take their place," wrote Salzinger (2001, p. B14). On the other hand, confirming a theory leads us toward underlying mechanisms; results explain why something happened.

We often read research results such as these:

- ESP is shown to work.
- Listening to Mozart improves spatial reasoning.
- People who go to church live longer.

When we see headlines like these, we should ask ourselves why these things might happen. What are the explanations? Does religion have anything to do with living longer, or is it the fact that people who go to church receive emotional reinforcement, network to learn about better physicians, get out of the house, are likely to walk to church, or any other number of explanations?

Salzinger (2001) suggested that we should spend more time thinking about what causes the results and the possible mechanisms, and identify the critical ingredient that allows results to be applied in other contexts.

Reference

Salzinger, K. (February 16, 2001). Scientists should look for basic causes, not just effects. *Chronicle of Higher Education, 157* (23), B14.

Of course, no single problem necessarily meets all the criteria perfectly. For example, some theoretical problems may have limited direct application. Theoretical problems should be directed toward issues that may ultimately prove useful to practitioners. By honestly answering the questions set forth by McCloy (1930), a practical evaluation of the selected problem is possible (see the sidebar Criteria for Selecting a Research Problem).

An intriguing way to develop a research problem is to see how experts develop problems. Snyder and Abernethy (1992), in their book *The Creative Side of Experimentation,* had well-established scholars in motor control, motor development, and sport psychology give first-person accounts of factors that influenced their career research programs. The editors also deduced themes that seemed to run through these scholars' research programs. Their analysis focused on such questions as these:

- What are common personal and professional characteristics of expert researchers?
- What types of experimentation do expert researchers perform?
- What strategies do expert experimenters use to enhance their ability to ask important questions?

In addition, Locke, Spirduso, and Silverman (2014) provided a comprehensive 20-step approach to help graduate students identify topics and develop a proposal.

Using Inductive and Deductive Reasoning

The means for identifying specific research problems come from two methods of reasoning: inductive and deductive. Figure 2.1 provides a schema of the inductive reasoning process. Individual observations are tied together into specific hypotheses, which are grouped into more general explanations that are united into theory. To move from the level of observations to that of theory requires many individual studies that test specific hypotheses. But even beyond the individual studies, someone must see how all the findings relate and then offer a theoretical explanation that encompasses all the individual findings.

CRITERIA FOR SELECTING A RESEARCH PROBLEM

In 1930, the first volume (issue no. 2) of *Research Quarterly* (now *Research Quarterly for Exercise and Sport*) had a paper by Dr. C.H. McCloy, one of our historically famous scholars. McCloy provided a list of questions by which a researcher can judge the quality and feasibility of a research problem. Those questions (listed here) are as valid and useful today as they were in 1930:

- Is the problem in the realm of research?
- Does it interest you?
- Does it possess unity?
- Is it worthwhile?
- Is it feasible?
- Is it timely?
- Can you attack the problem without prejudice?
- Are you prepared in the techniques to address the problem?

Our thanks to Jim Morrow, University of North Texas, for pointing out this information.

---

closed-loop theory— A theory of motor skill learning advanced by Adams (1971) that proposes that information received as feedback from a movement is compared with some internal reference of correctness.

An example of this process can be found in the motor learning and control area. Adams (1971) proposed a **closed-loop theory** of motor skill learning. Basically, the closed-loop theory proposes that information received as feedback from a movement is compared with some internal reference of correctness (assumed to be stored in memory). Then the discrepancies between the movement and the intended movement are noted. Finally, the next attempt at the movement is adjusted to approximate the movement goal more closely. Adams' theory ties together many previous observations about movement response. The theory was tightly reasoned but limited to slow-positioning responses. This limitation really makes it a more general explanation, according to figure 2.1.

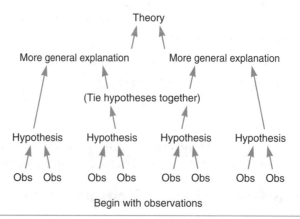

Figure 2.1 Inductive reasoning. Obs = observation.

Reprinted from R.L. Hoenes and B. Chissom, 1975, *A student guide for educational research* (Statesboro, GA: Vog Press), 22. By permission of Arlene Chissom.

Schmidt (1975) proposed a **schema theory** that extended Adams' reasoning to include more rapid types of movements, frequently called ballistic tasks. (Schema theory also deals with several other limitations of Adams' theory that are not important to this discussion.) The point is that schema theory proposed to unify two general explanations, one about slow movements and the other about ballistic (rapid) movements, under one theoretical explanation—clearly an example of inductive reasoning.

Reasoning must be careful, logical, and causal; otherwise, one of our examples of inappropriate induction may result:

> **A researcher spent several weeks training a cockroach to jump. The bug became well trained and would leap high in the air on the command "Jump." The researcher then began to manipulate his independent variable, which was to remove the bug's legs one at a time. After removing the first leg, the researcher said, "Jump," and the bug did. He then removed in turn the second, third, fourth, and fifth legs and said, "Jump." After each leg was removed, the bug jumped every time. After the researcher removed the sixth leg and gave the "Jump" command, the bug just lay there. The researcher's conclusion from this research was this: "When all the legs are removed from a cockroach, the bug becomes deaf." (Thomas, 1980, p. 267)**

Figure 2.2 presents a model of deductive reasoning. Deductive reasoning moves from a theoretical explanation of events to specific hypotheses that are tested against (or compared with) reality to evaluate whether the hypotheses are correct. Using the previously presented notions from his schema theory (to avoid explaining another theory), Schmidt advanced a hypothesis frequently called **variability of practice**. Essentially, this hypothesis (reasoned or deduced from the theory) says that practice of a variety of movement experiences (within a movement class), when compared with practicing a single movement, facilitates transfer to a new movement (still within the same class). Several researchers have tested this hypothesis, identified by deductive reasoning, and found it viable. In fact, within any given study, both inductive and deductive reasoning are useful. The total research setting was presented in chapter 1, figure 1.2. Review how the deductive and inductive processes operate; that is, at the beginning of a study, the researcher deduces hypotheses from relevant theories and concepts and induces hypotheses from relevant findings in other research.

schema theory—A theory of motor skill learning advanced by R.A. Schmidt (1975) as an extension of Adams' (1971) closed-loop theory. The theory proposed to unify two more general explanations under one theoretical explanation.

variability of practice—A tenet of motor skill learning advanced by R.A. Schmidt that states that the practice of a variety of movement experiences facilitates transfer to a new movement when compared with practicing a single movement.

Figure 2.2 Deductive reasoning. Hyp = hypothesis.

Reprinted from R.L. Hoenes and B. Chissom, 1975, *A student guide for educational research* (Statesboro, GA: Vog Press), 23. By permission of Arlene Chissom.

PURPOSE OF THE LITERATURE REVIEW

A major part of developing the research problem is reading what has already been published about the problem. Much research may already have been done on the problem in which you are interested. In other words, the problem has been pretty much "fished out." We hope this will not be the case when you start this phase of the research process. (In many instances, your major professor can steer you away from a saturated topic. We urge you to work closely with her or him, because doing so will make your task more efficient.) Whatever the topic, past research is invaluable in planning new research.

Browsing in a library confirms that information—and lots of it—exists. The dilemma lies in knowing how to locate and evaluate the information that you want and, ultimately, how to use the information after you have found it.

Reviews of literature serve several purposes. Frequently, they stimulate inductive reasoning. A scholar may seek to locate and synthesize all the relevant literature on a particular topic to develop a more general explanation or a theory to explain certain phenomena. For the dissertation or thesis, the review will be more focused, directly addressing the questions and methods for the study. An alternative way of analyzing the literature, mentioned in chapter 1, is meta-analysis (Schmidt & Hunter, 2014), which is discussed in detail in chapter 14.

> In literature reviews, studies can be related to one another based on similarities and differences in theoretical frameworks, problem statements, methodologies, and findings.

The major problem of literature reviews is how all those studies can be related to one another in an effective way. Most frequently, authors attempt to relate studies by similarities and differences in theoretical frameworks, problem statements, methodologies (participants, instruments, treatments, designs, and statistical analyses), and findings. Results are then determined by counting votes. For example, you would write, "From the eight studies with similar characteristics, five found no significant difference between the treatments; thus, this treatment has no consistent effect."

This procedure is most easily accomplished through use of a summary sheet, such as the one in table 2.1, which relates the frequency and intensity of exercise to the percentage of change in body fat. The conclusion from looking at these studies might be the following:

Exercising 20 min per day for 3 days per week for 10 to 14 weeks at 70% of the maximal heart rate produces moderate losses of body fat (4 to 5%). More frequent exercise bouts produce minimal increases, but less frequent or less intense exercise is substantially less effective in eliminating body fat.

Techniques of this type lend themselves to the development of the literature review around central themes or topics. This approach not only allows synthesis of the relevant findings but also makes the literature review interesting to read.

TABLE 2.1

Sample Form for Synthesizing Studies (Hypothetical Example)

Characteristics of studies					
Study	Problem statement	Participant description	Instrument	Procedure and design	Finding
Smith (1985)	Effects of exercise on body fat	30 college-age males	Underwater weighing	Exercise 3 d/wk at 70% of (220 − age) for 12 wk	4% reduction in body fat
Johnson (1978)	Effects of exercise on body fat	45 college-age males	Underwater weighing	Jog 3 d/wk at 70% or 50% of (220 − age) for 10 wk	5% for 70% gp 2% for 50% gp
Andrews (1989)	Effects of frequent and intense exercise on body fat	36 college-age males	Skinfold calipers	Jog 2, 4, 6 d/wk at 75% of (220 − age) for 12 wk	1% for 2 d 4% for 4 d 5% for 6 d
Mitchell (1980)	Effects of workload on body fat	24 high school males	Skinfold calipers	Pedal at 30, 45, 60 rpm with 2 KP resistance for 20 min, 3 d/wk for 14 wk	1% at 30 rpm 3% at 45 rpm 4% at 60 rpm

Identifying the Problem

As we have already discussed, the literature review is essential in identifying the specific problem. Often, the final statement of the problem, questions, and hypotheses cannot be written until the literature is reviewed and the study is situated in the previous literature. Of course, after locating a series of studies, the first task is to decide which studies are related to the topic area. This objective can frequently be accomplished by reading the abstract and, if necessary, some specific parts of the paper. After a few key studies are identified, a careful reading usually produces several ideas and unresolved questions. You will find it useful to discuss these questions with a professor or advanced graduate student from your area of specialization. Doing so can eliminate unproductive approaches or dead ends. After the problem is specified, an intensive library search begins.

Developing Hypotheses or Questions

Hypotheses are deduced from theory or induced from other empirical studies and real-world observations. These hypotheses are based on logical reasoning and, when predictive of the study's outcome, are labeled research hypotheses. For example, after spending a good deal of time at the grocery store, we are able to put forward this hypothesis for you to test: The shortest line is always the slowest. If you change lines, the one you left will speed up, and the one you entered will slow down. In qualitative studies, more general questions often serve the same purpose as hypotheses do for quantitative studies.

Developing the Method

Although considerable work is involved in identifying the problem and specifying hypotheses and questions, one of the more creative parts of research is developing the method to answer the hypotheses. If the method is planned and pilot-tested appropriately, the outcome of the study allows the hypotheses and questions to be evaluated. We believe that the researcher fails when the results of a study are blamed on methodological problems. Post hoc methodological blame results from lack of (or poor) planning and pilot work before undertaking the research.

The review of literature can be extremely helpful in identifying methods that have been successfully used to solve particular types of problems. Valuable elements from other studies may include the characteristics of the participants, data collection instruments and testing procedures, treatments, designs, and statistical analyses. All or parts or combinations of the previously used methods are helpful as the researcher plans the study, but these should not limit the researcher in designing the study. Creative methodology is a key to good hypothesis testing. But neither other scholars' research nor creativity ever replaces the need to conduct thorough pilot work.

> Elements from other studies can help identify methods for the study design. These may include characteristics of the participants, data collection instruments, and aspects of testing and statistical analysis.

BASIC LITERATURE SEARCH STRATEGIES

The prospect of beginning a literature search can sometimes be frightening or depressing. How and where do you begin? What kind of sequence or strategy should you use in finding relevant literature? What services does the library offer in your search?

Authors of research texts have advanced strategies for finding pertinent information on a topic. We know of no single right way of doing it. The search process depends considerably on your initial familiarity with the topic. In other words, if you have virtually no knowledge about a particular topic, your starting point and sequence will be different from that of someone already familiar with the literature.

In recent years a computer search has become a fundamental part of every literature search. Some people are inclined to begin by doing a computerized search. This strategy can prove fruitful, to be sure, but it is not foolproof. Gaining some familiarity with the topic and

secondary sources—
Sources of data in research in which authors have evaluated and summarized previous research.

primary sources—First-hand sources of data in research; original studies.

Primary sources, most commonly in the form of journal articles, are the most valuable for researchers.

search descriptors will aid in developing and refining the search. For students who are not well grounded in the topic, certain preliminary (general) sources may be helpful in locating **secondary sources** with which to become familiar with the topic at hand, and they prepare the student to acquire and understand **primary sources.**

An index, such as the *Education Index,* is a preliminary (not primary) source. An index can provide the researcher with books and articles that relate to the problem. Textbooks are valuable *secondary* sources that can give the reader an overview of the topic and what has been done in the way of research. In fact, secondary sources such as encyclopedias and scholarly books may be the starting point for the search, by which the student becomes aware of the problem in the first place. *Primary* sources are ultimately the most valuable for the researcher in that the information is firsthand. Most primary sources in a literature review are journal articles. Theses and dissertations are also primary sources, and those completed since 1997 are available online through ProQuest Dissertations & Theses Global.

STEPS IN THE LITERATURE SEARCH

You should follow six steps when reviewing the literature. Performing these steps ensures thoroughness and makes the search more productive.

Step 1: Write the Problem Statement

We discuss the formal writing of the problem statement in chapter 3. At this point, you are merely trying to specify what research questions you are asking. For example, a researcher wants to find out whether the student-teaching experience influences attitudes toward teaching. More specifically, the researcher wishes to examine the attitudes toward teaching before and after the student-teaching experience of students in a teacher preparation program in physical education. By carefully defining the research problem, the researcher can keep the literature search within reasonable limits. Remember, the problem statement, questions, and hypotheses may change based on the search. Write the statement as completely (but concisely) as you can at this time.

Step 2: Consult Secondary Sources

This step helps you gain an overview of the topic, but you can omit it if you are knowledgeable about the topic. Secondary sources such as textbooks and encyclopedias are helpful when students have limited knowledge about a topic and can profit from background information and a summary of previous research. A review paper on the topic of interest is especially valuable.

Encyclopedias

Encyclopedias provide an overview of information on research topics and summarize knowledge about subject areas. General encyclopedias provide broad information about an entire field. Specialized encyclopedias offer much narrower topics. Examples are the *Encyclopedia of Sport Sciences and Medicine; Encyclopedia of Physical Education, Fitness, and Sports; Encyclopedia of Educational Research;* and *Handbook of Research on Teaching.*

Because a rather lengthy period (years) often elapses from the time the authors submit their contributions until the publication date, you should be aware that the information in an encyclopedia is dated. Nevertheless, you can get important background information about a subject, become familiar with basic terms, and note references to some pertinent research journals.

Research Reviews

Reviews of research are an excellent source of information for three reasons:

- Some knowledgeable person has spent a great deal of time and effort in compiling the latest literature on the topic.

- The author has not only found the relevant literature but also critically reviewed and synthesized it into an integrated summary of what is known about the area.

- The reviewer often suggests areas of needed research, for which the graduate student may be profoundly grateful.

Some actual review publications are the *Annual Review of Medicine, Annual Review of Psychology, Review of Educational Research, Physiological Reviews, Psychological Review,* and *Exercise and Sport Sciences Reviews.* These sources, such as *Annual Review of Psychology* and *Physiological Reviews,* are available electronically as well as in print.

Many journals publish one or two review papers on a regular basis. For example, *Research Quarterly for Exercise and Sport* publishes a review paper from the annual C.H. McCloy Lecture given at the research consortium breakfast at the SHAPE America annual meeting. Thomas J. Templin and K.A.R. Richard's C.H. McCloy lecture in 2014, "Reflections on Socialization Into Physical Education: An Intergenerational Perspective," was published in the December 2014 issue (vol. 85, pp. 431-445) of *Research Quarterly for Exercise and Sport.* Occasionally, a journal publishes a special issue that consists entirely of review papers.

Step 3: Determine Descriptors

Descriptors are terms that help locate sources pertaining to a topic. For the topic of the effect of student teaching on attitudes toward teaching in physical education, obvious descriptors are attitude (toward teaching), changes in attitudes, student teaching, and physical education.

I'm having trouble organizing my literature review of Mickey Mouse.

© Vasiliy Koval/fotolia

Descriptors can be classified as major and minor. The combination of descriptors helps the researcher pinpoint pertinent related literature. Obviously, terms such as *attitude, student teaching,* and *physical education* by themselves are too broad. Each database has its own descriptors for topics. We discuss this further when we describe computer searching later in this chapter.

Step 4: Search Preliminary Sources

Use preliminary (general) sources to find primary sources through computer-aided searches. Preliminary sources primarily consist of abstracts and indexes. Preliminary sources that are helpful to researchers in physical education, exercise science, and sport science are described in the following sections.

Abstracts

Concise summaries of research studies are valuable sources of information. Abstracts of papers presented at research meetings are available at national, district, and most state conventions. Abstracts of symposia sponsored by the SHAPE America Research Consortium, free communications, and **poster sessions** presented at the national SHAPE America convention are published each year in *Research Quarterly for Exercise and Sport Supplement. Medicine & Science in Sports & Exercise* publishes a special supplement of abstracts each year for papers to be presented at the annual meeting of the American College of Sports Medicine.

Other abstract sources are *Dissertation Abstracts International,* which contains abstracts of dissertations from most colleges and universities in the United States and Canada, and *Index and Abstracts of Foreign Physical Education Literature,* which provides abstracts from journals outside the United States. Sources of abstracts in related fields include *Biological Abstracts, Psychological Abstracts, Sociological Abstracts,* and *Current Index to Journals in Education.*

Indexes

Several indexes provide references to magazine and journal articles concerning specific topics. Some general indexes commonly used in physical education, exercise science, and sport science include *Education Index, Reader's Guide to Periodical Literature, New York Times Index, Web of Science,* and *Physical Education Index.* Despite the title, this last source provides a comprehensive subject index to domestic and foreign periodicals in the fields of dance, health education, recreation, sports, physical therapy, and sports medicine. Researchers in the areas within physical education, exercise science, and sport science tend to use indexes that pertain to their topics of interest, such as *Index Medicus, PsycINFO, ERIC, Current Contents,* and *SPORTDiscus.* All these indexes are available electronically, the format preferred in most research libraries.

■ *Index Medicus.* Widely used in exercise science, *Index Medicus* provides access to more than 2,500 biomedical journals around the world. It is published monthly, and each issue has subject and author sections and a bibliography of medical reviews. *Index Medicus* can also be searched by computer through *MEDLINE.*

■ *PsycINFO.* A computer index in the field of behavioral science, *PsycINFO* selects keywords and lists appropriate titles, identifying authors and journals.

■ *ERIC.* The acronym *ERIC* stands for Educational Resources Information Center. It is the world's largest source of education information. The *ERIC* system collects, sorts, classifies, and stores thousands of documents on topics concerning education and related fields. Its basic indexes are *Resources in Education (RIE)* and *Current Index to Journals in Education (CIJE).* Besides containing the abstracts, *RIE* provides information on how you can obtain a document—whether you can purchase it on microfilm or as a pdf file, order an

poster session—A method of presenting research at a conference in which the author places summaries of his or her research on the wall or on a poster stand and answers questions from passersby.

ERIC copy, or request the original copy from the publisher. Both *RIE* and *CIJE* provide valuable assistance in the search for information on specific topics. In addition, *ERIC* produces a thesaurus containing thousands of index terms that can be used in locating references and in conducting a computer search.

■ *Current Contents.* This small weekly magazine, published by the Institute for Scientific Information, contains the table of contents of journals published recently within a general content area (e.g., social and behavioral sciences) and divides journals by subareas (e.g., psychology, education, rehabilitation, and special education). *Current Contents* indexes journal titles by topic and author and provides authors' addresses, so that you can obtain reprints. It also publishes a section on current books in each issue as well as weekly editions for life sciences; physical, chemical, and earth sciences; social and behavioral sciences; agriculture, biology, and environmental sciences; clinical medicine; engineering, technology, and applied sciences; and arts and humanities. *Current Contents* is now available electronically as *Web of Science* in most university libraries.

■ *SPORTDiscus.* This index is a valuable resource for both practical and research literature on sport, physical fitness, and physical education topics. Topics include sports medicine, exercise psychology, biomechanics, psychology, training, coaching, physical education and fitness, and other sport- and fitness-related topics. *SPORTDiscus* also provides comprehensive indexing of the *Microform Publications* collection of dissertations and theses of the International Institute for Sport and Human Performance. *SPORTDiscus* is available only electronically; there is no print equivalent. Coverage begins with 1975.

Bibliographies

Bibliographies list books and articles about specific topics. They come in many forms, depending on how the information is listed. All contain the authors, titles of books or articles, journal names, and publishing information. An **annotated bibliography** includes brief descriptions of the nature and scope of the articles or books.

annotated bibliography—A list of resources that provides a brief description of the nature and scope of each article or book.

The bibliography of a recent study on the topic in question is an invaluable aid to the researcher. Some authors have stated that one of the most valuable contributions of a dissertation is the review of literature and bibliography. But you cannot simply lift the literature review from a previous study. Just because someone else has reviewed pertinent sources does not relieve you of the responsibility to read and evaluate each source yourself. Keep in mind that (a) the previous author may have been careless and cited the source or sources incorrectly and (b) the previous author may have taken the results of a study out of context or from a point of view different from the original author's or your own.

We have found incorrect bibliographic entries on numerous occasions. A study by Stull, Christina, and Quinn (1991) revealed that of 973 citations in the 1988 and 1989 volumes of *Research Quarterly for Exercise and Sport,* 457 contained one or more errors, an error rate of 47%. Such findings are by no means unique to this journal. A 50% error rate was reported for *JAMA: The Journal of the American Medical Association* (Goodrich & Roland, 1977). Stull, Christina, and Quinn (1991) emphasized that every component in a bibliographic citation is important. Moreover, the ultimate responsibility for accuracy rests with the author. In the publishing process, mistakes can creep in, starting with copying the source, typing the article, revising the manuscript, and checking page proofs.

A good search strategy is to look for the most recent sources of information and then work backward. You will save much time by consulting the most recent studies, and you will profit from the searches of others. Some examples of bibliographies are the *Annotated Bibliography on Movement Education; Bibliography of Research Involving Female Subjects; Annotated Bibliography on Perceptual-Motor Development; Bibliography of Medical Reviews in Index Medicus;* and *Annotated Bibliography in Physical Education, Recreation, and Psychomotor Function of Mentally Retarded Persons.*

Library Information Systems

The traditional card catalog with little trays of cards containing bibliographic information by author and subject has all but disappeared. In fact, the main university library where one of the authors works now has the only card catalog on campus located in an exhibit that focuses on the history of the library. All university libraries have gone to a computerized catalog system.

Computerized catalogs abound. Usually, the searcher first selects the type of search from a menu, such as author, title, keyword, or call number. When the source is found, the full display for the reference includes author, title, publication information, all the index terms, and the call number. Libraries in colleges and universities can also be accessed on the Internet by faculty and students using personal computers. As you become familiar with the computer operations, the search process becomes much faster and more productive. Remember, if you have questions about any library operations, ask a librarian. They tend to be remarkably helpful and courteous.

Computer Searches

Computer service facilities can greatly expedite the literature search. Automated searching provides more effective and efficient access to indexes and information than does manual searching; manual searching still is possible with some databases, but it is recommended that you use the computer when you are ready to find primary sources. Databases relevant to areas in kinesiology are listed in the sidebar Some Databases Available for Computer Searches. The computer search covers many abstracting and indexing services in the sciences, humanities, and social sciences.

Before you begin your search, we strongly recommend that you take a little time to plan it. The amount of time saved and the articles found will make this preliminary step valuable. As you talk with your major professor and read preliminary and secondary sources, record the words that describe the topic and save them for the initial aspects of the search. This list might be long, but having these terms can help you adjust the scope of the search and venture into related areas.

Becoming familiar with multiple computerized databases will assist in your search. Most students doing a literature review or wanting to understand a topic use three or more databases. For example, a student in pedagogy will want to use Web of Science, ERIC, and ProQuest Dissertations & Theses Global because they cover different journals and the search protocols are slightly different. Your major professor and advanced graduate students can assist you in identifying databases that will be most helpful. If the library has orientation sessions for using databases, we strongly suggest that you attend. If you used computerized searches at another library or more than a couple of years earlier, you should take the time to learn about the databases and the interface at your university—systems change quickly, and each university system is slightly different. As you become familiar with each database, keep notes about how the database works (e.g., can it use wildcards to signify multiple possible related terms, or does it work better if you use words from the thesaurus?).

Most databases now permit users to save, e-mail, and print information. In addition, many show the history of changes during a session. Keeping notes on the files saved and changes made during a search is helpful. It may be particularly helpful to have a folder on your computer where you save the list of marked articles from the search. Before saving (or e-mailing), add a note to the comments section so that you can easily identify it when you go back to it later.

Adjusting the Scope of the Search

As mentioned before, in virtually all university libraries, a person can do the computer search from any computer connected to the Internet. The information can then be saved and printed with a great deal of information (e.g., abstract, keywords, references, the number of times the paper was cited by others). Remember, as noted earlier, that the key to a successful literature search is careful planning. Therefore, record your problem statement and then formulate cogent descriptors and keywords. If the database you are using has a thesaurus, use it. An effective way to find productive keywords or subject terms is to find one or more journal articles pertaining to your topic and see how they are indexed.

> Planning is key to a successful literature search. Record your problem statement and take time to formulate descriptors and keywords

Some Databases Available for Computer Searches

ERIC (EDUCATIONAL RESOURCES INFORMATION CENTER) (1966)

Descriptor guidelines: *Thesaurus of ERIC Descriptors*

Type of information: All topics concerning education from preschool to adult education are covered in areas of administration, curriculum, teaching, and learning. Published articles are accessed from *Current Index to Journals in Education,* and unpublished materials can be obtained from the *Education Resources* part of *ERIC.* These two databases can be searched simultaneously or separately. An *ERIC* computer search is also available on CD-ROM.

MEDLINE (1966)

Descriptor guidelines: *Medical Subject Headings*

Type of information: The vast coverage of journal articles in biomedicine, health care, gerontology, and other areas in *Index Medicus* can be accessed through *MEDLINE.*

PsycINFO (1967)

Descriptor guidelines: *Thesaurus of Psychological Index Terms*

Type of information: Journal articles, dissertations, and technical reports (book and book chapter citations have been available since 1992) in psychology and related behavioral sciences. This database comes from *Psychological Abstracts.*

Sociological Abstracts (1963)

Descriptor guidelines: *Thesaurus of Sociological Indexing Terms*

Type of information: This database contains comprehensive coverage of journal articles, conference publications, and books in social, developmental, and clinical psychology. Several database files are combined in this coverage.

Dissertation Abstracts Online (1861)

Descriptor guidelines: None. Keywords are listed by searcher.

Type of information: Dissertation abstracts and master's theses (since 1962) from *Dissertation Abstracts International* and *American Doctoral Dissertations.* It includes dissertations from nearly all U.S. doctorate-granting institutions and numerous dissertations from Canada and other countries. This is available electronically as ProQuest Dissertations & Theses Global. Dissertations completed during or after 1997 are available to download as pdf files.

ingentaconnect (1988)

Descriptor guidelines: None. Keywords or names are listed by searcher.

Type of information: Journal articles and journal tables of contents from over 27,000 journals with thousands of citations added daily. This fee-based service offers the opportunity to acquire a fax or pdf file of most articles, usually within 24 hours.

SPORTDiscus (1975)

Descriptor guidelines: None. Keywords are listed by searcher.

(continued)

Some Databases Available for Computer Searches *(continued)*

Type of information: Wide coverage of literature on sport, physical fitness, and physical education topics. Also, a comprehensive index of dissertations and theses plus citations on social and psychological aspects of sport and leisure research. This service is available online.

Web Of Science (1965)

Descriptor guidelines: Keywords, author's name, journal name.

Type of information: Web of Science is a searchable electronic database that is most often used through a library. Access to three large indexes of journals and other information is provided: *Science Citation Index* (1965), *Social Sciences Citation Index* (1965), and *Arts and Humanities Citation Index* (1975). Web of Science provides a list of all the journals that are covered (by general area, field of work, and alphabetically). Users can locate published papers, authors of papers, and journals. In addition, data about impact factors for various journals and citation rates for various papers and authors are searchable.

The scope of your search can be narrowed or broadened by using special words called Boolean operators. The two most common operators, or connectors, are the words *and* and *or*. To narrow the search, add another term with the word *and*. For example, in the proposed study on investigating changes in attitude toward teaching after the student-teaching experience, 2,237 items were listed under the descriptor "attitude change." For "attitude + teaching," 186 items were listed, and for "student teacher attitudes," 100 items were listed. When all were connected with the word *and*, 44 references were listed, which is a manageable number to examine. Figure 2.3 illustrates the Boolean logic with this example.

The word *or* broadens the search. Additional related terms can be connected with *or* so that the computer searches for more than one descriptor (see figure 2.4). For example, a researcher who seeks information about "practice teaching" (133 items) could broaden the search by using "practice teaching" (133 items) *or* "internship" (495 items) to obtain 628 related items.

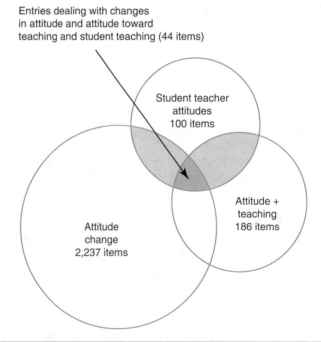

Figure 2.3 An illustration of the *and* connector to narrow a search.

Figure 2.4   An illustration of the *or* connector to broaden the search.

If the completed search produces one or more articles that are often cited or considered key to the topic, it is possible to identify where each was cited since publication. This task is accomplished by doing a cited reference search on Web of Science. This approach permits you to double-check and find articles that you may have missed. In addition, it identifies related articles that may use different keywords or descriptors.

You may prefer to have someone help you with your first search. In this case, you probably will consult with a librarian, and the search may have to be scheduled and carried out by library personnel. You will most likely be asked to complete a form on which the research statement, keywords, and databases are specified. Library personnel can be extremely helpful in selecting descriptors and in the entire literature search process. Search results can be saved and downloaded or e-mailed to you so that you can print them at your convenience. Although you may have to pay a fee for such searches, in most cases the resulting bibliography is well worth the expense, especially in light of its comprehensiveness and the savings in time and energy. We note, however, that if you plan to use the library frequently for professional or research reasons, it will be well worth the effort to learn how to do searches yourself.

Obtaining the Primary Sources

After you have a list of related references, you must obtain the actual studies and read them. Many electronic databases have abstracts in addition to the bibliographic information. The abstract is extremely helpful in making the decision about whether the article is worth retrieving. For most searches, reading the abstract will inform you whether the paper will be needed, and you can easily include this with the saved bibliographic reference with the citations you selected to look at further.

After you know which articles you want to retrieve, you need to obtain them. Virtually all journals are available in electronic form. Some journals have many years of electronic archives, whereas others have fewer years because they provided digital versions of papers. As you learn to use the databases available in your library, it is a good practice to learn how to retrieve journal articles electronically. Each library has a different way of doing this—everything from direct links from the database to having to search for each journal separately—and we recommend becoming proficient in finding and retrieving papers at the library you will use.

Your library may not have all the journals that are on the list of references. You must consult your library's information system to see whether your library carries a particular journal. If the article is from an older volume, you may have to retrieve the physical copy and make a photocopy. If the library does not have the journal or volume that you need, you should use interlibrary loan or possibly obtain a fax or pdf file of the article through the service *ingentaconnect*.

Table 2.2 lists journals that are often used by students studying physical activity. Each journal title is accompanied by its impact factor and 5-year impact factor from the *Journal Citation Reports* (Thomson Reuters, 2014). The impact factor is a standardized measure of how often articles from the journal have been cited over the previous 2 years. The 5-year impact factor is a similar measure for the previous 5 years. The higher the impact factor is, the more citations there have been per article published by the journal over the period.

TABLE 2.2

Journals With Impact Factors

Journal title	Impact factor	5-year impact factor
ACSM's Health & Fitness Journal	0.268	0.265
Adapted Physical Activity Quarterly	1.079	1.695
American Educational Research Journal	2.275	3.618
American Journal of Physical Medicine & Rehabilitation	2.012	2.088
American Journal of Physiology— Heart and Circulatory Physiology	4.012	4.046
American Journal of Sports Medicine	4.699	5.009
Applied Physiology, Nutrition, and Metabolism (Physiologie Appliquée, Nutrition et Métabolisme)	2.225	2.531
Archives of Physical Medicine and Rehabilitation	2.441	2.531
Aviation, Space, and Environmental Medicine	0.782	0.998
Biology of Sport	0.527	0.403
British Journal of Sports Medicine	4.171	4.340
Child Development	4.235	5.834
Clinical Biomechanics	1.880	2.589
Clinical Journal of Sports Medicine	2.012	2.413
Clinics in Sport Medicine	2.583	2.413
Current Sports Medicine Reports	1.600	1.486
Epidemiology	6.178	5.906
Ergonomics	1.608	1.025
European Journal of Applied Physiology	2.298	2.073
European Journal of Sport Science	1.314	1.516
Exercise Immunology Review	9.929	6.103
Exercise and Sport Sciences Reviews	4.818	5.504
Experimental Brain Research	2.168	2.619
Gait & Posture	2.299	2.985
High Altitude Medicine & Biology	1.818	1.945
Human Biology	1.522	1.380
Human Factors	1.290	1.964
Human Movement Science	2.027	2.478
International Journal of Sport Nutrition and Exercise Metabolism	1.982	2.309
International Journal of Sport Physiology and Performance	2.683	2.934
International Journal of Sport Psychology	0.453	1.177
International Journal of Sports Medicine	2.374	2.424
Isokinetics and Exercise Science	0.354	0.613
Journal of Aging and Physical Activity	1.411	2.110
Journal of Applied Biomechanics	0.904	0.994
Journal of Applied Physiology	3.434	4.193
Journal of Applied Sport Psychology	1.098	1.789
Journal of Athletic Training	1.509	2.455
Journal of Back and Musculoskeletal Rehabilitation	1.041	1.029
Journal of Biomechanics	2.496	3.141
Journal of Curriculum Studies	0.779	0.899
Journal of Electromyography and Kinesiology	1.725	2.264

Journal title	Impact factor	5-year impact factor
Journal of Experimental Psychology: Human Perception and Performance	3.105	3.636
Journal of Motor Behavior	1.406	1.841
Journal of Orthopaedic & Sports Physical Therapy	2.376	3.195
Journal of Orthopaedic Trauma	1.540	2.174
Journal of the Philosophy of Sport	0.250	0.329
Journal of Physiotherapy (formerly the Australian Journal of Physiotherapy)	2.894	2.972
Journal of Rehabilitation Medicine	1.895	2.552
Journal of Science and Medicine in Sport	3.079	3.493
Journal of Shoulder and Elbow Surgery	2.365	2.699
Journal of Sport and Exercise Psychology	2.593	3.787
Journal of Sport and Health Science	1.227	1.108
Journal of Sport Management	0.727	1.108
Journal of Sport Rehabilitation	0.855	1.261
Journal of Sports Medicine and Physical Fitness	0.757	1.056
Journal of Sports Sciences	2.095	2.751
Journal of Strength and Conditioning Research	1.858	2.400
Journal of Teaching in Physical Education	0.479	1.169
Journal of the International Society of Sports Medicine	1.500	1.987
The Knee	1.702	1.926
Knee Surgery, Sports Traumatology, Arthroscopy	2.837	2.862
Medicina Dello Sport	0.125	0.152
Medicine & Science in Sports & Exercise	4.459	5.206
Motor Control	1.453	1.660
Movement Disorders	5.634	5.251
Operative Techniques in Sports Medicine	0.211	0.254
Pediatric Exercise Science	1.613	2.028
Perceptual & Motor Skills	0.521	0.658
Physical Therapy	3.245	2.932
Physical Therapy in Sport	1.373	1.830
Physician and Sportsmedicine	1,490	1.308
PM&R	1.662	2.249
Psychology of Sport and Exercise	1.768	2.405
Quest	0.902	0.994
Research Quarterly for Exercise and Sport	1.261	1.794
Scandinavian Journal of Medicine & Science in Sports	3.174	4.193
Science & Sports	0.537	0.443
Sociology of Sport Journal	1.125	1.382
Sports Biomechanics	0.867	1.170
Sport, Education and Society	1.333	1.642
Sport Psychologist	0.933	1.382
Sports Medicine	5.320	6.697
Sports Medicine and Arthroscopy Review	1.977	2.662
Sportverletzung-Sportschaden	0.283	0.371
Strength and Conditioning Journal	0.772	0.758
Teaching and Teacher Education	1.607	1.612
Wilderness & Environmental Medicine	0.790	0.925

Adapted from Cardinal and Thomas 2005.

This simple and easily obtained metric may appear to be a valid way to evaluate journal quality. Many authorities, however, have criticized the use of impact factors to evaluate journals (e.g., see Joint Committee on Quantitative Assessment of Research, 2008; Kulinna, Scrabis-Fletcher, Kodish, Phillips, & Silverman, 2009; Lawrence, 2003; Sammarco, 2008; Silverman, Kulinna, & Phillips, 2014). Among the objections are that authors can cite themselves, that impact factors do not tell whether the citation was positive or negative, and that comparisons across fields are problematic because there are larger and smaller groups of researchers. In a recent paper, Larivière and Gingras (2009) did an extensive search and found that duplicate papers published in two journals had different citation rates. They suggested that some journals have a cumulative advantage in maintaining their impact factors. These and other bibliographic rating indexes are just one bit of information to consider in evaluating journals.

Using the Internet: The Good, the Bad, and the Ugly

A great deal of information may be found online when searching the Internet. The correctness and accuracy of information, however, are often difficult to verify. In fact, much information is wrong and misleading. You can search the Internet using keywords with one of the standard search engines. You will likely find many sites when you conduct the search. Separating the good and valid ones from the useless and biased ones is difficult. However, several sites consistently contain useful information on topics that students want to find.

- Internet Scout (http://scout.wisc.edu) is a joint project from the National Science Foundation and the University of Wisconsin at Madison. It includes a considerable amount of useful scientific information on the biological sciences; physical sciences; and math, engineering, and technology.

- SOSIG stands for the Social Science Information Gateway (www.ariadne.ac.uk/issue2/sosig) and connects to more than 50,000 social science websites in such areas as anthropology, business, economics, education, environmental sciences, European studies, geography, government, law, philosophy, politics, psychology, and sociology.

- The WWW Virtual Library (http://vlib.org) is organized using the Library of Congress subject headings, thus making it similar to your local college or university library. Headings include agriculture, business and economics, communications and media, computing and computer science, education, engineering, humanities and humanistic studies, international affairs, law, recreation, regional studies, social and behavioral sciences, and society.

Using your computer to search on a system such as Google Scholar is very productive. For example, if you enter the term *schema theory,* you will see Richard Schmidt's original paper published in *Psychological Review* in 1975. This paper has been cited over 2,000 times. If you look down the list of papers in your search, you will see many papers that cite schema theory. If you have connections to your university library, you can download many of these papers at no cost.

Personal Computers

We have already mentioned that many faculty and students can access library sources and conduct literature searches from electronic devices in their own living quarters or offices. Your library should have information that will tell you how to log on and what information will be required to complete a search from a remote location.

You should also be aware of the ability to store bibliographic entries, abstracts, and even reprints of studies in the form of pdf files. A number of commercial programs are available for this purpose (e.g., EndNote, ProCite, and Reference Manager). After the bibliographic information and notes have been entered, the information can be retrieved by the use of pertinent keywords, author names, or journal names. Additions, changes, and deletions are easily accomplished, and the number of items that can be stored is limited only by electronic storage space limits. Each new entry is automatically stored in alphabetical order, and the complete bibliography is always instantly accessible. Linking pdf files of articles directly to the notes is easy, and additional information can be found quickly. Many universities now make one or more of these programs available free or at a reduced cost. In addition, most libraries have classes to learn the program or consultants who can help if a problem is encountered. Storing information in the cloud is also an option (e.g., Amazon Cloud Drive, Bitcasa, Box, CertainSafe, Dropbox).

Other Library Services

The library services available depend mostly on the size of the institution. Larger schools generally provide more financial support for the library, although some relatively small institutions have excellent library resources that provide outstanding support for the institutional and research aspects of the school.

Besides the usual services of the library information system (the computerized card catalog, reference and circulation departments, bibliographic collections, and stack areas), libraries also offer resources and services such as copy services, newspaper rooms, government document sections, telephone directories, college catalogs, and many other special features and services. One such valuable service is the interlibrary loan, which enables you to get books, theses and dissertations, and copies of articles in journals that your library does not carry. An interlibrary loan usually takes a short time from the date it is requested, and most libraries are now delivering the requests as pdf files.

Universities typically have extensive collections of resources available on **microform**, a general term that embraces microfilm, microfiche, microcard, and microprint. Microforms are simply miniaturized photographic reproductions of the contents of a printed page. You must use special machines called readers to enlarge the information so that you can read it.

An obvious advantage of microforms is economy of space. In addition, they are useful for acquiring material that would otherwise not be available. Rare books and manuscripts and deteriorating materials can be preserved on microform. Some journals are on microform so that libraries can obtain them more economically. Collections of government documents are often contained on microform, as are *ERIC* publications. The library usually has a reader–printer that can make paper copies of microform material. Microform holdings are normally listed in the library's electronic card catalog system and periodical listings. Many microform collections (e.g., *ERIC*) are being converted to pdf files for easier access and storage.

Many libraries offer guided tours, short orientation courses, self-guided tours, and tutorials in the use of electronic resources. Get acquainted with your library. The hours spent will be the wisest investment of your time as a graduate student.

microform—A general term that encompasses microfilm, microfiche, and any form of data storage on which the pages of a book, journal, or newspaper are photographed and reduced in size.

Step 5: Read and Record the Literature

Collecting related literature is a major undertaking, but the next step is even more time consuming. You must read, understand, and record the relevant information from the literature, keeping in mind one of the many (anonymous) Murphy's laws: No matter how many years you save an item, you will never need it until after you have thrown it away.

The person who originated this quote was probably a researcher working on a literature review. You can count on the fact that if you throw away one note from your literature search, that paper will be cited incorrectly in your text or reference list. If you delete your notes on an article because you do not believe that it is relevant to your research, your major professor, a committee member, or the journal to which you submit the paper will request inclusion of the article. Then, when you go back to locate the article in the library, either it will have been ripped out of the journal, the whole journal will be missing, or a professor will have checked it out and never returned it (and the librarian will not reveal the professor's name). Therefore, when you find a particular paper, take careful and complete notes, including exact citation information, and use a bibliographic program to link to the pdf file so that you are prepared if you need to see the article again.

To learn more about understanding the literature you read, try to decipher the scientific phrases in the Deciphering Scientific Phrases sidebar. After you understand what each phrase really means, you should have little difficulty understanding authors of research articles. Seriously, though, as a researcher, you should note the following information from research studies that you read:

- Statement of the problem (and maybe hypotheses)
- Characteristics of the participants
- Instruments and tests used (including reliability and validity information if provided)
- Testing procedures
- Independent and dependent variables
- Treatments applied to participants (if an experimental study)
- Design and statistical analyses
- Findings
- Questions raised for further study
- Citations to other relevant studies that you have not located

When studies are particularly relevant to the proposed research, save a copy. If the article is not available electronically, photocopy it and write the complete citation on the title page if the journal does not provide this. Also, indicate in your electronic system or on a note card which studies are saved electronically on your hard drive and which are photocopied.

Another useful way to learn to read and understand related literature is to critique a few studies. Use the series of questions in the Criteria for Critiquing a Research Paper sidebar when critiquing a study. The sidebar Sample Form for a Critique is a form that we use for reporting critiques. A few critiquing attempts should aid you in focusing on the important information contained in research studies.

Reviewers can occasionally go overboard in their critiques, much like the critique of Schubert's *Unfinished Symphony* presented in the sidebar A Critique of Schubert's Productivity.

To summarize, by using a computerized system for note taking (or index cards if you prefer), you can record the important information about most studies and index it by topic. Always be sure to record the complete and correct citation with the appropriate citation style used by your institution (e.g., American Psychological Association [APA], *Index Medicus*). If using a computerized system, be sure to back up the data files and article pdf files on a stick drive or in the cloud.

Step 6: Write the Literature Review

After you have located and read the necessary information and have recorded the appropriate bibliographic data, you are ready to begin writing the literature review. The literature review has three basic parts:

1. Introduction
2. Body
3. Summary and conclusions

Deciphering Scientific Phrases

WHAT WAS SAID	WHAT WAS MEANT
It has long been known that . . .	I haven't bothered to look up the original reference but . . .
Of great theoretical and practical importance	Interesting to me
Although it has not been possible to provide definite answers to those questions ...	The experiment didn't work out, but I figured I could at least get a publication out of it.
The W-PO system was chosen as especially suitable to show the predicted behavior.	The researcher in the next lab had some already made up.
Three of the samples were chosen for detailed study.	The results on the others didn't make sense.
Accidentally strained during mounting	Dropped on the floor
Handled with extreme care throughout the experiment	Not dropped on the floor
Typical results are shown.	The best results are shown.
Agreement with the predicted curve is	
excellent	fair
good	poor
satisfactory	doubtful
fair	imaginary
It is suggested that . . .	I think . . .
It is believed that . . .	I think . . .
It may be that . . .	I think . . .
It is clear that much additional work will be required before a complete understanding . . .	I don't understand it . . .
Let me make one thing perfectly clear.	A snow job is coming.
Unfortunately, a quantitative theory to account for these results has not been formulated.	Neither I nor anybody else understands it.
Correct with an order of magnitude . . .	Wrong.
Thanks are due to Joe Glotz for assistance with the experiments and to John Doe for valuable discussion.	Glotz did the work, and Doe explained what it meant.

CRITERIA FOR CRITIQUING A RESEARCH PAPER

I. Overall impression (most important): Is the paper a significant contribution to knowledge about the area?

II. Introduction and literature review
   a. Is the research plan developed within a reasonable theoretical framework?
   b. Is current and relevant research cited and properly interpreted?
   c. Is the statement of the problem clear, concise, testable, and derived from the theory and research reviewed?

III. Methods
   a. Are relevant participant characteristics described, and are the participants appropriate for the research?
   b. Is the instrumentation appropriate?
   c. Are testing or treatment procedures described in sufficient detail?
   d. Are the statistical analyses and research design sufficient?

IV. Results
   a. Do the results evaluate the stated problems?
   b. Is the presentation of the results complete?
   c. Are the tables and figures appropriate?

V. Discussion
   a. Are the results discussed?
   b. Are the results related to the problem, theory, and previous findings?
   c. Is there excessive speculation?

VI. References
   a. Are all references in the correct format, and are they complete?
   b. Are all references cited in the text?
   c. Are all dates in the references correct, and do they match the text citation?

VII. Abstract
   a. Does it include a statement of the purpose; description of participants, instrumentation, and procedures; and a report of meaningful findings?
   b. Is the abstract the proper length?

VIII. General
   a. Are keywords provided?
   b. Are running heads provided?
   c. Does the paper use nonsexist language and provide for protection and appropriate labeling of human participants?

The introduction should explain the purpose of the review and its organization. The body of the review should be organized around important topics. Finally, the review should summarize important implications and suggest directions for future research. The purpose of the review is to demonstrate that your problem requires investigation and that you have considered the value of relevant research in developing your hypotheses and methods—that is, that you know and understand what other people have done and how that research relates to and supports what you plan to do.

Sample Form for a Critique

I. Basic information
   a. Name of journal
   b. Publisher of journal
   c. Number of articles in this issue
   d. Presence of publication guidelines
   e. Name of the editor
   f. Presence of a yearly index
   g. Writing style that is used (e.g., APA, *Index Medicus*)?
   h. Complete reference in APA style for article you will review

II. Summary of article

III. Critique of article
   a. Introduction and review
   b. Methods
   c. Results
   d. Discussion
   e. References
   f. Overall

IV. Photocopy of article

V. Standard critique rules
   a. It must be typed (double-spaced) in APA style.
   b. It should not be in any type of folder; pages should be stapled in the upper left-hand corner.
   c. A cover page should identify the course, purpose, and yourself.
   d. The critique should not exceed five typewritten pages exclusive of the cover page.

A Critique of Schubert's Productivity

A company CEO was given a ticket for a performance of Schubert's *Unfinished Symphony.* Unable to go, he passed the invitation to the company TQM (total quality management) coordinator. The next morning the CEO asked the TQM coordinator how he enjoyed the performance. Instead of voicing a few plausible observations, the TQM coordinator handed the CEO a memorandum that read as follows:

- For considerable periods, the oboe players had nothing to do. Their number should be reduced, and their work spread over the whole orchestra, thus avoiding peaks of inactivity.

- All 12 violins were playing identical notes. This seems unnecessary duplication, and the staff of this section should be drastically cut. If a large volume of sound is required, this could be obtained through the use of an amplifier.

- Much effort was involved in playing the demi-semiquavers. This seems an excessive refinement, and it is recommended that all notes be rounded up to the nearest semiquaver. If this were done, it should be possible to use trainees instead of craftsmen.

- No useful purpose is served by repeating with horns the passage that has already been handled by the strings. If all such redundant passages were eliminated, the concert could be reduced from 2 hours to 20 minutes.

In light of the preceding, one can only conclude that had Schubert given attention to these matters, he would probably have had the time to finish his symphony.

The introduction to the review (or to topical areas within the review) is important. If these paragraphs are not well done and interesting, the reader may skip the entire section. Attempt to attract the reader's attention by identifying in a provocative way the important points to be covered.

The body of the literature review requires considerable attention. Relevant research must be organized, synthesized, and written in a clear, concise, and interesting way. No unwritten law dictates that literature reviews must be boring and poorly written, although we suspect that some graduate students work from that assumption. Part of the problem stems from graduate students' belief that they must find a way to make their scientific writing complex and circuitous as opposed to simple and straightforward. Apparently, the rule is never to use a short and simple word when a longer, more complex one can be substituted. In table 2.3, we provide a useful aid to potential research writers. We strongly recommend Day and Gastel's (2006) book, which can easily be read in four to six hours. It provides many excellent and humorous examples that are valuable in writing for publication and for theses and dissertations.

TABLE 2.3

Examples of Avoiding Jargon When Writing

Jargon	Preferred usage
a considerable amount of	much
a majority of	most
a variety of	many
absolutely essential	essential
along the lines of	like
an order of magnitude faster	10 times faster
are of the same opinion	agree
as a consequence of	because
as a matter of fact	in fact (or leave out)
as a means of	to
as is the case	as happens
as to	about (or leave out)
at an earlier date	previously
at the present time	now
based on the fact that	because
basic fundamentals	basics
benchmarking	measuring
by means of	by, with
is capable of	can
consensus of opinion	consensus
definitely proved	proved
despite the fact that	although
drive standards-based niches	assess
due to the fact that	because
during the course of	during, while
end result	result
entirely eliminate	eliminate
exhibits a tendency to	tends
fabricate	make
fewer in number	fewer

Jargon	Preferred usage
first of all	first
for the duration of	during
for the purpose of	for
for the reason that	since, because
from the point of view of	for
give consideration to	consider
give rise to	cause
has a requirement for	needs
has the capability of	can
if at all possible	if possible
if conditions are such that	if, when
in a number of cases	some
in a satisfactory manner	satisfactorily
in a timely matter	quickly
in a very real sense	in a sense (or leave out)
in all cases	always, invariably
in case	if
in connection with	about, concerning
in many cases	often
in my opinion it is not an unjustifiable assumption that	I think (or leave out)
in order to	to
in respect to	about
in some cases	sometimes
in terms of	about
in the course of	during, while
in the event that	if
in view of the fact that	because, since
inasmuch as	for, as
initiate	begin, start
is defined as	is
it has been reported by Smith	Smith reported
it is apparent that	apparently
it is believed that	I think
it is clear that	clearly
it is clear that much additional work will be required before a complete understanding	I do not understand it
it is evident that a produced b	a produced b
it is of interest to note that	(leave out)
it is often the case that	often
it may be that	I think
it may, however, be noted that	but
it should be noted that	note that (or leave out)
it was observed in the course of the experiments that	we observed
it would thus appear that	apparently
makes an attempt	attempts
many different	many
necessitates the inclusion of	needs, requires

(continued)

Table 2.3 *(continued)*

Jargon	Preferred usage
needless to say	(leave out, and consider leaving out whatever follows it)
notwithstanding	even though
of great theoretical and practical importance	useful
on a daily basis	daily
on the basis of	by
on the grounds that	since, because
on the part of	by, among, for
our attention has been called to the fact that	we belatedly discovered that
past history	past
perform	do
pertaining to	about
postpone until later	postpone
potentialities	chances
prior to	before
protein determinations were performed	proteins were determined
provides guidance for	guides
quite unique	unique
rather interesting	interesting
red in color	red
referred to as	called
resultant effect	result
scaled back	cut
sufficient	enough
take into consideration	consider
terminate	end
the opinion is advanced that	I think
the question as to whether	whether
the reason is because	because
there can be little doubt that this is	this probably is
there is reason to believe	I think
this result would seem to indicate	this result indicates
throughout the whole of the experiment	throughout the experiment
through the use of	by, with
upon	on
utilize	use
was of the opinion that	believed
we have insufficient knowledge	we do not know
we wish to thank	we thank
without further delay	now
with reference to	about (or leave out)
with regard to	concerning, about (or leave out)
with the possible exception of	except
with the result that	so that

Based on R.A. Day and B. Gastel, 2006, *How to write and publish a scientific paper,* 6th ed. (Westport, CT: Greenwood Press).

Besides removing as much jargon as possible (using Day and Gastel's suggestions), you should be clear and to the point. We advocate the KISS principle (Keep it simple, stupid) as a basic tenet for writing. Many grammatical errors can be avoided by using simple declarative sentences. The list in the sidebar Ten Commandments of Good Writing—Plus a Few Others rephrases the commandments of good writing from a 1968 Council of Biology Editors newsletter. The first 10 items are Day's, and we included a few of our own.

Proper syntax (the way words and phrases are put together) is the secret to successful writing. Some examples of improper syntax and other unclear writing are highlighted in the sidebar Examples of Unclear Writing.

As mentioned previously, the literature review should be organized around important topics. These topics serve as subheadings in the paper to direct the reader's attention. The best way to organize the topics and the information within topics is to develop an outline. The more carefully the outline is planned, the easier the writing will be. A good task is to select a review paper from a journal or from a thesis or dissertation review of literature and reconstruct the outline that the author must have used. In looking at older theses and dissertations, we find that the literature review tends to be a historical account, often presented in chronological order. We suggest that you not select one of these older studies, because this style is cumbersome and usually poorly synthesized. Locke and colleagues (2014) provided detailed advice for preparing and writing the literature review.

To write a good literature review, you should write as you like to read. No one wants to read abstracts of study after study presented in chronological order. A more interesting and readable approach is to present a concept and then discuss the findings about that concept, documenting them by references to the research reports related to them. In this way, consensus and controversy can be identified and discussed. More relevant and important studies can be presented in detail, and several studies with the same outcome can be covered in one sentence.

Ten Commandments of Good Writing—Plus a Few Others

1. Each pronoun should agree with their antecedent.
2. Just between you and i, case is important.
3. A preposition is a poor word to end a sentence with. (Incidentally, did you hear about the streetwalker who violated a grammatical rule? She unwittingly approached a plainclothes officer, and her proposition ended with a sentence.)
4. Verbs has to agree with their subjects.
5. Don't use no double negatives.
6. A writer mustn't shift your point of view.
7. When dangling, don't use participles.
8. Join clauses good, like a conjunction should.
9. Don't write a run-on sentence it is difficult when you got to punctuate it so it makes sense when the reader reads what you wrote.
10. About sentence fragments.
11. Don't use commas, which aren't necessary.
12. Its' important to use apostrophe's right.
13. Check to see if you any words out.
14. As far as incomplete constructions, they are wrong.
15. Last but not least, lay off cliches.

Adapted from Day 1983.

Examples of Unclear Writing

UNCLEAR QUESTIONS TO CHILDREN AND THEIR ANSWERS

- Name the four seasons. Salt, pepper, mustard, and vinegar.
- How can you delay milk turning sour? Keep it in the cow.
- What is the fibula? A small lie.
- What does *varicose* mean? Nearby.
- How is dew formed? The sun shines and makes the leaves perspire.
- What do fish do in school? They take debate.
- What do you see in Los Angeles when the fog lifts? UCLA.

SENTENCES WITH IMPROPER SYNTAX

- The patient was referred to the hospital for repair of a hernia by a social worker.
- As a baboon who grew up wild in the jungle, I realized that Wiki had special nutritional needs.
- No one was injured in the blast, which was attributed to a buildup of gas by one town official.
- Table 1 contains a summary of responses pertaining to suicide and death by means of a questionnaire. (We are aware that some questionnaires can be ambiguous, irrelevant, and trivial, but we had no idea they could be fatal.)

In a thesis or dissertation, the two most important aspects of the literature review are criticism and completeness. The studies should not simply be presented relative to a topic. The theoretical, methodological, and interpretative aspects of the research should be criticized not study by study, but rather across studies. This criticism demonstrates your grasp of the issues and identifies problems that should be overcome in the study you are planning. Frequently, the problems identified by criticism of the literature provide justification for your research.

Completeness (not in the sense of the length of the review but rather of reference completeness) is the other important aspect of the literature review. You should demonstrate to your committee that you have located, read, and understood all the related literature. Many studies may be redundant and need only appropriate citing, but they must be cited. The thesis or dissertation is your passport to graduation because it demonstrates your competence; therefore, never fail to be thorough. This advice, however, applies only to the thesis or dissertation. Writing for publication or using alternate thesis or dissertation formats does not require emphasizing the completeness of the literature cited (note the use of the word *cited,* not *read*). Journals have limited space and usually want the introduction and literature review to be integrated and relatively short.

> Insightful criticism and completeness of references are the most important aspects of the literature review.

SUMMARY

Identifying and formulating a researchable problem are often difficult tasks, especially for the novice researcher. This chapter offered suggestions to help you find suitable topics. Inductive and deductive reasoning for formulating research hypotheses were discussed. Inductive reasoning moves from observations to specific hypotheses to a more general theoretical model. Deductive reasoning moves from a theoretical explanation to specific hypotheses to be tested.

The steps in a literature search include writing the problem statement, consulting secondary sources, determining descriptors, searching preliminary sources to locate primary sources, reading and recording the literature, and writing the review.

There are no shortcuts to locating, reading, and indexing the literature and then writing the literature review. If you follow our suggestions, you can do it more effectively, but much hard work is still required. A good scholar is careful and thorough. Do not depend on what others report, because they are often incorrect. Look it up yourself.

No one can just sit down and write a good literature review. A careful plan is necessary. First, outline what you propose to write, write it, and then write it again. When you are convinced that the review represents your best effort, have a knowledgeable graduate student or faculty member read it. Then welcome their suggestions. Next, have a friend who is not as knowledgeable read it. If your friend can understand it, then your review is probably in good shape. Of course, your research methods professor will find something wrong or at least something that he or she thinks should be different. Just remember that professors feel obligated to find errors in graduate students' work.

Check Your Understanding

These exercises will help you locate, synthesize, organize, critique, and write a literature review. Suggestions from this chapter will help you complete these exercises.

1. Critique a research study in your area of interest. Use the questions in the sidebar Criteria for Critiquing a Research Paper. Report the critique using the sidebar Sample Form for a Critique as your model.

2. Select a research paper from a journal. Construct the outline that the author probably used to write the introd uction and literature review (whether presented together or separately).

Presenting the Problem

You are entitled to your own opinions—but not your own facts.

—Daniel Patrick Moynihan

In a thesis or dissertation, the first section or chapter introduces the problem. Indeed, it is often titled "Introduction." Several sections in the introduction serve to convey the significance of the problem and set forth the dimensions of the study. This chapter discusses each of the following sections, which are frequently required in the first part of a thesis or dissertation:

- Title
- Introduction
- Problem statement
- Hypothesis
- Definitions
- Assumptions and limitations
- Significance

Not all thesis advisors subscribe to the same thesis format, because there is no universally accepted one. Moreover, the nature of the research problem causes the format to vary. For example, a historical study would not adhere to the same format used in an experimental study, and section titles might differ between descriptive studies and qualitative studies. We merely present sections, each with a purpose and with specific characteristics, typically found in the introduction.

CHOOSING THE TITLE

Although discussing the title first may seem logical, it might surprise you to learn that titles are often not determined until after the study has been written. But at the proposal meeting, you must have a title (although it may be provisional), so we discuss it first.

Some writers claim that there is a trend toward shortening titles (e.g., Day & Gastel, 2006). But an analysis of more than 10,000 dissertations in seven areas of education failed to demonstrate such a trend (Coorough & Nelson, 1997). Many titles are, in essence, the statement of the problem (some even include the methods section). Following is an example of a too-lengthy title. (*Note:* The examples we use to represent poor practices are fictional. Frequently, they have been suggested by actual studies, but any similarity to a real study is purely coincidental.)

An Investigation of a Survey and Analysis of the Influence of PL 94-142 on the Attitudes, Teaching Methodology, and Evaluative Techniques of Randomly Selected Male and Female Physical Education Teachers in Public High Schools in Cornfield County, State of Confusion

Such a title includes simply too much information. Day (1983) humorously addressed this problem by reporting a conversation between two students. When one asked whether the other had read a certain paper, the reply was, "Yes, I read the paper, but I haven't finished the title yet" (p. 10). A better title for the previous would be "PL 94-142's Influence on Physical Education Teachers' Attitudes, Methodology, and Evaluations."

The purpose of the title is to convey the content, but this purpose should be accomplished as succinctly as possible. For example, "The Twelve-Minute Swim as a Test for Aerobic Endurance in Swimming" (Jackson, 1978) is a good title because it tells the reader exactly what the study is about. It defines the purpose, which is the validation of the 12 min swim, and it delimits the study to the assessment of aerobic endurance for swimmers.

> A good title succinctly informs the reader of the study's content.

However, beware of going to the other extreme in striving for a short title. A title such as "Professional Preparation" is not particularly helpful. It does not include the field or the aspects of professional preparation studied. The key to the effectiveness of a short title is whether it reflects the contents of the study. A title that is specific is more easily indexed and retrieved through electronic databases and is more meaningful for a reader who is searching for literature on a certain topic.

Avoid superfluous words and phrases such as *An Investigation of, An Analysis of,* and *A Study of.* They simply increase the length of the title and contribute nothing to the description of the content. Consider this title: "A Study of Three Teaching Methods." Half the title consists of superfluous words: *A Study of.* The rest of the title is not specific enough to be indexed effectively: Teaching what? What methods?

Furthermore, always be aware of your audience. You can assume that your audience is reasonably familiar with the field, the accepted terminology, and the relevant problem areas. An outsider can question the significance of studies in any field. Some titles of supposedly scholarly works are downright humorous. Just recently we ran across the title "The Nature of Navel Fluff" from the journal *Medical Hypotheses* (Steinhauser, 2009), which suggested that lint might provide a cleaning function for the navel.

For a rousing good time, peruse the titles of theses and dissertations completed at a university in any given year. For example, we found "The Phospholipid Distribution in the Testes of the House Cricket." How weird can you get? We are joking, of course. The point is that people tend to criticize studies done in other disciplines simply because they are ignorant about those disciplines.

WRITING THE INTRODUCTION

The introductory portion of a thesis or research article should create interest in the problem. You use the introduction to persuade readers of the significance of the problem, provide background information, bring out areas of needed research, and then skillfully and logically proceed to the purpose of the study.

How to Write a Good Introduction

A good introduction requires literary skill because it should flow smoothly yet be reasonably brief. Be careful not to overwhelm the reader with technical jargon, because the reader must be able to understand the problem to gain an interest in the solution. Therefore, an important rule is this: Do not be too technical. A forceful, simple, and direct vocabulary is more effective for purposes of communication than scientific jargon and the worship of polysyllables. Day (1983, pp. 147–148) related a classic story of the pitfalls of scientific jargon:

> This reminds me of the plumber who wrote the Bureau of Standards saying that he had found hydrochloric acid good for cleaning out clogged drains. The Bureau wrote back, "The efficacy of hydrochloric acid is indisputable, but the corrosive residue is incompatible with metallic permanence." The plumber replied that he was glad the Bureau agreed. The Bureau tried again, writing, "We cannot assume responsibility for the production of toxic and noxious residues with hydrochloric acid and suggest that you use an alternative procedure." The plumber again said that he was glad the Bureau agreed with him. Finally, the Bureau wrote to the plumber, "Don't use hydrochloric acid. It eats the hell out of pipes."

Audience awareness is important. Again, you can assume that readers are reasonably informed about the topic (or they probably would not be reading it in the first place). But even an informed reader needs some refresher background information to understand the nature of the problem, to be sufficiently interested, and to appreciate your rationale for studying the problem. You must remember that your audience has not been as completely and recently immersed in this area of research as you have been.

The introductory paragraphs must create interest in the study; thus, your writing skill and knowledge of the topic are especially valuable in the introduction. The narrative should introduce the necessary background information quickly and explain the rationale behind the study. A smooth, unified, well-written introduction should lead to the problem statement with such clarity that the reader could state the purpose of the study before reading it. Sometimes, however, statistical findings may be incorrectly connected in the introduction, producing inappropriate facts: A recent study found that the average Australian walks about 900 miles per year and drinks approximately 22 gallons of beer per year, so Australians get about 41 miles per gallon—not bad!

> Write the introduction so that readers know the study's purpose before reading the problem statement. Introductions should also create interest, provide background information, and explain the study's rationale.

The following introductions were selected from research journals for their brevity of presentation and effectiveness. This is not to say that brevity in itself is a criterion, because some topics require more comprehensive introductions than do others. For example, studies developing or validating a theoretical model usually need longer introductions than do applied research topics. Furthermore, theses and dissertations (in the traditional format) almost always have longer introductions than journal articles simply because of the page-cost considerations in the latter.

Examples of Good Introductions

The following examples demonstrate some desirable features in an introduction; they include a general introduction, background information, a mention of gaps in the literature and areas of needed research, and a logical progression leading to the problem statement. After you have read them, see whether you can write the purpose for each study.

Example 3.1 (from Teramoto & Golding, 2009)

From *Research Quarterly for Exercise and Sport* Vol. 80, pgs. 138-145, Copyright 2009 by the American Alliance for Health, Physical Education, Recreation and Dance, 1900 Association Drive, Reston, VA 20191

> [General introduction] Research shows that participating in regular, vigorous physical activity is a key factor in reducing the risk of coronary heart disease (CHD) (Mazzeo et al., 1998). CHD is the leading cause of death in the United States (Rosamond et al., 2007). Physical inactivity is one of the CHD risk factors that can be modified (National Cholesterol Education Program [NCEP] Expert Panel, 2002).
>
> [Background information] Elevated plasma levels of cholesterol and triglycerides are associated with the development of CHD. . . . Regular physical activity can improve certain lipid levels associated with the risk of CHD. . . .

[Lead-in] Although the research on positive effects of regular physical activity on plasma lipids is extensive, few studies have monitored the changes in plasma lipid levels in individuals who exercise regularly for more than 10 years.

Example 3.2 (from Stodden, Langendorfer, & Roberton, 2009)

From *Research Quarterly for Exercise and Sport*, Vol. 80, pgs. 223-229, Copyright 2009 by the American Alliance for Health, Physical Education, Recreation and Dance, 1900 Association Drive, Reston, VA 20191.

[General introduction] Developing healthy lifestyles includes maintaining appropriate levels of health-related physical fitness . . . and physical activity. Maintaining physical fitness levels may reduce mortality risk and incidence of chronic diseases (Blair et al., 1995; CDC, 2001; Freedman, Dietz, Srinivasan, & Berenson, 1999; Centers for Disease Control and Prevention, 2001) and muscular strength and endurance . . . the functional limitations that lead to dependent care in the elderly . . . .

[Background information] While physical fitness in children, adolescents, and adults has been promoted (Blair et al., 1995; McKenzie et al., 2003; Okely, Booth, & Patterson, 2001; Sallis et al., 1997), relatively little is known about the effects of maintaining motor skills on physical fitness throughout the lifespan. Intermediate to high levels of competence in fundamental motor skills (FMS) required for successful participation in many sports and physical activities may be associated with higher levels of performance and health-related physical fitness.

[Lead-in] Overall, research on the relationship between motor skill competence and aspects of physical fitness is scanty.

STATING THE RESEARCH PROBLEM

The problem statement follows the introduction. We should point out that the literature review is often included in the introductory section and thus precedes the formal statement of the problem. If this is the case, then a brief problem statement should appear fairly soon in the introductory section before the literature review.

The problem statement in example 3.1 from Teramoto and Golding (2009) was to examine well-documented longitudinal effects of regular exercise on reducing the risk of CHD especially for older adults' maintenance of an independent lifestyle.

The problem statement in the Stodden, Langendorfer, and Roberton (2009) study was also obvious from the introduction—to evaluate the role of motor skill competence in health-related aspects of physical fitness.

Identifying the Variables

The problem statement should be succinct. When the study has several subpurposes, however, this goal is not always easily accomplished. The statement should identify the variables in the study, including the independent variable, the dependent variable, and the categorical variable (if any) for an experimental or quasi-experimental study. Usually, some **control variables** (which could possibly influence the results and are kept out of the study) can also be identified here.

The independent and dependent variables were mentioned in chapter 1. The independent variable is the experimental, or treatment, variable; it is the cause. The dependent variable is what is measured to assess the effects of the independent variable; it is the effect. A **categorical variable**, sometimes called a moderator variable (Tuckman, 1978), is a kind of independent variable except that it cannot be manipulated, such as age, race, or sex. A categorical variable is studied to determine whether its presence changes the cause-and-effect relationship of the independent and dependent variables.

control variable—A factor that could possibly influence the results and that is kept out of the study.

categorical variable—A kind of independent variable that cannot be manipulated, such as age, race, or sex; also called *moderator variable*.

Anshel and Marisi (1978) studied the effect of synchronous and asynchronous movement to music on endurance performance. One group performed an exercise in synchronization to background music, one group exercised with background music that was not synchronized to the pace of the exercise, and a third group exercised with no background music.

The independent variable was the background music condition. There were three levels of this variable: synchronous music, asynchronous music, and no music. The dependent variable was endurance performance, which was reflected by the amount of time the participant could exercise on a bicycle ergometer until exhaustion. In this study, the endurance performances of men and women under the synchronous, asynchronous, and no-music conditions were compared. The authors thus sought to determine whether men responded differently than women to the exercise conditions. Sex, then, represented a categorical variable. Not all studies have categorical variables.

The researcher decides which variables to manipulate and which variables to control. He or she can control the possible influence of some variable by keeping it out of the study. Thus, the researcher chooses not to assess a variable's possible effect on the relationship between the independent and dependent variables, so this variable is controlled. For example, suppose a researcher is comparing stress-reduction methods on the competitive state anxiety of gymnasts before dual meets. The gymnasts' years of competitive experience might have a bearing on their anxiety scores. The researchers have a choice. They can include this attribute as a categorical variable by requiring that half the participants have had a certain number of years of experience and that the other half have had less, or they can control the variable of experience by requiring that all participants have similar experience.

Researchers control the influence of variables by deciding which variables should be manipulated and which should be controlled.

The decision to include or exclude some variable depends on several considerations, such as whether the variable is closely related to the theoretical model and how likely it is that an interaction will be present. Practical considerations include the difficulty of making a variable a categorical variable or controlling it (such as the availability of participants having a particular trait) and the amount of control the researcher has over the experimental situation.

In the study of the effects of synchronized music on endurance (Anshel & Marisi, 1978), the factor of fitness level was controlled by giving all the participants a physical working capacity test. Then, on the basis of this test, each person exercised at a workload that would cause a heart rate of 170 beats per minute (bpm). Thus, even though the ergometer resistance settings would be different from person to person, all participants would be exercising at approximately the same relative workload; the differences in fitness were controlled in this manner. Another way of controlling fitness as a variable would be to test participants on a fitness test and select just those with a certain level of fitness.

Extraneous variables are factors that could affect the relationship between the independent and dependent variables but are not included or controlled. The possible influence of an extraneous variable is usually brought out in the discussion section. Anshel and Marisi (1978) speculated that some differences in the performances of men and women might be due to the women's reluctance to exhibit maximum effort in the presence of a male experimenter. Consequently, this would be an extraneous variable. (All the kinds of variables are discussed in more detail in chapter 18.)

extraneous variable—A factor that could affect the relationship between the independent and dependent variables but that is not included or controlled.

Rarely are the variables labeled as such in the problem statement. Occasionally, the researcher identifies the independent and dependent variables, but mostly these variables are just implied.

In summary, an effectively constructed introduction leads smoothly to the purpose of the study. This item is expressed as the statement of the problem and should be as clear and concise as the subpurposes, or variables, allow it to be.

Structuring the Problem Statement

To achieve clarity in the statement of the problem, a final but important aspect to consider is sentence structure, or syntax. For example, suppose a researcher conducted a study "to compare sprinters and distance runners on anaerobic power, as measured by velocity in running up a flight of stairs." Observe the difference in meaning if the researcher had worded the purpose as "to compare the anaerobic power of sprinters and distance runners while running up a flight of stairs." The researcher would have to be in good shape to make those comparisons while running up stairs. Another example of faulty syntax was the case in which the purpose of the study was "to assess gains in quadriceps strength in albino mice using electrical stimulation." Those mice had to be awfully clever to use electrical instruments.

PRESENTING THE RESEARCH HYPOTHESIS

After you have stated the research problem, you must present the hypothesis (or present a question for a qualitative study). The formulation of hypotheses was discussed in chapters 1 and 2. The discussion here is on the statement of the hypotheses and the distinction between research hypotheses and the null hypothesis. Remember that **research hypotheses** are the expected results. In the study by Anshel and Marisi (1978), a research hypothesis might be that endurance performance would be enhanced by exercising to synchronized music. The introduction produces a rationale for that hypothesis. Another hypothesis might be that exercising to asynchronous music would be more effective than exercising with no background music (because the pleasurable sensory stimuli blocks the unpleasant stimuli associated with the fatiguing exercise). As a further example, a researcher in cardiac rehabilitation might hypothesize that distance from the exercise center is more influential as a factor in patients' exercise adherence than the type of activities offered in the cardiac rehabilitation program. In the example given in chapter 1, a dance teacher hypothesized that the use of video in the instructional program would enhance the learning of dance skills. Here are a few humorous research hypotheses:

- A plastic bag will never open from the end you try first.
- Pressing harder on the remote control will help if the batteries are weak.
- No matter the color of the bubble bath soap, the bubbles are always white.
- Glue will not stick to the inside of the bottle.

In contrast, the **null hypothesis** is used primarily in the statistical test for the reliability of the results; the null hypothesis says that there are no differences among treatments (or no relationships among variables). For example, any observed difference or relationship is due simply to chance (see chapter 7). The null hypothesis is usually not the research hypothesis, and the research hypothesis is what usually is presented. Generally, the researcher expects one method to be better than others or anticipates a relationship between two variables. One does not embark on a study if nothing is expected to happen. On the other hand, a researcher sometimes hypothesizes that one method is just as good as another. For example, in the multitude of studies done in the 1950s and 1960s on isometric versus isotonic exercises, it was often hypothesized that the upstart isometric exercise was just as effective as the traditional isotonic exercise, if there was regular specific knowledge of results. In a study on the choice of recreational activities of mentally impaired children, Matthews (1979) showed that most research in this area, which reported differences between children with and without mental impairment, failed to consider socioeconomic status. Consequently, he hypothesized that there were no differences in the frequency of participation in recreational activities between children who were mildly mentally impaired and those who were not mentally impaired when socioeconomic status was held constant.

research hypothesis—A hypothesis deduced from theory or induced from empirical studies that is based on logical reasoning and predicts the outcome of the study.

null hypothesis—A hypothesis used primarily in the statistical test for the reliability of the results that says that there are no differences among treatments (or no relationships among variables).

Furthermore, sometimes the researcher does not expect differences in some aspects of the study but does expect a difference in others. A researcher might hypothesize that children of high aptitude in learning would do better with one style of teaching, whereas children of low aptitude would fare better with another style. In a study of age differences in the strategy for recall of movement (Thomas, Thomas, Lee, Testerman, & Ashy, 1983), the authors hypothesized that because location is automatically encoded in memory, there would be no real difference between younger and older children in remembering a location (where an event happened during a run). But they hypothesized that there would be a difference in remembering distance because the older child spontaneously uses a strategy for remembering whereas the younger child does not. The formulation of hypotheses is an important aspect of defining and delimiting the research problem.

OPERATIONALLY DEFINING TERMS

Another task in the preparation of the first section of a thesis or dissertation is operationally defining certain terms so that the researcher and the reader can adequately evaluate the results. The dependent variable must be operationally defined.

An **operational definition** describes an observable phenomenon, as opposed to a synonym definition or dictionary definition. To illustrate, a study such as Anshel and Marisi's (1978), which investigated the effects of music on forestalling fatigue, must operationally define fatigue. The author cannot use a synonym, such as *exhaustion,* because that idea is not concrete enough. We all might have our own ideas of what fatigue is, but if we are going to say that some independent variable affects fatigue, we must supply some observable evidence of changes in fatigue. Therefore, fatigue must be operationally defined. Anshel and Marisi did not use the term *fatigue,* but from their description of procedures we can infer its operational definition as being when the participant was unable to maintain the pedaling rate of 50 revolutions per minute for 10 consecutive seconds.

operational definition—An observable phenomenon that enables the researcher to test empirically whether the predicted outcomes can be supported.

Another researcher might define fatigue as the point at which maximal heart rate is achieved; still another might define fatigue as the point of maximal oxygen consumption. In all cases, though, the criterion must be observable.

A study dealing with dehydration must provide an operational definition, such as a loss of 5% of body weight. The term *obesity* in males could be defined as having 25% body fat. A study of teaching methods on learning must operationally define the term *learning.* The old definition "a change in behavior" is meaningless in providing evidence of learning. Learning might be demonstrated by five successful throwing performances or some other observable performance criterion.

You may not always agree with the investigator's definitions, but at least you know how a particular term is being used. A common mistake of novice researchers is thinking that every term needs to be defined. (We have seen master's students define terms not even used in their studies!) Consider an example of an unnecessary definition in a study of the effects of strength training on changes in self-concept. *Self-concept* would need to be defined (probably as represented by some scale), but *strength* would not. The strength-training program used would be described in the methods section. Basically, operational definitions are directly related to the research hypotheses because, if you predict that some treatment will produce some effect, you must define how that effectiveness will be manifested. Here are a few humorous operational definitions:

- Osteopornosis: A degenerate disease
- Arachnoleptic fit: The frantic dance performed just after walking through a spiderweb
- Caterpallor: The color you turn after finding half a worm in the fruit you are eating

BASIC ASSUMPTIONS, DELIMITATIONS, AND LIMITATIONS

Besides writing the introduction, stating the research problem and hypothesis, and operationally defining your terms, you may be required to outline the basic assumptions and limitations under which you performed your research.

Assumptions

Every study has certain fundamental premises that it could not proceed without. In other words, you must assume that certain conditions exist and that the behaviors in question can be observed and measured (along with other basic suppositions). A study in pedagogy that compares teaching methods must assume that the teachers involved are capable of promoting learning; if this assumption is not made, the study is worthless. Furthermore, in a learning study, the researcher must assume that the sample selection (e.g., random selection) results in a normal distribution with regard to learning capacity.

A study designed to assess an attitude toward exercise is based on the assumption that this attitude can be reliably demonstrated and measured. Furthermore, you can assume that the participants will respond truthfully, at least for the most part. If you cannot assume those things, you should not waste your time conducting the study.

Of course, the experimenter does everything possible to increase the credibility of the premises. The researcher takes great care in selecting measuring instruments, sampling, and gathering data concerning such things as standardized instructions and motivating techniques. Nevertheless, the researcher still must rely on certain basic assumptions.

Consider the following studies. Johnson (1979) investigated the effects of different levels of fatigue on visual recognition of previously learned material. Among his basic assumptions were that (a) the mental capacities of the participants were within the normal range for university students, (b) the participants understood the directions, (c) the mental task typified the types of mental tasks encountered in athletics, and (d) the physical demands of the task typified the levels of exertion commonly experienced by athletes.

Lane (1983) compared skinfold profiles of black and white girls and boys and tried to determine which skinfold sites best indicated total body fatness with regard to race, sex, and age. She assumed that (a) the skinfold caliper is a valid and reliable instrument for measuring subcutaneous fat, (b) skinfold measurements taken at the body sites indicate the subcutaneous fat stores in the limbs and trunk, and (c) the sum of all skinfolds represents a valid indication of body fatness.

In some physiological studies, the participants are instructed (and agree) to fast or refrain from smoking or drinking liquids for a specified period before testing. Obviously, unless the study is conducted in some type of prison environment, the experimenter cannot physically monitor the participants' activities. Consequently, a basic assumption is that the participants will follow instructions.

limitation—Shortcoming or influence that either cannot be controlled or is the result of the delimitations imposed by the investigator.

delimitation—A limitation imposed by the researcher in the scope of the study; a choice that the researcher makes to define a workable research problem.

Delimitations and Limitations

Every study has **limitations**. Limitations are shortcomings or influences that either cannot be controlled or are the results of the restrictions imposed by the investigator. Some limitations refer to the scope of the study, which is usually set by the researcher. These are often called **delimitations**. Kroll (1971) described delimitations as choices the experimenter makes to define a workable research problem, such as the use of one personality test in the assessment of personality characteristics. Moreover, in a study dealing with individual-sport athletes, the researcher may choose to restrict the selection of participants to athletes in just two or three sports, simply because all individual sports could not be included in one study. Thus, the researcher delimits the study. You probably notice that these delimitations resemble operational definitions. Although they are similar, they are not the same. For example, the size of the sample is a delimitation but is not included under operational definitions.

You can also see that basic assumptions are entwined with delimitations as well as with operational definitions. The researcher must proceed on the assumption that the restrictions imposed on the study will not be so confining as to destroy the external validity (generalizability) of the results.

Remember, theses or dissertations do not have one "correct" format. An examination of studies shows numerous variations in organization. Some studies have delimitations and limitations described in separate sections. Some use a combination heading, some list only one heading but include both in the description, and still others embed these in other sections. As with all aspects of format, much depends on how the advisor was taught. Graduate colleges and schools often allow great latitude in format as long as the study is internally consistent. You will find considerable differences in format even within the same department. The sidebar Limitations and Explanations presents a few humorous explanations of limitations.

In Kroll's (1971) example of delimiting the scope of the study to just two sports to represent individual-sport athletes, an automatic limitation is how well these sports represent all individual sports. Moreover, if the researcher is studying the personality traits of these athletes and delimits the measurement of personality to just one test, a limitation is present. Furthermore, at least one inherent limitation in all self-report instruments in which participants respond to questions about their behavior, likes, or interests is the truthfulness of the responses.

Thus, you can see that limitations accompany the basic assumptions to the extent that the assumptions fail to be justified. As with assumptions, the investigator tries as much as possible to reduce limitations that might stem from faulty procedures. In R.L. Johnson's 1979 study of fatigue effects on the visual recognition of previously learned material, the participants first had to learn the material. He established criteria for learning (operational definition) and tried to control for overlearning (i.e., differences in the level of learning).

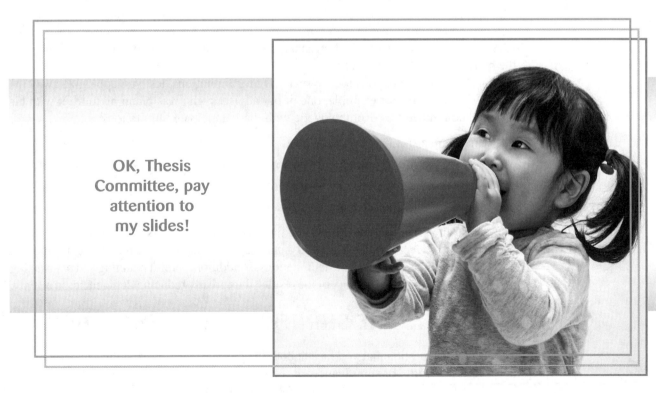

OK, Thesis Committee, pay attention to my slides!

Despite his efforts, however, he recognized the limitation of possible differences in the degree of learning, which could certainly influence recognition.

Limitations and Explanations

Limitations	Explanation
On a garbage truck	Yesterday's Meals on Wheels
On a plumber's truck	Repairing what your husband fixed
On an electrician's truck	Let us remove your shorts
On a maternity room door	Push, Push, Push
On a fence door	Salespeople welcome, dog food is expensive
In a dog vet's waiting room	Be right back—Sit, Stay
In a funeral home	Drive carefully, we'll wait
In an eye doctor's office	If you can't read this, you're at the right place

In her study of skinfold profiles, Lane (1983) had to delimit the study to a certain number of participants in one part of Baton Rouge, Louisiana. Consequently, a limitation was that the children were from one geographic location. She also recognized that changes in body fatness are associated with the onset of puberty, but she was unable to obtain data on puberty or other indices of maturation, and this therefore posed a limitation. Still another limitation was the inability to control possible influences on skinfold measurement, such as dehydration and other diurnal variations. Finally, because there are no internationally recognized standard body sites for skinfold measurement, generalizability may have been limited to the body sites used in this study.

You should not be overzealous in searching for limitations, lest you apologize away the worth of your study. For example, one of our advisees who was planning to meet with his proposal committee was overapologetic with these anticipated limitations:

- The sample size may be too small.
- The tests may not represent the parameter in question.
- The training sessions may be too short.
- The investigator lacks adequate measurement experience.

As a result, the proposal was revised and the method was reassessed.

No study is perfect. You must carefully analyze the delimitations to determine whether the resulting limitations outweigh the delimitations. In addition, careful planning and painstaking methodology increase the validity of the results, thus greatly reducing deficiencies in a study.

JUSTIFYING THE SIGNIFICANCE OF THE STUDY

The inevitable question that you face at both the proposal meeting and the final oral exams deals with the worth, or significance, of the study. The question may be asked in different ways, such as So what? or What good is it? or How is this of any importance to your field? Regardless of the manner in which the question is asked, you must answer it. Perhaps because of the inevitability of the question, most students are required to include a section titled Significance of the Study or sometimes Need for the Study, Importance of the Study, or Rationale for the Study.

Basic and Applied Research Revisited

The worth of the research study is judged largely by whether it is basic or applied research. In chapter 1 we explained that basic research does not have immediate social significance; it usually deals with theoretical problems and is conducted in a controlled laboratory setting. Applied research addresses immediate problems for the purpose of improving practice. It offers less control but ideally more real-world application. Consequently, basic and applied research cannot be evaluated by the same criteria.

The significance of a basic research study obviously depends on the purpose of the study, but usually the criteria focus on the extent to which the study contributes to the formulation or validation of some theory. The worth of applied research must be evaluated on the basis of its contribution to the solution of some immediate problem.

Writing the Section on the Significance of the Study

The significance section is often difficult to write, probably because the student thinks only in terms of the practicality of the study—how the results can be immediately used to improve some aspect of the profession. Kroll (1971) emphasized the importance of maintaining continuity of the significance section with the introduction. Too often the sections are written with different frames of reference instead of a continuous flow of thought. The significance section should focus on such things as contradictory findings of previous research, gaps in knowledge in particular areas, and the contribution the study might make to practice. Difficulties in measuring aspects of the phenomenon in question are sometimes emphasized. The rationale for verifying existing theories may be the focus of the section in some studies, whereas in others the practical application is the main concern. Generally, both theoretical and practical reasons are expressed, but the emphasis will vary according to the study.

Just as the length of the introduction varies, the length of the significance section varies considerably from study to study. A sample of a significance section from Lane (1983) illustrates an approach that focuses on some conflicts between previous research findings and present practice:

> Because the measurement of body composition has become an important aspect of physical fitness testing, the validity, reliability, and administrative feasibility of the measurements are of paramount concern. Adult formulas for estimating percent fat are not considered valid for children; thus skinfolds are in themselves used as measures of body composition.

> The *AAHPERD Health Related Physical Fitness Test Manual* (American Alliance for Health, Physical Education, Recreation and Dance, 1980) describes two skinfold measurements: the triceps and subscapular. Norms for the total of these two measures are provided for boys and girls ages 6 to 17 years. Abbreviated norms are also given for the triceps skinfolds only. Norms were taken from HES data (Johnston et al., 1974). There is minimal evidence for why these two body sites were selected, especially when one, the subscapular, poses some problem with regard to modesty. If the triceps and subscapular were selected to represent one site from the limbs and one from the trunk, are they the most predictive of total fatness as indicated by the sum of several skinfolds from the limbs and trunk? Furthermore, if some sites are equally predictive, the ease of administration needs to be considered.

> **Of major significance in this study is whether the skinfolds that best represent body fatness in white children are equally suitable for black children. Authors (Cronk & Roches, 1982; Harsha et al., 1974; Johnston et al., 1974) have reported differences in skinfold thicknesses between black and white children, yet the AAHPERD norms make no distinction. To be of value, norms must represent the population for which they are intended. Moreover, there may be greater differences in fatness between blacks and whites at different ages.**
>
> **Similarly, it may be that a different combination of skinfolds would be more valid for girls than for boys. It does not seem to be of any great administrative advantage to use the same sites for both sexes if other sites are equally valid indicators of total fatness.**

A final word of warning: In being asked the inevitable question in the final oral exam about the significance of the study, do not reply, "It was necessary to get my degree." The stony silence will be unnerving.

DIFFERENCES BETWEEN THE THESIS AND THE RESEARCH ARTICLE

A mere glance at research articles in journals reveals that a number of the sections found in the traditional thesis or dissertation described in this chapter are missing. At least two reasons account for this. The first is financial: Periodicals are concerned about publishing costs, so brevity is emphasized. Second, a kind of novice–expert ritual seems to be in operation. The novice is required to state the hypotheses explicitly, define terms, state assumptions, recognize limitations, and justify the worth of the study in writing. Certainly, these steps are all part of defining and delimiting the research problem, and it is undoubtedly a worthwhile experience to address each step formally.

Research journal authors, on the other hand, need not explain the step-by-step procedure they used in developing the problem. Typically, a research journal has an introduction that includes a short literature review. The length varies considerably, and some journals insist on brief introductions.

The purpose of the study is nearly always given but is usually not designated by a heading; rather, it is often the last sentence or so in the introduction. For example, in 30 articles in one volume of the *Research Quarterly for Exercise and Sport,* only one had a section titled "Purpose of the Study." The authors of 24 articles ended their introductions with sentences that began, "The purpose of this study was . . . ." Four indicated the purpose with sentences beginning, "This study was designed to . . ." or "The intent of the study was . . . ." One study did not state the purpose at all. In 29 cases, the authors and editors felt that the purpose was evident from the title and introduction.

Research hypotheses and questions are sometimes given but with little uniformity. Operational definitions, assumptions, limitations, and the significance of the study are rarely if ever stated in research articles. Apparently, for the "expert" researcher these steps were accomplished in the development of the problem and are understood or obvious, or both, and need not be stated. If the article is well written, you should be able to discern the operational definitions, the assumptions and limitations, and the independent, dependent, and categorical variables even though they are not specifically stated. Moreover, the significance of the study should be implicitly obvious if the author has written a good introduction.

There is growing movement to replace the traditional thesis or dissertation with one or more published research papers (e.g., Patton, 1990). Often, these papers are collected into a bound volume labeled as a thesis or dissertation, but some institutions accept one or more of these publications in lieu of a thesis or dissertation. In later chapters we provide a format for collecting published papers into a thesis or dissertation, a practice accepted by most university graduate schools.

SUMMARY

This chapter discussed the information that is typically presented in the first section or chapter of a thesis or dissertation (excluding the review of literature). First, we considered the length and substance of the study title. The importance of a good, short, descriptive title for indexing and searching the literature is sometimes overlooked.

The introduction of a research study often proves difficult to write. Conveying the significance of the study to the reader requires a great deal of thought, effort, and skill. If this section is poorly done, the reader may not bother to read the rest of the study.

The problem statement and the research hypotheses or questions commonly appear in most research studies, whether they are theses or dissertations, journal articles, or research grants. Operational definitions, assumptions, limitations and delimitations, and the significance of the study usually appear explicitly only in theses and dissertations. Their purpose is to help (or force) the researcher to define and delimit the research problem succinctly. Operational definitions specifically describe how certain terms (especially the dependent variables) are being used in the study. Assumptions identify the basic conditions that must be assumed to exist for the results to have credibility. Delimitations relate to the scope of the study imposed by the researcher, such as the number and characteristics of the participants, the treatment conditions, the dependent variables used, and the way the dependent variables are measured. Limitations are possible influences on the results that are consequences of the delimitations or that cannot be completely controlled.

The section on the significance of the study forces the researcher to address the inevitable question of the worth of the study. This section should be a continuation of the introduction in terms of the contextual flow. It usually calls attention to the relationship (and differences) between this study and previous ones, controversies and gaps in the literature, and the contribution that this study might make to the practitioner, to existing theoretical models, or to both.

Check Your Understanding

1. For each of the following brief descriptions of studies, write a title, the purpose or purposes, and three research hypotheses. The researchers assessed the following:

    a. Skill acquisition of three groups of fourth-grade boys and girls who had been taught by different teaching styles (A, B, and C).

    b. Self-concept of two groups of boys (a low-strength group and a high-strength group) before and after a strength-training program.

    c. Body composition (estimated percentage of fat) using the electrical impedance analysis method on participants at normal hydration and again after they had become dehydrated.

    d. Grade point averages of male and female athletes participating in major and minor (club) sports at large universities and small private colleges.

2. Locate five articles from research journals, and for each try to determine (a) the hypotheses, if not stated, (b) the independent variable or variables, (c) the dependent variable or variables, (d) an operational definition for the dependent variables, (e) at least two delimitations, (f) at least two limitations, and (g) a basic assumption.

4

Formulating the Method

Chapter 3 provided an overview of the introduction in the proposal for a thesis or dissertation. As already indicated, the journal format typically includes the literature review (see chapter 2) as part of the introduction. (If the chapter format is used, the literature review may appear as a separate chapter or as part of the introduction.) Regardless, after completing the introduction, the researcher must describe the methodology for the research. Typically, this section is labeled Methods. We give an overview of the four parts of the methods section here:

- Participants
- Instruments or apparatuses
- Procedures
- Design and analysis

For our purposes, let us assume that the journal format is used and that the literature review is included in the introduction of your thesis or dissertation, followed by the methods section. Much of the remainder of this book focuses on the methods.

- Important aspects of the study: the participants, instruments, procedures, design and analysis (described in this chapter)
- How to measure and analyze the results (part II)
- How to design the study (part III)
- How to write and present research (part IV)

The purpose of the methods section is to explain how the study was conducted. The standard rule is that the description should be thorough enough that a competent researcher could reproduce the study.

Although qualitative studies are formatted somewhat differently from quantitative studies, they also present a methods section describing the participants and how the data were obtained.

Competent researchers should be able to reproduce the study after reading the methods section.

However, detailed explanations of the settings in which data were collected as well as descriptions of the participants are often required. We present more methodological details on qualitative studies in chapter 19.

Science as we know it today grew out of the murky lore of the Middle Ages (e.g., sorcery and religious ritual).

> But while witches, priests, and chiefs were developing taller and taller hats, scientists worked out a method for determining the validity of their experimental results: They learned to ask, Are they reproducible?—that is, would anyone using the same materials and methods arrive at the same results? (Scherr, 1983, p. ix)

For example, if a 10 kg and a 5 kg iron ball were dropped from the Empire State Building by King Kong 40 years ago or by Arnold Schwarzenegger today, they should strike the ground simultaneously on both occasions.

HOW TO PRESENT METHODOLOGICAL DETAILS

Dissertations and theses differ considerably from published articles in terms of the methodological details provided. When using the journal format, however, you should place the additional materials in an appendix. Journals try to conserve space, but space is not an issue in a dissertation or thesis. Thus, whereas standard techniques in a journal article are referenced only to another published study (in an easily obtainable journal), a thesis or dissertation should provide considerably greater detail in the appendix. Note that we indicated that a technique could be referenced to an easily obtainable journal. When writing for publication, use common sense in this regard. Consider, for example, this citation:

> Farke, F.R., Frankenstein, C., & Frickenfrack, F. (1921). Flexion of the feet by foot fetish feet feelers. *Research Abnormal: Perception of Feet, 22,* 1–26.

By most standards, this citation would not be considered easily obtainable. Therefore, if you are in doubt, give the details of the study or technique.

Furthermore, because theses and dissertations have appendixes, much of the detail that would clutter and extend the methods section should be placed there. Examples include exact instructions to **participants**, samples of tests and answer sheets, diagrams and pictures of equipment, sample data-recording sheets, and informed-consent agreements.

participants—People who are used as subjects in a study. In APA style the term *participants* is used rather than *subjects*.

WHY PLANNING THE METHODS IS IMPORTANT

The purpose of planning the method is to eliminate any alternative or rival hypotheses. This statement really means that when you design the study correctly and the results are as predicted, the only explanation is what you did in the research. Using a previous example to illustrate, our hypothesis is: Shoe size and mathematics performance are positively related during elementary school. To test this hypothesis, we go to an elementary school, measure shoe sizes, and obtain standardized mathematics performance scores of the children in grades 1 through 5. When we plot these scores, they appear as in figure 4.1, each dot representing a single child. Moving from the dot to the x-axis shows shoe size, whereas the y-axis shows math performance. "Look!" we say, "We are correct. As shoe sizes get larger, the children's mathematics performance increases. Eureka! All we need to do is buy the children bigger shoes and their mathematics performance will improve." But wait a minute! We have overlooked two things. First, a rival explanation is present: Both shoe size and mathematics performance are related to age. That connection really explains the relationship. As the kids get older, their feet grow larger and they perform better on mathematics tests. Second, just because two things are related does not mean that one causes the other. Correlation does not prove causation. Obviously, we cannot improve children's math performances by buying them bigger shoes.

In research we want to use the **MAXICON principle**: Maximize true variance, or increase the odds that the real relationship or explanation will be discovered; minimize error variance, or reduce all the mistakes that could creep into the study to disguise the true relationship; and control extraneous variance, or make sure that rival hypotheses are not the real explanations of the relationship.

TWO PRINCIPLES FOR PLANNING EXPERIMENTS

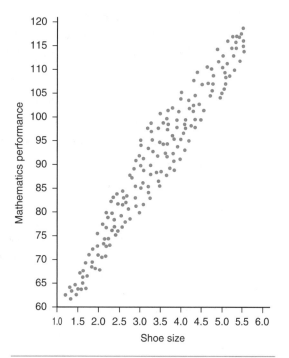

Figure 4.1 Relationship of mathematics performance and shoe size of children in grades 1 through 5.

> **MAXICON principle**—A method of controlling any explanation for the results except the hypothesis that the researcher intends to evaluate. This is done by maximizing true variance, minimizing error variance, and controlling extraneous variance.

In an interesting paper, Cohen (1990) put forward two principles that make good sense when planning experiments. The first is Less is more. Of course, this principle seldom applies to the number of participants in a study, but it does apply to other aspects. For example, graduate students want to conduct meaningful studies that address and solve important problems. To do this, they often plan complex studies with many independent and dependent variables. From one perspective, this approach is good: The world of physical activity is truly complex. Students frequently start with useful ideas to study, but the ideas become so cumbersome that the studies often fail because of sheer complexity. Carefully evaluate the number of independent and dependent variables that are practically and theoretically important to your study. Do not let anyone convince you (except your major professor, of course, who can convince you of anything) to add additional variables just to see what happens. This action complicates your study and causes all types of measurement and statistical problems.

This idea leads to the second principle from Cohen (1990): Simple is better. This statement is true from the design to the treatments to the analysis to displaying data to interpreting results. Keep your study straightforward so that when you find something, you can understand and interpret what it means. Understanding your data is an important concept. Although fancy statistical procedures are impressive and informative, there is no substitute for plotting data graphically and evaluating it carefully. Summary statistics (e.g., mean, standard deviation) are helpful and informative, but they are no substitute for looking at the distribution of the original data. Summary statistics may not show us what we really need to know—things that become evident when we look at a graphic display of the data.

> Evaluating and graphically plotting the data can enhance your understanding and improve your interpretation of the data.

DESCRIBING PARTICIPANTS

This section of the method of a thesis or dissertation describes how and why the participants were selected and which of their characteristics are pertinent to the study. These are questions to consider when selecting participants:

- Are participants with special characteristics necessary for your research?
    - Age (children, elderly)
    - Sex (females, males, or both)
    - Level of training (trained or untrained)
    - Level of performance (experts or novices)
    - Size (weight, fatness)
    - Special types (athletes, cyclists, runners)
- Can you obtain the necessary permission and cooperation from the participants?
- Can you find enough participants?

Of course, you want to select participants who will respond to the treatments and measures used in the study. For example, if you want to see the results of training a group of children in overhand throwing, selecting expert 12-year-old baseball pitchers as participants will not likely produce a change in measures of throwing outcome. An intense, long-term training program would be required to have any influence on these participants. Selecting children who are soccer players and have never played organized baseball would offer better odds for a training program to produce changes.

In experimental research, the interactions among participants, measures, and the nature of the treatment program are essential in allowing the treatment program to have a chance to work (Thomas, Lochbaum, Landers, & He, 1997). If you select participants who have high levels of physical fitness, subjecting them to a moderate training program will not produce changes in fitness. Also, participants high in physical fitness will have a small range of scores on a measure of cardiorespiratory endurance (e.g., $\dot{V}O_2$max). For example, you will not find a significant correlation between $\dot{V}O_2$max and marathon performance in world-class marathon runners. Their range of scores in $\dot{V}O_2$max is small, as is their range of scores in marathon performance times. Because the range of scores is small in both measures, no significant correlation will be found. This result does not mean that running performance

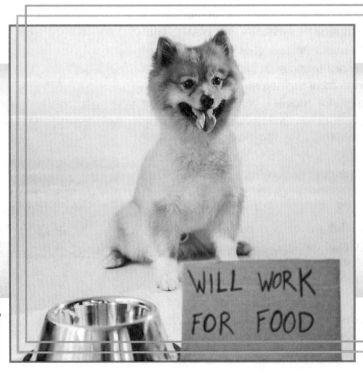

You may have to offer something to the participants to make being in the study worthwhile.

and cardiorespiratory endurance are not related. It means that you have restricted the range of participants' performance so much that the correlation cannot be exhibited. If you had selected moderately trained runners (e.g., women who jog three times per week for 40 min each time), a significant correlation would be found between running performance on a 5K run and $\dot{V}O_2max$. (We discuss procedures for selecting sample participants in chapter 6.)

What to Tell About the Participants

The exact number of participants should be given, as should any loss of participants during the time of study. In the proposal, some of this information may not be exact. For example, the following might describe the potential participants:

Participants: For this study, 48 males, ranging in age from 21 to 34 years, will be randomly selected from a group ($N = 147$) of well-trained distance runners ($\dot{V}O_2max = 60$ ml · kg · min^{-1} or higher) who have been competitive runners for at least 2 years. Participants will be randomly assigned to one of four groups ($n = 12$).

After the study is completed, details are available on the participants, so now this section might read as follows:

Participants: In this study, 48 males, ranging in age from 21 to 34 years, were randomly selected from a group ($N = 147$) of well-trained runners ($\dot{V}O_2max = 60$ ml · kg · min^{-1} or higher) who had been competitive runners for at least 2 years. The participants had the following characteristics (standard deviations in parentheses): age, $M = 26$ years (3.3); height, $M = 172.5$ cm (7.5); weight, $M = 66.9$ kg (8.7); and $\dot{V}O_2max$, $M = 65$ ml · kg · min^{-1} (4.2). Participants were randomly assigned to one of four groups ($n = 12$).

The participant characteristics listed are extremely pertinent in an exercise physiology study but not at all pertinent, for example, in a study of equipment used by children on the playground. The nature of the research dictates the participant characteristics of interest to the researcher. Carefully think through the characteristics that you will report in your research. Look at related studies for ideas about important characteristics to report.

The characteristics of participants that you identify and report must be clearly specified. Note in the example that well-trained runners were exactly defined; that is, their $\dot{V}O_2max$ must be equal to 60 ml · kg · min^{-1} or higher. When participants of different ages are to be used is another good example. Saying only that 7-, 9-, and 11-year-olds will be the participants is not sufficient. How wide is the age range for 7-year-olds? Is it ±1 month, ±6 months, or what? In the proposal, you may say that 7-, 9-, and 11-year-olds will be included in the study. At the time of testing, each age will be limited to a range of ±6 months. Then, when the thesis or dissertation is written, it may read as follows:

> **At each age level, 15 children were selected for this study. The mean ages are as follows (standard deviations in parentheses): the youngest group, 7.1 years (4.4 months); 9-year-olds, 9.2 years (3.9 months); and the oldest group, 11.2 years (4.1 months).**

A graduate student we know was in his committee meeting when a faculty member on his committee asked him to describe one of his participants. The student said, "He was 6 feet tall, had brown hair and a beard, and trained regularly by distance running." The really sharp faculty member then asked, "Was this a male or female?" to which the student was unable to respond because of laughter.

Protecting Participants

Most research in the study of physical activity deals with humans, often children, but it also includes animals. Therefore, the researcher must be concerned about any circumstances in the research setting or activity that could harm humans or animals. In chapter 5, Ethical

> Participants, measures, and treatment programs are interrelated. Be sure to choose participants who will respond to the treatment program and have a broad enough range of results when measured with the chosen techniques.

Issues in Research and Scholarship, we provide considerable details on what the researcher must do to protect both humans and animals used in research. Obtaining informed consent from humans and ensuring the protection and care of animals are particularly important.

DESCRIBING INSTRUMENTS

Information about the instruments, apparatuses, or tests used to collect data is used to generate the dependent variables in the study. Consider the following when selecting tests and instruments:

- What is the validity and reliability of the measures?
- How difficult is it to obtain the measures?
- Do you have access to the instruments, tests, or apparatuses needed?
- Do you know (or can you learn) how to administer the tests or use the equipment?
- Do you know how to evaluate participants' test performance?
- Will the tests, instruments, or apparatuses yield a reasonable range of scores for the participants you have selected?
- Will the participants be willing to spend whatever time is required for you to administer the tests or instruments?

Incorrect Metric Conversions

1 trillion microphones = 1 megaphone

1 million bicycles = 2 megacycles

2,000 mockingbirds = 2 kilomockingbirds

1/2 lavatory = 1 demijohn

1 millionth of a fish = 1 microfiche

453.6 graham crackers = 1 pound cake

10 rations = 1 decoration

10 millipedes = 1 centipede

10 monologues = 5 dialogues

For example, in a sport psychology study you are interested in how a group of university football players will be affected by a lecture on steroid use. In addition, you suspect that the players' attitudes might be modified by certain personality traits. So you select three tests—a steroid knowledge test, an attitude inventory about responsible drug use, and a trait personality measure—and administer them to all participants. The knowledge and attitude tests will probably be given before and after the lecture; and the trait personality test, only before the lecture (traits should not change, and this test is being used to stratify players in some way). In the instrument section, you should describe the three tests and probably put complete copies of each in the appendix (see chapter 5 about the ethical use of standardized tests). You also should describe the reliability (consistency) and validity (what the test measures) information that is available on each test with appropriate citations. Then you should explain the scoring sheets (place a sample in the appendix) and the scoring methods (but do not use inappropriate conversions for measurements like the ones in the sidebar Incorrect Metric Conversions).

Another example is a motor behavior study in which participants' reaction and movement times are measured under various conditions. In the instruments section, you should describe the testing apparatus and provide a diagram or picture. If the apparatus interfaces with a computer to control the testing situation and data collection, you should describe the computer (brand name and model) and how the interface was made. At least a description of how the computer program operates should be included in the appendix. You should explain the dependent variables generated for reaction and movement time and give reliability estimates for these characteristics. All the necessary information could be presented by the use of both the instrument (or apparatus) part of the methods section and the appendix, thus allowing the methods section to flow smoothly.

DESCRIBING PROCEDURES

In the procedures section, you should describe how the data were obtained, including all testing procedures for obtaining scores on the variables of interest. How tests were given and who gave them are important features. You should detail the setup of the testing situation and instructions given to the participants (although you may place some of this information in the appendix). If the study is experimental, then you should describe the treatments applied to the groups of participants.

Consider these points when planning procedures:

- Collecting the data
    - When? Where? How much time is required?
    - Do you have pilot data to demonstrate your skill and knowledge in using the tests and equipment and knowing how participants will respond?
    - Have you developed a scheme for data acquisition, recording, and scoring? (These are often computer controlled.)

- Planning the treatments (in quasi-experimental and experimental studies)
    - How long? How intense? How often?
    - How will participants' adherence to treatments be determined?
    - Do you have pilot data to show how participants will respond to the treatments and that you can administer these treatments?
    - Have you selected appropriate treatments for the type of participants to be used?

One of our favorite summaries of the problems encountered and solutions proposed is presented in the Errors in Experiments sidebar. These statements are extracted from an article by Martens (1973).

The procedures section contains most of the details that allow another researcher to replicate the study. (But these details must be useful, unlike the examples in the sidebar Statements in Procedures That Sometimes Get Twisted.) Tuckman (1978) outlined these details, which generally include the following:

- Specific order in which steps were undertaken
- Timing of the study (e.g., time taken for various procedures and time between procedures)
- Instructions given to participants
- Briefings, debriefings, and safeguards

Avoiding Methodological Faults

No single item in this book is more important than our advice to pilot all your procedures. Physical education, kinesiology, exercise science, and sport science have produced thousands of studies in which the discussions centered on methodological faults that caused the research to lack validity. It is worth repeating that placing post hoc blame on the methodology for inadequate results is unacceptable. Every thesis or dissertation proposal should present pilot work—Work undertaken to verify that you can correctly administer the tests and treatments for your study using appropriate participants.

pilot work that verifies that all instruments and procedures will function as specified on the type of participants for which the research is intended. In addition, you must demonstrate that you can use these procedures and apparatuses accurately and reliably.

Do not let your procedures become so complex that you end up like Santa Claus on Christmas Eve, with more to do than seems possible.

> According to our calculations, Santa has about 31 hours to work on Christmas (because of different time zones and the rotation of the earth). If he travels from east to west and has 91.8 million households to visit in the United States, Santa must make 822.6 visits per second, allowing him 1/1,000 of a second to park, hop out of his sleigh, slide down the chimney, fill the stockings, distribute the remaining presents under the tree, eat whatever snacks have been left, get back up the chimney, get back into the sleigh, and move on to the next house.

ERRORS IN EXPERIMENTS

Martens' method derives from the basic premise that

In people experiments, people errors increase disproportionately to the contact people have with people.

It is obvious that the most logical deduction from this premise is

To reduce people errors in people experiments, reduce the number of people.

Although this solution might be preferred for its elegant simplicity, its feasibility can be questioned. Therefore, the following alternative formulation warrants consideration:

The contact between people testers and people subjects in people experiments should be minimized, standardized, and randomized.

Reprinted, by permission, from R. Martens, 1973, "People errors in people experiments," *Quest* 20: 22.

Statements in Procedures That Sometimes Get Twisted

Dyslexics have more fnu.

Clones are people, two.

Cole's law is thinly sliced cabbage.

Does the name Pavlov ring a bell?

Energizer bunny arrested and charged with battery.

Staph only in the microbiology lab.

A blood test of 20 pessimists was always b-negative.

Without geometry, life is pointless.

When two egotists fight, it's an I for an I.

Santa's helpers are subordinate clauses.

Importance of Pilot Work

During our years as professors, editors, and researchers, we have seen abstracts of thousands of master's theses and doctoral dissertations. More than 75% of these research efforts are not publishable and do not contribute to theory or practice because of major methodological flaws that could have been easily corrected with pilot work. Sadly, this circumstance reflects negatively not only on the discipline and profession but also on the graduate students who conducted the research and the faculty who directed it. Yet nearly all the problems could have been corrected by increased knowledge of the topic, better research design, and pilot work on the procedures.

Graduate students frequently seek information about appropriate procedures from related literature (and they should). Procedures for intensity, frequency, and duration of experimental treatments are often readily available, as is information about testing instruments and procedures. Remember, however, that procedures in one area do not necessarily work well in another, as the example in the sidebar Research Procedures May Not Generalize illustrates.

> A study should be designed so that dependent variables change only when independent variables are changed.

DESCRIBING DESIGN AND ANALYSIS

Design is the key to controlling the outcomes from experimental and quasi-experimental research. The independent variables are manipulated to judge their effects on the dependent variable. A well-designed study is one in which the only explanation for change in the dependent variable is how the participants were treated (independent variable). The design and theory have enabled the researcher to eliminate all rival or alternative hypotheses. The design requires a section heading in the methods for experimental and quasi-experimental research. The plans for data analysis must also be reported. In most studies some type of statistical analysis is used, but there are exceptions (e.g., historical or qualitative research that uses other types of analysis).

Typically, the researcher explains the proposed application of the statistics. In nearly all cases, descriptive statistics are provided, such as means and standard deviations for each variable. If correlational techniques (relationships among variables) are used, then the variables to be correlated and the techniques are named. For example, a researcher might write the following: "The degree of relationship between two estimates of percentage fat will be established by using Pearson r to correlate the sum of three skinfolds with fat estimates from a Bod Pod." In experimental and quasi-experimental studies, descriptive statistics are provided for the dependent measures, and the statistics for establishing differences among groups are reported. A researcher might write the following: "A t test was used to determine whether youth league hockey players who watched professional games produced more violent actions during their games than youth league players who did not watch professional hockey games."

The major problem that graduate students encounter in the description of statistical techniques is the tendency to inform everyone of their knowledge of statistics. Of course, that is not much of a problem for the new graduate student. But if your program of studies is a research-oriented one in which you take several statistics courses, your attitude may change rapidly.

> Hiawatha, who at college majored in applied statistics, consequently felt entitled to instruct his fellow men on any subject whatsoever. (Kendall, 1959, p. 23)

The point is for you to describe your statistical analyses but not to instruct in their theoretical underpinnings and proper use.

> In terms of scientific progress, any statistical analysis whose purpose is not determined by theory, whose hypothesis and methods are not theoretically specified, or whose results are not related back to theory must be considered, like atheoretical fishing and model building, to be hobbies. (Serlin, 1987, p. 371)

Research Procedures May Not Generalize

Dr. I.M. Funded was a good life scientist who studied the biochemistry of exercise in a private research laboratory. He had also done several studies with a colleague in sport psychology to determine whether some biochemical responses he had found were factors in psychological responses to exercise. Thus, he had a firm grasp of some of the social science techniques as well as those of life science.

Unfortunately, Dr. Funded's funding ran out, and he lost his job. A friend of his was the superintendent of a large school district. Dr. Funded went to his friend, Dr. Elected, and said, "I am a good scientist well trained in problem-solving techniques. Surely, you must need someone like me in your administrative structure. In addition, I have an undergraduate degree in physical education, so I am certified to teach, although I never have." Dr. Elected agreed to hire him as his teaching effectiveness supervisor because the school system was having difficulty identifying good teaching. Dr. Elected thought that perhaps a scientist with good problem-solving skills and the ability to make careful measurements could find a solution.

Dr. Funded decided that his first task was to identify some good teachers so that he could determine their characteristics. He would use some of the techniques he had acquired from his colleague in sport psychology to identify good teachers. He had learned that questionnaires were effective in surveying large groups, but that interviews were more valid. Dr. Funded drew a random sample of 6 schools from the 40 in the district. Then, he randomly selected six teachers in each school and interviewed them. He used a direct interview question: "Are you a good teacher?" All 36 indicated that they were. So he went back to Dr. Elected, explained what he had done, and said, "You don't have a teaching problem. All of your teachers are good." (Of course, he noted that there could be some sampling error, but he was certain of his results.) Dr. Elected was not very happy with Dr. Funded's procedures and results and suggested that perhaps he needed more sophisticated techniques and strategies to identify good teachers.

Dr. Funded was slightly distraught but thought, "I have always questioned the techniques of those psychologists anyway. I will return to my life science techniques to determine the answer." He went back to the previously selected 36 teachers with a plan to draw blood, sample urine, and biopsy muscles (at four sites) once per week for four weeks. Immediately, 34 teachers said that they would not take part, but 2 who were triathletes agreed to participate. Dr. Funded noted that the participant mortality rate was about normal for biopsy studies, so the data should be generalizable. He collected all the data, did the correct chemical analyses, and reported back to Dr. Elected. He indicated that effective teachers had 84% slow-twitch fiber, higher-than-average amounts of hemoglobin per deciliter of blood, and a specific profile of catecholamines (epinephrine and norepinephrine) in the urine. In addition, good teachers trained for at least 100 miles (160 km) per week on the bicycle, 50 miles (80 km) running, and 7,500 meters swimming. Dr. Funded sat back smugly and said, "Techniques for the life sciences can be applied to solve many problems." Dr. Elected said, "You are fired."

ESTABLISHING CAUSE AND EFFECT

The establishment of cause and effect in an experimental study is much more than a statistical and design issue. The issue is one of logic. If the null hypothesis is not true, then what hypothesis is true? Of course, the scholar hopes the research hypothesis is true, but this is difficult to establish. In science the researcher seeks to explain that certain types of effects normally happen given specific circumstances or causes. For example, an effect may occur in the presence of something but not in its absence. Water boils in the presence of a high enough temperature but not in the absence of this specific temperature (given a specific air pressure).

People can (and do) differ in their opinions about what can be a cause-and-effect relationship or even whether such a relationship can exist. For example, whether you believe in universal laws, in destiny, in free will, or in an omnipotent deity is likely to influence your view of cause and effect.

Must causes be observable? If yes, then two criteria are needed for establishing cause and effect. First is the method of agreement. If an effect occurs when both A and B are present, and A and B have only C in common, then C is the likely cause (or at least part of it). Second is the method of disagreement. If the effect does not occur in E and F when C is the only common element absent, then C is the cause (or part of it). Thus, it is clear that the researcher's reasoning influences the establishment of cause and effect because the researcher's beliefs set the stage for what may be considered a cause or even whether one can exist. This suggests that the researcher's theoretical beliefs as well as the study's design and analysis are essential factors in establishing cause and effect (for a more detailed discussion of causation, see White, 1990).

Manipulation Effects

When using a treatment in experimental or quasi-experimental research, the participant is involved in the treatment on a regular basis. For example, if participants in a home exercise program are to exercise daily for at least 40 min, how can the researcher be sure they are doing so? Some sort of quantitative manipulation effect needs to be carried out to be certain. Maybe the researcher will ask each participant to wear a movement recording device every day during the training. Because this device records amount of movement, the researcher can determine whether a vigorous program of exercise was undertaken for the correct amount of time. Even a manipulation check such as the one described, however, has flaws: The participant could fail to exercise but ask someone else who was exercising to wear the device.

In any experimental study, researchers must have a plan as part of the methodology to check the manipulation of the independent variable in relation to the dependent variable. In a laboratory exercise study that involves one participant and one researcher, this is relatively easy. But as the research moves further into real-world settings, manipulation checks become increasingly important. In pedagogical, psychological, and sociological types of research related to physical activity, manipulation checks are nearly always essential to verify that participants perceived and responded to the treatments as the researcher intended.

Sometimes, manipulation checks can be done in a qualitative way by questioning participants either during or following the individual treatment sessions or after the treatment period is over. Although a quantitative manipulation check is preferred, a qualitative one is sometimes all that is possible or reasonable.

Fatal Flaws in Research

Every study proposal, particularly its methodology, should be carefully evaluated for fatal flaws. That is, the researcher should ask, Does this study lack a characteristic that will cause it to be rejected for publication regardless of the way it is conducted or the outcomes? The following are some questions that can help to identify fatal flaws:

- Are all hypotheses logical when considering theory and study characteristics?
- Are all the assumptions made about the study good ones?
- Are the "right" participants in sufficient numbers selected for the study, especially in relation to the treatments and measurements used?
- Are the treatments intense enough and of sufficient length to produce the desired changes (especially considering the participants selected)?
- Are all measurements valid and reliable as well as appropriate for the planned study, especially considering the characteristics of participants and treatment levels in experimental studies?
- Are data collection and storage procedures well planned and carefully done?
- Are all extraneous variables being controlled?

Even in famous studies, important errors occur. For example, Franklin Henry was one of the most famous researchers in kinesiology. His paper "Memory Drum Theory" (Henry & Rogers, 1960), which established the first real explanation for how movements are controlled, is the most highly cited paper ever to appear in *Research Quarterly for Exercise and Sport* (Cardinal & Thomas, 2005). Yet Fischman, Christina, and Anson (2008) reported a significant error in the method used. They suggested that this discovery provides yet another lesson about the need for replication in science.

INTERACTION OF PARTICIPANTS, MEASUREMENTS, AND TREATMENTS

Correlational studies use selected participants to evaluate the relationship among two or more measures, whereas experimental studies use selected participants to evaluate whether a specific treatment causes a specific outcome. In both instances, selecting the appropriate participants, measurements, and treatments is extremely important. In correlational studies, the measurements must assess the characteristics of interest and produce an appropriate range of scores on the measures; if they do not, then the researcher has reduced the chances of discovering important relationships. For example, if the purpose of the study is to discover the relationship between anxiety and motor performance, then participants must exhibit a range of both anxiety and motor performance. If all participants have low levels of anxiety and have low motor skill, little opportunity exists to find a relationship.

The same idea is true in an experimental study to discover the influence of weight training on jumping performance. If all the participants are regular weightlifters who work on leg strength and are excellent jumpers, a weight-training program will not increase jumping performance. Said another way, the participants must show a range of performance levels on the measures of interest in correlational studies, and the measures must capture the characteristics that the researcher hopes to assess. In experimental studies, the participants must be at the right level to respond to the treatments, the treatments must be intense enough and frequent enough to produce the changes that the researcher hopes to see, and the measuring instruments must be responsive to the treatment changes in the participants.

SUMMARY

This chapter provided an overview of the method for the research study. We identified the major parts as participants, instruments or apparatuses, procedures, and design and analysis. The major purposes of the four parts of the methods section are to eliminate alternative or rival hypotheses and to eliminate any explanation for the results except the hypothesis that the researcher intends to evaluate. The MAXICON principle shows the way to accomplish this: (a) Maximize the true or planned sources of variation, (b) minimize any error or unplanned sources of variation, and (c) control any extraneous sources of variation. In later chapters, we detail how to do this from the viewpoints of statistics and measurement (part II) and of design (part III). We explain the final sections of the thesis or dissertation in part IV.

Check Your Understanding

1. Find an experimental study in *Research Quarterly for Exercise and Sport* and critique the methods section. Comment on the degree to which the author provided sufficient information concerning the participants (and informed consent), instruments, procedures, and design and analysis.

2. Locate a survey study and compare and contrast its methods description with that in the study in problem 1.

5

Ethical Issues in Research and Scholarship

> When choosing between two evils, I always like to try the one I've never tried before.
>
> —Mae West

A
s a graduate student, you will encounter a number of ethical issues in research and scholarship. In this chapter, we draw your attention to many of these issues and provide a framework for discussion and decision making. But the choices are not always clear-cut. Most important to making good decisions is having good information and obtaining advice from trusted faculty members. The major topics presented include misconduct in science, working with faculty and other graduate students, and using humans and animals as participants in research.

SEVEN AREAS OF SCIENTIFIC DISHONESTY

The White House Office of Science and Technology Policy has defined *scientific misconduct* for all U.S. agencies:

> **Scientific misconduct is fabrication, falsification, or plagiarism in proposing, performing, or reviewing research, or in reporting research results.** (*Federal Register,* October 14, 1999)

In this section we discuss issues in scientific misconduct that are generally pertinent to all scholarly areas of physical activity. We have also supplied a list of references on scientific misconduct as well as some situations and questions about honesty in science.

Shore (1991) identified seven areas in which scientific dishonesty might occur; each is discussed in the following subsections. A 1993 issue of *Quest* (Thomas & Gill) included several thought-provoking articles on ethics in the study of physical activity, and Drowatzky (1996) is a valuable resource on the topic. For a recent list of sources and information on misconduct in science, go to the Office of Research Integrity at UC Riverside: http://research.ucr.edu/ori/rcr/resources.aspx or to the Wikipedia page on scientific misconduct at http://en.wikipedia.org/wiki/Scientific_misconduct.

Plagiarism

Plagiarism means using the ideas, writings, or drawings of others as your own. Of course, this is completely unacceptable in the research process (including writing). Plagiarism carries severe penalties at all institutions. A researcher who plagiarizes work carries a stigma for life in his or her profession. No potential outcome is worth the risk involved.

On occasion, a graduate student or faculty member can inadvertently be involved in plagiarism. This situation generally occurs on work that is coauthored. If one author plagiarizes material, the other could be equally punished even though he or she is unaware of the plagiarism. Although no means of protection is surefire (except not working with anyone else), never allow a paper with your name on it to be submitted (or revised) unless you have seen the complete paper in its final form.

In scientific writing, originality is also important. Common practice is to circulate manuscripts and drafts of papers among scholars (which are often shared with graduate students) who are known to be working in a specific area. If ideas, methods, findings, and so on, are borrowed from these, proper credit should always be given.

Fabrication and Falsification

Based on information in 2010 from the National Science Foundation (www.nsf.gov/statistics/seind10/c0/c0s6.htm), published peer review research articles increased from 460,000 (1988) to 760,000 (2008). Thus, it is not surprising to learn that scientists have occasionally been caught making up or altering research data. Of course, this action is completely unethical, and severe penalties are imposed on people who are caught. Pressure has been particularly intense in medical and health-related research because such research is often expensive, requires external funding, and involves risk. The temptation is great to make a little change here or there or to make up data because "I only need a few more participants, but I am running out of time." The odds of being detected in these types of actions are high, but even if you should get away with it, you will always know that you did it, and you will probably put other people at risk because of your actions.

Although graduate students and faculty may knowingly produce fraudulent research, established scholars are sometimes indirectly involved in scientific misconduct. This may occur when they work with other scientists who produce fraudulent data that follow the predicted outcomes (e.g., as in a funded grant in which the proposal suggests what outcomes are probable). In these instances, the established scholar sees exactly what he or she expects to see in the data. Because this result verifies the hypotheses, the data are assumed to be acceptable. One of the most famous cases involved Nobel laureate David Baltimore, who coauthored a paper with principal authors Thereza Imanishi-Kari and David Weaver that was published in *Cell* in April 1986. Baltimore checked the findings, but he saw in the data the expected outcomes and agreed to submit the paper. The fact that the data were not accurate subsequently led to Baltimore's resignation as president of Rockefeller University. Thus, although Baltimore was not the principal author of the paper, his career was seriously damaged.

Falsification can also occur with cited literature. Graduate students should be careful how they interpret what an author says. Work of other authors should not be "bent" to fit projected hypotheses. For that reason, graduate students should read original sources instead of relying on the interpretations of others, because those interpretations may not follow the original source closely.

Nonpublication of Data

Some data are not included because they do not support the desired outcome. This tactic is sometimes known as "cooking" data. A thin line separates the action of eliminating "bad" data from "cooking" data. Bad data should be caught, if possible, at the time of data acquisition. For example, if a test value seems too large or small and the researcher checks the instrument and finds it out of calibration, eliminating this bad data is good research practice. But deciding that a value is inappropriate when data are being analyzed and changing it is "cooking" data.

Another term applied to unusual data is **outlier**. Some people have called such data "outright liars," suggesting that the data are bad. But extreme values now are sometimes trimmed. Just because a score is extreme does not mean that it is based on bad data. Although extreme scores can create problems in data analysis, trimming them automatically is a poor practice (see chapter 10 on nonparametric analysis for one solution to this problem).

The most drastic instance in this category is the failure to publish results that do not support projected hypotheses. Journals are often accused of a publication bias, meaning that only significant results are published, but authors should publish the outcomes of solid research regardless of whether findings support projected hypotheses. The results from well-planned studies based on theory and previous empirical data have important meaning regardless of whether the predicted outcomes are found.

> **outlier**—An unrepresentative score; a score that lies outside of the normal scores.

Faulty Data-Gathering Procedures

A number of unethical activities can occur at the data-gathering stage of a research project. In particular, graduate students should be aware of issues such as these:

- Continuing with data collection from participants who are not meeting the requirement of the research (e.g., poor effort; failure to adhere to agreements about diet, exercise, rest)
- Malfunctioning equipment
- Inappropriate treatment of participants (e.g., failure to follow the guidelines from the human subjects committee)
- Recording data incorrectly

For example, a doctoral student whom we know was collecting data on running economy in a field setting. Participants returned several times to be video-recorded while repeating a run at varying stride lengths and rates. On the third day of testing, a male runner performed erratically during the run. When the experimenter questioned him, she learned that he had been out drinking with his buddies late on the previous evening and was hung over. Of course, she wisely discarded the data and scheduled him for another run several days later. Had she not noted the unusual nature of his performance and questioned him carefully, she would have included data that would probably have skewed her results because the participant was not adhering to previously agreed-upon conditions of the study.

Poor Data Storage and Retention

Data must be stored and maintained as originally recorded and not altered. All original records should be maintained so that the original data are available for examination. Federal agencies and most journals suggest that original data be maintained for at least three years following publication of the results.

Misleading Authorship

A major ethical issue among researchers involves joint research projects or, more specifically, the publication and presentation of joint research efforts. Generally, the order of authorship for presentations and publications should be based on the researchers' contributions to the project. The first, or senior, author is usually the researcher who developed the idea and the plan for the research. Second and third authors are normally listed in the order of their contributions (see Fine & Kurdek, 1993, for a detailed discussion and case studies).

Although this sounds easy enough, authorship decisions are difficult at times. Sometimes researchers make equal contributions and decide to flip a coin to determine who is listed first. Regardless, decide the order of authors at the beginning of a collaborative effort. This approach saves hard feelings later, when everyone may not agree on whose contribution was most important. If the contributions of various authors change over the course of the research project, discuss a change at that time.

A second issue is who should be an author (see Crase & Rosato, 1992, for a discussion of the changing nature of authorship). Studies occasionally have more authors than participants. In fact, sometimes the authors are also the participants. When you look at what participants must go through in some research studies, you can see why only a major professor's graduate students would allow such things to be done to them. Even then, they insist on being listed as authors as a reward. More seriously, the following two rules should help define authorship:

- **Authorship should involve only those who contribute directly to the research project.** This listing does not necessarily include the laboratory director or a graduate student's major professor. The only thing that we advocate in the chain letter in the sidebar Chain Letter to Increase Publications is the humor.

- **Technicians do not necessarily become joint authors.** Graduate students sometimes think that because they collect the data, they should be coauthors. Only when graduate students contribute to the planning, analysis, and writing of the research report are they entitled to be listed as coauthors. Even this rule does not apply to grants that pay graduate students for their work. Good major professors involve their graduate students in all aspects of their research programs; thus, these students frequently serve as coauthors and technicians.

For an interesting article on authorship, see http://sciencecareers.sciencemag.org/career_magazine/previous_issues/articles/2010_04_16/caredit.a1000039

Unacceptable Publication Practices

The final scientific dishonesty concern deals with joint publications—specifically, those between the major professor and the graduate student. Major professors do (and should) immediately begin to involve graduate students in their research programs. When this happens, the general guidelines that we suggested earlier apply. But two conflicting forces are at work. A professor's job is to foster and develop students' scholarly ability. At the same time, pressure is increasing on faculty to publish so that they can obtain the benefits of promotions, tenure, outside funding, and merit pay. Being the first (senior) author is a benefit in these endeavors. As a result, faculty members want to be selfless and assist students, but they also feel the pressure to publish. This issue may not affect senior faculty, but it is certainly significant for untenured assistant professors. As mentioned previously, there are no hard-and-fast rules other than that everyone agree before the research is undertaken.

The thesis or dissertation is a special case. By definition, the thesis or dissertation is how a graduate student demonstrates competence to receive a degree. Frequently, for the master's thesis, the major professor supplies the idea, design, and much of the writing and editing. In spite of this, we believe that it should be regarded as the student's work. The dissertation should always be regarded as the student's work, but second authorship for the major professor on either the thesis or the dissertation is acceptable under certain circumstances. The American Psychological Association (2010, pp. 18–20) has defined these circumstances adequately, and we recommend the use of their guidelines:

- Only second authorship is acceptable for the dissertation supervisor.
- Second authorship may be considered obligatory if the supervisor designates the primary variables, makes major interpretive contributions, or provides the database.

Chain Letter to Increase Publications

Dear Colleague:

We are sure that you are aware of the importance of publications in establishing yourself and procuring grants, awards, and well-paying academic positions or chairpersonships. We have devised a way your curriculum vitae can be greatly enhanced with very little effort.

This letter contains a list of names and addresses. Include the top two names as coauthors on your next scholarly paper. Then remove the top name and place your own name at the bottom of the list. Send the revised letter to five colleagues.

If these instructions are followed, by the time your name reaches the top of the list, you will have claim to coauthorship of 15,625 refereed publications. If you break this chain, your next 10 papers will be rejected as lacking in relevance to real-world behavior. Thus, you will be labeled as ecologically invalid by your peers.

Sincerely,

Jerry R. Thomas, Professor

Jack K. Nelson, Professor

List as coauthors:

Jerry R. Thomas

Jack K. Nelson

I.M. Published

U.R. Tenured

C.D. Raise

- Second authorship is a courtesy if the supervisor designates the general area of concern, is substantially involved in the development of the design and measurement procedures, or substantially contributes to the writing of the published report.

- Second authorship is not acceptable if the supervisor provides only encouragement, physical facilities, financial support, critiques, or editorial contributions.

- In all instances, agreement should be reviewed before the writing for publication is undertaken and at the time of the submission. If disagreements arise, they should be resolved by a third party using these guidelines.

Authors must also be careful about **dual publication**. Sometimes, this circumstance is legitimate; for example, a scientific paper published by one journal may be reprinted by another journal or in a book of readings (this should always be noted). Authors may not, however, publish the same paper in more than one copyrighted original research journal. But what constitutes the "same paper"? Can more than one paper be written from the same set of data? The line is rather hazy. For example, Thomas (1986, pp. iv–v) indicated that

> **frequently, new insights may be gained by evaluating previously reported data from a different perspective. However, reports of this type are always classed as research notes whether the reanalysis is undertaken by the original author or someone else. This does not mean that reports which use data from a number of studies (e.g., meta-analyses, power analyses) are classed as research notes.**

Generally, good scientific practice is to publish all the appropriate data in a single primary publication. For example, if both psychological and physiological data were collected as a result of a specific experiment on training, publishing these separately may not be appropriate.

dual publication—Having the same scientific paper published in more than one journal or other publication; this is generally unethical.

Frequently, the main finding of interest is the interaction between psychological and physiological responses. But in other cases, the volume of data may be so large as to prohibit an inclusive paper. Sometimes the papers can be published as a series; at other times they may be completely separate. Another example is a large-scale study in which a tremendous amount of information is collected (e.g., exercise epidemiological or pedagogical studies). Usually, data are selected from computer records (or videos) to answer a set of related questions for a research report. Researchers may then use a different part (or even an overlapping part) of the data set to address another set of questions. More than one legitimate publication may then result from the same data set, but the authors should identify that more than one paper has been produced. Researchers who do not follow these general types of rules may be viewed as lacking scientific objectivity in their work and certainly as lacking in modesty (Day & Gastel, 2006).

> A single primary publication should include either all appropriate data (e.g., psychological, physiological) or a notice that relevant data is published elsewhere.

Most research journals require that the author include a statement that the paper has not been previously published or submitted elsewhere while the journal is considering it. Papers published in one language may not be published as an original paper in a second language.

ETHICAL ISSUES REGARDING COPYRIGHT

Graduate students should be aware of copyright regulations and the concept of fair use as it applies to educational materials. Copyrighted material is often used in theses and dissertations, and this use is acceptable if it is fair and reasonable. Often, graduate students want to use a figure or table in a thesis or dissertation. If you use a table or figure from another source, you must seek permission (see the Sample Copyright Permission Letter sidebar) from the copyright holder (for published papers, this is usually the author but sometimes the research journal) and cite it appropriately (e.g., "used by permission of . . .").

The concept of fair use has four basic rules:

1. *Purpose.* Is the use to be commercial or educational? More leeway is given for educational use such as theses, dissertations, and published research papers.

2. *Nature.* Is copying expected? Copying a journal article for your personal use is expected and reasonable. Copying a complete book or standardized test, however, is not expected and is probably a violation of the fair-use concept.

3. *Amount.* How much is to be copied? The significant issue is the importance of the copied part.

4. *Effect.* How does copying affect the market for the document? Making a single copy of a journal article has little effect on the market for the journal, but copying a book (or maybe a book chapter) or a standardized test reduces royalties to the author and income for the publisher. That act is not fair use.

Few standard answers apply to fair use. Fair use is a flexible idea (or alternatively, a statement that cannot be interpreted rigidly). It's better to be safe than sorry when using material in your thesis or dissertation. If you have any doubt, seek permission. University Microfilms (which produces *Dissertation Abstracts International,* the national warehouse for dissertations) reports that about 15% of dissertations require follow-up on copyright issues.

MODEL FOR CONSIDERING SCIENTIFIC MISCONDUCT

Intention is often used as the basis for differentiating between scientific misconduct and mistakes. Drowatzky (1993, 1996) noted that the following model is often suggested:

Scientific misconduct → sanctions   Scientific mistakes → remedial activities

SAMPLE COPYRIGHT PERMISSION LETTER

Date

Permissions Editor (or name of an individual author)

Publisher (not needed if an individual)

Address

Dear _____:

I am preparing my thesis/dissertation, tentatively titled _____.

I would like permission to use the following material: _____

Title of article in journal, book, book chapter: _____

Author of article, book, or book chapter: _____

Title of journal or book (include volume and issue number of journal): _____
_____

Editor if edited book: _____

Year of publication: _____

Place of publication and name of publisher of book or journal: _____
_____

Copyright year and holder: _____

Page number(s) on which material appears: _____ ☐ to be reprinted ☐ to be adapted

A COPY OF THE MATERIAL IS ATTACHED.
Table, figure, or page number in my thesis or dissertation: _____

I request permission for nonexclusive rights in all languages throughout the world. I will, of course, cite a standard source line, including complete bibliographic data. If you have specific credit line requirements, please make them known in the space provided below.

_____

A duplicate copy of this form is for your files. Your prompt cooperation will be appreciated. If permission is granted, please sign the release below and return to:

YOUR SIGNATURE: _____

YOUR NAME AND ADDRESS: _____

....

Permission granted, Signature: _____

Address: _____
_____ Date: _____

Sanctions for Scientific Misconduct

Sanctions are often imposed on those who are fraudulent in their scientific work. Internal sanctions imposed by the researcher's university have included the following:

- Restriction of academic duties
- Termination of work on the project
- Reduction in professorial rank
- Fines to cover costs
- Separation from the university (either with or without loss of benefits)
- Salary freeze
- Promotion freeze
- Supervision of future grant submissions
- Verbal reprimands
- Letter of reprimand (either included or not included in the permanent record)
- Monitoring research with prior review of publications

Besides internal university sanctions for scientific fraud, sanctions may be imposed by the agency that has funded the research, scholarly journals that published the work, and related scholarly or professional groups. In recent years, external sanctions have included the following:

- Revocation of prior publications
- Letters to offended parties
- Prohibition from obtaining outside grants
- Discontinuance of service to outside agencies
- Release of information to agencies and professional organizations
- Referral to legal system for further actions
- Fines to cover costs

Responsibilities of Graduate Students

As a graduate student, you must become concerned with ethical issues. Of course, the issues are important for much more than just the biological science areas (such as exercise physiology, biomechanics, motor behavior, health promotion and exercise, and exercise psychology) or behavioral science areas (such as sport sociology, sport philosophy, sport history, sport psychology, and physical education pedagogy). Fraud, misrepresentation, inaccurate interpretation of data, plagiarism, unfair authorship issues, and unethical publications practices are problems that extend to any area of scholarship. Although these practices sometimes occur simply because the person is unethical, often pressures that exist in our system of higher education tempt researchers to behave unethically. Following are some examples of such pressures:

- The need to obtain external funding for research
- Pressure to publish scholarly findings
- The need to complete graduate degree work
- The desire to obtain rewards in higher education (e.g., promotion, merit raises)

Academic units should be encouraged to develop mechanisms for discussing ethical scholarship issues with graduate students. The approach might include seminars for graduate students focused on these issues. But some systematic means is needed to bring the issues to the forefront for discussion.

Philosophical Positions Underlying Ethical Issues

A person's basic philosophical position on ethical issues drives the decision-making process in research. Drowatzky (1993, 1996) summarized the ethical views that underpin the decision-making process:

- The individual is precious, and the individual's benefit takes precedence over society.
- Equality is of utmost importance, and everyone must be treated equally.
- Fairness is the overriding guide to ethics, and all decisions must be based on fairness.
- The welfare of society takes precedence over that of the individual, and all must be done for the benefit of society.
- Truth, defined as being true, genuine, and conforming to reality, is the basis for decision making.

Of course, several statements in this list are in direct conflict with each other and will lead to substantially different decisions depending on one's view. Discussing and evaluating these statements and what they imply about decisions in scholarship should enhance graduate students' understanding of important issues. A few more philosophical statements to ponder are listed in the sidebar Philosophical Statements to Think About.

Finally, reading some of the literature on fraud and misconduct in research is a sobering experience for anyone. A notable example is the special issue of *Ethics and Behavior* titled "Whistleblowing and Scientific Misconduct" (vol. 3, no. 1, 1993). This issue gives a fascinating account of the David Baltimore and Herbert Needleman cases, including overviews and responses by the whistleblowers and those accused of scientific misconduct. Drowatzky (1996) also offered numerous examples of misconduct and the problems associated with it.

WORKING WITH FACULTY

Ethical considerations among researchers and ethical factors in the graduate student–major professor relationship are two important topics (see Roberts, 1993, for a detailed discussion). Major professors should treat graduate students as colleagues. If we want our students to be scholars when they complete their graduate work, then we should treat them like scholars from the start, because graduate students do not become scholars on receipt of a degree. By the same token, graduate students must act like responsible scholars. This means producing careful, thorough, and high-quality work.

Selecting a Major Professor

Students should try to select major professors who share their views in their area of interest. Master's students frequently choose an institution based on location or the promise of financial aid. Doctoral students, on the other hand, should select an institution by evaluating the program's quality and the faculty in their area of specialization (see Hollingworth & Fassinger, 2002, for a discussion). A listing and electronic link to doctoral programs in physical activity is available at www.nationalacademyofkinesiology.org. Do not choose your major professor hastily. If you are already at the institution, carefully evaluate the specializations available in your interest areas. Ask questions about faculty and learn whether they publish in those areas. Read some of their publications and determine your interest. What financial support, such as laboratories and equipment, is given to those areas? Also, talk to fellow graduate students. Finally, talk with the faculty members to determine how effectively you will be able to work with them.

Philosophical Statements to Think About

1. The professor discovered that her theory of earthquakes was on shaky ground.
2. A backward poet writes inverse.
3. I planted some birdseed and a bird came up. What do I do with it?
4. I went to San Francisco and found someone's heart. Now what?
5. Do protons have mass? I didn't even know they were Catholic.
6. What is a free gift? Aren't all gifts free?
7. I used to be indecisive. Now I'm not sure.
8. How can there be self-help groups?
9. Is it my imagination, or do buffalo wings taste like chicken?

We advocate a mentor model for preparing graduate students (particularly doctoral students) in kinesiology, physical education, exercise science, and sport science. To become a good researcher (or a good clinician), a student needs a one-on-one student–faculty relationship. This means several things about graduate students and graduate faculty.

First, graduate students need to be full-time students to develop the research and clinical skills needed for success in research and teaching. They need to work with a mentor in an ongoing research program. This arrangement lends continuity to research efforts and pulls graduate students together into effective research teams. Theses and dissertation topics arise naturally from these types of settings. Additionally, more senior students become models and can offer assistance to novice graduate students. Expertise is acquired by watching experts, by working with them, and then by practicing the techniques acquired.

Conversely, for faculty members to be good mentors, they must have active research programs. Appropriate facilities and equipment must be available, and faculty members must have time to devote to research and graduate student mentoring. Potential graduate students should carefully investigate the situations into which they will place themselves, especially if they have a major interest in research (for a good description of mentors, see Newell, 1987).

If you are not yet enrolled at an institution, find out which institutions offer the specialties you are interested in. Request information from their graduate schools and departments and look at their websites. Read the appropriate journals (over the past 5 to 10 years) and see which faculty are publishing. After you narrow your list, explore financial support and plan a visit. Speak to the graduate coordinator for the department and the faculty in your area of interest. Sometimes you can meet faculty at conventions, such as those of SHAPE America (national or district meetings), the American College of Sports Medicine, the International Society of Biomechanics, and the North American Society for Psychology of Sport and Physical Activity.

After you select a major professor, you must select a committee. Normally, the master's or doctoral committee is selected in consultation with your major professor. The committee selected should be one that can contribute to the planning and evaluation of your work, not one that might be the easiest. It is preferable to wait a semester or quarter or two (if you can) before selecting a final committee. This interval gives you the opportunity to have several committee members as teachers and to evaluate whether you have common interests.

Changing Your Major Professor

What happens, however, if you have a major professor (or committee member) who is not ideal for you? First, evaluate the reason. You need not be best friends, but it is important that you and your major professor are striving for the same goals. Sometimes students' interests change. Sometimes people just cannot get along. If handled professionally, however, this situation should not be a problem. Go to your major professor and explain the situation as you perceive it and offer him or her an opportunity to respond. Of course, the conflict may be more personal. If so, use an objective and professional approach. If you cannot, or if this does not produce satisfactory results, the best recourse may be to seek the advice of the graduate coordinator or department chairperson.

PROTECTING HUMAN PARTICIPANTS

Most research in the study of physical activity deals with humans, often children. Therefore, the researcher must be concerned about any circumstances in the research setting or activity that could harm the participants. Harm should be interpreted to mean frightening, embarrassing, or negatively affecting the participants (Tuckman, 1978). Of course, researchers always run the risk of creating a problem. What must be balanced is the degree of risk, the participants' rights, and the potential value of the research in contributing to knowledge, to the development of technology, and to the improvement of people's lives.

> There will always be a risk of harm. Researchers should balance the degree of risk, rights of participants, and value of the study.

What Should Research Participants Expect?

Tuckman (1978) summarized the participants' rights that experimenters must consider:

- **The right to privacy or nonparticipation.** This item includes researchers' not asking for unnecessary information and their obtaining direct consent from adults and consent from parents for children (as well as the consent of the children themselves when appropriate).

- **The right to remain anonymous.** The researcher should explain that the study focuses on group data and that an ID number (rather than the participant's name) will be used to record data.

- **The right to confidentiality.** Participants should be told who will have access to original data by which the participants might be identified (keep this to as few people as possible).

- **The right to expect experimenter responsibility.** The experimenter should be well meaning and sensitive to human dignity. If the participant is not told the purpose of the study (or is misled), the participant must be informed immediately after the completion of testing.

Qualitative research (discussed in detail in chapter 19) lends itself to some potential ethical problems because of the close, personal interaction with participants. The researcher often spends a great deal of time with participants, getting to know them and asking them to share their thoughts and perceptions. Griffin and Templin (1989) raised ethical concerns about whether to share field notes, how to protect a participant's self-esteem without compromising accuracy in the research report when the two are in conflict, and what to do if you are told about (or observe) something illegal or immoral while collecting data. Locke, Spirduso, and Silverman (2014) discussed these issues further.

There are no easy answers to situational ethics in **fieldwork**. Qualitative research sometimes deals with so-called deviants, such as drug addicts and unlawful motorcycle gangs. Informed consent is impossible in some circumstances. Punch (1986, p. 36) made this point when he described his research with police. The patrol car that he was riding in was directed to a fight. As the police officers jumped out and started wrestling with the combatants, Punch wondered whether he was supposed to yell "Freeze," thrust his head between the entangled limbs, and, Miranda-like, chant out the rights of the participants. Similarly, when Powermaker (cited by Punch) came face to face with a lynch mob, should she have flashed her academic identity card and explained to the crowd the nature of her presence? By these two examples we are not implying that qualitative researchers are exempt from considerations such as informed consent and deception. We are simply pointing out that certain types of

> fieldwork—Methodology common in qualitative research in which data are gathered in natural settings.

qualitative research situations present special ethical problems. We invite you to read the discussion by Punch (1986) and consult some of the sources he cited concerning this issue.

People with disabilities present a special issue as research participants. A participant with a disability is protected under the U.S. Right to Privacy Act. Thus, institutions are prohibited from releasing the names of people with disabilities as potential research participants. The researcher must contact the institution about possible participants. The institution must then request permission from the participant or his or her parents to release the participant's name and the nature of the disability to the researcher. If the participant or parents approve, the institution allows the researcher to contact the participant or parents to seek approval for the research to be undertaken. Although this procedure is cumbersome and varies from state to state, people clearly have the right not to be cited in studies and labeled as disabled unless they so choose.

Informed Consent

Consideration must be given to the protection of human participants. Researchers are required to protect the rights and well-being of participants in their studies. The regulations detailing the procedures are published by the U.S. Department of Health and Human Services (45 CFR 46.101). Most institutions regulate this protection in two ways. First, researchers are required to complete some type of form describing their research. Training on obtaining permission for use of human subjects is typically online, as are the forms to use. The procedures and forms vary somewhat from institution to institution but are easy to complete and submit.

The researcher completes this form online, attaches an abstract, and has it approved before beginning any work, including pilot work. The source of approval may vary. For example, some institutions require that all forms be approved by a central committee. Other institutions delegate approval for so-called standard types of research to a lower level (e.g., a college committee).

Typically, institutions require that you include your informed-consent form with your application to conduct human-participant research. If the participants are minors, then you must obtain their parents' permission and the children's assent if they are old enough to understand. Every university has slightly different guidelines for research with human subjects and for obtaining consent. You should get these guidelines early and make sure your informed-consent forms conform to them.

Following are basic elements of informed consent as specified by the advisory committee of *Research Quarterly for Exercise and Sport* (Thomas, 1983, p. 221):

- A fair explanation of the procedures to be followed, including an identification of those that are experimental.
- A description of the attendant discomforts and risks.
- A description of the benefits to be expected.
- A disclosure of appropriate alternative procedures that would be advantageous for the participant.
- An offer to answer any inquiries concerning the procedures.
- An instruction that the participant is free to withdraw consent and discontinue participation in the project or activity at any time. In addition, the agreement should contain no exculpatory language through which the participant is made to waive, or appear to waive, any legal right or to release the institution or its agents from liability or negligence.

The researcher is required to comply with any institutional guidelines for the protection of human participants and for informed consent. A description of this compliance should be included under the participant heading in the methods section of the thesis or dissertation.

> Federal law requires researchers to protect participants' rights and well-being.

See no evil, hear no evil, speak no evil—can I ignore this ethical issue?

© Thomas Perkins/fotolia

Most journals also require a statement with regard to this issue. The form used for informed consent is normally placed in an appendix of the thesis or dissertation.

PROTECTING ANIMAL SUBJECTS

Matt (1993), in her article "Ethical Issues in Animal Research," pointed out that this issue is not new, having been discussed in Europe for over 400 years and in the United States for over 100 years. The ground rules were established long ago when Descartes indicated that it was justifiable to use animals in research because they cannot reason and are therefore lower in the order of things than humans (Matt, 1993). Bentham (1970), however, said that the issue is not whether animals can think and reason but whether they perceive pain and suffer.

As Matt (1993) argued, far fewer animals are used in research than are slaughtered for food, held in zoos, and killed as unwanted pets in animal shelters. In fact, animal studies may have more stringent criteria for approval than studies with humans. Institutional review boards typically require that investigators demonstrate that animal studies add significant knowledge to the literature and are not replications, a requirement not placed on studies using humans as participants.

If animals are well treated, is their use in research justified? Matt (1993, pp. 46–47) said yes if the purposes of the research fall into one of five categories:

(1) drug testing, such as the development and testing of AIDS drugs; (2) animal models of disease, such as the development of animal models of arthritis, diabetes, iron deficiency, auto-immune dysfunction, and aging; (3) basic research, focused on examining and elucidating mechanisms at a level of definition not possible in human models; (4) education of undergraduate and graduate students in laboratories and lectures, with experience and information gained from the use of animal models; and (5) development of surgical techniques, used extensively in the training of medical students and the testing of new surgical devices and procedures.

A careful consideration of these categories suggests few suitable alternatives (but see Zelaznik, 1993, for a discussion).

If animals are used as subjects for research studies in kinesiology, institutions require adherence to *Guide for the Care and Use of Laboratory Animals,* published by the U.S.

Department of Health and Human Services, as detailed in the Animal Welfare Act (PL 89-544, PL 91-979, and PL 04-279). Most institutions also support the rules and procedures for recommended care of laboratory animals outlined by the American Association for Accreditation of Laboratory Animal Care.

All these documents recognize that for advancements to be made in human and animal research, animals must be used. These animals must be well tended, and if their use results in the animals being incapacitated or sacrificed, this procedure must be done humanely.

> Although research advancement requires animal testing, animals must be well cared for and treated humanely.

SUMMARY

This chapter addressed ethical issues that affect graduate students in their research and scholarly activities. We identified ethical issues and set the stage for you to think about and discuss these values as they influence your graduate and scholarly life.

Points that typically arise in the area of scientific misconduct include plagiarism, fabrication and falsification of data, nonpublication of data, data-gathering problems, data storage and retention issues, authorship controversies, and publication practices. We discussed copyright issues in research and publication. We also presented the model most frequently used to deal with scientific misconduct and some of the internal and external sanctions that have been imposed on people found guilty of scientific misconduct.

We discussed ethical and procedural issues in working relationships. How should graduate students select a major professor and committee members? Should students seek new major professors or committee members if they are unable to work effectively or amicably with their original choices?

Finally, we discussed ethics and procedures in the use of human and animal participants. This topic included remembering participants' rights, protecting human and animal subjects, and obtaining informed consent from human participants.

Check Your Understanding

1. Go online to your university's research office and find the instructions and forms for requesting the use of humans in research. Prepare a summary of the procedures for class discussion.

2. Our thanks go to Dr. Philip Martin, professor of kinesiology at Iowa State University, for developing the following case studies and allowing us to use them. Names and institutions used should not be taken seriously; we are only trying to bring humor to our friends.

Plagiarism

- What is plagiarism? Can you recognize it when you see it?

Case study. In preparing her thesis introduction, Graduate Assistant Christina periodically takes multiple sentences verbatim from some of her sources (her attitude is, "I couldn't have written it better myself!").

- Is she wrong to do this?
- If she provides a reference to her sources at the end of her paragraph, is she still wrong?

Fabrication or Falsification of Data

Case study. Professor Wade has strength-training data on 20 elderly participants. As he was madly processing data to meet the ACSM abstract deadline, Professor Wade realized that his sample did not show a significant increase in strength. Being disappointed about this finding, he examined his data more closely and noticed that 15 participants appeared to increase strength substantially whereas 5 participants actually showed strength declines. Professor Wade concluded that these 5 participants must not have adhered to the training program, and he decided to drop them from the study. Using data on only the remaining 15 participants, he could now demonstrate statistically significant improvements in strength, and he wrote his abstract based on those 15 participants.

- Did Professor Wade act ethically?
- What is an outlier? How do you define an outlier?
- How long should you keep raw data available for others to review following publication?
- Are you obligated to provide your raw data for examination on request?

Authorship of Presentations and Publications

Case study. Professor Martin is well known for her research on the effects of exercise on bone density. In 2006 she was awarded a five-year research grant from NIH to study bone density. Graduate Assistant (GA) Jackson began work under Professor Martin's mentorship in 2012 and was immediately assigned the task of running one of the experiments on bone density that was outlined in the grant proposal. Professor Martin could never be found in the laboratory for data collection, but she regularly held lab meetings during which she discussed progress on data collection with GA Jackson. After completing data collection, GA Jackson organized and presented the data to Professor Martin, who was pleased with the data. Professor Martin then assigned GA Jackson the task of drafting a manuscript based on the data. After several iterations, both were pleased with the final product and concluded that the manuscript was ready for submission. Unfortunately, they had had no conversations about authorship.

- Should the authorship be Martin and Jackson or Jackson and Martin?
- Professor Martin indicated to GA Jackson during the manuscript preparation phase that she would like to include GAs Powers, Cauraugh, Stelmach, and Thomas (the other four GAs in Martin's lab) as coauthors because they had been involved in other aspects of the funded research (after all, they needed publications on their curricula vitae to secure prestigious postdoc positions).
- Is this request by Professor Martin reasonable? As a side note, an interesting exercise is to compute some oddball descriptors for manuscripts with long lists of authors. Consider (1) the ratio of words to author, (2) the ratio of lines to author, and (3) the ratio of participants to author, and then consider just how significant each author's contributions might be.
- If the project for which GA Jackson collected the data was for his dissertation topic (an offshoot of Professor Martin's funded research line), should the authorship be Martin and Jackson, Jackson and Martin, or just Jackson?
- If Professor Martin hired Technician Magill to assist GA Jackson with data processing (for the original scenario), should the authorship be Martin and Jackson; Martin, Jackson, and Magill; Jackson and Martin; or Jackson, Martin, and Magill?
- How important is having some agreement about authorship at the beginning of the writing process?

Other Publication and Presentation Issues

Case study. Professor Sharp, an expert in forensic biomechanics, presented a research paper on the relationship between footprint spacing and body size at the 2010 annual meeting of American College of Sports Medicine (ACSM). Shortly after the meeting, during a moment of free time, he saw an announcement in one of his research magazines for a meeting titled The Science of Forensic Biomechanics, sponsored by the Society of Police Detectives. He submitted the same abstract used at the ACSM annual meeting and subsequently presented the same research paper at the SFB meeting.

- Right or wrong?

Case study. Professor Sharp submitted a manuscript titled "The Relationship Between Footprint Spacing and Body Size" to the *Journal of Biomechanics* (*JoB*). It was accepted and scheduled to be published in the March 2016 issue. He also had an abstract on the same subject accepted for presentation at the May 2016 ACSM meeting.

- Is there anything wrong with presenting published data?

Case study. Professor Sharp wanted to change jobs and needed to strengthen his curriculum vitae. He submitted his manuscript titled "The Relationship Between Footprint Spacing and Body Size" to the *Journal of Biomechanics.* To ensure publication of his research, Professor Sharp decided to submit the same manuscript to the *Journal of Forensic Science* while his *JoB* manuscript was still under review (after all, two chances are better than one).

 ■   Right or wrong?

Case study. Professor Sharp submitted his manuscript titled "The Relationship Between Footprint Spacing and Body Size" to the *Journal of Biomechanics.* He then used the same data but interpreted them from a somewhat different perspective. He prepared another manuscript titled "The Relationship Between Footprint Spacing and Walking Speed" and submitted it to the *Journal of Forensic Science* while his *JoB* manuscript was still under review.

 ■   Did Professor Sharp do anything wrong?

Case study. Assistant Professor Roberts, a new faculty member at the University of Reallycold, has just completed a broad, multidisciplinary dissertation on the benefits of endurance training on psychological and physiological markers of health and well-being and biomechanical aspects of the running pattern. In each of three major areas (psychology, physiology, and biomechanics), she has four major dependent variables. Knowing that she needs a healthy publication record when she reaches the promotion and tenure review in five years, she decides to create 12 publications from her dissertation research ("The Benefits of Endurance Training on . . .").

 ■   How do you react to her strategy for dealing with the dictum Publish or Perish?

Case study. Assistant Professor French, a new faculty member at the University of South Columbia, completed his doctorate at Big Time University under the direction of Dr. Samoht. His dissertation topic fit within the general scope of research activities being pursued by Dr. Samoht but reflected a unique focus, one that Dr. Samoht had not considered before French's work as a doctoral student. French submits a manuscript to *Medicine & Science in Sports & Exercise* listing himself as the sole author of the paper and the University of South Columbia as his affiliation.

 ■   Has French behaved ethically?
 ■   Has he given appropriate credit to Professor Samoht or Big Time University?

Case study. Professor Conan Barbarian is an icon in the field of gerontology. He is the director of the Institute of Gerontological Research (IGR) at Jellystone University, a highly funded research lab in which several faculty, postdocs, and graduate students work. Professor Barbarian requires that he be listed as an author on all manuscripts based on research completed at the IGR.

 ■   Is Professor Barbarian justified in his demand, or is this an example of ego gone wild?

Changing Your Major Professor

Case study. Graduate Assistant (GA) Lee has a strong interest in the mechanical behavior of muscle and was accepted into the exercise science doctoral program to study with Professor Silverman, an expert on muscle mechanics. After one year in the program, the chemistry between GA Lee and Professor Silverman is not great. In addition, GA Lee notices that one of his GA colleagues is getting some travel support from her mentor, Professor Moran, an expert on muscle energetics. GA Lee wants to continue to study muscle mechanics but thinks that he would like to do so under Professor Moran's direction rather than Professor Silverman's.

 ■   Should GA Lee pursue a mentor change? If so, how should he go about doing so? What are GA Lee's obligations to Professor Silverman?

Juggling Multiple Job Offers

Case study. Postdoc Gallagher is pursuing several tenure-track job opportunities in kinesiology, following years of training as a graduate student and postdoctoral fellow. Her first choice is a faculty position at the University of Minnesota. UM has an excellent reputation in kinesiology, the job description seems to match Gallagher's interests well, superb lab facilities are already in place, and her family lives within a few hours' drive in Wisconsin. While Gallagher waits to hear from UM, she gets an invitation for an interview at Gator University in central Florida. She accepts the invitation, and the interview goes well.

She likes the faculty at Gator and thinks that there is good potential to build a solid research program at Gator despite the fact that the kinesiology program there and the university are not considered top tier. Gator calls and offers her a faculty position with a modest startup package. She is not excited about living in Florida (she is not fond of gators, mosquitoes, and humidity), but no other offers are pending, and time is getting short in the job search process. Thus, she accepts in writing the faculty offer from Gator. Two days after mailing her response to Gator, Gallagher receives an invitation to interview at the University of Minnesota.

- ■ What should she do?

Assume for the moment that Gallagher accepted the invitation from UM to interview. While visiting UM, she is confronted by Professor Smith, who has a good friend and colleague at Gator. Professor Smith heard from his Gator colleague during a friendly telephone conversation that Gallagher had interviewed at Gator and had accepted their job offer.

- ■ How would you expect the faculties at UM and Gator to react to this situation?
- ■ If Gallagher had not yet received a written offer from Gator University but had only orally accepted an offer from Gator (i.e., she was waiting for the official offer to reach her by mail), would the situation be any different?

Statistical and Measurement Concepts in Research

> The statistics on sanity are that 1 out of 4 is suffering from some sort of mental illness. Look at the 3 people sitting around you in class—If they are OK, then it's you.
>
> —Anonymous

In the following six chapters, we present some basic statistical and measurement techniques that are frequently used in physical activity research. We give more attention here to the basic statistical techniques than to the more complex methods. The underlying concepts of statistical techniques have been emphasized rather than any derivation of formulas or extensive computations. Because an understanding of how the basic statistical techniques work facilitates a grasp of the more advanced procedures, we have provided the computational procedures for most of the basic statistics and examples of their use. We use the procedures from Statistical Package for the Social Sciences (SPSS) to analyze data on 30 professional golf players from the 2008 season of the PGA Tour as examples.

Chapter 6 discusses the need for statistics. We describe types of sampling procedures and summarize the basic statistics used in describing data, such as distributions, measures of central tendency, and measures of variation. Statistics can reveal two things about data: significance and meaningfulness.

Chapter 7 introduces the concepts of probability, effect size, and—most important—power. Using power analysis to evaluate studies you read and studies you plan is essential.

Chapter 8 pertains to relationships among variables. We review correlational techniques, such as the Pearson r for the relationship between only two variables. We explain partial correlation as a technique by which one can determine the correlation between two variables while holding the influence of additional variables constant. The use of correlation for prediction is discussed when using more than one variable to predict a criterion (multiple regression, logistic regression, and discriminant function analysis). Finally, we briefly present multivariate techniques of correlation: canonical correlation, factor analysis, and structural modeling.

Chapter 9 focuses on statistical techniques for comparing treatment effects on groups, such as different training methods or different samples. The simplest comparison of differences is between two groups: the *t* test. Next, we describe analysis of variance as a means of testing the significance of the differences among two or more groups. We also discuss the use of factorial analysis of variance, by which two or more independent variables can be compared. The use of repeated measures is common in our field, and we describe how they are analyzed. In addition, we provide an overview of multivariate techniques in this area, including discriminant analysis and multivariate analysis of variance.

Chapter 10 provides information on nonparametric techniques for data analysis. These are procedures in which the data fail to meet one or more of the basic assumptions of parametric techniques described in chapters 8 and 9. We present an approach to nonparametrics that parallels the procedures described in chapters 8 and 9. Thus, you do not need to learn a new set of statistics to use the techniques.

Finally, chapter 11 reviews many of the measurement issues that apply when conducting research. We focus on the validity and reliability of dependent variables. We present a short overview of measurement issues regarding data collected about physical performance, although complete coverage of this topic is not possible. In much research on physical activity, dependent variables may be affective and knowledge measures. Thus, we discuss these types of dependent variables.

After reading the six chapters in part II, you will not be a statistician or a measurement expert (unless you were one before you started). But if you read and study these chapters carefully and perhaps explore some of the references further, you should be able to comprehend the statistical analysis and measurement techniques of most research studies.

6

Becoming Acquainted With Statistical Concepts

Anything is possible if you do not know what you are talking about.

—From Lachlan's Laws (Scottish Highland philosopher)

The idea of learning statistics frightens many people. If you are intimidated, you need not be. Statistics is one of the few ways that data can be reported uniformly to allow relevant, accurate conclusions and comparisons to be made. Statistics are methodical, logical, and necessary, not random, inconsistent, or terrifying. Our approach to teaching statistics in this book is to acquaint you with the basic concepts and give you a working knowledge; it is not our purpose to make you a statistician, especially given this well-known saying: There are liars, damn liars, and statisticians.

WHY WE NEED STATISTICS

Remember that throughout the process in which you conceive, plan, execute, and write up research, it is your informed judgment as a scientist on which you must rely, and this holds as much for the statistical aspects of the work as it does for all the others (Cohen, 1990, p. 1310).

The variance associated with statistical procedures is:

Total (observed) variance = true variance + error variance

Thus, all statistical estimates are one of two arrangements of the variance formula:

Significance tests (t, F, and X^2) = true variance ÷ error variance

Correlation (r^2, R^2, % variance accounted for) = true variance ÷ total variance

Parametric statistical tests (chapters 8 and 9) assume that the distribution of data is normal (e.g., it forms a normal curve) for appropriate use of the t and F tables to test significance. Nonparametric statistical tests (see chapter 10) do not require the normality assumptions because comparisons are to the X^2 table.

Statistics is simply an objective means of interpreting a collection of observations. Various statistical techniques are necessary for describing the characteristics of data, testing relationships between sets of data, and testing the differences among sets of data. For example, if height and a standing-long-jump score were measured for each student in a seventh-grade class, you could sum all the heights and then divide the sum by the number of students. The result (a statistic to represent the average height) is the **mean**, $M = \Sigma X / N$, where Σ means sum, X = each student's height, and N = number of students; read this as "the sum of all Xs divided by N." The mean (M) describes the average height in the class; it is a single characteristic that represents the data.

An example of testing relationships between sets of data would be to measure the degree of association between height and the scores on the standing long jump. You might hypothesize that taller people can jump farther than shorter people. By plotting the scores (figure 6.1), you can see that taller people generally do jump farther than shorter people. But note that the relationship is not perfect. If it were, the scores would begin in the lower left-hand corner of the figure and proceed diagonally in a straight line toward the upper right-hand corner. One measure of the degree of association between two variables is called the **Pearson product moment coefficient of correlation** (also known as Pearson r, interclass correlation, or simple correlation). When two variables are unrelated, their correlation is

mean—A statistical measure of central tendency that is the average score of a group of scores.

Pearson product moment coefficient of correlation—The most commonly used method of computing correlation between two variables; also called *interclass correlation*, *simple correlation*, or *Pearson* r.

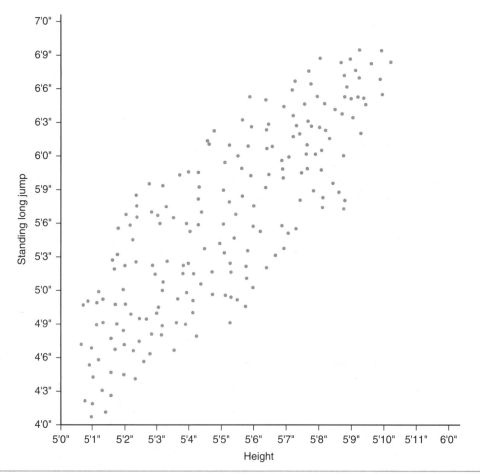

Figure 6.1 Relationship of height to standing long jump.

approximately zero. In figure 6.1 the two variables (height and standing-long-jump scores) have a moderately positive correlation (r is probably between .40 and .60). Relationships and correlation are discussed in greater detail in chapter 8, but for now you should see that researchers frequently want to investigate the relationships between variables.

Besides descriptive and correlational techniques, a third category of statistical techniques is used to measure differences among groups. Suppose you believe that weight training of the legs will increase the distance a person can jump. You divide a seventh-grade class into two groups and have one group participate for 8 weeks in a weight-training program designed to develop leg strength. The other students continue their regular activities. You want to know whether the independent variable (weight training versus regular activity) produces a change in the dependent variable (standing-long-jump score). Therefore, you measure the two groups' scores in the standing long jump at the end of the 8 weeks (treatment period) and compare their average performances. Here, a statistical technique to assess differences between two independent groups, a *t* test, would be used. By calculating t and comparing it with a value from a t table, you can judge whether the two groups were significantly different on their average long-jump scores. Ways for assessing differences among groups are discussed in chapter 9.

t **test**—A statistical technique to assess differences between two groups.

USE OF COMPUTERS IN STATISTICAL ANALYSIS

Computers are extremely helpful in the calculation of statistics. Computers do not make the mistakes that can occur in hand calculations, and they are many times faster. Desktop or notebook computers are most frequently used for statistical analysis in laboratories, offices, and homes. The computer and its attachments (disk drives, monitors, printers, flash drives, and so on) are called **hardware**, and the computer programming is called **software**. Numerous software programs have been written to calculate statistics.

Two popular software packages for use on computers are the Statistical Package for the Social Sciences (SPSS) and the Statistical Analysis System (SAS). Most colleges and universities have one or both of these packages, which can be used on a personal computer (with proper hardware and software). Your computer center has information about available equipment and services. Many computer centers provide user services or user consultant centers where you can get advice about the hardware and software available and instructions in how to use it.

Most institutions teach statistics courses in which these software packages are used. Statistics departments or other departments may also offer consulting services to help with the use of statistics for research projects. You should investigate the services your institution offers. Senior graduate students and faculty can also advise you on available services.

hardware—The physical units of a computer, such as the printed circuit boards, monitor, keyboard, disk drive, and printer.

software—The programs or instructions used to make computers function in the desired manner.

DESCRIPTION AND INFERENCE ARE NOT STATISTICAL TECHNIQUES

At the beginning of this chapter, we stated that statistical techniques allow the description of data characteristics, the testing of relationships, and the testing of differences. When we discuss description and **inference**, though, we are not discussing statistical techniques, although those two words are sometimes confused with statistical techniques. This confusion is the result of saying that correlations describe relationships and that cause and effect are inferred by techniques for testing differences between groups. These statements are not necessarily true. Any statistic describes the **sample** of participants for which it was calculated. If the sample of participants represents some larger group, then the findings can be inferred (or generalized) to the larger group. The statistic used, however, has nothing to do

inference—The generalization of results to some larger group.

sample—A group of participants, treatments, or situations selected from a larger population.

with inference. The method of selecting the sample, procedures, and context is what does or does not allow inference.

WAYS TO SELECT A SAMPLE

The sample is the group of participants, treatments, and situations on which the study is conducted. A key issue is how these samples are selected. In the following sections we discuss the types of sampling typically used in designing studies and using statistical analysis.

Random Selection

population—The larger group from which a sample is taken.

random numbers table—A table in which numbers are arranged in two-digit (or greater) sets so that any combination of rows or columns is unrelated.

The sample of participants might be randomly selected from some larger group, or a **population**. For example, if your college or university has 10,000 students, you could randomly select 200 for a study. You would assign each of the 10,000 students a numeric ID. The first numeric ID would be 0000, the second 0001, the third 0002, up through the last (the 10,000th), who would be 9999. Then a **random numbers table** (see table 1 in the appendix) would be used to select the sample. The numbers in this table are arranged in two-digit sets so that any combination of rows or columns is unrelated. In this instance (IDs 0000–9999), you need to select 200 four-digit numbers. Because the rows and columns are unrelated, you can choose any type of systematic strategy to go through the table. Enter the table at random (close your eyes and put your finger on the page). Suppose the place of entering the table is the sixth column of a two-digit number on row 8. The numeric ID here is 3020 (including columns 6 and 7 to get a four-digit number). You select the student with that numeric ID (3020). From here you can go down, across, or diagonally. Going down, to the next row, the four-digit combination of columns 6 and 7 is 9695, then 2953, and so on, until you have selected 200 students. After completing this column, go to the next four-digit column. Column 7 is not used because it was included with column 6 to yield a four-digit number.

The purpose of a random sample is to infer that the findings apply to the larger population.

The system used in the random numbers table is not the only one. You can use any systematic way of going through the table. Of course, the purpose of all this is to select a sample randomly so that it represents the larger population; that is, the findings in the sample can be inferred back to the larger population. From a statistical viewpoint, inference means that a characteristic, relationship, or difference found in the sample is likely also to be present in the population from which the sample was selected.

Computer programs can also be used to generate a set of random numbers. You tell the program the population size and how many random numbers to select.

Stratified Random Sampling

stratified random sampling—A method of stratifying a population on some characteristic before random selection of the sample.

In **stratified random sampling**, the population is divided (stratified) on some characteristic before random selection of the sample. Returning to the previous example, the selection was 200 students from a population of 10,000; suppose your college is 30% freshmen, 30% sophomores, 20% juniors, and 20% seniors. You could stratify on class before random selection to make sure the sample is exact in terms of class representation. Here, you would randomly select 60 students from the 3,000 freshmen, 60 from the 3,000 sophomores, 40 from the 2,000 juniors, and 40 from the 2,000 seniors. This procedure still yields a total sample of 200.

Stratified random sampling might be particularly appropriate for survey or interview research. Suppose you suspect that attitudes toward exercise participation change over the college years. You might use a stratified random sampling technique for interviewing 200 college students to test this hypothesis. Another example would be to develop normative data on a physical fitness test for grades 4 through 8 in a school district. Because performance would be related to age, you should stratify the population by age before randomly selecting the sample on which to collect normative data.

Systematic Sampling

If the population from which the sample is to be selected is very large, assigning a numeric ID to each potential participant is time-consuming. Suppose you want to sample a town with a population of 50,000 concerning the need for new sport facilities. One approach would be to use systematic sampling from the telephone book. You might decide to call a sample of 500 people. To do so, you would select every 100th name in the phone book (50,000 / 500 = 100). Of course, you are assuming that the telephone book represents the population or, said another way, that everyone you need to sample has a listed telephone number. This turned out to be a poor assumption in the 1948 presidential election (Dewey vs. Truman). The pollsters had predicted that Dewey would win by a substantial margin, but they had sampled from telephone books in key areas. Unfortunately for Dewey, many people without telephones voted for Truman. His victory was called an upset, but it was an upset only because of poor sampling procedures. Systematic sampling can yield a good sample and should be equivalent to random sampling if the sample is fairly large. Researchers, however, generally prefer random sampling.

Random Assignment

In experimental research, groups are formed within the sample. The issue here is not how the sample is selected but how the groups are formed within the sample. Chapter 18 discusses experimental research and true experimental designs. All true designs require that the groups within the sample be randomly assigned or randomized. Although this requirement has nothing to do with selecting the sample, the procedures used for random assignment are the same. Each person in the sample is given a numeric ID. If the sample has 30 participants, the numeric IDs range from 00 to 29. In this case, suppose that three equal groups ($n = 10$) are to be formed. Enter a table of random numbers; the first numeric ID encountered between 00 and 29 goes in group 1, the second in group 2, and so on, until each group has 10 participants. This process allows you to assume that the groups are equivalent at the beginning of the experiment, which is one of several important features of good experimental design that is intended to establish cause and effect.

Computer programs are also available to randomize groups. You supply the sample size and the number of groups and decide whether the groups are to have an equal number of participants. The computer program then randomly assigns participants.

JUSTIFYING POST HOC EXPLANATIONS

Frequently, the sample for research is not randomly selected; rather, the researcher attempts a post hoc justification that the sample represents some larger group. A typical example is showing that the sample does not differ in average age, racial balance, or socioeconomic status from some larger group. Of course, the purpose is to allow the findings on the sample to be generalized to the larger group. A post hoc attempt at generalization may be better than nothing, but it is not the equivalent of random selection, which allows the assumption that the sample does not differ from the population on the characteristics measured (as well as any other characteristics). In a post hoc justification, only the characteristics measured can be compared. Whether those are the ones that really matter is open to speculation.

Post hoc justification is also used to compare intact groups, or groups within the sample that are not randomly formed. However, in this case, the justification is that because the groups did not differ on certain measured characteristics before the study began, they can be judged equivalent. Of course, the same point applies: Are the groups different on some unmeasured characteristic that affects the results? This question cannot be answered satisfactorily. But, as before, a good post hoc justification of equivalence does add strength to comparisons of intact groups.

DIFFICULTY OF RANDOM SAMPLING AND ASSIGNMENT: HOW GOOD DOES IT HAVE TO BE?

In many studies of physical activity, random sampling procedures are not possible. For example, when comparing experts with novice performers, typically, neither group is randomly sampled, nor are group memberships (expert and novice) randomly formed. The same is true when studying trained versus untrained runners or cyclists. Often, we are interested in comparing the responses of different age groups, ethnic groups, and sexes to training in skill or exercise. Obviously, these groups are seldom randomly selected and cannot be randomly formed. In many studies, sampling is not done at all; researchers are happy to have any participants who volunteer.

Sampling also applies to the treatments used for different groups of participants. How are the treatments selected? Do they represent some population of potential treatments? What about the situational context under which participants are tested or receive the experimental treatment? Are these sampled in some way? Do they represent the potential situations?

The real answers to these questions about sampling are that we seldom randomly sample a population at any level: participants, treatments, or situations. Yet we hope to be able to use statistical techniques based on sampling assumptions and to infer that what we find applies to some larger group than the participants used in the study. What is really needed is a sample that is "good enough for our purpose" (Kruskal & Mosteller, 1979, p. 259). This concept is extremely important for research in physical activity. If we cannot disregard the strict requirements of random sampling that allow study outcomes to be generalized to some similar group, we will never be able to generalize beyond the characteristics of any particular study—location, age, race, sex, time, and so on. Strictly speaking, there is no basis for the findings from a sample to be similar to a population that differs in any way from the sample (even one hour after the sampling). "A good-enough principle of sampling, however, would allow generalizations to any population for which the sample is representative enough" (Serlin, 1987, p. 366).

For the sample of participants, treatments, and situations to be good enough to generalize, it must be selected on some theoretical basis. For example, if the theory proposes that the cardiovascular system responds to training in a specific way in all untrained adults, using volunteer untrained undergraduates in a training study may be acceptable. "It is only on the basis of theory that one decides whether the experimental results can be generalized to the responder population, to the stimulus or ecological population, or both" (Serlin, 1987, p. 367). The best possible generalization statement is to say that the findings may be plausible in other participants, treatments, and situations, depending on their similarity to the study characteristics.

> A good sample leads to a generalization statement that states that it is plausible that the findings apply to a broader population.

UNIT OF ANALYSIS

A question related to sampling and the statistical analyses discussed in subsequent chapters is labeled the **unit of analysis**. This concept refers to what can be considered the most basic unit from which data can be produced. We typically think of this as the individual participant; for example, we measure the change in the level of physical activity of a specific elementary school child.

The unit of analysis concept is also connected to the level of the treatment given. For example, consider a large study about how a specific program of intervention in physical

> **unit of analysis**—The concept, related to sampling and statistical analysis, that refers to what is considered the most basic unit from which data can be produced.

activity influences students in 16 elementary schools within a school district. We would like to randomly assign students within elementary schools to the two levels of intervention. The experimental treatment is a physical education class that focuses on all children being physically active and understanding why total physical activity (outside and inside school programs) is important to health. The control condition is children who are continuing their regular physical education program (assume a program that stresses competitive games and sports more than total physical activity). As you might guess, assigning students at random to these two conditions is extremely difficult. How can different programs be offered at the same time to children within a single class?

The researcher decides that children cannot be randomly assigned to the two treatments, but that classes can. Fortunately for our researcher, each elementary school has two classes at each grade level (planning clean experiments is easy when you make up the plan as we did; the task is much harder in the real world where situations are not so accommodating). So within an elementary school, one of the two classes at each grade level is randomly assigned to either the experimental or control treatment. In this instance, the unit of analysis is the class, not the individual child, because the class is what has been randomly assigned. Thus, a mean physical activity score is calculated for each class rather than for each child, and the class score is used in the statistical analysis. You can see how drastically that modification influences the number of data points for the statistical analysis. If we consider each child as the unit of analysis, and each grade of K through 5 has two classes of 25 children each, then the total sample size for a school (n) would be 300 and the total sample size (N) for the 16 schools would be 4,800. But if we consider the class as the unit of analysis, then the sample size is 12 for each of the 16 elementary schools, giving a total sample size (N) of 192.

However, consider that giving the two treatments within an elementary school might confound the data; that is, students receiving one treatment might well talk with students receiving the other even though they are in different classes. Maybe the children receiving the control treatment decide to exercise at home because a friend in another class (receiving the experimental treatment) is doing so. This possibility contaminates the data and increases variability. To avoid this, the researcher might decide to assign the 16 schools randomly to each of the two treatments. Now the unit of analysis becomes the school. The experimental treatment has an n of 8, and the control treatment has an n of 8.

You can see that deciding the appropriate unit of analysis drastically influences the sample size for statistical analysis (in the previous example, it ranges from a low of 16 to a high of 4,800). As you will see in chapter 7, appropriate statistical analysis demands statistical **power**. Power is influenced by sample size; larger samples generally have more power to detect relationships or differences. In planning studies in which intact units (classes or schools in the previous example) are assigned to a treatment, the effect of the appropriate unit of analysis on power should be considered early. Otherwise, during analysis the researcher will be faced with two bad choices—using the wrong unit of analysis and inflating N, or having to collect further data to have sufficient power to determine whether the treatment was effective.

power (statistical)—The probability of rejecting a false null hypothesis.

The unit of analysis significantly affects the sample size.

MEASURES OF CENTRAL TENDENCY AND VARIABILITY

Some of the more easily understood statistical and mathematical calculations are those that find the **central tendency** and **variability** of scores. When you have a group of scores, one number may be used to represent the group. That number is generally the mean, median, or mode. These terms are ways of expressing central tendency. Within the group of scores, each individual score differs to some degree from the central tendency score. The degree of

central tendency (measure of)—A single score that best represents all the scores.

variability—The degree of difference between each individual score and the central tendency score.

difference is the variability of the score. Two terms that describe the variability of the scores are **standard deviation** and **variance**.

Central Tendency Scores

The statistic for the central tendency score with which you are probably familiar is the mean (*M*), or average:

$$M = \Sigma X / N \tag{6.1}$$

Thus, if you have the numbers 6, 5, 10, 2, 5, 8, 5, 1, and 3, then

$$M = (6 + 5 + 10 + 2 + 5 + 8 + 5 + 1 + 3) / 9 = 45 / 9 = 5$$

The number 5 is the mean and represents this series of numbers.

Sometimes, the mean may not be the most representative or characteristic score. Suppose we replace the number 10 in the previous set of scores with the number 46. The mean is now 9, a number larger than all but one of the scores. That number is not representative because one score (46) made the average high. In this case, another measure of central tendency is more useful. The **median** is defined as the value in the middle; the middle value is the value that occurs in the place (*N* + 1) / 2 when the values are put in order. In our example, arrange the numbers from lowest to highest—1, 2, 3, 5, 5, 5, 6, 8, 46—then count (9 + 1) / 2 = 5 places from the first score; the median is the value 5, which is a much more representative score. If *N* is an even number, the median value may be a decimal. For example, for scores 1, 2, 3, 4, the median is 2.5.

Most often, you are interested in the mean of a group of scores. You may occasionally be interested in the median or perhaps in another measure of central tendency, the **mode**, which is defined as the most frequently occurring score. In the previous example, the mode is also 5, because it occurs three times. Some groups of scores may have more than one mode.

Variability Scores

Another characteristic of a group of scores is the variability. An estimate of the variability, or spread, of the scores can be calculated as the standard deviation (*s*):

$$s = \sqrt{\Sigma (X - M)^2 / (N - 1)} \tag{6.2}$$

This formula translates as follows. Calculate the mean by equation 6.1, subtract the mean from each person's score (*X* − *M*), square the answer, sum (Σ) the squared scores, divide by the number of scores minus one (*N* − 1), and take the square root of the answer. Table 6.1 provides an example.

The mean and standard deviation together are good descriptions of a set of scores. If the standard deviation is large, the mean may not be a good representation. Roughly 68% of a set of scores fall within ±1*s*, about 95% of the scores fall within ±2*s*, and about 99% of the scores fall within ±3*s*. This distribution of scores is called a normal distribution (discussed later in the chapter).

Equation 6.2 was used to help you understand the meaning of the standard deviation. For use with a hand calculator, equation 6.3 is simpler:

TABLE 6.1

Calculations of Mean and Standard Deviation

X	X−M	(X−M)²
6	1	1
5	0	0
10	5	25
2	−3	9
5	0	0
8	3	9
5	0	0
1	−4	16
3	−2	4
Σ = 45	0	64

$$M = \Sigma X / N = 45 / 9 = 5$$

$$s = \sqrt{\Sigma (X - M)^2 / (N - 1)} = \sqrt{64 / 8} = \sqrt{8} = 2.83$$

$$s = \sqrt{ N\sum X^2 - \left(\sum X\right)^2 / \left[N(N-1) \right]} \qquad (6.3)$$

One final point for later consideration is that the square of the standard deviation is called the variance, or s^2.

Range of Scores

Sometimes the range of scores (highest and lowest) may also be reported, particularly when the median is used rather than the mean. The median and the mean may be used in connection with each other. For example, 15 people might be given 10 blocks of 10 trials (100 total trials) on a reaction-time task. The experimenter may decide to use the median reaction time of a person's 10 trials as the most representative score in each block. Thus, each person would have 10 median scores, one for each of the 10 trial blocks. Here, the range of scores from which the median was selected should be reported. Both the mean and the standard deviation should be reported for the 15 participants' median scores at trial block 1, trial block 2, and so on. Thus, the range is reported for the selection of the median of each trial block, whereas the standard deviation is reported for the mean of each trial block.

Confidence Intervals

Confidence intervals are an effective technique used by researchers to help interpret a variety of statistics such as means, medians, and correlations. They are also used in hypothesis testing. A confidence interval provides an expected upper and lower limit for a statistic at a specified probability level, usually either 95% or 99%. The size, or length, of a confidence interval is affected by the size of the sample, the homogeneity of values within the sample, and the level of confidence selected by the researcher. Confidence intervals are based on the fact that any statistic possesses sampling error. This error relates to how well the statistic represents the target population. When we compute a mean for a sample, we are making an estimate of the mean of the target population. A confidence interval provides a band within which the estimate of the population mean is likely to fall instead of a single point.

> Confidence intervals should be used because statistics vary in how well they represent target populations.

A confidence interval (CI) of a statistic such as the mean employs the following information:

CI = observed statistic ± (standard error × specified confidence level value) (6.4)

We will construct a confidence interval for a mean of a sample with the following characteristics: $n = 30$, $M = 40$, $s = 8$.

The observed statistic is the mean ($M = 40$). The **standard error** represents the variability of the sampling distribution of a statistic. Imagine that instead of one sample of $n = 30$, you had selected 100 samples of this size. The means for all the samples would not be the same; in fact, they would start to approximate a normal distribution. If you calculated the standard deviation of all of these means, you would have an estimate of the standard error. Fortunately, we do not have to draw hundreds of samples to calculate standard error; we simply divide our sample standard deviation by the square root of the sample size. Do not ask why; just accept it. (Have we ever lied to you before?) So, in our example, the standard error of the mean is $s_M = 8 / \sqrt{30} = 1.46$.

> **standard error**—The variability of the sampling distribution.

The last piece of information we need to construct a confidence interval is the value for the specified confidence level. This is easy: We just look it up in a table. Recall from the discussion of the normal curve that 95% of the scores are contained within approximately ±2s. Actually, it is 1.96s instead of 2.00, and 99% of the scores are encompassed by ±2.576s instead of 3.00 (see table 2 in the appendix). Hence, if we knew that our sample had a normal distribution, we would use 1.96 for the 95% confidence level value or 2.58 for 99%.

OK, I've put in all the data but this computer will not calculate the mean.

But because we cannot assume that our sample is normally distributed, we should use a distribution table that takes into account the size of the sample. So we use the *t* distribution in table 5 in the appendix. The first column is labeled *df* (degrees of freedom). For our purposes here, this just means $n - 1$, or $30 - 1 = 29$. We are going to use the values under the two-tailed test (to reflect the spread above and below the mean). The value for the 95% confidence level is under the .05 column. Thus, reading across from 29 *df,* we see that the value to use is 2.045. If we were using the 99% level of confidence, we would use the .01 column, and the value would be 2.756.

We can now construct the confidence interval (95% confidence) using equation 6.4:

CI = sample mean ± (standard error × confidence level table value)

$$= 40 \pm (1.46 \times 2.045) = 40 \pm 2.99$$

We subtract and add 2.99 to 40 to obtain a confidence interval of 37.01 to 42.99. Thus, we are 95% confident that this interval includes the target population mean.

Confidence intervals are also used in hypothesis testing, such as applying the null hypothesis to the difference between two or more means. If we were comparing the means of two groups, the difference between the two means would be the observed statistic. The standard error would be the standard error of the difference (which combines the standard errors of the two means). The table value would be taken from the *t* table (appendix table 5) using *df* of $(n_1 - 1) + (n_2 - 1)$ or $(N - 2)$ for the selected level of confidence (e.g., 95%). The calculated confidence interval would give an estimate of the range within which the true difference between these two population means would fall 95% of the time.

Frequency Distribution and the Stem-and-Leaf Display

A common technique for summarizing data is to produce a picture (called a histogram) of the distribution of scores by means of a **frequency distribution**. A simple frequency distribution just lists all the scores and the frequency of each score. If there is a wide range of values, a grouped frequency distribution is used in which scores are grouped into small ranges called **frequency intervals**. For example, figure 6.2*a* is a histogram of the scoring averages of 30 selected PGA players in 2008. The scoring averages of players are on the *x*-axis (bottom), and the frequency of each score is represented by the height of the column on the *y*-axis. You can see by looking at the figure that 6 players averaged about 70.25 strokes per round. Note that a replication of the normal curve is placed over the actual distribution so that the two can be visually compared. One drawback to a grouped frequency distribution is that information is lost; that is, a reader does not know the exact score of each individual within a given interval.

frequency distribution—A distribution of scores including the frequency with which they occur.

frequency intervals—Small ranges of scores within a frequency distribution into which scores are grouped.

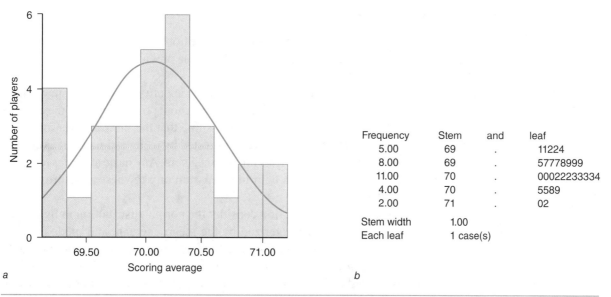

a   *b*

Figure 6.2   *(a)* Histogram of scoring averages of 30 selected PGA Tour players in 2008. *(b)* Stem-and-leaf display of scoring averages of 30 selected PGA Tour players in 2008.

Another effective way to provide information about the distribution of a set of data is called a **stem-and-leaf display**. This representation is similar to a grouped frequency distribution, but no information is lost.

Figure 6.2*b* provides a stem-and-leaf distribution of the same golf data. You can observe each player's actual scoring average—note that 11 players are listed by the first score of 70. Reading to the right, 3 players average 70.0, 3 are at 70.2, 4 are at 70.3, and 1 is at 70.4. When the intervals are ordered from lowest to highest, you can turn the page on its side and see a type of graph similar to the histogram. This method helps you visualize the normality of the distribution.

stem-and-leaf display—A method of organizing raw scores by which score intervals are shown on the left side of a vertical line and individual scores falling into each interval are shown on the right side.

BASIC CONCEPTS OF STATISTICAL TECHNIQUES

Besides measures of central tendency and variability, other, slightly more complicated statistical techniques can be used. Before we explain each in detail, however, you must understand some general information about statistical techniques.

Two Categories of Statistical Tests

The two general categories of statistical tests are **parametric** and **nonparametric**. Using the various tests in each category requires meeting the assumptions for those tests. The first category, parametric statistical tests, has three assumptions about the distribution of the data:

- The population from which the sample is drawn is normally distributed on the variable of interest.
- The samples drawn from a population have the same variances on the variable of interest.
- The observations are independent.

Certain parametric techniques have additional assumptions. The second category, nonparametric statistics, is called **distribution free** because the previous assumptions need not be met.

Whenever the assumptions are met, parametric statistics are often said to have more power, although there is some debate on this issue. Having more power increases the chances of rejecting a false null hypothesis. You frequently assume that the three criteria for use of parametric statistics are met. The assumptions can be tested by using estimates of **skewness** and **kurtosis**. (Only the meaning of these tests is explained here. Any basic statistics textbook provides considerably more detail. For a helpful discussion on skewness and kurtosis, see Newell and Hancock, 1984.)

To understand skewness and kurtosis, first consider the normal distribution in figure 6.3. This is a **normal curve**, which is characterized by the mean, median, and mode being at the same point (center of the distribution). In addition, $\pm 1s$ from the mean includes 68% of the scores, $\pm 2s$ from the mean includes 95% of the scores, and $\pm 3s$ includes 99% of the scores. Thus, data distributed as in figure 6.3 would meet the three assumptions for use of parametric techniques. Skewness of the distribution describes the direction of the hump of the curve (labeled A in figure 6.4) and the nature of the tails of the curve (labeled B and C). If the hump (A) is shifted to the left and the long tail (B) is shifted to the right (figure 6.4a), the skewness is positive. If the shift of the hump (A) is to the right and the long tail (C) is to the left (figure 6.4b), the skewness is negative. Kurtosis describes the vertical aspect of the curve, such as whether the curve is more or less peaked than the normal curve. Figure 6.5a shows a more peaked curve, and figure 6.5b shows a flatter curve.

Table 2 in the appendix shows a unit normal distribution (z) for a normal curve. The column z shows the location of the mean. When the mean is in the center of the distribution, its z is equal to .00; thus, .50 (50%) of the distribution is beyond (to the right of) the mean, leaving .50 (50%) of the distribution as a remainder (to the left of the mean). As the mean of the distribution moves to the right in a normal curve (say, to a z of $+1s$), .8413 (84%) of the distribution is to the left of the mean (remainder) and .1587 is to the right of (beyond) the mean. This table allows you to determine the percentage of the normal distribution included by the mean plus any fraction of a standard deviation. Suppose you want to know what percentage of the distribution would be included by the mean plus one-half (.50) of a standard deviation. Using table 2 you can see that it would be .6915 (remainder), or 69%.

For chapters 8 and 9, consider that the three basic assumptions for parametric statistical tests have been met. This is done for two reasons. First, the assumptions are very robust to

parametric statistical test—A test based on data assumptions of normal distribution, equal variance, and independence of observations.

nonparametric statistical test—Any of a number of statistical techniques used when the data do not meet the assumptions required to perform parametric tests.

distribution free—A term used to describe nonparametric statistical tests, because the data distribution requirements for a parametric test do not have to be met.

skewness—A description of the direction of the hump of the curve of the data distribution and the nature of the tails of the curve.

kurtosis—A description of the vertical characteristic of the curve showing the data distribution, such as whether the curve is more peaked or flatter than the normal curve.

normal curve—A distribution of data in which the mean, median, and mode are at the same point (center of the distribution) and in which $\pm 1s$ from the mean includes 68% of the scores, $\pm 2s$ from the mean includes 95% of the scores, and $\pm 3s$ includes 99% of the scores.

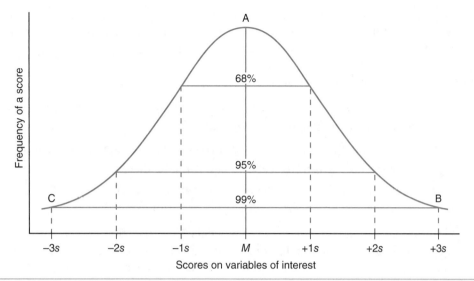

Figure 6.3   The normal curve.

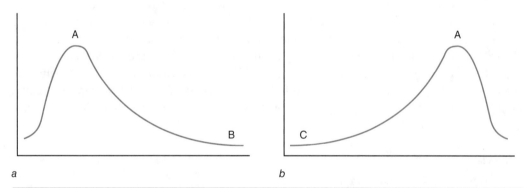

Figure 6.4   Skewed curves: *(a)* positive skewness; *(b)* negative skewness.

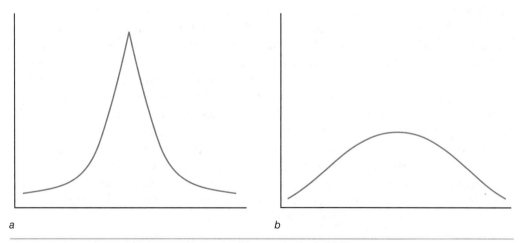

Figure 6.5   Curves with abnormal kurtosis: *(a)* more peaked; *(b)* flatter.

violations, meaning that the outcome of the statistical test is relatively accurate even with severe violations of the assumptions. Second, most of the research in physical activity uses parametric tests.

What Statistical Techniques Tell About Data

The statistical techniques presented in the next four chapters answer the following two questions about the data to which they are applied:

1. Is the effect or relationship of interest statistically significant? In other words, if the research is repeated, will the effect or relationship be there again (is it reliable)?

2. How strong (or meaningful) is the effect or relationship of interest? This question refers to the magnitude, or size, of the effect or relationship.

Two facts are important about these two statements. First, question 1 is usually answered before question 2, because the strength of the relationship or effect may be of minimal interest until it is known to be reliable (significant). Second, question 2 is always of interest if the effect or relationship is significant and occasionally of interest without a significant effect. Sometimes, in elating over the significance of effects and relationships, researchers lose sight of the need to look at the strength, or meaningfulness, of these relationships. This point is particularly true in research in which differences among groups are compared. The experimenter frequently forgets that relatively small differences can be significant. Finding a significant difference means only that the differences are reliable or that the same answer is likely to be obtained if the research is repeated. The experimenter then needs to look at the size of the differences to interpret whether the findings are meaningful. For each technique presented in the next three chapters, we first look at whether the relationship or effect is significant (reliable). Then we suggest ways to evaluate the strength (meaningfulness) of the relationship or effect.

Distinction Between Correlation and Differences in Statistical Techniques

Statistical techniques can be usefully divided into two categories: (a) statistical techniques used to test relationships between or among variables in one group of participants (regression or correlation) and (b) techniques used to test differences between or among groups of participants (*t* tests and analysis of variance). This division is, strictly speaking, inaccurate because both sets of techniques are based on the general linear model and involve only different ways of entering data and manipulating variance components. An introduction to research methods, however, is neither the place to reform the world of statistics nor the place to confuse you. Thus, the techniques are considered as two distinct groups. Chapter 8 discusses relationships among variables, and chapter 9 discusses differences among groups. Learning the simple calculations underlying the easier techniques and then building on them helps you understand the more complex ones. Do not panic because statistics involves manipulating numbers. You can escape from this section with a reasonable grasp of how, why, and when the various statistical techniques are used in the study of physical activity.

Remember that correlation between two variables does not indicate causation (recall the discussion in chapter 4). Causation is not determined from any statistic or correlation. Cause and effect are established by theory, logic, and the total experimental situation, of which statistics is a part. As summarized by Pedhazur (1982, p. 579),

> **"Correlation is no proof of causation."** Nor does any other index prove causation, regardless of whether the index was derived from data collected in experimental or in nonexperimental research. Covariations or correlations among variables may be suggestive of causal linkages. Nevertheless, an explanatory scheme is not arrived at on the basis of the data, but rather on the basis of knowledge, theoretical formulations and assumptions, and logical analyses. It is the explanatory scheme of the researcher that determines the type of analysis to be applied to data, and not the other way around.

For example, Descartes has been credited with the logical statement, "I think, therefore I am." But the famous philosopher Edsall Murphy "recognized it as a syllogism with an unstated major premise" (Morgenstern, 1983, p. 112):

A nonexistent object cannot think (major premise).

I think (minor premise).

Therefore, I am (conclusion).

Murphy was not satisfied with this and so tried to find deeper meaning and a better logical analysis.

Q: How can you be sure that you exist?

A: I think.

Q: How can you be sure that you are thinking?

A: I cannot, but I do think that I think.

Q: Does that make you sure that you exist?

A: I think so.

This exchange should make it clear that Descartes went too far. He ought to have said, "I think I think, therefore I am," or possibly, "I think I think, therefore I think I am, I think." You do not really exist unless others are aware of your existence. Murphy proclaimed, "I stink, therefore I am" (Morgenstern, 1983, p. 112).

DATA FOR USE IN THE REMAINING STATISTICAL CHAPTERS

As we indicated in the introduction to part II, we will use a standard data set throughout the next several chapters. The data are from the PGA website, and we selected data from the top 10 money winners (no, Tiger Woods is not on the list; remember the knee operation in 2008), money winners 21 through 30, and money winners 41 through 50. The data by player characteristics are in table 6.2—player's name, money won, world rank, scoring average, driving distance, driving accuracy, greens in regulation, putts per round, scrambling, sand saves, player rank in three groups (coded 1 = top 10, 2 = 21–30, and 3 = 41–50), U.S. (coded 1) versus international player (coded 2).

TABLE 6.2

PGA Data for 2008 on Top 10, 21–30, and 41–50 Players Based on Money Won

Name	Money won	World rank	Average score	Drive distance	Drive accuracy	Greens in regulation	Putts per round	Scrambling	Sand saves	Grouping by money	United States vs. International
Singh	6601094.00	5.00	69.58	297.80	59.45	68.45	29.47	58.92	45.11	1.00	2.00
Mickelson	5188875.00	3.00	69.17	295.70	55.27	65.81	28.74	60.42	62.50	1.00	1.00
Garcia	4858224.00	2.00	69.12	294.60	59.39	67.06	29.61	57.59	57.02	1.00	2.00
Perry	4663794.00	14.00	69.83	296.00	61.97	67.47	29.25	57.57	50.00	1.00	1.00
Kim	4656265.00	11.00	69.28	300.90	58.34	65.78	28.85	59.32	50.35	1.00	1.00
Villegas	4422641.00	7.00	69.49	293.30	58.15	64.60	28.97	53.52	54.61	1.00	2.00
Harrington	4313551.00	4.00	69.28	296.30	59.37	60.67	28.04	61.02	58.06	1.00	2.00
Cink	3979301.00	16.00	70.02	296.90	55.27	66.94	29.16	55.60	51.13	1.00	1.00
Leonard	3943542.00	23.00	69.77	281.40	67.72	66.61	28.85	60.07	55.17	1.00	1.00
Allenby	3606700.00	27.00	69.73	291.70	65.64	70.40	30.07	55.26	46.49	1.00	2.00
Els	2537290.00	9.00	69.90	296.00	56.88	61.33	29.28	56.61	54.37	2.00	2.00
Pettersson	2512538.00	62.00	70.24	286.00	59.87	63.54	28.80	59.00	53.13	2.00	2.00
Appleby	2484630.00	38.00	69.73	290.90	58.19	61.90	28.55	60.24	56.30	2.00	2.00
Stricker	2438304.00	15.00	69.98	283.60	56.25	63.81	28.76	61.83	52.34	2.00	1.00
Campbell	2404770.00	65.00	70.04	289.90	65.68	68.44	29.50	54.68	43.41	2.00	1.00
Weekley	2398751.00	48.00	70.41	291.70	64.75	67.87	30.19	57.08	50.39	2.00	1.00
Trahan	2304368.00	74.00	70.33	291.30	65.31	66.25	29.52	55.69	42.48	2.00	1.00
Ames	2285707.00	37.00	69.99	283.80	62.72	65.04	28.99	58.72	50.76	2.00	2.00
Duke	2238885.00	81.00	70.38	284.90	62.27	64.80	28.79	57.82	50.96	2.00	1.00
Hart	2218817.00	54.00	70.59	275.50	61.18	66.11	28.83	61.12	63.71	2.00	1.00
Thompson	1869329.00	156.00	71.04	295.40	66.85	63.14	29.75	57.75	48.78	3.00	1.00
D. Johnson	1789895.00	143.00	71.22	309.70	53.05	63.71	29.40	51.58	42.48	3.00	1.00
Perez	1756038.00	82.00	70.08	294.20	63.27	66.91	29.38	56.20	55.80	3.00	1.00
Mayfair	1750683.00	92.00	70.51	284.00	72.16	68.12	29.84	60.07	54.60	3.00	1.00
Clark	1722030.00	28.00	70.31	281.10	64.98	63.61	29.97	58.32	54.55	3.00	2.00
Bryant	1719153.00	85.00	70.21	279.50	73.87	64.88	29.47	59.18	48.31	3.00	1.00
Pampling	1702952.00	56.00	70.81	288.00	65.65	63.61	29.31	56.11	53.46	3.00	2.00
Love	1695237.00	80.00	70.30	301.30	58.22	64.74	29.50	55.56	50.00	3.00	1.00
Baddeley	1665587.00	47.00	70.20	290.30	59.45	62.02	28.38	58.52	52.00	3.00	2.00
Kelly	1652400.00	102.00	70.94	277.00	66.21	63.36	29.39	57.76	48.68	3.00	1.00

SUMMARY

Statistics are used to describe data, to determine relationships among variables, and to test for differences among groups. In this chapter, we have tried to make the point that the type of statistics used does not determine whether findings can be generalized; rather, it is sampling that permits (or limits) inference. Whenever possible, random sampling is the method of choice, but in behavioral research the more important question may be whether the sample is good enough. In some types of research, such as surveys, stratified random sampling is desirable so that the study represents certain segments of a population. In experimental research, random assignment of participants to groups is definitely desirable so that the researcher can assume equivalence at the beginning of the experiment.

We began our coverage of statistical techniques with basic concepts such as measures of central tendency, variability, and normal distribution. Remember that statistics can do two things: establish significance and assess meaningfulness. Significance means that a relationship or difference is reliable—that you could expect it to happen again if the study were repeated. Meaningfulness refers to the importance of the results.

Check Your Understanding

The following scores represent the number of pull-ups attained by 15 students:

8	4	13
11	6	2
6	7	5
5	0	4
8	6	5

1. Use equation 6.1 to calculate the mean.
2. Use equation 6.2 to calculate the standard deviation.
3. Use equation 6.3 to calculate the standard deviation, and see whether the answer is the same as that derived from equation 6.2.
4. Rearrange the scores from low to high. Use the formula $(N + 1) / 2$ to find the median point. Count up from the bottom that number of values to locate the median score.
5. Make a list of 50 names. Using the table of random numbers (table 1 in the appendix), randomly select 24 subjects and then randomly assign them to two groups of 12 each.
6. Construct a stem-and-leaf display using intervals of 10 scores (e.g., 10–19, 20–29, and so on) for the following 30 scores:

50	42	64	18	41	30	48	68	21	48
43	27	51	42	62	53	45	31	13	58
60	35	28	46	36	56	39	46	25	49

7. Consult equation 6.4 and construct a confidence interval for the mean of a sample ($M = 50$, $s = 6$, $n = 100$). Use the 95% confidence level. Interpret the results.

Statistical Issues in Research Planning and Evaluation

> **The probability of being watched is directly proportional to the stupidity of your act.**
>
> —From Lachlan's Laws (Scottish Highland philosopher)

To plan your own study or evaluate a study by someone else, you need to understand four concepts and their interrelationships: alpha, power, sample size, and effect size. In this chapter, we present these concepts and show how to use them in research planning and evaluation.

PROBABILITY

Another concept that deals with statistical techniques is **probability**, which asks what the odds are that certain things will happen. You encounter probability in everyday events. What are the chances that it will rain? You hear from a weather report that the probability of rain is 90%. You wonder whether this means that it will rain in 90% of the places or, more likely, that the chances are 90% that it will rain where you are, especially if you are planning to play golf or tennis.

A concept of probability related to statistics is called **equally likely events**. For example, if you roll a die, the chances of the numbers from 1 to 6 occurring are equally likely (i.e., 1 in 6, unless you are playing craps in Las Vegas). Another pertinent approach to probability involves **relative frequency**. To illustrate, suppose you toss a coin 100 times. You would expect heads 50 times and tails 50 times; the probability of either result is one-half, or .50. When you toss, however, you may get heads 48 times, or .48. This is the relative frequency. You might perform 100 tosses 10 times and never get .50, but the relative frequency would be distributed closely around .50, and you would still assume the probability as .50.

In a statistical test, you sample from a population of participants and events. You use probability statements to describe the confidence that you place in the statistical findings. Frequently, you encounter a statistical test followed by a probability statement such as *p*

probability—The odds that a certain event will occur.

equally likely events—A concept of probability in which the chances of one event occurring are the same as the chances of another event occurring.

relative frequency—A concept of probability concerning the comparative likelihood of two or more events occurring.

< .05. This interpretation is that a difference or relationship of this size would be expected less than 5 times in 100 as a result of chance.

Alpha

In research, the test statistic is compared with a probability table for that statistic, which tells you what the chance occurrence is. The experimenter may establish an acceptable level of chance occurrence (called **alpha**, α) before the study. This level of chance occurrence can vary from low to high but can never be eliminated. For any given study, the probability of the findings being due to chance always exists, or, to quote Holten's Homily, "The only time to be positive is when you are positive you are wrong."

In behavioral research, alpha (probability of chance occurrence) is frequently set at .05 or .01 (the odds that the findings are due to chance are either 5 in 100 or 1 in 100). There is nothing magical about .05 or .01. These values are used to control for a **type I error**. In a study, the experimenter may make two types of error. A type I error is to reject the null hypothesis when the null hypothesis is true. For example, a researcher concludes that there is a difference between two methods of training, but there really is not. A **type II error** is not to reject the null hypothesis when the null hypothesis is false. For example, a researcher concludes that there is no difference between the two training methods, but there really is a difference. Figure 7.1 is called a **truth table**, which displays type I and type II errors. As you can see, to accept a true null hypothesis or reject a false one is the correct decision. You control for type I errors by setting alpha. For example, if alpha is set at .05, then if 100 experiments are conducted, a true null hypothesis of no difference or no relationship would be rejected on only 5 occasions. Although the chances for error still exist, the experimenter has specified them exactly by establishing alpha before the study.

alpha (α)—A level of probability (of chance occurrence) set by the experimenter before the study; sometimes referred to as *level of significance.*

type I error—Rejection of the null hypothesis when the null hypothesis is true.

type II error—Acceptance of the null hypothesis when the null hypothesis is false.

truth table—A graphic representation of correct and incorrect decisions regarding type I and type II errors.

	H_0 true	H_0 false
Accept	Correct decision	Type II error (β)
Reject	Type I error (α)	Correct decision

Figure 7.1 Truth table for the null hypothesis (H_0).

Adapted from Kirk 1995.

To some extent the issue is this: If you had to make an error, which type of error would you be willing to make? The level of alpha reflects the type of error that you are willing to make. In other words, is it more important that you avoid concluding that one training method is better than another when it really is not (type I), or is it more important that you avoid the conclusion that one method is not any better than the other when it really is (type II)? For example, in a study of the effect of a cancer drug, the experimenter would not want to accept the null hypothesis of no effect if there were any chance that the drug worked. Thus, the experimenter might set alpha at .30 even though the odds of making a type I error would be inflated. The experimenter is making sure that the drug has every opportunity to show its effectiveness. On the other hand, setting alpha at a very low level, for example, .001, greatly decreases the odds of making a type I error and consequently increases the difficulty of detecting a real difference (a type II error).

We cannot tell you where to set alpha, but we can say that the levels .05 or .01 are widely accepted in the scientific community. If alpha is to be moved up or down, be sure to justify your decision, but consider this: "Surely God loves the .06 nearly as much as the .05" (Rosnow & Rosenthal, 1989, p. 1277).

Even when experimenters set alpha at a specific level (e.g., .05) before the research, they often report the probability of a chance occurrence for the specific effects of the study at the level it occurred (e.g., $p = .012$). This procedure is appropriate (and recommended),

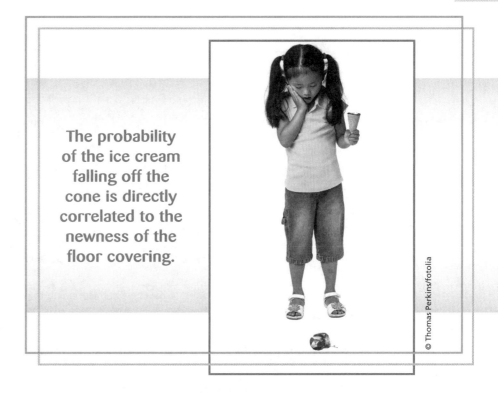

because the researchers are only demonstrating to what degree the level of probability exceeded the specified level.

It is debatable whether alpha should be specified before the research and the findings reported at the alpha that has been specified. Some argue that results are either significant at the specified alpha or they are not (not much is in between). The condition is like being pregnant—one either is or is not. Comparisons with alpha work the same way. Alpha is established as a criterion, and either the results meet the criterion or they do not. Although experimenters sometimes report borderline significance (if alpha is .05), they label from .051 to .10 as borderline.

The probability of the ice cream falling off the cone is directly correlated to the newness of the floor covering.

© Thomas Perkins/fotolia

A sounder approach may be to report the exact level of probability (e.g., $p = .024$) associated with the test statistics (e.g., r, t) and then to estimate the meaningfulness of the difference or relationship. Using the statistical information (significance and meaningfulness), the researcher should interpret the findings within the theory and hypotheses that have been put forward. Rather than making the decision a statistical one, this approach places the responsibility of decision making where it belongs: on the researcher who has placed the study within a theoretical model and has considered related research. Much of the criticism of statistics has revolved around the blind application of statistical techniques to data without appropriate interpretation of the outcomes. For excellent articles on the use of common sense with statistical outcomes, read Cohen (1990) and Serlin (1987).

Beta

Although the magnitude of a type I error is specified by alpha, you may also make a type II error, the magnitude of which is determined by **beta (β)**. By looking at figure 7.2, you can see the overlap of the score distribution on the dependent variable for x (the sampling distribution if the null hypothesis is true) and y (the sampling distribution if the null hypothesis is false). By specifying alpha, you indicate that the mean of y (given a certain distribution) must be at a specified distance from the mean of x before the null hypothesis is rejected. But if the mean of y falls anywhere between the mean of x and the specified y, you could be making a type II error (beta); that is, you do not reject the null hypothesis when, in fact, there is a true difference. As you can see, there is a relationship between alpha and beta; for example, as alpha is set increasingly smaller, beta becomes larger.

beta (β)—The magnitude of a type II error.

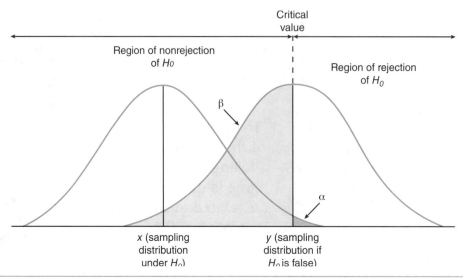

Figure 7.2   Regions under the normal curve corresponding to probabilities of making type I and type II errors.

MEANINGFULNESS (EFFECT SIZE)

Besides reporting the significance of the findings, scholars need to be concerned about the **meaningfulness** of the outcomes of their research. The meaningfulness of a difference between two means can be estimated in many ways, but the one that has gained the most attention is **effect size** (suggested by Cohen, 1969). You may be familiar with using effect size (ES) in meta-analysis (if not, you will be soon; see chapter 14 on research synthesis). The formula for ES is

$$ES = (M_1 - M_2) / s \tag{7.1}$$

This formula subtracts the mean of one group (M_1) from the mean of a second group (M_2) and divides the difference by the standard deviation. That places the difference between the means in the common metric called *standard deviation units,* which can be compared to guidelines for behavioral research suggested by Cohen (1969): 0.2 or less is a small ES, about 0.5 is a moderate ES, and 0.8 or more is a large ES. For a more detailed discussion of the use of effect size in physical activity research, see Thomas, Salazar, and Landers (1991).

Many authors (e.g., Cohen, 1990; Serlin, 1987; Thomas, 2014; Thomas, Salazar, & Landers, 1991) have indicated the need to report some estimate of meaningfulness with all tests of significance.

POWER

Power is the probability of rejecting the null hypothesis when the null hypothesis is false (e.g., detecting a real difference), or the probability of making a correct decision. Power ranges from 0 to 1. The greater the power is, the more likely you are to detect a real difference or relationship. Thus, power increases the odds of rejecting a false null hypothesis. Of course, to a degree, in behavioral research, *the null hypothesis is always false!*

What this statement reflects is that in behavioral research the means of two groups are never the same. Thus, if enough participants are obtained (one way of obtaining power), any two means can be declared significantly different. Remember what that means. If you do it again, you will get about the same answer. The more interesting questions in behavioral research are the following:

meaningfulness—The importance or practical significance of an effect or relationship.

effect size—The standardized value that is the difference between the means divided by the standard deviation, also called *delta.*

- How large a difference is important in theory or practice?
- How many participants are needed to declare an important difference as significant?

Understanding the concept of power can answer these questions. If a researcher can identify the size of an important effect through previous research or even simply estimate an effect size (e.g., 0.5 is a moderate ES) and establish how much power is acceptable (e.g., a common estimate in the behavioral science is 0.8), then the size of the sample needed for a study can be estimated.

Figures 7.3 and 7.4 show the relationship among sample size (y-axis), power (x-axis), and effect size (ES) curves when alpha is either .05 or .01.

Figure 7.3   Effect size curves when alpha is .05 for a two-tailed test.

HOW MANY SUBJECTS?: STATISTICAL POWER ANALYSIS IN RESEARCH by C. Kraemer and S. Thiemann. Copyright 1987 by SAGE PUBLICATIONS INC BOOKS. Reproduced with permission of SAGE PUBLICATIONS INC BOOKS in the format Text book via Copyright Clearance Center.

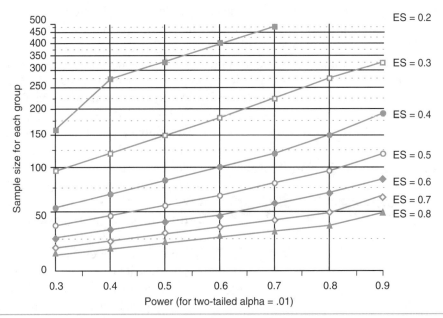

Figure 7.4   Effect size curves when alpha is .01 for a two-tailed test.

HOW MANY SUBJECTS?: STATISTICAL POWER ANALYSIS IN RESEARCH by C. Kraemer and S. Thiemann. Copyright 1987 by SAGE PUBLICATIONS INC BOOKS. Reproduced with permission of SAGE PUBLICATIONS INC BOOKS in the format Text book via Copyright Clearance Center.

Consider the following example. An investigator plans a study that will have two randomly formed groups, but she does not know how many participants are needed for each group to detect a meaningful difference between treatments. However, there are several related studies, and the investigator has calculated an average ES = 0.7 (using equation 6.1 in chapter 6) favoring the experimental group from the outcomes in these studies. The investigator decides to set alpha = .05 and wants to keep beta at four times the level of alpha (thus, beta = .20) because Cohen (1988) suggested that in the behavioral sciences, the seriousness of type I to type II error should be in a 4-to-1 ratio. Because power is 1 − beta (1.0 − 0.2 = 0.8), power is set at .8 (often recommended as an appropriate power in behavioral research; Green, S.B., 1991, p. 502). When this information is known—alpha, ES, and power—then the number of participants needed in each of the two groups can be estimated from figure 7.3. Read the 0.7 ES curve to where it passes the x-axis (power) at .8. Then read across to the y-axis (sample size), and note that 30 people would be needed in each group. Note how this relationship works: As the number of people in each group is reduced, power is reduced (at the same ES).

In figure 7.4 (alpha = .01), note that for the same level of power (.8) and ES (0.7), the number of people per group increases from 30 (from figure 7.3, alpha = .05) to 50. So if all else stays the same but a more stringent alpha is used (e.g., .05 to .01), a greater number of participants are required to detect a significant difference. Hence, with a more stringent (lower) alpha (e.g., .001) power is reduced, making it more difficult to detect a significant difference. A larger alpha such as .10 or .20 increases power; that is, it lessens the chance of making a type II error. Of course, it increases the chance of making a type I error (proclaiming that a difference is present when there is none). As stated before, the researcher has to decide which type of error is more important to avoid. A practical guideline to follow for applied research is that if the costs (in money, time, and so on) are the same for both treatments, go for power: Find out whether there is a difference. But if one treatment costs more than the other, avoid the type I error. You do not want to adopt a more expensive treatment (method, program, instrument) if it is not *significantly* better.

The size of the sample is extremely influential on power. Power increases with increased n. Table 7.1 illustrates this statement using an ES of 0.5 and alpha at .05. With a very small number of participants per group, for example, n = 10, the power would be .20 (off the chart in figure 7.3). This is only a 20% chance of detecting a real difference. Conversely, with a very large sample size, such as n = 100, the power is actually about .95, which is approaching certainty of finding a real difference.

Keep in mind the relationships among alpha, sample size, and ES in planning a study. If you have access to only a small number of participants, then you need to have a really large ES or use a larger alpha, or both. Do not just blindly specify the .05 alpha if detecting a real difference is the main issue. Use a higher one, such as .20 or even .30. This approach is extremely pertinent in pilot studies.

In chapter 9 we discuss some of the things you can do to increase the ES. Sometimes, however, in planning a study, you may not be able to determine an expected ES from the literature. Here you can benefit from a pilot study to try to estimate the ES.

Our approach in chapter 9 is more post hoc, because it involves reporting the effect size and variance accounted for in significant findings and interpreting whether this is a meaningful effect. The a priori procedures described previously are more desirable

> The relationships among alpha, sample size, and ES are important. For example, studies with a small sample size require a larger ES or alpha, or both.

TABLE 7.1

Relationship of Sample Size and Power (ES = 0.5, alpha = .05)

n	Power
10	.20*
20	.30
50	.70
75	.85
100	.95*

*These values are from the chart in figure 7.3.

but not always applicable. (For a more detailed discussion of important factors associated with establishing significance levels and determining power, see Franks & Huck, 1986, or Thomas, Lochbaum, et al., 1997.)

Finally, a summary of all this is well presented in the short poem by Rosenthal in the sidebar Achieving Power (1991, p. 221).

To summarize, knowledge of four concepts is needed for planning and evaluating any quantitative study: alpha, power, sample size, and effect size.

- **Alpha** establishes the acceptable magnitude of type I error (i.e., the chances of rejecting a true null hypothesis) and the level of significance selected. It is typically an arbitrary value—often .05 or .01 in the behavioral and biological areas of physical activity.

- **Power** represents the chances of rejecting a false null hypothesis. It is based on the magnitude of beta (i.e., the chances of making a type II error, accepting a false null hypothesis). Beta is typically set at 4 × alpha, so if alpha is .05, then beta is .20. Power is typically established as 1 – beta or .80 in the behavioral and biological areas of physical activity.

- **Sample size** is the number of participants in the study being evaluated or planned.

- **Effect size** is the outcome of a study typically expressed in standard deviation units.

$$ES = (M_1 - M_2) / s \tag{7.2}$$

R squared (r^2) is also a form of effect size and can be expressed in standard deviation units using the formula (Rosenthal, 1994)

$$ES = [r / \sqrt{(1 - r^2)}] \, [\sqrt{(df \, (n_1 + n_2)) / n_1 \, n_2}] \tag{7.3}$$

Most often in evaluating published studies, all the information to determine the four concepts is included, although everything might not be in final form (e.g., the effect size can be calculated from the means and standard deviations or converted from r).

Achieving Power

I. THE PROBLEM

Oh, *F* is large and *p* is small
That's why we are walking tall.
What it means we need not mull
Just so we reject the null.
Or Chi-square large and *p* near nil
Results like that, they fill the bill.
What if meaning requires a poll?
Never mind, we're on a roll!
The message we have learned too well?
Significance! That rings the bell!

II. THE IMPLICATIONS

The moral of our little tale?
That we mortals may be frail.
When we feel a *p* near zero
Makes us out to be a hero.
But tell us then is it too late?
Can we perhaps avoid our fate?
Replace that wish to null-reject
Report the *size* of the effect.
That may not insure our glory
But at least it tells a story
That is just the kind of yield
Needed to advance our field.

USING INFORMATION IN THE CONTEXT OF THE STUDY

context—The interre-
lationships found in the
real-world setting. (See
also chapter 12, which
addresses the use of con-
text.)

The item missing in these four concepts that is used to evaluate all research is the one of greatest interest: **context**. How do the findings from this study fit into the context of theory and practice? In fact, the question of greatest interest to the reader of research (which should make it of considerable interest to the researcher) is one that often goes unanswered: Are effect sizes for significant findings large enough to be meaningful when interpreted within the context of the study, or for the application of findings to other related samples, or for planning a related study?

In planning research, the experimenter often has either previous findings or pilot research from which appropriate effect sizes of important variables can be estimated. If the experimenter sets alpha at .05 and power at .80, then the question becomes, How many participants would be needed to detect an effect of a specific magnitude if the treatment works or the relationship is present? Charts and tables in books (see figure 7.3) as well as computer programs can be used to estimate the number of participants needed, given that the experimenter has established or estimated the other three components. Numerous online power calculators will do a power analysis for you (just put the term *statistical power* in your browser).

One of the major benefits of using effect sizes in estimating the meaningfulness of data is that they are not directly sensitive to sample size. Certainly, effect sizes have some sensitivity to sample size because the size of the standard deviation is influenced by the sample size. Effect sizes are based on either the difference between the means (divided by the standard deviation) or the size of the correlation. Table 7.2 provides the magnitude of effect sizes as they reflect how two treatment groups overlap in their scores. Note that when an effect size is 0, the distribution of the scores of the two groups overlaps completely and the midpoints are the same (50% of participants' scores on either side of the mean). As effect sizes

TABLE 7.2

Interpretation of Effect Size Using the Distribution From the Normal Curve

Effect size	Percentage of experimental group difference above control group mean	Percentage – 50
0.0	50	0
0.1	54	4
0.2	**58**	**8**
0.3	62	12
0.4	65	15
0.5	**69**	**19**
0.6	73	23
0.8	**79**	**29**
1.0	84	34
1.2	88	38
1.5	93	43
2.0	98	48

Adapted from McNamara 1994.

become larger, there is a net gain so that a greater percentage of the participants' scores in one group (typically the experimental) exceeds the midpoint (50%) of the distribution of the other group's scores (control). Effect sizes are often interpreted by their absolute size. In table 7.2 a small effect size (often labeled 0.2 or less) suggests that the experimental group has a net gain (as a result of the treatments) of 8% over the control group; that is, 58% of the distribution of the experimental group's scores is beyond the midpoint (50%) of the control group's scores. A moderate effect size (often labeled about 0.5) shows a net gain of 19% over the midpoint of the control group's scores, whereas a large effect size (often labeled 0.8 and above) has a net gain of 29%. This is an excellent way of describing the influence of changes in an experimental group compared with a control group. That is, if the experimental and control groups begin at similar points on the dependent variable, what percentage of experimental participants' scores are above the midpoint (50%) in the distribution of the control group's scores? Thus, the larger the effect size, the less the overlap between the distribution of scores in the two groups.

Finally, as we move into the next three chapters on statistics, remember that general linear model statistical analyses are of value only if the sample is neither too small nor too large. In very small samples, a single unusual value can substantially influence the results. Moreover, within- and between-participant variation (the error variance) tends to be large, which causes the error variance in tests of significance to be large, resulting in few significant findings. On the other extreme of sample size, statistics has little value for very large samples because nearly any difference or relationship is significant. Recall that the error variance in significance tests is divided by the degrees of freedom, a calculation based on sample size. Thus, you will notice that large public health studies tend to discuss data analyses such as risk factors, percentage of the sample with a characteristic (prevalence, incidence, mortality rate), and odds ratios rather than significance (see chapter 17 by B.E. Ainsworth and C.E. Matthews in this text).

In chapter 3 we discussed the research and null hypotheses. Cohen (1990, 1994) and Hagen (1997) provided interesting insights into the value (or lack thereof) of the null hypothesis as a statistical test. But in the behavioral and biological sciences of physical activity dealing mostly with human participants, at some level the null hypothesis is nearly always false (Cohen, 1990; but see Hagen, 1997, for a different view). Remember, conceptually, the null hypothesis comes from the development of statistical tables that researchers use to compare their test statistic (e.g., F ratios). These tables are based on population estimates of how often randomly selected samples might differ if there really is no difference. Thus, $p = .05$ means that two randomly selected samples of a certain size (n) will differ by chance only 5 times in 100. We use these tabled values for comparison with our calculated statistical values (e.g., F ratios). Two groups of humans, however, are hardly ever exactly alike on a measured variable, and if there are enough participants, p will be less than .05. Testing the null hypothesis is the wrong concept. The correct questions are What is a meaningful difference (effect size) within the context of theory and application and how many participants are required to reject a difference this large at $p < .05$ and power $> .80$? Of course, $p < .05$ and power $> .80$ are arbitrary values.

Remember, statistical routines are just that: computer routines into which numbers are input and that provide standardized output based on those numbers. Just as the numbers we assign to variables do not know where they come from (i.e., someone has developed the procedures for assigning values to certain levels of characteristics), statistical computer programs do not know where the numbers come from, nor do they care. You as the researcher must verify that the numbers are good ones. You must verify that the numbers (data) meet the assumptions used in the development of the tables to which they are being compared (e.g., normal distribution of data for the F table or the use of the chi-square table for nonnormal distributions). This judgment is the very basis for evaluating data and making decisions between parametric and nonparametric statistical analyses (more on this in chapter 10).

Context is what matters with regard to meaningfulness. You must ask yourself, Within the context of what I do, does an effect of this size matter? The answer nearly always depends on who you are and what you are doing (and practically never on whether $p = .05$ or $.01$). Thus, having a significant (reliable) effect is a necessary, but not sufficient, condition in statistics. To meet the criteria of being both necessary and sufficient, the effect must be significant and meaningful within the context of its use. Said another way,

> An effect that is significant and meaningful is necessary and sufficient.

- estimates of significance are driven by sample size,
- estimates of meaningfulness are driven by the size of the difference, and
- context is driven by how the findings will be used.

CONTEXT: THE KEY TO MEANING

Here is a simple example of context. Suppose we told you that we have developed (and advertised on TV) a program of physical training, the Runflex System, that will improve your speed in the 100 m dash by 100 ms. This system requires that you attach cables to your feet from flexible rods, lie on your back, and flex the rods by pulling your knees toward your chest. This system is guaranteed to work in 6 weeks or you get your money back. Testing by a major university laboratory in exercise physiology shows that the system does produce an average improvement of 100 ms in speed in the 100 m dash. Would you be willing to send us three easy payments of $99.95 for this system? Right now you are saying to yourself, Are these people nuts? I don't even run 100 m dashes. Suppose, however, that you are the second fastest 100 m sprinter in the world, and the fastest sprinter runs 100 ms faster than you do. Would you buy our Runflex apparatus now? *Context matters!*

Quantitative researchers often do not provide appropriate contexts for their studies, especially the important findings. Some context may be provided in the introduction to papers when the rationale for the research is developed, but even then it is sometimes difficult to understand why a study is important and how it fits in the scheme of things. In the discussion section we often tend to let the data speak for themselves with a minimal attempt at explaining why the findings are important, how they support (or do not support) a theory, why the findings add to previous knowledge (or dispute it), what the logical subsequent research questions are, and why practitioners should care about the outcomes.

Qualitative researchers are typically much stronger than quantitative researchers at placing findings in context (see Biddle, Markland, Gilbourne, Chatzisarants, & Sparkes, 2001, for a good discussion). Of course, this difference occurs because context drives qualitative researchers' research plans, data collection, data analyses, and interpretations of findings. We are not suggesting a switch to qualitative research by quantitative scholars; we have learned much, and much is yet to be learned, from quantitative research. (In addition, we are too old to learn all the techniques.) But quantitative researchers can gain considerable benefit by learning to use context as the basis for planning and interpreting research, particularly important outcomes. Mixed-methods research takes advantage of the strength of both quantitative and qualitative methodologies. We discuss mixed methods in chapter 20 and recommend that you give them considerable thought as you plan your research.

SUMMARY

In this chapter, you learned the interrelationships among alpha, power, sample size, and effect size. Appropriate use of this information is the most important aspect of planning your own study or evaluating someone else's. Placing this information in the context in which you plan the research or plan to use the results allows others to interpret and use your research findings appropriately.

Check Your Understanding

1. Locate a data-based paper from a research journal. Answer the following questions about this paper:

   a. What probability levels did the researchers use to test their hypotheses?

   b. What was the sample size for the group or groups?

   c. What was the effect size for some aspect of the study?

   d. How would you estimate power using your answers to a, b, and c?

2. Draw a figure (such as figure 7.2) with two normal curves (one for an experimental group and one for a control group) that reflects an effect size of 0.2. Do this again for an effect size of 0.5. Repeat for an effect size of 0.8.

8

Relationships Among Variables

When you are dissatisfied and would like to go back to youth, think of Algebra.

—Will Rogers

I n chapter 6, we promised that after we had presented some basic information to help you understand statistical techniques, we would begin to explain in detail some specific techniques. We begin with correlation.

Correlation is a statistical technique used to determine the relationship between two or more variables. In this chapter, we discuss several types of correlation, the reliability and meaningfulness of correlation coefficients, and the use of correlation for predictions, including partial and semipartial correlations, multiple regression, logistic regression, and discriminant function analysis. Finally, we briefly overview multivariate forms of correlation: canonical correlation, factor analysis, and structural modeling.

> **correlation**—A statistical technique used to determine the relationship between two or more variables.

WHAT CORRELATIONAL RESEARCH INVESTIGATES

Often, a researcher is interested in the degree of relationship between, or correlation of, performances, such as the relationship between performances on a distance run and a **step test** as measures of cardiorespiratory fitness. Sometimes, an investigator wishes to establish the relationship between traits and behavior, such as how personality characteristics relate to participation in high-risk recreational activities. Still other correlational research problems might involve relationships between anthropometric measurements, such as skinfold thicknesses and fat percentages as determined by underwater weighing. Here, the researcher may wish to predict fat percentage from the skinfold measurements.

> **step test**—A test used to measure cardiorespiratory fitness involving the measurement of pulse rate after stepping up and down on a bench.

Correlation may involve two variables, such as the relationship between height and weight. It may involve three or more variables, such as when a researcher investigates the relationship between a criterion (dependent variable) such as cardiorespiratory fitness and two or more predictor variables (independent variables) such as body weight, fat percentage, speed, and muscular endurance. This technique is multiple correlation. Another technique, canonical correlation, establishes the relationships between two or more dependent variables and two or more independent variables. Factor analysis uses correlations among a number of variables to try to identify the underlying relationships or factors. Finally, structural modeling provides evidence for how variables may directly or indirectly influence other variables.

UNDERSTANDING THE NATURE OF CORRELATION

The **coefficient of correlation** is a quantitative value of the relationship between two or more variables. The correlation coefficient can range from .00 to 1.00 in either a positive or negative direction. Thus, perfect correlation is 1.00 (either +1.00 or –1.00), and no relationship at all is .00 (see the sidebar Correlation: Perfect and Not So Perfect for examples of perfect correlations and examples of no correlation).

Positive Correlation

A **positive correlation** exists when a small value for one variable is associated with a small value for another variable and a large value for one variable is associated with a large value for another. Strength and body weight are positively correlated: Heavier people are generally stronger than lighter people. (The correlation is not perfect because some lighter people are stronger than some heavier people and weaker than some who weigh even less.)

Figure 8.1 is a graphic illustration of a perfect positive correlation involving boys. Notice that Bill's body weight is 70 lb (32 kg) and his strength measure is 150 lb (68 kg). Dick's data are weight = 80 lb (36 kg) and strength = 175 lb (79 kg); the increase continues through Tom's data, where weight = 110 lb (50 kg) and strength = 250 lb (113 kg). Thus, when the scores are plotted, they form a perfectly straight diagonal line. This is perfect correlation ($r = 1.00$). The relative positions of the boys' pairs of scores are identical in the two distributions. In other words, each boy is the same relative distance from the mean of each set of scores. Common sense tells us that perfect correlation does not exist in human traits, abilities, and performances because of so-called people variability and other influences.

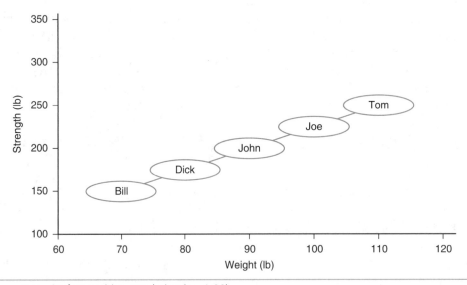

Figure 8.1   Perfect positive correlation ($r = 1.00$).

Correlation: Perfect and Not So Perfect

QUOTES FROM ANONYMOUS FAMOUS PEOPLE ABOUT PERFECT CORRELATION ($r = 1.00$)

Smoking kills. If you are killed, you've lost a very important part of your life.

Traditionally, most of Australia's imports come from overseas.

It's bad luck to be superstitious.

Things are more like they are now than they ever were before.

The police are not here to create disorder but to preserve disorder.

The Internet is a great way to get on the net.

The president has kept all the promises he intends to keep.

China is a big country inhabited by many Chinese.

That low-down scoundrel deserves to be kicked to death by a jackass, and I'm just the one to do it.

It's like déjà vu all over again.

The loss of life will be irreplaceable.

STUDENTS' STATEMENTS THAT ARE UNCORRELATED TO TEACHING PERFORMANCE

Sometimes students write interesting things on the teaching evaluations for our research methods course. Here are a few of the statements that have an $r = .00$ to our teaching.

1. This class was a religious experience for me. I had to take it all on faith.
2. Textbook makes a satisfying "thud" when dropped on the floor.
3. Have you ever fallen asleep in one class and awakened in another?
4. Help! I've fallen asleep and can't wake up.
5. He teaches like Speedy Gonzalez on a caffeine high.
6. I'm learning by osmosis; I sleep with my head on the textbook.
7. Class is a great stress reliever. I was so confused I forgot who I was.
8. This class kept me out of trouble from 2:30 to 4:00 pm on Tuesday and Thursday.

Figure 8.2 illustrates a more realistic relationship using the PGA data presented at the end of chapter 6. The data are from 30 PGA players in 2008, and the correlation is between the percentage of greens hit in regulation (x-axis) and the number of putts per round (y-axis). The correlation is $r = .549$, which is significant at $p = .002$ with 28 degrees of freedom (number of participants, $N - 2$, called df). As can be observed, the data points (each point represents one person—read to the x-axis for the percentage of greens in regulation and to the y-axis for putts per round) generally progress from the lower left to the upper right representing a relationship but not a perfect one.

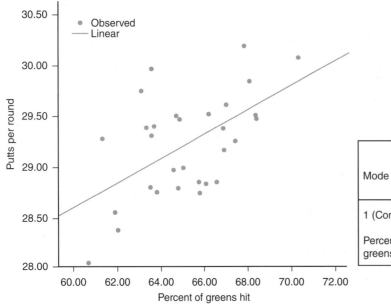

Mode 1	Unstandardized coefficients		Standardized coefficients
	B	Standard error	Beta
1 (Constant)	21.506	2.219	
Percent of greens hit	.118	.034	.549

M(PPR) = 29.22, SD = 0.51

M(GIR) = 65.23, SD = 2.35

Figure 8.2 More realistic relationship between PGA golfers in percentage of greens hit in regulation (*x*-axis) and putts per round (*y*-axis). Correlation is $r(28) = .549$, $p = .002$.

negative correlation—A relationship between two variables in which a small value for the first variable is associated with a large value for the second variable, and a large value for the first variable is associated with a small value for the second variable.

Negative Correlation

In figure 8.3, we have plotted the PGA data for driving distance versus driving accuracy. The correlation is $r = -.594$, $p < .001$, $df = 28$. This means that the correlation is negative—driving the ball a greater distance is negatively related to driving the ball into the fairway. This relationship is shown in figure 8.3, in which the general pattern of the data points is from upper left to lower right. This is a **negative correlation**. A perfect negative correlation would be a straight diagonal line at a 45° angle (the upper left corner of the graph to the lower right corner). Figure 8.3 depicts a negative correlation of a moderate degree ($r = -.594$), but an upper-left-to-lower-right pattern is still apparent.

Patterns of Relationships

Figure 8.4 is a hypothetical example of four patterns of relationships between two variables. Figure 8.4*a* is a positive relationship as previously described. Figure 8.4*b* is a negative relationship, also previously described. When virtually no relationship exists between variables, the correlation is .00, as shown in figure 8.4*c*. This example denotes independence between sets of scores. The plotted scores exhibit no discernible pattern at all. Finally, two variables may not have a linear relationship but can still be related, as in figure 8.4*d*, which shows a curvilinear relationship. The interpretation of the reliability and meaningfulness of correlations is explained later in this chapter.

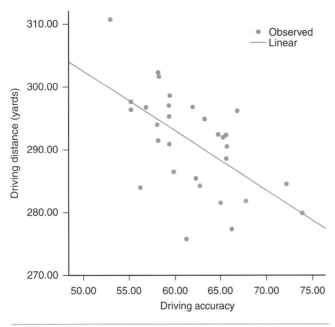

Figure 8.3 Plot of the relationship between driving accuracy (*x*-axis) and driving distance (*y*-axis) for the 30 PGA golfers. Correlation is $r(28) = .594$, $p < .001$.

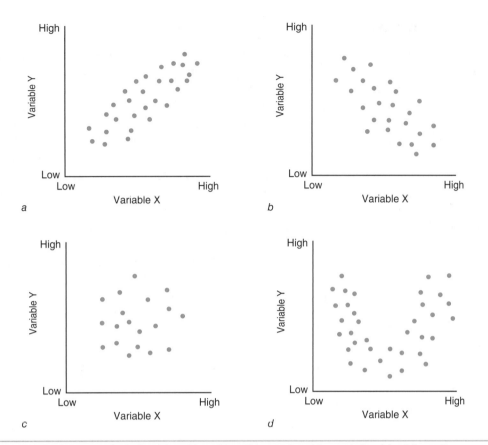

Figure 8.4   Patterns of relationships: *(a)* positive, *(b)* negative, *(c)* no relationship, *(d)* curvilinear.

Correlation and Causation

At this point we must stress again that a correlation between two variables does not mean that one variable causes the other. In chapter 4 we used an example of a research hypothesis (albeit a dumb one) that achievement in math could be improved by buying a child larger shoes. This hypothesis resulted from a correlation between math achievement scores and shoe size in elementary school children (with age not controlled). This example helps to drive home the point that correlation does not mean causation.

In Ziv's 1988 study on the effectiveness of teaching and learning with humor, the experimental group participants were told a story to illustrate the fact that correlations do not show a causal effect. In the story, aliens from another planet, who were invisible to earthlings, decided to study the differences between fat people and thin people. One alien observed the coffee-drinking behavior of fat and thin people at a cafeteria over an extended period. The alien then analyzed the observations statistically and decided that there was a correlation between coffee-drinking behavior and body weight. Thin people tended to drink their coffee with sugar, whereas fat people mostly drank theirs with an artificial sweetener. The alien concluded from this study that sugar makes humans thin, whereas artificial sweeteners make humans fat.

We are not saying that one variable cannot be the cause of another; in fact, using our example of a negative correlation, it is likely that driving the golf ball farther does cause a decrease in driving accuracy (as determined by the percentage of fairways that the player hits), but establishing that relationship does not establish cause and effect. For cause and effect to occur, the two variables must be correlated. Correlation is a necessary but not sufficient condition for causation. The only way causation can be shown is with an experimental study in which an independent variable can be manipulated to bring about an effect.

Pearson Product Moment Correlation

Several times in the preceding discussion we used the symbol r. This symbol denotes the **Pearson product moment coefficient of correlation**. This type of correlation has one criterion (or dependent) variable and one predictor (or independent) variable. Thus, every participant has two scores, such as driving distance and driving accuracy. An important assumption for the use of r is that the relationship between the variables is expected to be linear; that is, that a straight line is the best model of the relationship. When that is not true (e.g., figure 8.4d), r is an inappropriate way to analyze the data. If you were to apply the statistical formula to calculate r to the data in figure 8.4d, the outcome would be an r of about .00, indicating no relationship. Yet we can clearly see from the plot that a relationship does exist; it is just not a linear one.

The computation of the correlation coefficient involves the relative distances of the scores from the two means of the distributions. The computations can be accomplished with a number of formulas; we present just one. This formula is sometimes called the computer method because it involves operations similar to those performed by a computer. The formula appears large and imposing but actually consists of only three operations:

1. Sum each set of scores.

2. Square and sum each set of scores.

3. Multiply each pair of scores and obtain the cumulative sum of these products.

The formula is

$$r = \frac{N\sum XY - (\sum X)(\sum Y)}{\sqrt{N\sum X^2 - (\sum X)^2}\sqrt{N\sum Y^2 - (\sum Y)^2}}$$

(8.1)

To illustrate, we will again use the PGA data but just those from the top 10 players (example 8.1). Note in figure 8.5 that the correlation for the top 10 players is quite high, $r(8) = .925$ (see table 8.1 for SPSS output on this correlation). In the example you can see the hand calculations of this r using equation 8.1. Note that N refers to the number of paired scores, not the total number of scores.

Example 8.1

Known Values

2008 Data and Calculation of r for Greens Hit in Regulation and Putts per Round for Top 10 PGA Players

Player	Greens hit in regulation (X)	X^2	Putts per round (Y)	Y^2	XY
Singh	68.45	4,685.4	29.47	868.5	2,017.3
Mickelson	65.81	4,331.0	28.74	826.0	1,891.4
Garcia	67.06	4,498.4	29.61	876.8	1,985.9
Perry	67.47	4,552.2	29.25	855.6	1,973.6
Kim	65.78	4,327.0	28.85	832.3	1,897.8
Villegas	64.60	4,173.2	28.97	839.3	1,871.5
Harrington	60.67	3,680.8	28.04	786.2	1,701.3
Cink	66.94	4,481.0	29.16	850.3	1,952.1
Leonard	66.61	4,436.9	28.85	832.3	1,921.7
Allenby	70.40	4,956.2	30.07	904.2	2,116.9

Number of paired scores:	$N = 10$
Sum of greens hit in regulation	$\Sigma X = 663.8$
Sum of putts per round	$\Sigma Y = 291.0$
Sum of greens hit in regulation squared	$\Sigma X^2 = 44{,}121.98$
Sum of putts per round squared	$\Sigma Y^2 = 8{,}471.44$
Sum of greens hit × putts	$\Sigma XY = 19{,}329.4$

Working It Out (Equation 8.1)

$$r = \frac{N\sum XY - (\sum X)(\sum Y)}{\sqrt{N\sum X^2 - (\sum X)^2}\sqrt{N\sum Y^2 - (\sum Y)^2}}$$

$$r = \frac{(10)19{,}329.4 - (663.8)(291.0)}{\sqrt{10(44{,}121.98) - (663.8)^2}\sqrt{10(8{,}471.44) - (291.0)^2}}$$

$$r = \frac{193{,}294 - 193{,}165}{\sqrt{589.36}\sqrt{33}} = \frac{129.0}{139.5} = .925$$

TABLE 8.1

Correlation Between Greens Hit in Regulation and Putts per Round for Top 10 PGA Players

Model summary			
R	R squared	Adjusted R squared	Standard error of the estimate
.925	.855	.837	.223

The independent variable is greens hit in regulation.

For the $(\Sigma X)^2$ and the $(\Sigma Y)^2$ values, the sums of the raw scores for X and Y are then squared. Notice that these values are not the same as the ΣX^2 and ΣY^2 values, for which X and Y are squared and then summed. The ΣXY (the sum of the cross products of the X and Y scores) determines the direction of the correlation—that is, whether it is positive or negative.

In a correlation problem that simply determines the relationship between two variables, it does not matter which one is X and which is Y. If the investigator wants to predict one score from the other, then Y designates the criterion (dependent) variable (what is being predicted) and X the predictor (independent) variable. In this example, putts per round would be predicted from greens hit in regulation; it would not make much sense to predict in the other direction because hitting the green occurs before putting in golf. Prediction equations are discussed later in this section.

The relationship between the distance that a person could run in 12 min and heart rate after exercise would result in a negative

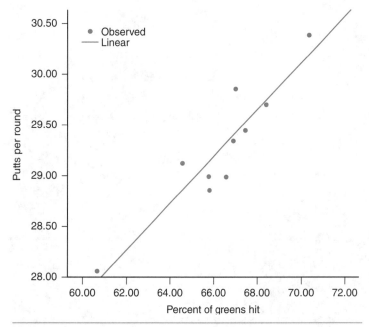

Figure 8.5   Correlation between greens hit in regulation and putts per round for top 10 PGA players in 2000.

correlation. This result would occur because a greater distance covered is good and a lower heart rate after exercise is good. A person who has good cardiorespiratory endurance would have a high score on one test (the run) and a low score on the other. So whether the correlation value is negative or positive depends not only on the relationship but also on the scoring direction of the X and Y variables.

WHAT THE COEFFICIENT OF CORRELATION MEANS

So far, we have dealt with the direction of correlation (positive or negative) and the calculation of r. An obvious question that arises is What does a coefficient of correlation mean in terms of being high or low, satisfactory or unsatisfactory? This seemingly simple question is not so simple to answer.

Interpreting the Reliability of r

First, there are several ways of interpreting r. One criterion is its reliability, or **significance**. Does it represent a real relationship? That is, if the study were repeated, what is the probability of finding a similar relationship? For this statistical criterion of significance, simply consult a table. In using the table, select the desired level of significance, such as the .05 level, and then find the appropriate degrees of freedom (*df,* which is based on the number of participants corrected for sample bias), which, for r, is equal to $N - 2$. Table 3 in the appendix contains the necessary correlation coefficients for significance at the .05 and .01 levels. Refer to the example of the correlation between greens hit in regulation and putts per round ($r = .925$). The degrees of freedom are $N - 2 = 10 - 2 = 8$ (remember, the variable N in correlation refers to the number of pairs of scores). When reading the table at 8 *df,* we see that a correlation of .632 is necessary for significance of a two-tailed test at the .05 level (and .765 at the .01 level). Therefore, we would have to conclude that our correlation of .925 is

Relationships are all unique—and the future of some may be hard to predict.

significant at $p < .01$. (We explain whether to find the correlation values under the column for a one-tailed or two-tailed test in the section Interpreting t in chapter 9.)

Another glance at appendix table 3 reveals a couple of obvious facts. The correlation needed for significance decreases with increased numbers of participants (*df*). In our example, we had only 10 participants (or pairs of scores). Very low correlation coefficients can be significant if you have a large sample of participants. At the .05 level, $r = .38$ is significant with 25 *df*, $r = .27$ is significant with 50 *df*, and $r = .195$ is significant with 100 *df*. With 1,000 *df*, a correlation as low as .08 is significant at the .01 level.

The second observation to note from the table is that a higher correlation is required for significance at the .01 level than at the .05 level. This statement should make sense. Remember, chapter 7 stated that the .05 level means that if 100 experiments were conducted, the null hypothesis (that there is no relationship) would be rejected incorrectly, just by chance, on 5 of the 100 occasions. At the .01 level, we would expect a relationship of this magnitude because of chance less than once in 100 experiments. Therefore, the test of significance at the .01 level is more stringent than at the .05 level, so a higher correlation is required for significance at the .01 level. When you use most computer programs to calculate statistics such as r, you are given the exact significance level in the printout.

Interpreting the Meaningfulness of *r*

The interpretation of a correlation for statistical significance is important, but because of the vast influence of sample size, this criterion is not always meaningful. As chapter 7 explained, statistics can answer two questions about data: Are the effects reliable? Are they meaningful?

The most commonly used criterion for interpreting the meaningfulness of the correlation coefficient is the **coefficient of determination (r^2)**. In this method, the portion of common association of the factors that influence the two variables is determined. In other words, the coefficient of determination indicates the portion of the total variance in one measure that can be explained, or accounted for, by the variance in the other measure.

Figure 8.6 offers a visual depiction of this idea. Circle A represents the variance in one variable, and circle B represents the variance in a second variable. At the left is a relationship in which $r = .00$ (no overlap between the two); thus, $r^2 = .00$. At the right, $r = .71$; thus, $r^2 = .50$ (considerable overlap between the two).

For example, the standing long jump and the vertical jump are common tests of explosive power. The tests are so commonly used that we tend to think of them as interchangeable; that is, as measuring the same thing. Yet correlations between the two tests usually range between .70 and .80. Hence, the coefficients of determination range from .49 ($.70^2$) to .64 ($.80^2$). Usually, the coefficient of determination is expressed as the percentage of variation. Thus, $.70^2 = .49 = 49\%$ and $.80^2 = .64 = 64\%$.

For a correlation of .70 between the standing long jump and the vertical jump, only about half (49%) of the variance (or influences) in one test is associated with the other. Both tests involve explosive force of the legs with

coefficient of determination (r^2)—The squared correlation coefficient; used in interpreting the meaningfulness of correlations.

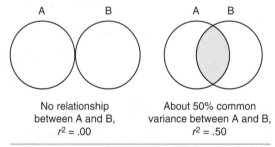

Figure 8.6 Example of shared variance between variables.

some flexion and extension of the trunk and swinging of the arms. Both are influenced by body weight in that the participant must propel his or her body through space, both involve the ability to prepare psychologically and physiologically to generate explosive force, both involve **relative strength**, and so on. These factors are held in common in the two tests. If $r = .80$, then 64% of the performance in one test is associated with, or explained by, the factors involved in the performance of the other test.

relative strength—The measure of a person's ability to exert maximal force in relation to his or her size.

But what about the unexplained variance ($1.0 - r^2$)? A correlation of .70 is made up of 49% common (explained) variance and 51% ($1.00 - .49$) error (unexplained) variance. What are some factors unique to each test? We cannot explain these differences fully, but some of the factors could be that (a) the standing long jump requires that the body be propelled forward and upward, whereas the vertical jump is only upward; (b) the scoring of the vertical jump neutralizes a person's height because standing reach is subtracted from jumping reach, but in the standing long jump, the taller person may have some advantage; and (c) perhaps more skill (coordination) is involved in the vertical jump because the person must jump, turn, and then touch the wall. This explanation is not intended to be any sort of mechanical analysis of the two tests. These are simply suggestions of some possible factors of common association, or explained variance, and some factors that might be unexplained or unique to each test of explosive power.

When we use the coefficient of determination to interpret correlation coefficients, it becomes apparent that a rather substantial relationship is needed to account for a great amount of common variance. A correlation of .71 is required to account for just half the variance in the other test, and a correlation of .90 accounts for only 81%. In some of the standardized tests used to predict academic success, correlations are generally quite low, often around .40. You can see by the coefficient of determination that a correlation of .40 accounts for only 16% of the factors contributing to academic success; therefore, unexplained variance is substantial. Still, these measures are often used rigorously as the criterion for admission to academic programs. Of course, the use of multiple predictors can greatly improve the estimate for success. The sizes of correlations can also be compared using the coefficient of determination. A correlation of .90 is not three times larger than a correlation of .30; it is nine times larger ($.30^2 = .09$, or 9%, and $.90^2 = .81$, or 81%).

Interpreting the correlation coefficient is further complicated because whether a correlation is good or inadequate depends on the purpose of the correlation. For example, if we are looking at the reliability (repeatability) of a test, a much higher correlation is needed than if we are determining simply whether a relationship exists between two variables. A correlation of .60 would not be acceptable for the relationship between two similar versions of an exercise knowledge test, but a correlation of .60 between exercise knowledge and exercise behavior would be noteworthy.

Z Transformation of r

Occasionally, a researcher wants to determine the average of two or more correlations. Trying to average the coefficients themselves is statistically unsound because the sampling distribution of coefficients of correlation is not normally distributed. In fact, the higher the correlation, in either a positive or a negative direction, the more skewed the distribution becomes. The most satisfactory method of approximating normality of a sampling distribution of linear relationships is by transforming coefficients of correlation to Z values. This is often called the **Fisher Z transformation**. (This Z should not be confused with the z used to refer to the height of the ordinate in the area of the normal curve.)

Fisher Z transformation—A method of approximating normality of a sampling distribution of linear relationship by transforming coefficients of correlation to Z values.

The transformation procedure involves the use of natural logarithms, but we need not use Fisher's formula to calculate the transformations (these conversions have been done for us in table 4 in the appendix). We simply consult the table and locate the corresponding Z value for any particular correlation coefficient.

Suppose, for example, that we obtained correlations between maximal oxygen consumption and a distance run (e.g., an 8 min run–walk) on four groups of participants of different ages. We would like to combine these sample correlations to obtain a valid and reliable estimate of the relationship between these two measures of cardiorespiratory endurance. The data for the following steps are shown in table 8.2.

TABLE 8.2

Average of Correlation Coefficients by Use of the *Z* Transformation

Age group	N	r	Z	$N - 3$	Weighted Z
13–14	30	.69	0.85	27	22.95
15–16	44	.85	1.26	41	51.66
17–18	38	.70	0.87	35	30.45
19–20	35	.77	1.02	32	32.64
				135	137.70

1. First, convert each correlation to a *Z* value using appendix table 4. For example, the correlation of .69 for the 13- and 14-year-olds has a corresponding *Z* value of 0.85; the next correlation of .85 for the 15- and 16-year-olds has a *Z* value of 1.26; and so on.

2. The *Z* values are then weighted by multiplying them by the degrees of freedom for each sample, which in this process is $N - 3$. So, for the 13- and 14-year-olds, the *Z* value of 0.85 is multiplied by 27 for a weighted *Z* value of 22.95. We do the same for the other three samples.

3. The weighted *Z* values are summed, and the mean weighted *Z* value is calculated by dividing by the sum of degrees of freedom ($N - 3$): 137.70 / 135 = 1.02.

4. The mean weighted *Z* value is converted back to a mean correlation by consulting appendix table 4 again. We see that the corresponding correlation for a *Z* value of 1.02 is .77.

Some authors declare that to average correlations by the Z-transformation technique, you must first establish that there are no significant differences among the four correlations. A comparison for differences could be made using a chi-square test of the weighted *Z* values (chi square is discussed in chapter 10). Other statisticians contend that averaging coefficients of correlation is permissible as long as the average correlation is not interpreted in terms of confidence intervals.

The *Z* transformation is also used for statistical tests (such as those for the significance of the correlation coefficient) and for determining the significance of the difference between two correlation coefficients. After reading chapter 9, you may wish to consult a statistics text, such as that by Mattson (1981), for further discussion on the use of the *Z* transformation for these procedures.

> The greater the correlation between the variables, the more accurately one can be predicted from the other.

USING CORRELATION FOR PREDICTION

We have stated several times that one purpose of correlation is prediction. College entrance examinations are used to predict success. Sometimes we try to predict a criterion such as fat percentage by the use of skinfold measurements or maximal oxygen consumption by a distance run. In studies of this type, the predictor variables (skinfold measurements) are less time-consuming, less expensive, and more feasible for mass testing than the criterion variable; thus, a **prediction**, or regression, **equation** is developed.

Prediction is based on correlation. The higher the relationship is between two variables, the more accurately you can predict one from the other. If the correlation were perfect, which it never is with humans, you could predict with complete accuracy.

> **prediction equation**—A formula to predict some criterion (e.g., some measure of performance) based on the relationship between the predictor variable(s) and the criterion; also called *regression equation*.

Working With Regression Equations

Of course, we do not encounter perfect relationships in the real world. However, in introducing the concept of prediction (regression) equations, it is often advantageous to begin with a hypothetical example of a perfect relationship.

Verducci (1980) provided one of the best examples in introducing the regression equation concerning monthly salary and annual income. If there are no other sources of income, we can predict with complete accuracy the annual income of, for example, teachers simply by multiplying their monthly salaries by 12. Figure 8.7 illustrates this perfect relationship. By plotting the monthly salary (the X, or predictor variable), the predicted annual income (the Y, or criterion variable) can be obtained. Thus, if we know that a teacher (e.g., Ms. Brooks) earns a monthly salary of $1,750, we can easily plot this on the graph where $1,750 on the horizontal (x) axis (the **abscissa**) intercepts the vertical (y) axis (the **ordinate**) at $21,000. The equation for prediction ($\hat{Y}$, predicted annual income) is thus $\hat{Y} = 12X$. In this example, the teacher's monthly salary (X) is inserted in the formula: $\hat{Y} = 12(1,750) = 21,000$.

Next, suppose that all teachers got an annual supplement of $1,000 for coaching or supervising cheerleaders or some other extracurricular activity. Now, the formula becomes $\hat{Y} = 1,000 + 12X$. Ms. Brooks' annual income is predicted as follows: $\hat{Y} = 1,000 + 12(1,750) = 22,000$. All teachers' annual incomes could be predicted in the same manner. This formula is the general formula for a straight line (remember back to middle school where you learned this) and is expressed as follows:

$$\hat{Y} = a + bX \tag{8.2}$$

where $\hat{Y}$ = the predicted score, or criterion; a = the intercept; b = the slope of the regression line; and X = the predictor.

In this example, the b factor was ascertained by common sense because we know that there are 12 months in a year. The slope of the line (b) signifies the amount of change in Y that accompanies a change of 1 unit of X. Therefore, any X unit (monthly salary) is multiplied by 12 to obtain the $\hat{Y}$ value. In actual regression problems, we do not intuitively know what b is, so we must calculate it by this formula:

$$b = r(s_Y / s_X) \tag{8.3}$$

where r = the correlation between X and Y, s_Y = the standard deviation of Y, and s_X = the standard deviation of X.

In our previous example, application of equation 8.3 uses these data:

X (monthly salary)	Y (annual income)
$M_X = 1,700$	$M_Y = 21,400$ (including $1,000 bonus)
$s_X = 141.42$	$s_Y = 1,697.06$
$r = 1.00$	

Therefore, $b = 1.00(1,697.06 / 141.42) = 12$. The a in the regression formula indicates the intercept of the regression line on the y-axis. In other words, a is the value of Y when X is zero. On a graph, if you extend the regression line sufficiently, you can see where the regression line intercepts Y. The a is a constant because it is added to each of the calculated bX values. Again, in our example, we know that this constant is 1,000. In other words, this is the value of Y even if there were no monthly salary (X). But to calculate the a value, you must first calculate b. Then use the following formula:

$$a = M_Y - bM_X \tag{8.4}$$

where a = the constant (or intercept), M_Y = the mean of the Y scores, b = the slope of the regression line, and M_X = the mean of the X scores. In our example, $a = 21,400 - 12(1,700) = 1,000$. Then the final regression equation is $\hat{Y} = a + bX$, or $Y = 1,000 + 12X$.

abscissa—The horizontal, or x, axis of a graph.

ordinate—The vertical, or y, axis of a graph.

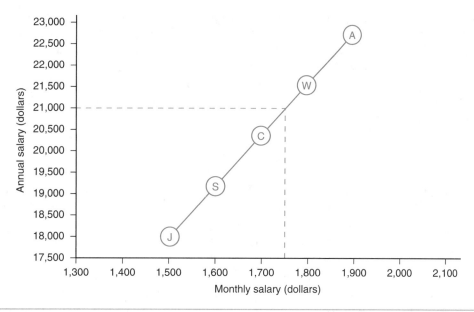

Figure 8.7   Plotting monthly and annual salaries with perfect *r*. *Note:* Letters refer to teachers' initials.

Next, let us use a more practical example in which the correlation is not 1.00. We can use the data from figure 8.2, in which the correlation between putts per round and greens in regulation was .55. The means and standard deviations are as follows:

X (greens in regulation)	*Y* (putts per round)
$M_X = 65.23$	$M_Y = 29.22$
$s_X = 2.35$	$s_Y = 0.51$
$r = .55$	

First, we calculate *b* as follows from equation 8.3 (see figure 8.2 for SPSS printout; data are within rounding error):

$$b = r(s_Y / s_X) = .55(0.51 / 2.35) = 0.12$$

Then, *a* is calculated as follows from equation 8.4:

$$a = M_Y - bM_X = 29.22 - 0.12(65.23) = 21.4$$

The regression formula (equation 8.2, $Y = a + bX$) becomes

$$\hat{Y} = 21.4 + 0.12X$$

For any PGA player's greens in regulation (GIR), we can calculate the predicted putts per round (PPR). For example, a player who hits 60% of the GIR would have a predicted PPR of $\hat{Y} = 21.4 + 0.12(60) = 28.6$. The main difference between this example and that of monthly and annual salaries is that the latter had no error of prediction because the correlation was 1.00. When we predicted PPR from GIR, however, the correlation was less than 1.00, so an error of prediction is present.

Line of Best Fit

Figure 8.2 shows a straight line connecting the GIR and PPR scores. This **line of best fit** is used to predict *Y* from the *X* scores.

This line passes through the intersection of the *X* and *Y* means. Note again our previous caution: This type of correlation is useful only if a straight line is the best fit to the data.

line of best fit—The calculated regression line that results in the smallest sum of squares of the vertical distances of every point from the line.

This assumption that a straight line fits the data applies to all the statistical techniques that we discuss in chapters 8, 9, and 10, as does the assumption that the data are normally distributed.

Figure 8.2 shows this line of best fit. You can readily see that the scores do not fall on the straight line, as they did with the perfect correlation example. These differences between predicted and actual Y scores represent the errors of prediction and are called **residual scores**. If we compute all the residual scores, the mean is zero, and the standard deviation is the **standard error of prediction** ($s_{Y \cdot X}$).

A simpler way of obtaining the standard error of prediction is to use this formula:

$$s_{Y \cdot X} = s_Y \sqrt{1 - r^2} \tag{8.5}$$

The standard error of prediction is interpreted the same way as the standard deviation. In other words, the predicted value (PPR) of the player in our example, plus or minus the standard error of prediction, occurs approximately 68 times out of 100.

The larger the correlation is, the smaller the error of prediction is. Also, the smaller the standard deviation of the criterion is, the smaller the error is.

The line of best fit is sometimes called the least-squares method. This means that the calculated regression line is one for which the sum of squares of the vertical distances of every point from the line is minimal. We will not develop this point here. Sum of squares is discussed in chapter 9.

PARTIAL CORRELATION

The correlation between two variables is sometimes misleading. Interpretation may be difficult when there is little or no correlation between the variables other than that caused by their common dependence on a third variable.

For example, many attributes increase regularly with age from 6 to 18 years, such as height, weight, strength, mental performance, vocabulary, and reading skills. Over a wide age range, the correlation between any two of these measures will almost certainly be positive and will probably be high because of the common maturity factor with which they are highly correlated. In fact, the correlation may drop to zero if the variability caused by age differences is eliminated. We can control this age factor in one of two ways. We can select only children of the same age, or we can partial out (remove the influence of something or hold something constant) the effects of age statistically by holding it constant.

The symbol for partial correlation is $r_{12 \cdot 3}$, which means the correlation between variables 1 and 2 with variable 3 held constant (we could hold any number of variables constant, e.g., $r_{12 \cdot 345}$). The calculation of partial correlation among three variables is simple. Let us refer to the previous correlation of shoe size and achievement in mathematics. This is a good example of **spurious correlation**, which means that the correlation between the two variables is due to the common influence of another variable (age or maturing). When the effect of the third variable (age) is removed, the correlation between shoe size and achievement in mathematics diminishes or vanishes completely. We label the three variables as follows: 1 = math achievement, 2 = shoe size, and 3 = age. Then, $r_{12 \cdot 3}$ is the partial correlation between variables 1 and 2 with 3 held constant. Suppose the correlation coefficients between the three variables were $r_{12} = .80$; $r_{13} = .90$; and $r_{23} = .88$. The formula for $r_{12 \cdot 3}$ is

$$\begin{aligned} r_{12 \cdot 3} &= \frac{r_{12} - r_{13}r_{23}}{\sqrt{1 - r_{13}^2}\sqrt{1 - r_{23}^2}} \\ &= \frac{.80 - .90 \times .88}{\sqrt{1 - .90^2}\sqrt{1 - .88^2}} \\ &= .039 \end{aligned} \tag{8.6}$$

Thus, we see that correlation between math achievement and shoe size drops to about zero when age is held constant.

The primary value of partial correlation is to develop a multiple-regression equation with two or more predictor variables. In the selection process, when a new variable is "stepped

residual scores—The difference between the predicted and actual scores that represents the error of prediction.

standard error of prediction—The computation of the standard deviation of all of the residual scores of a population; the amount of error expected in a prediction. Also called *standard error of estimate*.

spurious correlation—A relationship in which the correlation between two variables is due primarily to the common influence of another variable.

in," its correlation with the criterion is determined with the effects of the preceding variable held constant. The size and the sign of a partial correlation may be different from the zero-order (two-variable) correlation between the same variables.

SEMIPARTIAL CORRELATION

In the previous section on partial correlation, the effects of a third variable on the relationship between two other variables were eliminated by using the formula for partial correlation. In other words, in $r_{12 \cdot 3}$, the relationship of variable 3 to the correlation of variables 1 and 2, is removed. In some situations, the investigator may wish to remove the effects of a variable from only one of the variables being correlated. This is called **semipartial correlation**. The symbol is $r_{1(2 \cdot 3)}$, which indicates that the relationship between variables 1 and 2 is determined after the influence of variable 3 on variable 2 has been eliminated.

Suppose, for example, that a researcher is studying the relationship among perceived exertion (PE, a person's feelings of how hard he or she is working), heart rate (HR), and workload (WL). Obviously, WL is going to be correlated with HR. The researcher wants to investigate the relationship between PE and HR while controlling for WL. Regular partial correlation will show this relationship, but it will also remove the effects of WL on the relationship between PE and HR. But the researcher does not want to remove the effects of WL on the relationship of PE and HR; he wants to remove only the effects of WL on HR. In other words, the main interest is in the net effect of HR on PE after the influence of WL has been removed. Thus, in semipartial correlation, the effect of WL is removed from HR but not from PE.

> **semipartial correlation**—A technique in which just one variable is partialed out (i.e., effects are removed) from two variables in a correlation.

PROCEDURES FOR MULTIPLE REGRESSION

Multiple regression involves one dependent variable (usually a criterion of some sort) and two or more predictor variables (independent variables). The use of more than one predictor variable usually increases the accuracy of prediction. This should be self-evident. If you wished to predict scoring average on the PGA Tour, you would expect to get a more accurate prediction by using several skill results (e.g., driving distance, greens in regulation, putts per round) rather than by using only one.

The multiple correlation coefficient (R) indicates the relationship between the criterion and a weighted sum of the predictor variables. It follows then that R^2 represents the amount of the variance of the criterion that is explained or accounted for by the combined predictors. This is similar to the coefficient of determination (r^2), which was discussed earlier with regard to the common association between two variables. Now, however, we have the amount of association between one variable (the criterion) and a weighted combination of variables.

We wish to find the best combination of variables that gives the most accurate prediction of the criterion. Therefore, we are interested in knowing how much each of the predictors contributes to the total explained variance. Another way to say this is that we want to find the variables that most reduce the prediction errors. From a practical standpoint, in terms of the time and effort involved in obtaining measures of the predictor variables, it is desirable to find the fewest number of predictors that account for most of the variance of the criterion. Several selection procedures can be used for this purpose. (For additional information on multiple regression, see Cohen & Cohen, 1983, and Pedhazur, 1982.)

In deciding which predictors (or sets of predictors) to use, standard statistical packages offer several options. Table 8.3 includes the most common procedures. One option is simply to use all the predictor variables. But the fewer that can be used while still obtaining a good prediction, the more economical and valuable the prediction equation is. Sometimes researchers specify the order and combinations of predictor variables. This method, called hierarchical regression, should be based on theory and previous empirical evidence. The remaining four procedures listed in table 8.3 allow the program to select the order of predictor variables and the number to be used based on statistical criteria. All are designed to end up with fewer predictor variables than the full model.

> **multiple regression**—A model used for predicting a criterion from two or more independent, or predictor, variables.

> R represents the multiple correlation coefficient, which indicates the relationship between the criterion (a dependent variable) and a weighted sum of the predictor variables (independent variables).

TABLE 8.3

Selection Procedures in Multiple Regression

Procedure	How it works
Full model	All prediction variables are included in the model.
Hierarchical	Prediction variables are entered as blocks (e.g., two together) and are determined in advance.
Forward	The predictor with the highest correlation is entered first; then the next highest is entered after accounting for the relationship with the first predictor.
Backward	All variables are entered; then the one with the least contribution is removed first, and then the one with the next least contribution.
Stepwise	A combination of forward and backward procedures in which checks are made at each step to see whether a previously entered predictor should be removed.
Maximum R squared	The program calculates the best variable combination of two, three, or more predictors.

Example 8.2 provides an example of multiple regression using several of the golf skill variables to predict scoring average (from the complete data set in table 6.2). The means and standard deviations of the criterion (scoring average) and the predictors (driving distance, driving accuracy, greens in regulation, putts per round, scrambling, and sand saves) are in the top part of the table. The procedure uses the backward regression model—all the variables are included at the beginning (model 1 in the table), and a variable that is unimportant (not a good predictor) is removed at each step until all the remaining variables are important (contribute to the prediction model). In this example, model 2 has removed driving accuracy, and model 3 (the final model) has removed scrambling. Thus, model 3 uses the combination of sand saves, greens in regulation, driving distance, and putts per round to predict scoring average. The four predictors in a linear composite (each weighted for its contribution) significantly predicts $F(4,25) = 5.26$, $p = .003$, scoring average, $R = .676$. This linear composite of four aspects of golf performance accounts for 45.7% of the variance (R^2) in scoring average. The prediction equation is in the table. But because the linear composite of predictors accounts for only 45.7% of the variance in scoring average, the prediction may not be a useful one.

Example 8.2

Known Values

Multiple Regression Using Golf Skill Variables to Predict Scoring Average (Backward Multiple Regression)

Variable	Descriptive statistics		
	Mean	Standard deviation	N
Scoring average	70.1	0.5	30
Driving distance	290.6	7.8	30
Driving accuracy	61.9	4.9	30
Greens hit in regulation	65.2	2.3	30
Putts per round	29.2	0.5	30
Scrambling	57.8	2.4	30
Sand saves	51.9	5.1	30

Model	Model summary (d)	
	R	R^2
1	.703 (a)	.494
2	.703 (b)	.494
3	.676 (c)	.457

a. Predictors = sand saves, driving accuracy, greens in regulation, scrambling, driving distance, putts per round

b. Predictors = sand saves, greens in regulation, scrambling, driving distance, putts per round

c. Predictors = sand saves, greens in regulation, driving distance, putts per round

d. Criterion variable = scoring average

$F(4,25) = 5.26$, $p = .003$

Prediction Equation:

$\hat{Y} = 70.9 - (0.022)$ driving distance $- (0.121)$ greens in regulation $+ (0.529)$ putts per round $- (0.37)$ sand saves

Multiple-Regression Prediction Equations

The prediction equation resulting from multiple regression is basically that of the two-variable regression model, $\hat{Y} = a + bX$. The only difference is that there is more than one X variable; thus, the equation is

$$\hat{Y} = a + b_1X_1 + b_2X_2 + \ldots + b_iX_i \tag{8.7}$$

We will not delve into the formula for the calculation of the a and the bs for the selected variables. As we said before, researchers use computers for multiple-regression problems.

Some Problems Associated With Multiple Regression

The basic procedure to evaluate multiple regression is the same as that used in regression with only two variables: the size of the correlation. The higher the correlation, the more accurate the prediction. But some other factors should be mentioned.

One limitation of prediction relates to generalizability. Regression equations developed with a particular sample often lose considerable accuracy when applied to others. This loss of accuracy in prediction is called **shrinkage**. The term **population specificity** also relates to this phenomenon. Shrinkage and the use of cross-validation to improve generalizability are further discussed in chapter 11. We need to recognize that as researchers seek more accuracy through selection procedures (forward, backward, stepwise, maximum R^2) that capitalize on specific characteristics of the sample, they are less able to generalize to other populations. For example, a formula for predicting percentage of body fat from skinfold measurements developed with adult men would lose a great deal of accuracy if used for adolescents. Thus, the researcher should select the sample carefully with regard to the population for which the results are to be generalized.

In prediction studies the number of participants in the sample should be sufficiently large. Usually, the larger the sample, the more likely it will be to represent the population from which it is drawn. Another problem with small samples in multiple-regression studies is that the correlation may be spuriously high. A direct relationship exists between the correlation and the ratio of the number of participants to the number of variables. In fact, the degree to

shrinkage—The tendency for the validity to decrease when the prediction formula is used with a new sample.

population specificity—A phenomenon whereby a regression equation that was developed with a particular sample loses considerable accuracy when applied to others.

which the expected value of R^2 exceeds zero when it is zero in the population depends on two things: the size of the sample (n) and the number of variables (k). More precisely, it is the ratio $(k-1)/(n-1)$. To illustrate, suppose you read a study in which $R^2 = .90$. Impressive, right? The results would be meaningless, however, if the study had only 40 participants and 30 variables, because we could expect an R^2 of .74 just based on chance alone: $R^2 = (k-1)/(n-1) = (30-1)/(40-1) = .74$. You should be aware of this relationship between the number of participants and the number of variables when reading research that uses multiple regression. In the most extreme case, you can see that having the same number of variables as participants, $(k-1)/(n-1)$ yields an R^2 of 1.00! A participant-to-variable ratio of 10:1 or higher is often recommended.

LOGISTIC REGRESSION

Logistic regression is useful when the outcome (criterion) variable is in two categories (1 or 2, yes or no, female or male) and the predictor variable is continuous. (e.g., running speed, strength, weight). In the logistic regression curve shown in figure 8.8, body weight in pounds is used to predict whether the middle school child is female or male in a sample of 100 (50 females and 50 males). On the x-axis is body weight (60 pounds to 120 pounds), and on the y-axis (shown in the middle of the graph for ease of viewing) is the probability (from 0.00 to 1.00) of the child being male. Just as in regression, if the curved prediction line is read down to the x-axis, it represents the average weight at any point. When the line is read across to the y-axis, it represents the probability (odds ratio) of the child being male. However, the regression line in logistic regression is nonlinear. As you can see, the probability (or odds ratio) of the child being male increases with weight.

DISCRIMINANT FUNCTION ANALYSIS

Discriminant function analysis (DFA) is another type of prediction equation, but in this case the variable being predicted is group membership. Each of the methods listed in table 8.3 can be used to predict group membership when the groups are independent of one another. When employing DFA, the researcher uses data (a variety of predictor variables and the group each individual actually is in) to create an equation that can be used when group membership information is not available or does not exist. The researcher can examine, based on the best predictor variables, which group an individual is predicted to be a member of and the overall success of predictions (which predictions the equation got right).

As an example of the use of DFA, let's assume that a researcher wants to predict which undergraduate kinesiology majors will go to graduate school, which will go to professional school, and which will not go beyond the bachelor's degree. She has information on graduates for a 5-year period and a variety of variables that she believes would help create a prediction equation (e.g., undergraduate GPA, GPA in the major, a motivation questionnaire from the junior year, an indication of future plans). She puts all the information into a computer and completes a stepwise DFA that will indicate the best predictors and how good the predictions are. If the best equation predicts around one-third of graduates correctly, it is not a good prediction because that

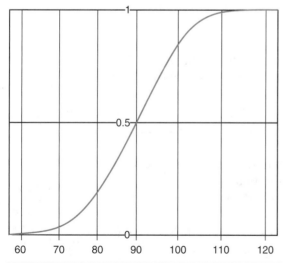

Figure 8.8 Logistic regression curve using body weight in pounds to predict the sex of middle school children.

could be obtained by chance. If, however, the prediction is much better, the equation can be considered valuable for predicting postgraduate education plans.

MULTIVARIATE FORMS OF CORRELATION

We have been discussing forms of correlation that are typically labeled as univariate, meaning that there are one or more predictor variables (independent variables) but only one criterion variable (dependent variable). In this short section we introduce the multivariate forms of correlation. Our purpose is to provide enough information that you can read a paper using one of these procedures and understand what has been done.

Canonical Correlation

Canonical correlation is an extension of multiple correlation (several predictors and one criterion) to an analysis that has several predictors and several criteria, represented by the symbol R_c. In multiple correlation, a linear composite is formed of the predictor variables that maximally predict the single criterion variable. In canonical correlation, two linear composites are formed: one of the predictor variables and one of the criterion variables. These two composites are formed to maximize the relationship between them, and this relationship is represented by R_c. R_c^2 interpreted just like R_c^2 from multiple correlation; it represents the shared variance between the two linear composites of variables.

For example, McPherson and Thomas (1989) studied expert and novice tennis players at two age levels (10–11 and 12–13). They measured three predictor variables (tennis knowledge, tennis serve, and tennis ground strokes). They then video-recorded the children playing tennis and coded the videos for three criterion variables of tennis play: control (being in position), decisions (choice of what stroke to use and where to hit the ball), and execution (doing what was intended). They used canonical correlation to evaluate the relationship between the linear composites of the three predictor and three criterion variables. They reported a significant canonical correlation, $R_c = .79$, $F(6,70) = 7.58$, $p < .01$ (the F ratio and the notation that describes it are discussed in chapter 9). When they evaluated the standardized canonical coefficients (such as the standardized beta weights in multiple correlation) and multiple-regression follow-ups, they found that tennis knowledge was most closely related to making good decisions during game play, whereas tennis ground strokes were most closely related to tennis execution during game play.

Factor Analysis

Many performance variables and characteristics are used to describe human behavior. Often, reducing a large set of performance and characteristic measures to a more manageable structure is useful. We have already discussed the likelihood that two performance measures might to some extent assess the same underlying characteristic. This represents the degree to which they are correlated. **Factor analysis** is an approach used to reduce a set of correlated measures to a smaller number of latent or hidden variables (Tinsley & Tinsley, 1987). Numerous procedures are grouped under the general topic of factor analysis. We do not discuss the techniques in detail, but we do provide a general explanation that will help you read and understand studies that use factor analysis. For good explanations and examples of factor analysis, read Tinsley and Tinsley (1987) and McDonald (1999).

Factor analysis is performed on data from a group of individuals on whom a series of measurements have been taken. The researcher usually wants to describe a reduced number of underlying constructs and possibly select the one or two best measures of each construct. Factor analysis begins by calculating the intercorrelations of all the measures used (the correlation between all possible pairs of variables; thus, if eight measures were used, the

factor analysis—A statistical technique used to reduce a set of data by grouping similar variables into basic components (factors).

correlation would be determined between variables 1 and 2, variables 1 and 3, and so on, for a total of 28 correlations). The goal of factor analysis is to discover the factors (underlying or hidden constructs) that best explain a group of measurements and describe the relation of each measure to the factor or underlying construct. The two general types of factor analysis are (a) exploratory, in which many variables are reduced to an underlying set, and (b) confirmatory, which either supports or does not support a structure proposed from theory. Confirmatory factor analysis is the more useful technique.

For example, Marsh, Marco, and Aþçý (2002) examined whether the Physical Self-Description Questionnaire (PSDQ) had similar factor structures when used with cross-cultural samples—high school students in Australia, Spain, and Turkey. They translated the instrument into Spanish and Turkish and used back-translation techniques to assess that the translations produced a similar instrument in the new language. Children in each country then completed the instrument, and factor structures and fit statistics (such as root mean square error of approximation [RMSEA] and the relative noncentrality index [RNI]) were compared. The research supported the use of the PSDQ in a variety of cultural environments. As we note in chapter 11, confirmatory factor analysis is often used to validate scores from attitude instruments or other questionnaires, and this study extended that into multiple countries.

Structural Equation Modeling

linear structural relations (LISREL)—A statistical approach used to establish relationships and examine the structural equations model.

Path analysis and **linear structural relations (LISREL)** are structural, or causal, modeling techniques that are used to explain the way certain characteristics relate to one another and attempt to imply cause. You should remember from the discussion in chapter 4 and earlier in this chapter that cause and effect are not a statistical result but a logical one. That is, if the experimenter can argue theoretically that changing a certain characteristic should result in a specific change in behavior, and if the actual experiments (and statistical analysis) support this hypothesis, then a particular independent variable is often inferred to cause a dependent variable. Of course, this is true only if all other possible influences have been controlled.

The way variables influence one another is not always clear; for example, $X \to Y \to Z$, or $X \to Y \to Z$. In the first case, X influences Y, which in turn influences Z, but in the second case, Y influences both X and Z. This is a simple example. Path analysis and LISREL allow a more complex modeling of the way variables influence one another. But all you can say about these models is that they are either consistent or inconsistent with the data and hypotheses. Whether they imply cause and effect depends on other things (e.g., control of all other variables, careful treatments, logical hypotheses, valid theories). For a useful overview and practical explanation of structural equation modeling, read Schumacker and Lomax (2004).

A good way to gain a basic understanding of LISREL is through an example from the physical education literature. Greenockle, Lee, and Lomax (1990) were interested in describing the relationship between student characteristics and exercise behavior. Figure 8.9 shows the relationship that they found in a sample of 10 existing high school physical education classes in which the teacher agreed to teach a specific physical fitness unit. The researchers measured four variables of the students' background: past activity level, perception of father's activity level, perception of mother's activity level, and knowledge of physical fitness. Two measures of attitude toward exercise were taken from a valid and reliable questionnaire. Two subjective norm measures, which indicated participants' perceptions of how others felt about the participants' exercise behavior, were obtained from the same questionnaire. One measure of intention to exercise was obtained. Five exercise behaviors were measured: jogging, fitness test score, percentage of participation, heart rate, and perceived exertion.

The technique of LISREL was used because it allows related measures to be grouped (as in factor analysis) into components (e.g., exercise behavior), and it shows the relationship among the components in terms of magnitude and direction. The estimates of each compo-

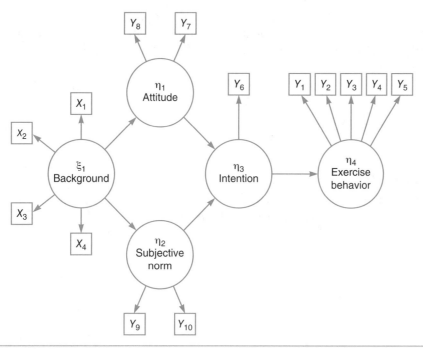

Figure 8.9 LISREL model.

"The relationship between selected student characteristics and activity patterns in a required high school physical education class," K.M. Greenockle, A. A. Lee, R. Lomax, *Research Quarterly for Exercise and Sport,* 61(1): 559-69, 1990 Taylor and Francis, reprinted by permission of the publisher (Taylor & Francis Ltd, www.tandfonline.com).

nent are grouped together in a linear equation (using their factor loadings). Then, a series of general linear equations (as in multiple regression) shows the relationships among the components. Model-fitting indices are used to assess the nature of the fit. The results of the study by Greenockle and colleagues (1990) showed "the prediction of exercise behavior by attitude and subjective norm to be significantly mediated by intention" (p. 59). You can see, then, that structural modeling techniques provide a way to evaluate the complex relationships that exist in real-world data. By establishing the directions of certain relationships, researcher can make stronger inferences about which characteristics are likely to cause other characteristics.

SUMMARY

We have explored some statistical techniques for determining relationships among variables. The simplest type of correlation is the Pearson r correlation, which describes the relationship between two variables. We introduced linear regression, which can be used to predict one variable from another. Correlation is interpreted for significance (reliability) and for meaningfulness (r^2), which indicates the portion of the total variance in one measure that can be explained or accounted for by the other measure.

Partial correlation is a procedure in which a correlation between two variables is obtained while the influence of one or more other variables is held constant. Semipartial correlation removes the influence of a third variable on only one of the two variables being correlated. Partial correlation (or semipartial correlation) is used in multiple correlation and in developing multiple-regression formulas. Logistic regression allows calculation of the odds ratio in the criterion variable as determined by a predictor variable. Discriminant analysis uses a group of variables to predict group membership.

In multiple regression two or more predictor (independent) variables are used to predict the criterion variable. The most efficient weighted linear composite of predictor variables is determined through such techniques as forward selection, backward selection, stepwise selection, and maximum R^2.

Finally, we provided a brief overview of three multivariate correlational procedures: canonical correlation, factor analysis, and structural equation modeling. An example of the use of each technique was presented.

Check Your Understanding

1. What is the correlation (r) between fat deposits on two body sites? Variable X is the triceps skinfold measure, and variable Y is the suprailiac skinfold measure.

X	Y
16	9
17	12
17	10
15	9
14	8
11	6
11	5
12	5
13	6
14	5
4	1
7	4
12	7
7	1
10	3

2. Using table 3 in the appendix, determine whether the correlation obtained in problem 1 is significant at the .05 level. What size would the correlation need to be for significance at the .01 level if you had 30 participants? What is the percentage of common variance between the two skinfold measurements in problem 1?

3. Determine the regression formula (equation 8.2) to predict maximal oxygen consumption ($\dot{V}O_2max$) from scores on a 12 min run. The information that you need follows:

X (12 min run)	Y ($\dot{V}O_2max$)
$M_X = 2{,}853$ m	$M_Y = 52.6$ ml · kg · min⁻¹
$s_X = 305$ m	$s_Y = 6.3$ ml · kg · min⁻¹
$r = .79$	

4. Using the prediction formula developed in problem 3, what is the predicted $\dot{V}O_2max$ for a participant who ran 2,954 m in 12 min? For a participant who ran 2,688 m in 12 min?

5. What is the standard error of prediction (equation 8.5) for the prediction formula in problem 3? How would you interpret the predicted $\dot{V}O_2max$ for the participants in problem 4?

9

Differences Among Groups

It's frustrating when you know all the answers but nobody bothers to ask you the questions.

—From Lachlan's Laws (Scottish Highland philosopher)

Statistical techniques are used for describing and finding relationships among variables, as we discussed in chapters 6 and 8. They are also used to detect differences among groups. The latter techniques are most frequently used for data analysis in experimental and quasi-experimental research. They enable us to evaluate the effects of an independent variable (cause or treatment) or categorical variable (sex, age, race, and so on) on a dependent variable (effect, outcome). Remember, however, that the techniques described in this chapter are not used in isolation to establish cause and effect but only to evaluate the influence of the independent variable. Cause and effect are not established by statistics but by theory, logic, and the total nature of the experimental situation.

HOW STATISTICS TEST DIFFERENCES

In experimental research, the levels of the independent variable may be established by the experimenter. For example, the experiment might involve the investigation of the effects of intensity of training on cardiorespiratory endurance. Thus, intensity of training is the independent variable (or treatment factor), whereas a measure of cardiorespiratory endurance is the dependent variable. Intensity of training could have any number of levels. If it were evaluated as a percentage of maximal oxygen consumption ($\dot{V}O_2max$), then it could be 30%, 40%, 50%, and so forth. The investigator would choose the number and the intensity of levels. In a simple experiment, the independent variable might be two levels of intensity of training; for example, 40% and 70% of $\dot{V}O_2max$. The length of each session (e.g., 30 min), frequency (e.g., three times per week), and number of weeks of training (e.g., 12) are controlled (equal for both groups). The dependent variable could be the distance that a person runs in 12 min.

The purpose of the statistical test is to evaluate the null hypothesis at a specific level of probability (e.g., $p < .05$). In other words, do the two levels of treatment differ significantly

> Statistics can only evaluate the influence of independent variables; they cannot establish cause and effect.

($p < .05$) so that these differences are not attributable to a chance occurrence more than 5 times in 100? The statistical test is always of the null hypothesis. All that statistics can do is reject or fail to reject the null hypothesis. Statistics cannot accept the research hypothesis. Only logical reasoning, good experimental design, and appropriate theorizing can do so. Statistics can determine only whether the groups are different, not why they are different.

In using logical techniques to infer cause and effect after finding significant differences, you must be careful to consider all possibilities. For example, we might propose the theorem that all odd numbers are prime numbers (this example is from Ronen et al., cited in Scherr 1983, p. 146). You know that prime numbers are those that can be divided only by 1 and by themselves. Thus, 1 is a prime number, 3 is a prime number, 5 is a prime number, 7 is a prime number, and so on. Using the induction technique of reasoning, every odd number is a prime number. In Ronen and colleagues' example, a small number of levels of the independent variable (prime numbers) were sampled, and an error was made in inferring that all levels of the independent variable were the same.

When you use statistics that test differences among groups, you want to establish not only whether the groups are significantly different but also the strength of the association between the independent and dependent variables, or the size of the difference between two groups. The t and the F ratios are used throughout this chapter to determine whether groups are significantly different. The meaningfulness of the differences is estimated by effect size (ES). Effect size (recall the discussion in chapter 7) is the standardized difference between two groups and is also used as an estimate of meaningfulness. In ANOVA techniques, R^2 is also used to establish meaningfulness. R^2 is the percentage of variance in the dependent variable accounted for by the independent variable.

The uses of the t and the F distributions as presented in this chapter have four assumptions (in addition to the assumptions for parametric statistics presented in chapter 6; Kirk, 1982, is a good book to read for more information):

- Observations are drawn from normally distributed populations.
- Observations represent random samples from populations.
- The numerator and denominator are estimates of the same population variance.
- The numerator and denominator of F (or t) ratios are independent.

Although t and F tests are robust to (only slightly influenced by) violations of these assumptions, the assumptions still are not trivial. You should be sensitive to their presence and to the fact that violations affect the probability levels that may be obtained in connection with t and F ratios.

TYPES OF t TESTS

We discuss three types of t tests: a t test between a sample and a population mean, a t test for independent groups, and a t test for dependent groups.

t Test Between a Sample and a Population Mean

First, we may want to know whether a sample of students differs from a larger population. For example, suppose that for a standardized knowledge test on physical fitness, the mean is 76 for a large population of college freshmen. When tested, a fitness class ($n = 30$) that you are teaching has a mean of 81 and a standard deviation of 9. Do your class members have significantly more knowledge about physical fitness than do the members of a typical college freshman class?

R^2 (R squared)—A method of interpreting the meaningfulness of the strength of the relationship between the independent and dependent variables; the proportion of total variance that is due to the treatments.

To establish whether groups are significantly different, remember two things: the significance of the association between the dependent and independent variables, and the size of the difference between the groups.

The t test is a test of the null hypothesis, which states that there is no difference between the sample mean (M) and the population mean (μ), or $M - \mu = 0$. Equation 9.1 is the t test between a sample and a population mean:

$$t = \frac{M - \mu}{s_M / \sqrt{n}} \tag{9.1}$$

where s_M = the standard deviation for the sample mean and n = the number of observations in the sample. Example 9.1 shows this formula applied to the means and standard deviation of the example for the standardized fitness knowledge test.

Example 9.1

Known Values

Population	$N = 10{,}000$
Fitness class	$n = 30$
Sample mean	$M = 81$
Population mean	$\mu = 76$
Standard deviation	$s_M = 9$

Working It Out (Equation 9.1)

$$t = \frac{M - \mu}{s_M / \sqrt{n}}$$

$$t = \frac{81 - 76}{9 / \sqrt{30}} = \frac{5}{1.64} = 3.05$$

Is the value of 3.05 significant? To find out, you will need to check table 5 in the appendix. To use the table, first work out the degrees of freedom (df) for the t test. Degrees of freedom are based on the number of participants with a correction for bias:

$$df = n - 1 \tag{9.2}$$

In our example, the degrees of freedom are $30 - 1$, or 29. Degrees of freedom are used to enter a t table to determine whether the calculated t is as large as or greater than the table t value. Note that across the top of appendix table 5 are probability levels. We want to know whether our t value is significant at $p < .05$, the level of significance that we set. Read across to the .05 level. Now read down the left side (df) to the row for the degrees of freedom that you calculated ($df = 29$). Read where the $df = 29$ row and the .05 column intersect. Is the calculated value ($t = 3.05$) larger than this value (2.045)? Yes, it is. So t is significant at $p < .05$. Thus, our sample class is reliably (significantly) different from the population average on the fitness test.

Independent t Test

The previous t test, applied to determine whether a sample differs from a population, is not used frequently. The most frequently used t test determines whether two sample means differ reliably from each other. This test is called an **independent t test**.

Suppose that we return to our example at the beginning of this chapter: Do two groups, training at different levels of intensity (40% or 70% of $\dot{V}O_2max$, 30 min per day, 3 days per week for 12 weeks), differ from each other on a measure of cardiorespiratory endurance (12 min run)? Let us further assume that 30 participants were randomly assigned to form two groups of 15 each.

independent t test—The most frequently used test to determine whether two sample means differ reliably from each other.

Equation 9.3 is the *t*-test formula for two independent samples:

$$t = \frac{M_1 - M_2}{\sqrt{s_1^2 / n_1 + s_2^2 / n_2}} \tag{9.3}$$

Equation 9.4 is the version of the *t*-test formula most easily performed with a calculator

$$t = \frac{M_1 - M_2}{\sqrt{\dfrac{\left[\Sigma X_1^2 - (\Sigma X_1)^2 / n_1 + \Sigma X_2^2 - (\Sigma X_2)^2 / n_2\right] \cdot (1 / n_1 + 1 / n_2)}{n_1 + n_2 - 2}}} \tag{9.4}$$

The degrees of freedom for an independent *t* test are calculated as follows:

$$df = n_1 + n_2 - 2 \tag{9.5}$$

In our example, $df = 15 + 15 - 2 = 28$ (or $N - 2 = 30 - 2 = 28$). Example 9.2 shows the results.

Estimating the Meaningfulness of Treatments

How meaningful is the effect in our comparison of training intensities? Or, more simply, is the increase in cardiorespiratory endurance of running an additional 548 m (3,004 – 2,456) worth the additional work of training at 70% of $\dot{V}O_2$max as compared with 40% of $\dot{V}O_2$max? Given the total variation in running performance of the two groups, what we really want to know is how much of this variation is accounted for by (associated with) the difference in the two levels of the independent variable (70% vs. 40%).

Effect Size

To estimate the degree to which the treatment influenced the outcome, use effect size (ES), the standardized difference between the means. Equation 9.6 (given previously as equation 7.1) is a way to estimate effect size (this concept was discussed in chapter 7 and is also used in meta-analysis, discussed in chapter 14):

$$ES = (M_1 - M_2) / s \tag{9.6}$$

where M_1 = the mean of one group or level of treatment, M_2 = the mean of a second group or level of treatment, and s = the standard deviation. The question is what standard deviation should be used. Considerable controversy exists over the answer to this question. Some statisticians believe that if there is a control group, its standard deviation should be used. If there is no control group, then the pooled standard deviation (equation 9.7) should be used. Some advocate the use of the pooled standard deviation on all occasions. Either method can be defended, but when there is no clear control group (as in the example used here), we recommend that you use the pooled standard deviation

$$s_p = \sqrt{\frac{s_1^2 (n_1 - 1) + s_2^2 (n_2 - 1)}{n_1 + n_2 - 2}} \tag{9.7}$$

where s_p = the pooled standard deviation, s_1^2 = the variance of group 1, s_2^2 = the variance of group 2, n_1 = the number of participants in group 1, and n_2 = the number of participants in group 2.

Effect size can be interpreted as follows: An ES of 0.8 or greater is large, an ES around 0.5 is moderate, and an ES of 0.2 or less is small. Thus, the ES calculated in example 9.2 of 5.0 is a large value and would typically be judged as a meaningful treatment effect.

Example 9.2

Known Values

	Group 1	Group 2
Mean distance run	$M_1 = 3,004$ m	$M_2 = 2,456$ m
Standard deviation	$S_1 = 114$ m	$S_2 = 103$ m
Number of subjects	$n_1 = 15$	$n_2 = 15$

Working It Out (Equations 9.6, 9.7, and 9.3)

$$s_p = \sqrt{\frac{s_1^2(n_1-1)+s_2^2(n_2-1)}{n_1+n_2-2}}$$

$$s_p = \sqrt{\frac{(114)^2(15-1)+(103)^2(15-1)}{15+15-2}} = 108.64$$

$$ES = (M_1 - M_2)/s$$

$$ES = \frac{3,004-2,456}{108.64} = 5.0$$

$$t = \frac{3,004-2,456}{\sqrt{(114^2/15)+(103^2/15)}}$$

$$t(28) = 13.73$$

$$p. < .005$$

The independent t test is a common statistical technique in research studies. Sometimes it is the main technique used, and sometimes it is just one of several statistical tests in a study. Sachtleben, Berg, Cheatham, Felix, and Hofschire (1997) compared serum lipoprotein and apolipoprotein levels in a group of weight trainers who used anabolic steroids with those in a group of weight trainers who did not use steroids. A number of comparisons were made. Here is a sample of how the authors reported the results of the t tests: "The fasting HDL-C level in the nonusers group was significantly higher than the off-cycle users group, $t(21) = 4.66, p < .007$" (p. 112). This statement is how the result of a t test is reported in APA style. The number in parentheses (21) is the df, which is $N - 2$. So we know there were 23 people in the two groups. The computed t was 4.66, and this was significant at $p = .007$. The authors used the Bonferroni procedure to adjust the alpha level. We discuss that procedure later in the chapter. Sachtleben and colleagues (1997) used ES to report meaningfulness. For the t we quoted earlier, the ES was 1.2, which would be considered large.

Checking for Homogeneity of Variance

All techniques for comparisons between groups assume that the variances (standard deviation squared) between the groups are equivalent. Although mild violations of this assumption do not present major problems, serious violations are more likely if group sizes are not approximately equal. Formulas given here and used in most computer programs allow unequal group sizes. The homogeneity assumption should be checked, however, if group sizes are very different or even when variances are very different (these techniques are not presented here but are covered in basic statistical texts).

Dependent t Test

We have now considered use of the t test to evaluate whether a sample differs from a population and whether two independent samples differ from each other. A third application is called a **dependent t test**, which is used when the two groups of scores are related in some manner. Usually, the relationship takes one of two forms:

dependent t test—A test of the significance of differences between the means of two sets of scores that are related, such as when the same participants are measured on two occasions.

■ Two groups of participants are matched on one or more characteristics and thus are no longer independent.

■ One group of participants is tested twice on the same variable, and the experimenter is interested in the change between the two tests.

The formula for a dependent t test is

$$t = \frac{M_{post} - M_{pre}}{\sqrt{\left[\left(s_{post}^2 + s_{pre}^2\right) - \left(2r_{pp} \cdot s_{post} \cdot s_{pre}\right)\right] / (N-1)}} \tag{9.8}$$

Notice that the top part of this formula is the same as the independent t-test equation (9.3). In addition, the first term in the bottom (within the parentheses at the left) is also similar. An amount is subtracted, however, from the bottom (error term) of the t-test formula. This term is read as "two times the correlation between the pre- and posttest (r_{pp}) times the standard deviation for the posttest times the standard deviation for the pretest." The independent t test (equation 9.3) assumes that the two groups of participants are independent. In this case, the same participants are tested twice (pretest and posttest). Thus, we adjust the error term of the t test downward (make it smaller) by taking into account the relationship (r) between the pretest and posttest adjusted by their standard deviations. The degrees of freedom for the dependent t test are

$$df = N - 1 \tag{9.9}$$

where $N =$ the number of *paired* observations. Equation 9.8 is cumbersome to compute because the correlation (r) between the pre- and posttest must be calculated. Thus, the raw-score formula is much easier to use:

$$t = \frac{\Sigma D}{\sqrt{\left[N\Sigma D^2 - (\Sigma D)^2\right] / (N-1)}} \tag{9.10}$$

where $D =$ the posttest minus the pretest for each participant and $N =$ the number of paired observations. Let us work out an example. Ten dancers are given a jump-and-reach test (the difference between the height on a wall they can reach and touch while standing and the height they can jump and touch). Then they take part in 10 weeks of dance activity that involves leaps and jumps 3 days per week. The dancers are again given the jump-and-reach test after the 10 weeks. (This is not proposed to be a real experiment, just a simple example illustrating the statistical technique.) Our research hypothesis is that the 10 weeks of dance experience will improve jumping skills as reflected by a change in jump-and-reach scores. The null hypothesis (H_0) is that the difference between the pretest and posttest of jumping is not significantly different from zero, $H_0 = M_{post} - M_{pre} = 0$. Example 9.3 shows how to calculate the dependent t value.

Example 9.3

Known Values

Subject	Pretest score (cm)	Posttest score (cm)	Posttest–pretest D	D^2
1	12	16	4	16
2	15	21	6	36
3	13	15	2	4
4	20	22	2	4
5	21	21	0	0
6	19	23	4	16
7	14	16	2	4
8	17	18	1	1
9	16	22	6	36
10	18	23	5	25

Sum of pretest scores	$\Sigma_{pre} = 165$ cm
Sum of posttest scores	$\Sigma_{post} = 197$ cm
Sum of D	$\Sigma D = 32$
Sum of D^2	$\Sigma D^2 = 142$
Number of paired observations	$N = 10$
Posttest mean	$M_{post} = 19.7$
Pretest mean	$M_{pre} = 16.5$

Working It Out (Equation 9.10)

$$t = \frac{\Sigma D}{\sqrt{\left[N \Sigma D^2 - (\Sigma D)^2\right] / (N-1)}}$$

$$t = \frac{32}{\sqrt{\dfrac{10(142) - (32)^2}{10-1}}} = \frac{32}{\sqrt{\dfrac{1,420 - 1,024}{9}}} = \frac{32}{6.63} = 4.83$$

The results indicate that the posttest mean (19.7 cm) was significantly better than the pretest mean (16.5 cm), $t(9) = 4.83$, $p < .05$. The null hypothesis can be rejected, and if everything else has been properly controlled in the experiment, we can conclude that the dance training produced a reliable increase in the height of jumping performance of 3.2 cm.

We could estimate the magnitude of the effect by dividing the average gain by the pretest mean, as in example 9.4.

Example 9.4

Known Values

Differences between post- and pretest means	$M_D = 3.2$
Mean of pretest scores	$M_{pre} = 16.5$

Working It Out

$$\text{Magnitude of increase} = \frac{M_D}{M_{pre}} = \frac{3.2}{16.5} = 0.194 = 19.4\%$$

The gain is 19.4% of the pretest, nearly a 20% improvement. Although this way of estimating the effect is rather crude, it does suffice. On the other hand, we could estimate ES for the pretest-to-posttest change by subtracting the pretest M from the posttest M and dividing by the pretest standard deviation. In our example this would be ES = 3.2 / 3.03 = 1.06, a large (and meaningful) effect size.

An example of a dependent t test in the literature is by Kokkonen, Nelson, and Cornwell (1998), who investigated the influence of acute stretching on maximal strength performance. The same participants were tested before and after stretching, which necessitated a dependent t test. One of the comparisons was reported thus: "The ST exercises altered sit-and-reach performance such that the poststretching sit-and-reach scores were significantly, $t(29) = 11.1$, $p < .05$, . . . increased 16% over the initial sit-and-reach scores" (p. 413).

INTERPRETING *t*

You have now learned to perform the calculations that determine differences among groups. What do the results mean? Are they significant? Does it matter whether they are or are not? We answer these questions in this section by explaining the difference between one-tailed and two-tailed *t* tests and discussing the aspects of the *t* test that influence power in research.

One-Tailed Versus Two-Tailed *t* Tests

At this point, refer again to table 5 in the appendix (the table of *t* values to which you compare the values calculated). Remember, you decide the alpha (probability) level (we have been using .05), calculate the degrees of freedom for *t*, and read the table *t* value at the intersection of the column (probability) and row (*df*). If your calculated *t* exceeds the table value, it is significant at the specified alpha and degrees of freedom. This table is used for a **two-tailed *t* test** because we assume that the difference between the two means could favor either mean. Sometimes we might hypothesize that group 1 will be better than group 2 or, at worst, no poorer than group 2. In this case, the test is a **one-tailed *t* test**; that is, it can go in only one direction. Then, when looking at appendix table 5 for a one-tailed *t* test, the .05 level is located in the .10 column for the two-tailed test, and the value for the .01 level for the one-tailed test is in the .02 column for the two-tailed test. Generally, in behavioral research, however, we are not so sure of our results that we can employ the one-tailed *t* table.

two-tailed *t* test—A test that assumes that the difference between the two means could favor either group.

one-tailed *t* test—A test that assumes that the difference between the two means lies in one direction only.

t Tests and Power in Research

In chapter 7, power was defined as the probability of rejecting the null hypothesis when the null hypothesis is false. Obtaining power in research is desirable, because the odds of rejecting a false null hypothesis are increased. The independent *t* test is used here to explain three ways to obtain power (besides setting the alpha level). These ways apply to all types of experimental research.

Consider the formula for the independent *t* test:

$$t = \frac{M_1 - M_2}{\sqrt{\dfrac{s_1^2}{n_1} + \dfrac{s_2^2}{n_2}}} \quad \begin{matrix}(1)\\(2)\\(3)\end{matrix} \qquad\qquad (9.11)$$

Note that we have placed numbers (1, 2, 3) beside the three horizontal levels of this formula. These three levels represent what can be manipulated to increase or decrease power.

The first level ($M_1 - M_2$) gives power if we can increase the difference between M_1 and M_2. You should see that if the second and third levels remain the same, a larger difference between the means increases the size of *t*, which increases the odds of rejecting the null hypothesis and thus increases power. How can the difference between the means be increased? The logical answer is by applying stronger, more concentrated treatments. One example would be to use a 12-week treatment instead of a 6-week treatment to give the treatment a better chance to show its effect. The 12-week treatment should move the means of the experimental and control groups farther apart than the 6-week treatment does. The result will be less overlap in their distributions.

The second level of the independent *t* formula is s_1^2 and s_2^2, or the variances (s^2) for each of the two groups. Recall that the standard deviation represents the spread of the scores about the mean. If this spread becomes smaller (scores distributed more closely about the mean), the variance is also smaller. If the variance term is smaller and the first and the third levels remain the same, the denominator of the *t* test becomes smaller and *t* becomes larger, thus increasing the odds of rejecting the null hypothesis and increasing power. How can the standard deviation and thereby the variance be made smaller? The answer is to apply the treatments more consistently. The more consistently the treatments are applied to each participant, the more similar they become in response to the dependent variable. This groups the distribution more tightly around the means, thus reducing the standard deviation (and thereby the variance).

Finally, the third level (n_1, n_2) is the number of participants in each group. If n_1 and n_2 are increased and the first and second levels remain the same, the denominator becomes smaller (note that n is divided into s^2) and t becomes larger, thus increasing the odds of rejecting the null hypothesis and obtaining power. Obviously, n_1 and n_2 can be increased by placing additional participants in each group.

Of course, power can also be influenced by varying the alpha (i.e., if alpha is set at .10 as opposed to .05, power is increased). But in doing this, we increase the risk of rejecting a true null hypothesis (i.e., we increase the chance of making a type I error).

Keep in mind that power can be influenced by your choice of design and the statistical techniques you use. We point this out as we discuss various statistical tests, but a simple example using tests that we have already addressed makes the point. We indicated that a dependent t test is used when you have either matched pairs or the same participants being measured twice. The dependent t test is used because the independent t assumes that the groups are composed of different participants and that the error term is attributed to sampling error. The dependent t reduces this error. Another way of looking at this is that we increase power by using the dependent t when the situation warrants it. To illustrate, if you were to calculate an independent t test on the scores in example 9.3, you would find the t to be 2.32, whereas the dependent t was 4.83. So even with fewer degrees of freedom for the dependent t, we increased power by obtaining a higher t with that statistical test. We will see that we can also increase power by using techniques such as analysis of covariance and factorial analysis of variance.

> To obtain power (and increase the odds of rejecting a false null hypothesis), consistently administer strong treatments, vary alpha, or use a larger sample size. Power can also be increased by using appropriate research design and statistical analysis.

In summary, obtaining power is desirable because it increases the odds of rejecting a false null hypothesis. Power can be obtained by using strong treatments, administering those treatments consistently, using as many participants as feasible, varying alpha, or using an appropriate research design and statistical analysis. Remember, however, that a second question always arises, even in the most powerful experiments. After the null hypothesis is rejected, the strength (meaningfulness) of the effects must be evaluated.

> A significant t ratio means that true variance significantly exceeds error variance.

The t ratio has a numerator and a denominator. From a theoretical point of view, the numerator is regarded as **true variance**, or the real difference between the means. The denominator is considered **error variance**, or variation about the mean. Thus, the t ratio is

$$t = \frac{\text{true variance}}{\text{error variance}}$$

> **true variance**—The portion of the differences in scores that is (theoretically) real.

where true variance = $M_1 - M_2$ and error variance = $\sqrt{s_1^2/n_1 + s_2^2/n_2}$. If no real differences exist between the groups, true variance = error variance, or the ratio between the two is true variance / error variance = 1.0. When a significant t ratio is found, we are really saying that true variance exceeds error variance to a significant degree. The amount by which the t ratio must exceed 1.0 for significance depends on the number of participants (df) and the alpha level established.

> **error variance**—The portion of the scores that is attributed to participant variability.

The estimate of the strength of the relationship (R^2) between the independent and dependent variables is represented by the ratio of true variance to total variance.

$$R^2 = \frac{\text{true variance}}{\text{total variance}}$$

Thus, R^2 represents the proportion of the total variance that is due to the treatments (true variance).

Effect size is also an estimate of the strength or meaningfulness of the group differences or treatments. Effect size places the difference between the means in standard deviation units, $(M_1 - M_2) / s$. Figure 9.1 shows how the normal distribution of two groups differs for two different effect sizes, 0.5 and 1.0. In figure 9.1*a* the standardized difference between the group means is ES = 0.5; that is, the mean of group 2 falls 0.5 standard deviation to the right of the mean of group 1. (If the mean of group 2 fell to the left, the ES would be –0.5.) Figure 9.1*b* shows the distribution when ES = 1.0; that is, the mean of group 2 is 1 standard deviation to the right of group 1. Notice that there is less overlap between the two groups' distribution of scores in figure 9.1*b* than in figure 9.1*a*. Said another way, when scores on the dependent variable were grouped by the independent variable in figure 9.1*b*, the means were farther apart and there was less distribution overlap than in figure 9.1*a*.

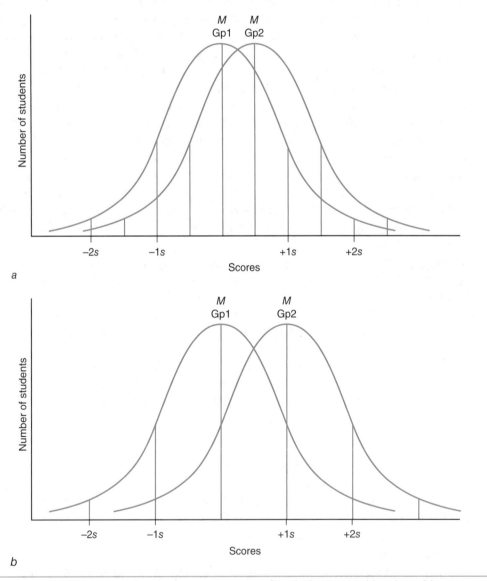

Figure 9.1   Distribution of scores for groups 1 and 2 for which ES = 0.5 *(a)* and 1.0 *(b)*.

Effect size is sometimes interpreted as a percentile change attributed to the treatment. For example, in figure 9.1*b*, the treatment group (group 2) mean was 1.0*s* higher than the control group mean. If we consult table 2 in the appendix, we see that a *z* of 1.00 (which is a distance of 1*s* above the mean) shows that only .1587 (16%) of the scores are higher than

1.0. In other words, the percentile rank for such a score is 84 (100 – 16). Consequently, in interpreting an effect size of 1.0, we can infer that the treatment improved average performance by 34 percentile points (i.e., treatment group = 84, control group = 50; 84 – 50 = 34).

RELATIONSHIP OF *t* AND *r*

As we mentioned previously, our separation of statistical techniques into the two categories of relationships among variables (chapter 8) and differences among groups (this chapter) is artificial because both sets of techniques are based on the general linear model. A brief demonstration with *t* and *r* should make the point. But the idea can be extended into the more sophisticated techniques discussed in chapter 8 and later in this chapter. Example 9.5 shows the relationship between *t* and *r*. Group 1 has a set of scores (dependent variable) for five people, as does group 2.

Example 9.5

Known Values

Group 1		Group 2	
Participant	Dependent variable	Participant	Dependent variable
A	1	F	6
B	2	G	7
C	3	H	8
D	4	I	9
E	5	J	10

Sum of group 1	$\Sigma_1 = 15$
Mean of group 1	$M_1 = 3$
Standard deviation of group 1	$s_1 = 1.58$
Sum of group 2	$\Sigma_2 = 40$
Mean of group 2	$M_2 = 8$
Standard deviation of group 2	$s_2 = 1.58$

Working It Out

1. Conduct an independent *t* test (equation 9.3) and test for significance.

$$ t = \frac{3-8}{\sqrt{\dfrac{1.58^2}{5} + \dfrac{1.58^2}{5}}} = 5.0^* \qquad df = (n_1 + n_2) - 2 = 8 $$
$$ t(8) = 5.0, p < .05 $$

*The sign is not important for this example.

2. Assign each participant a dummy code that stands for his or her group.

Group 1		Group 2	
Participant	Dummy code	Participant	Dummy code
A	1	F	0
B	1	G	0
C	1	H	0
D	1	I	0
E	1	J	0

(continued)

Example 9.5 *(continued)*

3. Apply the correlation formula (equation 8.1). If we treat the dummy-coded variable as X and the dependent variable as Y and ignore group membership (10 participants with two variables), then the correlation formula used earlier can be applied to the data.

$$r = \frac{N\Sigma XY - (\Sigma X)(\Sigma Y)}{\sqrt{N\Sigma X^2 - (\Sigma X)^2}\sqrt{N\Sigma Y^2 - (\Sigma Y)^2}}$$

$$= \frac{10(15) - 5(55)}{\sqrt{10(5) - 25}\sqrt{10(385) - 3{,}025}} = \frac{125}{144} = .87^*$$

*The sign is not important for this example.

4. Apply a t test to r. In this example $t = 5.0$, the same as the t test done on the group means (within rounding error).

$$t = \sqrt{\frac{r^2}{(1-r^2)/(N-2)}} = \sqrt{\frac{.87^2}{(1-.87^2)/(10-2)}} = \sqrt{\frac{.757}{.030}} = 5.0$$

The point is that r represents the relationship between the independent and dependent variables, and a t test can be applied to r (see step 4 in example 9.5) to evaluate the reliability (significance) of the relationship.

There are two sources of variance: true variance and error variance (true variance + error variance = total variance). The t test is the ratio of true variance to error variance, whereas r is the square root of the proportion of total variance accounted for by true variance. To get t from r only means manipulating the variance components in a slightly different way. This is because all parametric correlational and differences-among-groups techniques are based on the general linear model. This result can easily be shown to exist in the more advanced statistical techniques. (For a thorough treatment of this topic, see Pedhazur, 1982.)

You need to understand this basic concept because it is becoming increasingly common for researchers to use regression techniques to analyze experimental data. These data have traditionally been analyzed using the techniques discussed in this chapter. We have, however, demonstrated that what is traditional is not required. What is important is that the data be appropriately analyzed to answer the following questions:

■ Are the groups significantly different?

■ Does the independent variable account for a meaningful proportion of the variance in the dependent variable?

Significance is always evaluated as the ratio of true variance to error variance, whereas the percentage of variance accounted for is always the ratio of true variance to total variance.

ANALYSIS OF VARIANCE

Using t tests is a good way to determine differences between two groups. Often, though, experimenters work with more than two groups. A method of determining differences among them is needed in those cases. This section explains how analysis of variance is used to detect differences among two or more groups.

Simple Analysis of Variance

analysis of variance (ANOVA)—A test that allows the evaluation of the null hypothesis between two or more group means.

The concept of simple (sometimes called one-way but seldom considered simple by graduate students) **analysis of variance (ANOVA)** is an extension of the independent t test. In fact, t is just a special case of simple ANOVA in which there are two groups. Simple ANOVA allows the evaluation of the null hypothesis among two or more group means with the restriction that the groups represent levels of the same independent variable.

We are going to use an example from the PGA golf data (chapter 6, table 6.2) for ANOVA. We can compare three groupings of player performance (based on money won) with the dependent variable of scoring average. Table 9.1 shows the means, standard deviations, and 95% confidence intervals (CIs) for each group. Also displayed is the ANOVA, showing that the three groups are significantly different, $F(2,27) = 24.48$, $p < .0001$. The variance accounted for can be estimated by using $SS_{Between} / SS_{Total} = 5.444 / 8.445 = .64$; in other words, 64% of the variance in scoring average is accounted for by the grouping of players based on money won.

TABLE 9.1

ANOVA Among Three Performance Groups (Top 10, 21–30, and 41–50) With Scoring Average as the Dependent Variable

One-way ANOVA (also called simple ANOVA)

Scoring average	N	Descriptives			95% confidence interval for mean		Minimum	Maximum
		Mean	Standard deviation	Standard error	Lower bound	Upper bound		
1.00	10	69.5270	.30811	.09743	69.3066	69.7474	69.12	70.02
2.00	10	70.1590	.27066	.08559	69.9654	70.3526	69.73	70.59
3.00	10	70.5620	.40661	.12858	70.2711	70.8529	70.08	71.22
Total	30	70.0827	.53964	.09852	69.8812	70.2842	69.12	71.22

ANOVA

Scoring average	Sum of squares	df	MS	F	Significance
Between groups	5.444	2	2.722	24.482	.0001
Within groups	3.002	27	.111		
Total	8.445	29			

Why not do a *t* test between groups 1 and 2, a second *t* test between groups 1 and 3, and a third *t* test between groups 2 and 3? The reason is that this approach would violate an assumption concerning the established alpha level (let it be $p = .05$). The .05 level means that there is 1 chance in 20 of a difference because of sampling error, assuming that the groups of participants on which the statistical tests are done are from independent random samples. In our case, this assumption is not true. Each group has been used in two comparisons (e.g., 1 vs. 2 and 1 vs. 3) rather than only one. Thus, we have increased the chances of making a type I error (i.e., alpha is no longer .05). Making this type of comparison, in which the same group's mean is used more than once, is an example of increasing the experimentwise error rate (discussed later in this chapter). Simple ANOVA allows all three group means to be compared simultaneously, thus keeping alpha at the designated level of .05.

Calculating Simple ANOVA

Table 9.2 provides the formulas for calculating simple ANOVA and the *F* ratio. This method, the so-called ABC method, is simple:

$A = \Sigma X^2$: Square each participant's score, sum these squared scores (regardless of which group the participant is in), and set the total equal to A.

$B = (\Sigma X)^2 / N$: Sum all participants' scores (regardless of group), square the sum, divide by the total number of participants, and set the answer equal to B.

$C = (\Sigma X_1)^2 / n_1 + (\Sigma X_2)^2 / n_2 + \ldots + (\Sigma X_k)^2 / n_k$. Sum all scores in group 1, square the sum, and divide by the number of participants in group 1; do the same for the scores in group 2, and so on for however many groups (k) there are. Then add all the group sums and set the answer to C.

Next, fill in the summary table for ANOVA using A, B, and C. Thus, the between-groups (true variance) **sum of squares** (SS) is equal to C – B; the between-groups degrees of freedom (df) is the number of groups minus one (k – 1); the between-groups variance or mean square (MS_B) is the between-groups sum of squares divided by the between-groups degrees of freedom. The same is done within groups (error variance) and then for the total. The F ratio is MS_B / MS_W (i.e., the ratio of true variance to error variance).

Example 9.6 shows how the formulas in table 9.2 are used. The scores for groups 1, 2, and 3 are the sums of two judges' skill ratings for a particular series of movements. The groups are randomly formed of 15 junior high school students. Group 1 was taught with video, and the teacher made individual corrections while the student viewed the video. Group 2 was taught with video, but the teacher made only general group corrections while students viewed the video. The teacher taught group 3 without the benefit of video. From looking at the formulas in table 9.2, you should be able to see how each number in this example was calculated.

sum of squares—A measure of variability of scores; the sum of the squared deviations from the mean of the scores.

TABLE 9.2

Formulas for Calculating Simple ANOVA

$$A = \sum X^2$$

$$B = \frac{\left(\sum X\right)^2}{N}$$

$$C = \frac{\left(\sum X_1\right)^2}{n_1} + \frac{\left(\sum X_2\right)^2}{n_2} + \ldots + \frac{\left(\sum X_k\right)^2}{n_k}$$

Summary Table for ANOVA				
Source	*SS*	*df*	*MS*	*F*
Between (true)	C – B	k – 1	(C – B) / (k – 1)	MS_B / MS_W
Within (error)	A – C	N – k	(A – C) / (N – k)	
Total	A – B	N – 1		

Note: X = a participant's score, N = total number of participants, n = number of participants in a group, k = number of groups, SS = sum of squares, df = degrees of freedom, MS = mean square.

Example 9.6

Known Values

	Group 1		Group 2		Group 3	
	X	X^2	X	X^2	X	X^2
	12	144	9	81	6	36
	10	100	7	49	7	49
	11	121	6	36	2	4
	7	49	9	81	3	9
	10	100	4	16	2	4
Σ	50	514	35	263	20	102
M	10	—	7	—	4	—

Working It Out

$A = \sum X^2 = 514 + 263 + 102 = 879$

$B = \left(\sum X\right)^2 / N = (50 + 35 + 20)^2 / 15 = (105)^2 / 15 = 11{,}025 / 15 = 735$

$C = \left(\sum X_1\right)^2 / n_1 + \left(\sum X_2\right)^2 / n_2 + \left(\sum X_3\right)^2 / n_3 = (50)^2 / 5 + (35)^2 / 5 + (20)^2 / 5$

$\quad = 2{,}500 / 5 + 1{,}225 / 5 + 400 / 5 = 825$

	Summary Table for ANOVA			
Source	SS	df	MS	F
Between	90	2	45	10.00*
Within	54	12	4.5	
Total	144	14		

*$p < .05$

Your main interest in example 9.6, after you make sure that you understand how the numbers were obtained, is the F ratio of 10.00. Table 6 in the appendix contains F values for the .05 and .01 levels of significance. Although the numbers in the table are obtained the same way as in the t table, you use the table in a slightly different way. The F ratio is obtained by dividing MS_B by MS_W. Note in example 9.6 that the term MS_B (the numerator of the F ratio) has 2 df ($k - 1 = 2$), and the term MS_W (the denominator of the F ratio) has 12 df ($N - k = 12$). Notice also that appendix table 6 has degrees of freedom across the top (numerator) and down the left-hand column (denominator). For our F of 10.00, read down the 2 df column to the 12 df row; there are two numbers where the row and the column intersect. The top number (3.88) in dark print is the table F value for the .05 level, whereas the bottom number (6.93) in light print is the table F value for the .01 level. If our alpha had been established as .05, you can see that our F value of 10.00 is larger than the table value of 3.88 at the .05 level (it is also larger than the value of 6.93 at the .01 level). So our F is significant and could be written in the text of an article as $F(2,12) = 10.00$, $p < .01$ (read "F with 2 and 12 degrees of freedom equals 10.00 and is significant at less than the .01 level").

Buchowski, Darud, Chen, and Sun (1998) used a one-way (simple) ANOVA to compare work efficiency (WE) of aerobics instructors and noninstructors during an aerobics routine and found a significant difference: Instructors had the higher WE, $F(1,22) = 13.01, p < .002$. The APA notation tells us that there were two groups with 24 total participants, because the degrees of freedom in the parentheses (1,22) are first for the between-groups variation ($k - 1$) and then for the within-groups variation ($N - k$). The calculated F was 13.01 and was significant at the $p < .002$ level.

Follow-Up Testing

In our example of ANOVA with three groups, we now know that significant differences exist among the three group means ($M_1 = 10$, $M_2 = 7$, and $M_3 = 4$). But we do not know whether all three groups differ (e.g., whether groups 1 and 2 differ from group 3 but not from each other). So we next perform a follow-up test. One way to do this is to use t tests between groups 1 and 2, groups 1 and 3, and groups 2 and 3. But the same problem that we discussed earlier exists with alpha (type I error is increased). Several follow-up tests protect the experimentwise error rate (type I error, discussed later in this chapter). These methods include those of Scheffé, Tukey, Newman-Keuls, Duncan, and several others (see Toothaker, 1991, for both conceptual explanations and calculations of the various techniques). Each test is calculated in a slightly different way, but they all are conceptually similar to the t test in that they identify which pairs of groups differ from each other. The Scheffé method is the most conservative (which means that it identifies fewer significant differences), followed by Tukey's method. Duncan is the most liberal, identifying more significant differences. Newman-Keuls' method falls between the extremes but has some other problems.

For our purpose, one example should suffice, so we explain the use of the Scheffé method of making multiple comparisons among means. Scheffé (1953) is the most widely recognized of all multiple-comparison techniques (Toothaker, 1991). Because we believe that researchers should be conservative when using behavioral data, we generally recommend the Scheffé method for follow-up testing. Other multiple-comparison techniques, however, are appropriate for various situations.

The Scheffé technique has a constant critical value for the follow-up comparison of all means when the F ratio from a simple ANOVA (or a main effect in a factorial ANOVA, discussed next) is significant. Scheffé controls type I error (alpha inflation) for any number of appropriate comparisons. Equation 9.12 is the formula for the critical value (CV, required size of the difference) for significance using the Scheffé technique:

$$CV = \sqrt{(k-1)F_{\alpha;\, k-1,\, N-k}}\, \sqrt{2\left(MS_w\, /\, n\right)} \tag{9.12}$$

where k = the number of means to be compared, F_α = the table F ratio (from appendix table 6) for the selected alpha (e.g., .05) given dfs of $k - 1$ (for between groups) and $N - k$ (for within groups). For the second half of the equation, MS_w is obtained from the ANOVA summary table (SS_w / df_w), and n = the number in a single group.

Example 9.7 shows the Scheffé technique applied to the previous simple ANOVA shown in example 9.6. You find the F ratio in appendix table 6 at the intersection of the column df = 2 ($k - 1 = 3 - 1 = 2$) and the df = 12 row ($N - k = 15 - 3 = 12$) from the dfs in the summary ANOVA in table 9.1; the table F ratio is 3.88. Solving the left part of formula 9.12 in example 9.7, we find that the Scheffé value for significance at the .05 level is 2.79. For the second part of formula 9.12, we obtain $MS_w = 4.5$ (from table 9.1) and the n for each group is 5. Thus, $\sqrt{2(4.5\, /\, 5)} = 1.34$. Hence, formula 9.12 becomes

CV = (2.79)(1.34) = 3.74

This value is the size of the difference between means needed in order to be significant at $p < .05$.

Example 9.7 arranges the means from highest (10) to lowest (4). The observed difference for each of the comparisons is computed; then each observed difference is compared with the needed difference (CV) for significance. By following these steps, you can see that only the comparison between groups 1 and 3 was significant at $p < .05$. We could conclude that the techniques used in group 1 were significantly better than those used in group 3 but not significantly better than those used in group 2. Also, the techniques used in group 2 were not significantly better than those used in group 3.

Example 9.7

Working It Out

1. Order the means from highest (10) to lowest (4).
2. Compute the differences between all the means.

Group 1 ($M = 10$)	Group 2 ($M = 7$)	Group 3 ($M = 4$)	Observed difference
10	7		3
10		4	6
	7	4	3

3. Compute the critical value (difference needed) for significance at alpha $= .05$ (equation 9.11):

$$CV = \sqrt{(k-1)F_{\alpha;\, k-1,\, N-k}}\, \sqrt{2\left(MS_w / n\right)}$$

$$= \sqrt{(3-1)3.88}\, \sqrt{2(4.5/5)}$$

$$= (2.79)(1.34) = 3.74$$

4. Compare the CV of 3.74 with each observed difference to determine which difference or differences are significant. In this case, only the difference between groups 1 and 3 is as large as or larger than the CV of 3.74.

The researcher may also use **planned comparisons** to test for differences among groups. Planned comparisons are planned (i.e., testable hypotheses are developed) before the experiment. Thus, an experimenter might postulate a test between two groups before the experiment because in theory this comparison is important and should be significant. But the number of planned comparisons in an experiment should be small ($k - 1$, where k = the number of groups or treatment levels of an independent variable) relative to the total number of possible comparisons.

planned comparison—A comparison among groups that is planned before the experiment, rather than as a follow-up of a test such as ANOVA.

Determining the Meaningfulness of Results

Now that we know that F is significant for our example study and have followed it up to see which groups differ, we should answer our second question: What percentage of variance is accounted for by our treatments, or how meaningful are our results? One way to get a quick idea is to refer to table 9.1 and divide true variance by total variance: (SS_{true} / SS_{Total}) = 90 / 144 = .625, or 62.5% of the variance is accounted for by the treatments. Although this method is sufficient for a quick estimate, it is biased. The more accurate way is to use the following formula from Tolson (1980):

$$\omega^2 = \frac{\left[F(k-1)\right] - (k-1)}{\left[F(k-1)\right] + (N-k) + 1} \tag{9.13}$$

where F = the F ratio, k = the number of groups, and N = the total number of participants. If we do this with the data from example 9.6, we have

$$\omega^2 = \frac{\left[10.00(3-1)\right]-(3-1)}{10.00(3-1)+(15-3)+1} = \frac{18}{33} = .545$$

Thus, ω^2 indicates that 54.5% of the total variance is accounted for by the treatments.

Putting our statistics together, we could say that the treatment was significant, $F(2,12) = 10.00$, $p < .01$, and accounted for a meaningful proportion of the variance, $\omega^2 = 54.5\%$. In addition, a follow-up Scheffé test indicates that group 1 had the best performance and was significantly different ($p < .05$) from group 3. However, group 1 was not significantly different from group 2, nor was group 2 significantly different from group 3.

Summarizing Simple ANOVA

One final point to recall before leaving our discussion of simple ANOVA is that t is a special case of F when there are only two levels of the independent variable (two groups). In fact, this relationship ($t^2 = F$) is exact, within rounding error (see example 9.8).

Example 9.8

Known Values

	Control group			Experimental group		
	X	$(X-M)^2$	X^2	X	$(X-M)^2$	X^2
	2	.09	4	8	4	64
	4	2.89	16	7	1	49
	3	.49	9	5	1	25
	3	.49	9	4	4	16
	2	.09	4	7	1	49
	1	1.69	1	5	1	25
	1	1.69	1	6	0	36
Σ	16	7.43	44	42	12	264
M	2.3	—	—	6.0	—	—
S	1.11	—	—	1.41	—	—

Working It Out

1. Find t (equation 9.3).

$$t = \frac{M_1 - M_2}{\sqrt{\left(s_1^2/n_1\right)+\left(s_2^2/n_2\right)}} = \frac{2.3-6}{\sqrt{\left(1.11^2/7\right)+\left(1.41^2/7\right)}} = \frac{3.7}{0.68} = 5.44^*$$

*The sign is not important to report.

2. Compute the simple ANOVA and complete the summary table.

$$A = \sum X^2 = 44 + 264 = 308$$

$$B = \left(\sum X\right)^2/N = (58)^2/14 = 240.29$$

$$C = \left(\sum X_1\right)^2/n_1 + \left(\sum X_2\right)^2/n_2 = (16)^2/7 + (42)^2/7 = 36.57 + 252 = 288.57$$

	Summary Table for ANOVA			
Source	SS	df	MS	F
Between	48.28	1	48.28	29.80
Within	19.43	12	1.62	
Total	67.71	13		

$t^2 = F$ $(5.44)^2 = 29.6 \cong 29.8$ (within rounding error)

In other words, there is little need for t because F will handle two or more groups. But t remains in use because it was developed first, being the simplest case of F. Also, t is convenient and can be computed by hand because M and s are already calculated for each group to serve as descriptive statistics.

Factorial ANOVA

Up to this point, examples of two (t) or more levels (simple ANOVA) of one independent variable have been discussed. In fact, in our examples so far, all other independent variables have been controlled except the single independent variable to be manipulated and its effect on a dependent variable. This is called the law of the independent variable. But, in fact, we can manipulate more than one independent variable and statistically evaluate the effects on a dependent variable. This procedure is called **factorial ANOVA**, which is used when there is more than one factor, or independent variable. As you read the literature, you will see that many more studies use factorial ANOVA than use one-way, or simple, ANOVA. Theoretically, a factorial ANOVA may have any number of factors (two or more) and any number of levels within a factor (two or more), but we seldom encounter ANOVA with more than three factors. This is another good place to apply the KISS principle (Keep it simple, stupid).

factorial ANOVA—An analysis of variance in which there is more than one independent variable.

Components of Factorial ANOVA

For the purpose of explaining, we consider the simplest form of the factorial ANOVA, which uses only two independent variables and only two levels of each variable. This design, which is designated a 2×2 ANOVA, has two main effects and one interaction. **Main effects** are tests of each independent variable when the other is disregarded (and controlled). Look at table 9.3 and note that the first independent variable (IV_1) has two levels, labeled A_1 and A_2. In our example this IV represents the intensity of training: high intensity and low intensity. The second independent variable (IV_2) represents the level of fitness of the participants: low fitness (B_1) and high fitness (B_2). The numbers in the cells represent mean attitude scores of each of the four groups toward the type of training (high intensity or low intensity). Thus, we have one group of high-fitness people who trained under a high-intensity program and one group of high-fitness people who trained with a low-intensity program. The low-fitness participants were also split into the two intensity programs.

main effects—Tests of each independent variable when all other independent variables are held constant.

TABLE 9.3

Factorial (2 × 2) ANOVA Model

		IV$_2$ (fitness)		
		B_1 (low fit)	B_2 (high fit)	
IV$_1$ (intensity)	A_1 (high int)	$M = 10$	$= 40$	$M_{A1} = 25$
	A_2 (low int)	$M = 30$	$M = 20$	$M_{A2} = 25$
		$M_{B1} = 20$	$M_{B2} = 30$	

Summary table for ANOVA				
Source	SS*	df	MS	F
A (intensity)	—	$A - 1 = 2 - 1 = 1$	SS_A / df_A	MS_A / MS_{error}
B (fitness)	—	$B - 1 = 2 - 1 = 1$	SS_B / df_B	MS_B / MS_{error}
AB (interaction)	—	$(A - 1)(B - 1) = 1$	SS_{AB} / df_{AB}	MS_{AB} / MS_{error}
Error	—	$(N - 1) - [(A - 1) + (B - 1) + (A - 1)(B - 1)]$	SS_{error} / df_{error}	
Total	—	$N - 1$		

*For simplicity, formulas for computing sums of square are not given here.

We can test the main effect of IV_1 (level of intensity) by comparing the row means (M_{A1} and M_{A2}), because IV_2 (B_1 and B_2)—fitness level of the participants—is equally represented at each level of A. Actually, a sum of squares with $A - 1$ df and mean squares for the between-group and within-group sources of variance are computed. Thus, we can compute an F for main effect A (intensity).

The same holds true for IV_2, fitness level, by comparing the column means (M_{B1} and M_{B2}). You can see that the two levels of A (IV_1), intensity of training, are equally represented in the two levels of B. Therefore, the main effect of the attitudes of high- and low-fitness participants toward exercise can be tested by the F ratio for B.

In a study of this type, the main interest usually lies in the interaction. We want to know whether attitude toward exercise programs of high or low intensity (factor A) depends on (is influenced by) the fitness level of the participants (factor B). This effect is tested by an F ratio for the interaction ($A \times B$) of the two IVs, which evaluates the four cell means: M_{A1B1}, M_{A1B2}, M_{A2B1}, and M_{A2B2}. Unless some special circumstance exists, interest in testing the main effects is usually limited by a significant interaction, which means that what happens in one independent variable depends on the level of the other. Thus, normally it makes little sense to evaluate main effects when the interaction is significant.

This particular factorial ANOVA is labeled a 2 (intensity of training) × 2 (level of fitness) ANOVA (read "2-by-2 ANOVA"). The true variance can be divided into three parts:

- True variance because of A (intensity of training)
- True variance because of B (level of fitness)
- True variance because of the interaction of A and B

Each of these true variance components is tested against (divided by) error variance to form the three F ratios for this ANOVA. Each of these Fs has its own set of degrees of freedom so that it can be checked for significance in the F table (table 6 in the appendix).

Main effects (of more than two levels) can be followed up with a Scheffé test, but no follow-up is required for an independent variable with only two levels. If the F is significant, you only have to see which has the higher M. If the F for the interaction is significant, meaning that attitude toward high- or low-intensity training depends on the fitness level of the participants, it will be reflected in a plot, as shown in figure 9.2. You can see in table 9.3 that there is no mean difference for the main effect of intensity of training (both A_1 and A_2 are the same, $M = 25$). The plot in figure 9.2, however, reflects a significant interaction, because the mean attitude scores of the low-fitness group toward the low-intensity training was higher than their attitude toward high-intensity training, whereas the opposite was shown for the high-fitness participants. They preferred the high-intensity program.

This example shows how power is increased by using a particular type of statistical test. If we had just used a t test or simple ANOVA, we would not have found any difference in attitude toward the two levels of intensity (both means were identical, $M = 25$). When we added another factor (fitness level), however, we were able to discern differences in attitude that were dependent on the level of fitness of the participants.

As you can see, all we have done to follow up the significant interaction is to describe the plot in figure 9.2 verbally. Considerable disagreement exists among researchers about how to follow up significant interactions. Some researchers use a multiple-comparison-of-means test (such as Scheffé) to contrast the interaction cell means. These multiple-comparison tests, however, were developed

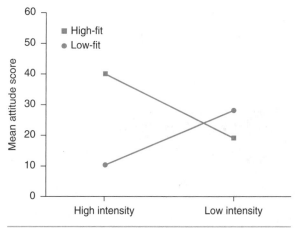

Figure 9.2   Plot of the interaction for a two-way factorial ANOVA.

for contrasting levels within an independent variable and not for cell means across two or more independent variables; thus, their use may be inappropriate. Other researchers use a follow-up called a test of simple main effects. If an interaction is to be followed up with a statistical test, using simple main effects via the Scheffé procedure is the preferred technique.

Our preference for evaluating an interaction is to do as we have done previously—plot the interaction and describe it. This takes into account the true nature of an interaction; that is, what happens in one independent variable depends on the other. But you are likely to encounter all these ways of testing interactions (and probably some others) as you read the research literature. Just remember that the researcher is trying to show you how the two or more independent variables interact. Also remember that follow-ups on main effects are usually unnecessary (or at least the interpretation must be qualified) when the interaction is significant.

Figure 9.3 shows a nonsignificant interaction. In this case, both groups preferred the same type of program over the other; hence, the lines are parallel. Significant interactions show deviations from parallel (as was shown in figure 9.2). The lines do not have to cross to reflect a significant interaction. Figure 9.4 shows a significant interaction in which those in the high-fitness group liked both forms of exercise equally, but there was a decided difference in preference in the members of the low-fitness group, who preferred the low-intensity program over the high-intensity program.

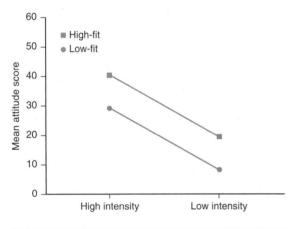

Figure 9.3 Plot of a nonsignificant interaction (parallel lines) in which both fitness groups prefer low-intensity programs over high-intensity programs.

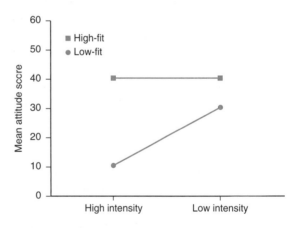

Figure 9.4 Plot of a significant interaction of no difference in high-fit people's preferences but a significant difference in intensity preference among low-fit people.

Determining the Meaningfulness of Factorial Analysis Results

Having answered the first statistical question about a factorial ANOVA (are the effects significant?), we now turn to the meaningfulness of the effects, or what percentage of the variance in the dependent variable is accounted for by the independent variables and their interaction? The three formulas that follow (from Tolson, 1980) provide the test for each component of ANOVA:

$$\omega_A^2 = \frac{(p-1)F_A - (p-1)}{(p-1)F_A + (q-1)F_B + (p-1)(q-1)F_{AB} + (N-pq)+1}$$

$$\omega_B^2 = \frac{(q-1)F_B - (q-1)}{(p-1)F_A + (q-1)F_B + (p-1)(q-1)F_{AB} + (N-pq)+1} \quad (9.14)$$

$$\omega_{AB}^2 = \frac{(p-1)(q-1)F_{AB} - (p-1)(q-1)}{(p-1)F_A + (q-1)F_B + (p-1)(q-1)F_{AB} + (N-pq)+1}$$

where p = the number of levels of A and q = the number of levels of B. These formulas involve the proportion of true variance to total variance. Note that the total variance (the denominator) is the same for all three proportions.

Each ω^2 represents the percentage of variance accounted for by that component of the ANOVA model. The ω^2_s can be summed to estimate the percentage of the total variance that is true variance. Of course, effect sizes can be calculated according to the previous equations (9.6 and 9.7) for any two means for which you want to compare main effects.

Summarizing Factorial ANOVA

Frequently, the levels of an independent variable are categorical (also called classified) rather than random. For example, we could do a simple ANOVA in which the levels (or groups) of the independent variable involve novice and expert tennis players, or the attitude toward intensity of training held by high-fitness participants as compared with low-fitness participants, as in our example in table 9.3. That is, we are interested in whether preference toward type of training programs affected high-fitness and low-fitness participants differently. Participants are randomly assigned to the levels of the first independent variable (intensity level of program); they cannot be randomly assigned to the categorical variable of fitness level. Common categorical variables are sex and age. A study might look at the effects of the levels of a treatment (independent variable) to which male and female, or 6-, 9-, and 12-year-old, participants are randomly assigned. Clearly, the interest in this type of factorial ANOVA is the following interaction: Does the effectiveness of treatment differ according to the sex (or age) of the participant?

Examples of factorial ANOVA are plentiful in the research literature. McPherson (1999) conducted a study in which she compared expert and novice tennis players from three age groups on various performance skills. The data were analyzed with 3 (age levels) × 2 (expertise levels) factorial ANOVA. Omega squared (ω^2) was used to interpret meaningfulness. The following is a sample description of the results of the factorial ANOVA: "Analysis of forceful game executions indicated significant main effects for age, $F(2,46) = 7.6$, $p < .002$, $\omega^2 = .10$, and expertise, $F(1,46) = 93.0$, $p < .00001$, $\omega^2 = .48$. The interaction, $F(2,46) = 0.9$, $p > .05$, was not significant" (p. 238). The significant main effects for age were followed up using the Scheffé method.

Repeated-Measures ANOVA

Much of the research in the study of physical activity involves studies that measure the same dependent variable more than once. For example, a study in sport psychology might investigate whether athletes' **state anxiety** (how nervous they feel at the time) differs before and after a game. Thus, a state-anxiety inventory would be given just before and immediately after the game. The question of interest is Does state anxiety change significantly after the game? A dependent t test could be used to see whether significant changes occurred. This is the simplest case of **repeated-measures ANOVA**.

Another study might measure participants on the dependent variable on several occasions. Suppose we want to know whether the distance that children (between ages 6 and 8) throw a ball using an overhand pattern increases over time. We decide to measure the children's distance throw every 3 months during the 2 years of the study. Thus, each child's throwing distance was assessed nine times. We now have nine repeated measures (at the beginning and in eight 3-month intervals) on the same children. We would use a simple ANOVA with repeated measures on a single factor. Basically, the repeated measures are used as nine levels of the independent variable, which is time (24 months).

The most frequent use of repeated measures involves a factorial ANOVA in which one or more of the factors (independent variables) are repeated measures. An example is an investigation of the effects of knowledge of results (KR) on skilled motor performance. Three

state anxiety—An immediate emotional state of apprehension and tension in response to a specific situation.

repeated-measures ANOVA—An analysis of scores for the same individuals on successive occasions, such as a series of test trials; also called *split-plot ANOVA or subject × trials ANOVA*.

groups of participants (three levels of the independent variable, KR) receive different types of KR (no KR, whether they were short or long of the target, and the number of centimeters they were short or long of the target). The task is to position a handle that slides back and forth on a track (called a linear slide) as close to the target as possible. But the participants are blindfolded, so they cannot see the target. They have only verbal KR to correct their estimates of where the target is on the track.

In this type of study, participants are usually given multiple trials (assume 30 trials in this example) so that the effects of the type of KR can be judged. The score on each trial is distance from the target in centimeters. This type of study is frequently analyzed as a two-way factorial ANOVA with repeated measures on the second factor. Thus, a 3 (levels of KR) × 30 (trials) ANOVA with distance error as the dependent variable is used to analyze the data. The first independent variable (level of KR) is a true one (three groups are randomly formed). The second independent variable (30 trials) is repeated measures. Sometimes this ANOVA is called a two-way factorial ANOVA with one between-subjects factor (levels of KR) and one within-subjects factor (30 repeated trials). Although an F ratio is calculated for each independent variable, the major focus is usually on the interaction. For example, do the groups change at different rates across the trials?

> Repeated-measures designs identify and separate individual differences from the error term, increase power, require fewer participants, and study a phenomenon across time.

Advantages of Repeated-Measures ANOVA

Repeated-measures designs have three advantages (Pedhazur, 1982). First, they provide the experimenter the opportunity to control for individual differences among participants, probably the largest source of variation in most studies. In between-subjects designs (completely randomized), the variation among people goes into the error term. Of course, this tends to reduce the F ratio unless it is offset by a large N. Remember that the error term of the F ratio is calculated by dividing the variation among participants by the degrees of freedom (based on the number of participants). In repeated-measures designs, variation from individual differences can be identified and separated from the error term, thereby reducing it and increasing power. As you can deduce from the first advantage, repeated-measures designs

Significant differences may be in the eye of the beholder.

are more economical because fewer participants are required. Finally, repeated-measures designs allow the study of a phenomenon across time. This feature is particularly important in studies of change in, for example, learning, fatigue, forgetting, performance, and aging.

Problems of Repeated-Measures ANOVA

Several problems adversely affect repeated-measures designs, including the following:

- Carryover effects. Treatments given earlier influence treatments given later.
- Practice effects. Participants improve at the task (dependent variables) as a result of repeated trials in addition to the treatment (also called the testing effect).
- Fatigue. Participants' performance is adversely influenced by fatigue (or boredom).
- Sensitization. Participants' awareness of the treatment is heightened because of repeated exposure.

Note that some of these problems may be the variables of interest in repeated-measures designs. Carryover effects may be of interest to a researcher of learning, whereas increased fatigue over trials may intrigue an exercise physiologist.

The tricky part of repeated-measures designs involves how to analyze the data statistically. We have already mentioned ANOVA models with repeated measures. Unfortunately, repeated-measures ANOVA has an assumption beyond the ones we have given for all previous techniques. This assumption is called **sphericity**: The repeated measures, "when transformed by a set of orthonormal weights, are uncorrelated with each other and have equal variances" (Schutz & Gessaroli, 1987, p. 134). If the design has a between-subjects factor, the pooled data (across all participants) must exhibit sphericity. How well the data meet these assumptions is best estimated by a statistic called epsilon (ε). Epsilon ranges from 1.0 (perfect sphericity, the assumption is met) to 0.0 (complete violation). An epsilon above .75 is desirable in repeated-measures experiments. Most of the widely used statistical packages (SPSS and SAS) have repeated-measures functions that provide both estimates of epsilon and tests that should be used to evaluate the F ratio for the repeated-measures factors in the designs. The failure to meet the sphericity assumption results in an increase in type I error; that is, the alpha level may be considerably larger than the researcher intended. Several statisticians (Davidson, 1972; Harris, 1985; Morrow & Frankiewicz, 1979) have suggested that multivariate techniques are more appropriate methods of analysis. Two additional points are important (Pedhazur, 1982):

- When the assumptions are met, ANOVA with repeated measures is more powerful than the multivariate tests.
- If the number of participants is small, only the ANOVA repeated-measures test can be used.

If you are contemplating conducting a study that uses a repeated-measures design, an additional source of reading should be helpful to you. Schutz and Gessaroli (1987) provide a tutorial on the use of repeated measures for univariate and multivariate data that gives a sound rationale for making decisions as well as specific examples of how the data may be analyzed and evaluated.

Components of Repeated-Measures ANOVA

Table 9.4 provides the sources of variation in a one-way ANOVA with repeated measures. This kind of analysis is also called subject × trials ANOVA, within-subjects ANOVA, or two-way ANOVA with one subject per cell.

The values we used to arrive at the statistics in table 9.4 are the same as in example 9.6. Recall that in example 9.6, 15 participants were randomly assigned to treatment group 1,

sphericity—An assumption that repeated measures are uncorrelated and have equal variance.

2, or 3. Suppose that, in a different study, five participants were each measured on three trials of a task, (or were given different treatments) and we coincidentally got the same 15 measurements as in example 9.6. In effect, rather than having 15 participants in groups 1, 2, and 3, we now have 5 participants who were measured on trials 1, 2, and 3 (or treatments 1, 2, and 3). The known values for table 9.4 would look identical to those in example 9.6, except that the columns could be labeled Trial 1, Trial 2, and Trial 3 (or Treatments 1, 2, and 3) instead of Group 1, Group 2, and Group 3. We are using the same 15 values so that when table 9.4 and example 9.6 are compared, you can see why the repeated-measures design results in increased economy. Notice that the sum of squares and degrees of freedom are the same. Also, the between-groups effect in example 9.6 and trials effect in table 9.4 are the same. But the sum of squares for the within-group effect (error) in example 9.6 is divided into two components in the repeated-measures analysis in table 9.4. The residual effect (estimating error) has a sum of squares of 28 with 8 df, and the subjects effect has a sum of squares of 26 with 4 df. This results in an F ratio for the repeated-measures ANOVA (in table 9.4) that is larger than the F ratio for the simple ANOVA (in example 9.6), despite the repeated-measures ANOVA having only one-third as many participants. The sum of squares for subjects in table 9.4 is not tested and simply represents the normal variation among subjects. Thus, across the three trials, the participants' mean performance decreased significantly (trial 1 = 10, trial 2 = 7, and trial 3 = 4). This analysis has all the strengths and weaknesses of repeated-measures designs previously discussed.

TABLE 9.4

Summary of Repeated-Measures ANOVA

Source	Summary table for ANOVA			
Source	SS	df	MS	F
Trials (T)	90	$(T-1) = 2$	45.0	12.86*
Subjects (S)	26	$(S-1) = 4$	6.5	
Residual (error)	28	$(S-1)(T-1) = 8$	3.5	
Total	144	$N-1 = 14$		

*$p < .05$.

Note: $\varepsilon = 1.00$; no adjustment of dfs needed.

The epsilon estimate from this analysis is shown at the bottom of table 9.4. This estimate was obtained by running a repeated-measures computer program from SPSS. If epsilon is less than 1.00, the degrees of freedom is adjusted according to the formula $df \times \varepsilon =$ adjusted df. This is done for the degrees of freedom in both the numerator and the denominator of any F ratio that includes the repeated-measures factor.

A conservative approach to this adjustment (called the **Geisser/Greenhouse correction**) can be done as follows (Stamm & Safrit, 1975):

$$\theta = 1\,(k-1) \tag{9.15}$$

where $k =$ the number of repeated measures. Then we multiply the degrees of freedom for trials by this value, $\theta\,(k-1)$, as well as the degrees of freedom for error, $\theta\,(n-1)(k-1)$. These adjusted degrees of freedom (to the nearest whole df) are used to look up the F ratio in the F table. If the F ratio is significant under the conservative test, then the effect likely is a real one. This procedure was advocated several years ago, and with the tests of repeated-measures designs now available, a computer program that provides the epsilon estimate is more appropriate.

Geisser/Greenhouse correction—A conservative approach to the adjustment of the epsilon estimate in repeated-measures ANOVA that calculates adjusted degrees of freedom to find an F ratio to determine significance.

It is probably safe to say that most ANOVA models in experimental research reported in national journals in our field use factorial ANOVA with repeated measures on one or more of the factors. An example of such a study is by Weiss, McCullagh, Smith, and Berlant (1998), who examined the role of peer mastery and coping models on children's swimming skills, fear, and self-efficacy. If the sphericity assumption was violated ($\varepsilon < .75$), the authors used a multivariate repeated-measures method. If sphericity was not violated, they used a 3 (peer model type) × 3 (assessment period) factorial ANOVA with repeated measures on the last factor.

ANALYSIS OF COVARIANCE

analysis of covariance (ANCOVA)—A combination of regression and ANOVA that statistically adjusts the dependent variable for some distractor variable called the *covariate*.

covariate—A distractor variable that is statistically controlled in ANCOVA and MANCOVA.

Analysis of covariance (ANCOVA) is a combination of regression and ANOVA. The technique is used to adjust the dependent variable for some distractor variable (called the **covariate**), which is some variable that could affect the treatments.

Using ANCOVA

Suppose we want to evaluate the effects of a training program to develop leg power on the time required to run 50 m. We know that reaction time (RT) influences the 50 m dash time because those who begin more quickly after the start signal have an advantage. We form two groups, measure RT in each group, train one with our power development program, use the other as a control group, and measure each participant's time for running 50 m. ANCOVA might be used in this study to analyze the data. There is an independent variable with two levels (power training and control), a dependent variable (50 m dash time), and an important distractor variable, or covariate (RT).

Analysis of a covariance is a two-step process in which an adjustment is first made for the 50 m dash score of each runner according to his or her RT. A correlation (r) is calculated between RT and the time in the 50 m dash. The resulting prediction equation, 50 m dash time $= a + (b)$RT (the familiar formula $a + bX$), is used to calculate each runner's predicted 50 m dash time (Y'). The difference between the actual 50 m dash time (Y) and the predicted time (Y') is called the residual ($Y - Y'$). A simple ANOVA is then calculated using each participant's residual score as the dependent variable (1 df for within-group sums of squares is lost because of the correlation). This allows an evaluation of 50 m dash speed with RT controlled.

Analysis of covariance can be used in factorial situations and with more than one covariate. The results are evaluated as in ANOVA except that one or more distractor variables are controlled. Also, ANCOVA is frequently used in situations in which a pretest is given, some treatment is applied, and then a posttest is given. The pretest is used as the covariate in this type of analysis. Note that in the preceding section on repeated-measures ANOVA, we indicated that this same situation could be analyzed by repeated measures. In addition, ANCOVA is used when comparing intact groups because the groups' performances (dependent variable) can be adjusted for distractor variables (covariates) on which they differ. If ANCOVA is warranted, it can increase the power of the F test (making it easier to detect a difference).

Limitations of ANCOVA

Although ANCOVA may seem to be the answer to many problems, it does have limitations. In particular, its use to adjust final performance for initial differences can result in misleading interpretations (Lord, 1969). In addition, if the correlations between the covariate and the dependent variable are not equal across the treatment groups, standard ANCOVA (there are nonstandard ANCOVA techniques) is inappropriate.

Turner and Martinek (1999) tested the validity of a technique called the games for understanding model in teaching field hockey. One group of participants was taught with this method, one group was taught with a technique approach to instruction, and one group served as a control group. For some of the dependent variables (e.g., ball control and passing decision making), ANCOVA was deemed appropriate with the pretest as the covariate.

EXPERIMENTWISE ERROR RATE

Sometimes researchers make several comparisons of dependent variables using the same participants. Usually, a multivariate technique (discussed in the next section) is the appropriate solution, but when dependent variables are combinations of other dependent variables (e.g., cardiac output is heart rate × stroke volume), a multivariate model using all three dependent variables is inappropriate. (This book is not the appropriate place to explain why. For more detail, see Thomas, 1977.) Thus, an ANOVA among three groups might be calculated separately for each dependent measure (i.e., three ANOVAs). The problem is that this procedure results in increasing the alpha that has been established for the experiment. One of two solutions is appropriate for adjusting alpha. The first, called the Bonferroni technique, simply divides the alpha level by the number of comparisons to be made:

$$\alpha_{EW} = \alpha / c \tag{9.16}$$

where α_W = alpha corrected for the experimentwise error rate, α = alpha, and c = the number of comparisons. If, for example, $\alpha = .05$ and $c = 3$, then the alpha for each comparison is .05 / 3 = .017. This means that the F ratio would have to reach an alpha of .017 to be declared significant.

Turner and Martinek (1999) used the Bonferroni adjustment for establishing significance for each of their dependent variables. Sachtleben and colleagues (1997) made seven comparisons using the t test. The authors used the Bonferroni procedure to adjust the alpha level. With alpha at .05, the adjusted alpha was .05 / 7 = .007. Consequently, to be significant, the t tests had to reach $\alpha < .007$.

The second option is to leave the overall alpha at .05 but to calculate the upper limit that the alpha might be:

$$\alpha_{UL} = 1 - (1 - \alpha)^k \tag{9.17}$$

where α_{UL} = alpha (upper limit) and k = the number of groups. Again, using the example with three comparisons and alpha at .05, $\alpha_{UL} = 1 - (1 - .05)^3 = .14$. Thus, the hypotheses are really being tested somewhere between an alpha of .05 if the dependent variables are perfectly correlated and an alpha of .14 if they are independent. When researchers make multiple comparisons using the same participants, they should either adjust alpha to the experimentwise error rate or at least report the upper limit on their alpha.

UNDERSTANDING MULTIVARIATE TECHNIQUES

Up to this point we have discussed experimental research examples involving one or more independent variables but only one dependent variable. Multivariate cases have one or more independent variables and two or more dependent variables. For example, it seems likely that when independent variables are manipulated, they influence more than one thing. The multivariate case allows more than one dependent variable. Using techniques that allow only one dependent variable (called univariate techniques) repeatedly when several dependent variables are present increases the experimentwise error rate (sometimes called probability pyramiding) in the same way as doing multiple t tests instead of simple ANOVA when there are more than two groups. But in some instances, using univariate techniques in a study that has multiple dependent variables is acceptable or the only choice. For example, as with all research, theory should drive decision making. You might not want to include a theoretically important dependent variable in a multivariate analysis when less theoretically important dependent variables could disguise its importance. Also, a multivariate technique is sometimes just not feasible because the study has a small number of participants. Following are multivariate techniques often used in experimental studies:

- Discriminant analysis
- Multivariate ANOVA (MANOVA) and special cases with repeated-measures designs
- Multivariate ANCOVA (MANCOVA)

Remember that the general linear model still underlies all the techniques, and we are still attempting to learn two things: Are we evaluating something significant (reliable)? How meaningful are significant findings?

Discriminant Analysis

We use discriminant analysis when we have one independent variable (two or more levels) and two or more dependent variables. The technique combines multiple regression and simple ANOVA. In effect, discriminant analysis uses a combination of the dependent variables to predict or discriminate among the levels of the independent variable, which in this case is group membership. In the discussion of multiple regression in chapter 8, we used several predictor variables in a linear combination to predict a criterion variable. In essence, discriminant analysis does the same thing except that several dependent variables are used in a linear combination to predict the group to which a participant belongs. This prediction of group membership is the equivalent of discriminating among the groups (recall how t could be calculated by r). The same methods used in multiple regression to identify the important predictors are used in discriminant analysis. These include forward, backward, and stepwise selection techniques. For greater detail, as well as a useful and practical description of discriminant analysis, read Betz (1987).

Forward, Backward, and Stepwise Selection

As mentioned in chapter 8, the forward selection technique enters the dependent variables in the order of their importance; that is, the dependent variable that contributes the most to separation of the groups (discriminates among or predicts group membership the best) is entered first. By correlation techniques, the effect of the first dependent variable on all others is removed, and the dependent variable that contributes the next greatest amount to the separation of the groups is entered at step 2. This procedure continues until all dependent variables have been entered or until some criterion for stopping the process (established by the researcher) is met.

The backward selection procedure is similar except that all the dependent variables are entered and the one contributing the least to group separation is removed. This continues until the only variables remaining are those that contribute significantly to the separation of the groups.

The stepwise technique is similar to forward selection except that at each step all the dependent variables are evaluated to see whether each still contributes to group separation. If a dependent variable does not contribute, it is stepped out (removed) from the linear combination, just as in multiple regression.

Example of Discriminant Analysis

Tew and Wood (1980) conducted a study in which American university varsity football players were classified into three groups: (a) offensive and defensive backs, (b) offensive and defensive linemen, and (c) linebackers and receivers. Data were collected for 28 athletes in each group ($N = 84$) on seven variables: 40 yd dash, 12 min run, shuttle run, vertical jump, standing long jump, bench press, and squat. Discriminant analysis was applied to determine how many of the seven dependent variables were needed to separate (predict) the three groups of football players.

Two criteria were set for the statistical computer program: to include the variable with the largest F ratio at each step and to stop when no remaining variable had an F significant at $p < .05$. The bench press was entered first because it had the largest univariate F. The 40 yd dash was selected next, followed by the vertical jump. At this point, none of the four remaining dependent variables had significant Fs, so the program provided the overall test of the linear composite of the dependent variables' (bench press, 40 yd dash, and vertical jump) ability to separate the three groupings of players. Squared canonical correlation, which is cumulative at each step, provides an estimation of meaningfulness: the percentage of variance accounted for (20% with the first variable, 34% with the first and second variables, and 35% with all three). The other four dependent variables were not selected because they failed to improve the ability to separate the three groups of players. This means that the characteristics underlying performance in the three dependent variables included were similar to the characteristics underlying performance in the four not included.

Discriminant analysis may be followed up with univariate techniques to determine which groups differed from one another on each of the dependent variables selected. There are several ways to approach this follow-up, but for simplicity, you could perform the Scheffé test among the three groups on the first dependent variable. Then you could use ANCOVA among the three groups on the second dependent variable using the first dependent variable as a covariate. This would provide adjusted means of the second dependent variable (means of the second dependent variable corrected for the first). A Scheffé test could be done among the adjusted means using the adjusted mean square for error. This procedure, called a *stepdown F technique*, is continued through each dependent variable using the previously stepped-in dependent variables as covariates. There are additional ways to follow up discriminant analysis.

Summarizing Discriminant Analysis

Tew and Wood (1980) used discriminant analysis in a situation in which the three groups were intact (i.e., not randomly formed). This is a common application of discriminant analysis. Discriminant analysis, however, does not overcome the need to form the groups randomly if determination of cause and effect is the purpose of the research.

Multivariate Analysis of Variance (MANOVA)

From an intuitive point of view, multivariate analysis of variance (MANOVA) is a straightforward extension of ANOVA. The only difference is that the F tests for the independent variables and interactions are based on an optimal linear composite of several dependent variables. There is no need to consider simple MANOVA here, because this is discriminant analysis (one independent variable with two or more levels and two or more dependent variables). For a good practical discussion, definition, and example of MANOVA, see Haase and Ellis (1987).

Using MANOVA

The mathematics is complex for factorial MANOVA, but the idea is simple. An optimal combination (linear composite) of the dependent variables is made that maximally accounts for (predicts) the variance associated with the independent variables. The variance associated with each independent variable is then separated out (as in ANOVA), and each of the independent variables and interactions on the optimal linear composite is tested. The associated F and degrees of freedom for each test are interpreted in the same way as for ANOVA. There are several ways to obtain F in MANOVA: Wilks' lambda, Pillai's trace, Hotelling's trace, and Roy's greatest characteristic root. We point this out only because authors sometimes identify how the MANOVA Fs were obtained. For your purposes, these distinctions are not important: Just remember that the MANOVA F ratios are similar to ANOVA F ratios.

After MANOVA has been used, a significant linear composite of dependent variables is identified that separates the levels of the independent variable or variables. Then, the important question usually is Which of the dependent variables contribute significantly to this separation? One of many ways to answer this question is to use discriminant analysis and the stepdown F procedures discussed in the previous section as follow-up techniques. This method works well for the main effects but not so well for interactions. Interactions in MANOVA are most frequently handled by calculating factorial ANOVA for each dependent variable, although this procedure fails to consider the relationships among the dependent variables.

Example of MANOVA

As an example of the use and follow-up of MANOVA, consider an experiment reported by French and Thomas (1987). One aspect of this study was to evaluate the influence of two groupings of age level (8- to 10-year-olds and 11- to 12-year-olds) and level of expertise (expert and novice players within each age level) on basketball knowledge and performance. A basketball knowledge test and two basketball skill tests (shooting and dribbling) were given to all the children. Following is French and Thomas's description of these results:

> A 2 × 2 (age league × expert/novice) MANOVA was conducted on the scores of the knowledge test and both skill tests. The results of the MANOVA indicated significant main effects for age league, $F(3,50) = 5.81$, $p < .01$, and expert/novice, $F(3,50) = 28.01$, $p < .01$, but no significant interaction. These main effects were followed up by a stepdown procedure using a forward selection discriminant analysis. The alpha level used as a basis for stepping in variables was set at .05. The discriminant analysis for age league revealed that knowledge was stepped in first, $F(1,54) = 8.31$, $p < .01$. Neither skill test was entered. Older children ($M = 79.5$) possessed more knowledge than younger children ($M = 64.9$). The discriminant analysis for expert/novice revealed that shooting was stepped in first, $F(1,54) = 61.40$, $p < .01$; knowledge was stepped in second, $F(1,53) = 5.51$, $p < .05$; and dribbling was not entered. Child experts ($M = 47.2$) performed significantly better than novices ($M = 25.7$) in shooting skills. The adjusted means for knowledge showed that child experts ($M = 77.1$) possessed more basketball knowledge than novices ($M = 64.2$). (p. 22)

If more than two levels had existed within an independent variable in the MANOVA (i.e., suppose three age-league levels had been used), then following the discriminant analysis, there would have been three means (one for each age-league level) to test among them for significance. It would be appropriate to use a Scheffé test for this follow-up, just as we did for ANOVA.

Multivariate Analysis of Covariance (MANCOVA)

Conceptually, multivariate analysis of covariance (MANCOVA) represents the same extension of ANCOVA that MANOVA did for ANOVA. In MANCOVA there are one or more independent variables, two or more dependent variables, and one or more covariates. Recall the earlier explanation of ANCOVA. A variable was used to adjust the dependent variable by correlation, and then ANOVA was applied to the adjusted dependent variable. In MANCOVA each dependent variable is adjusted for one or more covariates, and then an optimal linear composite that best discriminates among the independent variables is formed from the adjusted dependent variables, just as in MANOVA. Follow-up procedures are the same as in MANOVA, except that the linear composite of adjusted dependent variables is used. This technique is seldom used and typically used incorrectly. Normally, there is no reason to adjust a dependent variable for a linear composite of covariates. We have purposely not provided an example of MANCOVA use from the physical activity literature. MANCOVA is a fine technique if you understand what it does and you specifically want to test for that. But many researchers misunderstand MANCOVA and do not use it properly. Of course, the same is true of ANCOVA; for a good discussion of the issues, see A.C. Porter and Raudenbush (1987).

Repeated Measures With Multiple Dependent Variables

Rather frequently, experiments have multiple dependent variables that are measured more than once (over time). For example, in an exercise adherence study, both physiological and psychological variables might be measured once a week during a 15-week training program. If there were two training groups (different levels of training) and a control group, each with 15 participants, all of whom were measured every week (i.e., 15 times) on two psychological and three physiological measures, we would have a design that is 3 levels of exercise × 15 trials (3 × 15) for 5 dependent variables. This design offers several options for analysis.

We could do five 3 × 15 ANOVAs with repeated measures on the second factor. Here, we would follow the repeated-measures procedures described earlier in this chapter. But we would be inflating the alpha by doing multiple analyses on the same subjects. Of course, the alpha could be adjusted by the Bonferroni technique ($\alpha = .05 / 5 = .01$), but this fails to take into account the relationships among the dependent variables, which might be substantial and of considerable interest. An excellent tutorial on how to handle this problem is provided by Schutz and Gessaroli (1987). The following brief discussion is taken from their study, but if you are using this design and analysis, you should read their complete study and example.

Two options are available for this analysis: multivariate mixed-model (MMM) analysis and doubly multivariate (DM) analysis. Which is used depends on the assumptions that your data meet. Using the study previously described (3 levels of exercise × 15 trials for 5 dependent variables), the MMM analysis treats the independent variable (levels of exercise) as a true multivariate case by forming a linear composite of the five dependent variables to discriminate among the levels of the independent variable. If this composite is significant, it can be followed up by the stepdown F procedure (for an alternative procedure, see Schutz & Gessaroli, 1987). For the repeated-measures factor (and interaction), a linear composite is formed for each trial, and the linear composite at each trial is treated as a regular repeated-measures analysis. This means that the sphericity assumption must be met as previously described, and the epsilon can be used to test this assumption with the same standards previously described. The interpretation of the resulting F ratio for main effects for groups, trials, and the group × trial interaction are the same as in other designs. The question usually posed here is Do the groups change at different rates across the trials on the linear composite of dependent variables? This analysis is preferred if the sphericity assumption can be met because most authors believe that it offers more power. This assumption is difficult to meet, however, especially if there are more than two dependent variables measured on more than three trials.

The DM analysis does not require that the sphericity assumption be met. The analysis is the same for the independent variable of exercise. In the repeated-measures part of the analysis, however, a linear composite is formed not only of the dependent variables at each trial but also of the 15 trials (which themselves become a linear composite); thus the name *doubly multivariate*. The interpretations of the Fs for the two main effects and interaction remain essentially the same, but follow-ups become more complex.

McCullagh and Meyer (1997) compared four methods of providing information (physical practice with feedback, learning model with model feedback, correct model with model feedback, and learning model without model feedback) on learning correct form in the free-weight squat lift. There were two dependent variables (outcome and form) and five trials. A MANOVA with repeated measures was used to analyze the data. Univariate ANOVAs and post hoc comparisons were done as follow-ups for significant Fs.

SUMMARY

This chapter presented techniques used in situations in which differences among groups are the focus of attention. These techniques range from the simplest situation of two levels of an independent variable with one dependent variable to complex multivariate techniques involving multiple independent variables, multiple dependent variables, and multiple trials. These techniques can be categorized as follows:

■ A t test is used to determine how a group differs from a population, how two groups differ, how one group changes from one occasion to the next, and how several means differ (the Scheffé or some other multiple comparison test).

■ ANOVA shows differences among the levels of one independent variable (simple ANOVA), among the levels of two or more independent variables (factorial ANOVA), and among levels of independent variables when there is a distractor variable or covariate (ANCOVA).

■ Multivariate analysis of variance (MANOVA) is used when there is more than one dependent variable. Discriminant analysis is the simplest form of MANOVA (when there is only one independent variable and two or more dependent variables). More complex MANOVA involves two or more independent variables and two or more dependent variables.

■ MANCOVA is an extension of MANOVA in which there is one or more covariates.

■ Repeated-measures MANOVA is used when one or more of the independent variables are repeated measures.

Table 9.5 provides an overall look at the techniques presented in this chapter and chapter 8 (relationships among variables) and their appropriate use. Note that there are techniques for relationships among variables that parallel each technique for differences among groups. In fact, the relationship between t and r that was demonstrated earlier in this chapter holds, because ANOVA techniques are the equivalent of multiple regression. Each technique evaluates our two basic questions: Is the effect or relationship significant? Is the effect or relationship meaningful?

TABLE 9.5

Comparison of Statistical Techniques From Chapters 8 and 9

Description	Relationships among variables	Differences among groups
1 IV (2 levels) → 1 dv		Independent t test
1 predictor → 1 criterion	Pearson r	
2 or more IVs → 1 dv		Factorial ANOVA
2 or more predictors → 1 criterion	Multiple regression	
1 IV (2 or more levels) → 2 or more dvs		Discriminant analysis
2 or more IVs → 2 or more dvs		MANOVA
2 or more predictors → 2 or more criteria	Canonical correlation	

IV = independent variable; dv = dependent variable.

Some ideas presented here are complex and may not be easily understood in one reading. We provide some suggested readings and problems in this chapter that should be helpful. If you do not feel confident in your understanding of this material, reread the chapter, do the problems, and consult some of the suggested readings. This review is important because parts III and IV assume that you understand part II.

Check Your Understanding

1. Critique the statistical part of a study that uses an independent t test. Calculate ω^2 for t in this study.
2. Find a study that uses a multiple comparison test (Newman-Keuls, Duncan, or Scheffé) as a follow-up to ANOVA. Critique the use of this test.
3. A researcher wishes to compare the effectiveness of three desk arrangements (rows, clusters, and circles) on classroom behavior (frequencies of observed on-task behaviors). Elementary school children ($N = 18$) were randomly assigned to the three desk-arrangement groups for instruction. The M, s, and n for each group are as follows:

Group	M	S	n
Rows	6.0	1.4	6
Clusters	6.8	1.9	6
Circles	9.5	1.0	6

Here is a partial summary of the ANOVA:

Source	SS	df	MS	F
Between	40.1			
Within	34.3			
Total	74.4			

a. Complete the table. Use table 9.1 and example 9.6 for help.
b. Test the F for significance at the .05 level using appendix table 6
c. Compute ω^2 for meaningfulness.
d. Use the Scheffé method to determine where the differences are among the three groups (consult example 9.7).
e. Write a short paragraph interpreting the results of your data analysis.

4. Critique the statistical part of a study that uses a two-way factorial ANOVA. Calculate ω^2 for each factor and the interaction.
5. Locate a study that used MANOVA in its data analysis. Identify the independent and dependent variables.

10

Nonparametric Techniques

> If you have always done it that way, it is probably wrong.
>
> —Charles Kettering

In the preceding chapters, various parametric statistics were described. Recall that parametric statistics make assumptions about the normality and homogeneity of the variance of the distribution. Another category of statistics is called nonparametric statistics. This category is also referred to as distribution-free statistics because no assumptions are made about the distribution of scores. Nonparametric statistics are versatile because they can deal with ranked scores and categories. This feature can be a definite advantage when dealing with variables that do not lend themselves to precise interval-type or ratio-type data (which are more likely to meet parametric assumptions), such as categories of responses on questionnaires and affective behavior rating instruments. Data from quantitative research are frequently numerical counts of events that can be analyzed with nonparametric statistics.

The main drawback to nonparametric statistics that has often been voiced is that they are less powerful than parametric statistics. As you recall, power refers to the ability of a statistical test to reject a null hypothesis that is false. It should be pointed out that there is no agreement regarding the supposed power advantage of parametric tests (Harwell, 1990; Thomas, Nelson, & Thomas, 1999). Another drawback in the use of nonparametric tests has been the lack of statistical software for the more complex statistical tests, such as multivariate cases.

In chapter 6, we provided you with a set of procedures to judge the normality of data so that you can determine whether to use parametric or nonparametric approaches. These included looking at the distribution of the data (e.g., using a stem-and-leaf or histogram plot to evaluate whether the data fit a normal curve) and evaluating the skewness and kurtosis of the distribution. Deciding whether data meet the normality assumption is difficult (Thomas, Nelson, & Thomas, 1999). For example, how normal do data have to be to use parametric techniques? On numerous occasions a researcher's data will not meet the assumptions of normality. Micceri (1989), for example, maintained that much of the data in education and psychology are moderately or largely nonnormal, and thus nonparametric tests should be considered. Moreover, sometimes the only scores available are frequencies of occurrence or ranks (which often are not normally distributed), in which case the researcher should use nonparametric tests.

Because nonparametric statistics can deal with ranked scores and categories, they can be useful when working with data that do not meet parametric assumptions for normal distribution. Questionnaire responses and behavior rating techniques are two examples of data that can be analyzed using nonparametric statistics.

In this chapter, we present two categories of nonparametric analyses. First is which is used to analyze the frequency of responses that are in categories, as in these examples:

- How many highly fit children are girls and how many are boys?
- How many former athletes participate in recreational sports after age 30?
- How many nonathletes participate in recreational sports after age 30?
- How many highly fit women regularly participate in swimming, running, or cycling?

Second, we present a standard approach to analyzing ranked data when the ranks are not normally distributed. To use any of the general linear model (GLM) techniques described in chapters 8 and 9, the data must be normally distributed, and the data points must fit a straight line. Following are examples of data that might not meet the normality assumption:

- Questionnaire responses of *strongly agree, agree, neutral, disagree,* and *strongly disagree* can be considered ranked data; the responses are ranked from 1 (*strongly agree*) to 5 (*strongly disagree*).
- Data such as time, velocity, acceleration, and counts (e.g., how many push-ups you can do) may not be normally distributed).

In these instances, the use of parametric correlational and ANOVA procedures (including *t* tests) would be inappropriate because the assumptions for using the *r, t,* and *F* tables are not met.

nominal measure—A method of classifying data into categories, such as sex, age, grade level, or treatment group.

chi square— statistical test of the significance of the discrepancy between the observed and the expected results.

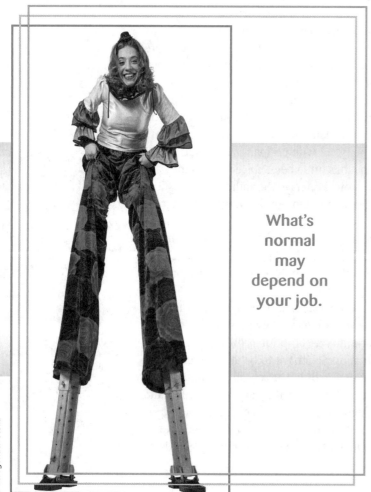

What's normal may depend on your job.

© aerogondo/fotolia

In this chapter, we provide a set of procedures for rank-order data that are parallel to the parametric procedures of correlation and ANOVA (including *t* tests). The ideas underlying these procedures are identical to the ones presented in chapters 8 and 9. The procedures use the same computer programs (e.g., SPSS, SAS) to analyze the ranked data, but instead of using the *r, t,* and *F* tables, one calculation is made (a statistic called *L*) that is then compared to a chi-square table, which does not require a normality assumption.

CHI SQUARE: TESTING THE OBSERVED VERSUS THE EXPECTED

Data are often sorted into categories, such as sex, age, grade level, treatment group, or some other **nominal** (categorical) **measure**. A researcher is sometimes interested in evaluating whether the number of cases in each category is different from what would be expected on the basis of chance, some known source of information (such as census data), or some other rational hypothesis about the distribution of cases in the population. **Chi square** provides a statistical test of the significance of the discrepancy between the observed and the expected results.

The formula for chi square is

$$\chi^2 = \Sigma[(O - E)^2 / E]$$ (10.1)

where O = the observed frequency and E = the expected frequency. Thus, the expected frequency in each category—often labeled cells, as in a table divided into four equal boxes—is subtracted from the observed (or obtained) frequency. This difference is then squared (which means that all differences will be positive), and these values are divided by the expected frequency for their respective categories and then summed.

For example, tennis coach Roger Rabbitfoot believes in jinxes. He believes that one court at his university is definitely unlucky for his team. He has kept records of matches on four courts over the years and is convinced that his team has lost significantly more matches on court 4 than on any other. Roger has set out to prove his point by comparing the number of losses on each of the four courts. The same number of matches has been played on each court, and his team has lost 120 individual matches on these courts during this period. Theoretically, it would be expected that each court would have one fourth, or 30, of the losses. The observed frequencies and the expected frequencies are shown in example 10.1.

Example 10.1

Known Values

		Court number				
		1	2	3	4	Total
Observed number of losses	$O =$	24	34	22	40	120
Expected number of losses	$E =$	30	30	30	30	120

Working It Out (Equation 10.1)

	Court number			
	1	2	3	4
$(O - E)$	–6	+4	–8	+10
$(O - E)^2$	36	16	64	100
$(O - E)^2 / E$	1.20	0.53	2.13	3.33
$\chi^2 = \Sigma[(O - E)^2 / E] = 7.19$				

The resulting chi square is then interpreted for significance by consulting table 7 in the appendix. Because there are four courts (or cells), the number of degrees of freedom (df) is $c - 1 = 3$. The researcher finds the critical value for 3 df, which is 7.82 for the .05 probability level. Roger's calculated value of 7.19 is less than this, so the null hypothesis that there were no differences among the four courts in the number of losses is not rejected. The observed differences could be attributed to chance. Roger probably still believes in his heart that he is right (called cardiac research).

In some cases, the expected frequencies for classifications can be obtained from existing sources of information, as in the following example. A new assistant professor, Nancy Niceperson, is assigned to teach the large sections of an introductory kinesiology course. After a few semesters, the department chairperson hears rumors that Dr. Niceperson is too lenient in her grading practices, so he uses chi square to compare the grades of her 240 students with the department's prescribed normal-curve grade distribution: 3.5% As and Fs, 24% Bs and Ds, and 45% Cs. If Dr. Niceperson were adhering to this normal distribution, she would be expected to have given 8 As and 8 Fs, (3.5% × 240), 58 Bs and 58 Ds (24% × 240), and 108 Cs (45% × 240). The observed frequencies (Dr. Niceperson's grades) and the expected frequencies (the department's mandated distribution) are compared using chi square in example 10.2.

Example 10.2

Known Values

		Grade					
		A (3.5%)	B (24%)	C (45%)	D (24%)	F (3.5%)	Total
Observed number of grades given	$O =$	21	75	114	28	2	240
Expected number of grades given	$E =$	8	58	108	58	8	240

Working It Out (Equation 10.1)

	Grade				
	A	B	C	D	F
$(O - E)$	13	17	6	−30	−6
$(O - E)^2$	169	289	36	900	36
$(O - E)^2 / E$	21.13	4.98	.33	15.52	4.50
$\chi^2 = \Sigma[(O - E)^2 / E] = 46.46$					

The chairperson, who is always fair, does not want to make a bad decision, so he decides to use the .01 level of probability. Table 7 in the appendix shows that for 4 *df* (there are five grades, or cells), a chi square of 13.28 is needed for significance at the .01 level. The obtained chi square of 46.46 exceeds the value, indicating a significant deviation from the expected grade distribution. Clearly, Dr. Niceperson's grades show too many As and Bs and too few Ds and Fs. The chairperson does the only fair thing by firing her on the spot.

Contingency Table

Often, a problem involves two or more categories of occurrences and two or more groups (a two-way classification). A common example is in the analysis of the results of questionnaires or attitude inventories in which there are several categories of responses (e.g., *agree, no opinion, disagree*) and two or more groups of respondents (e.g., exercise adherents and nonadherents). This type of two-way classification is called a **contingency table**.

To illustrate, suppose a group of athletes and a group of nonathletes respond to the following statement on a fair play inventory: "A baseball player who traps a fly ball between the ground and his glove should tell the umpire that he did not catch it." In the previous examples of one-way classification, the expected frequencies were determined by some type of rational hypothesis or source of information. In a contingency table, the expected values are computed from the marginal totals. Example 10.3 shows the participants' responses.

contingency table—A two-way classification of occurrences and groups that is used for computing the significance of the differences between observed and expected scores.

Example 10.3

Known Values

Observed responses	Agree	No opinion	Disagree	Total
Athletes	30	46	124	200
Nonathletes	114	80	56	250
Total	144	126	180	450

Working It Out

1. **Find the expected values (column total $\times$ row total) / N**

Expected responses	Agree	No opinion	Disagree	Total
Athletes	144 x 200 /450 = 64	126 x 200 /450 = 56	180 x 200 /450 = 80	200
Nonathletes	144 x 250 /450 = 80	126 x 250 /450 = 70	180 x 250 /450 = 100	250
Total	144	126	180	450

2. **Compute χ^2 (equation 10.1)**

Responses	$O - E$	$(O - E)^2$	$(O - E)^2 / E$
Athletes agree	−34	1,156	18.06
Nonathletes agree	34	1,156	14.45
Athletes no opinion	−10	100	1.79
Nonathletes no opinion	10	100	1.43
Athletes disagree	44	1,936	24.20
Nonathletes disagree	−44	1,936	19.36
			$\chi^2 = 79.29$

A total of 144 respondents agreed with the statement. Because there are 450 people in all, 144 / 450, or 32% of the total group, agreed with the statement. Thus, if there were no difference between athletes and nonathletes on fair play (the null hypothesis) as reflected by this statement, then 32% of the athletes ($0.32 \times 200 = 64$) and 32% of the nonathletes ($0.32 \times 250 = 80$) would be the expected frequencies for these two cells.

A much faster method of calculating these expected frequencies is simply to multiply the column total by the row total for each cell and divide by the total number (N), as we worked out in step 1 of example 10.3. Chi square is computed in the same manner as in the examples of one-way classification.

The degrees of freedom for a contingency table is $(r - 1)(c - 1)$, where r stands for rows and c for columns. Here, we have two rows and three columns, so $df = (2 - 1)(3 - 1) = 2$. As before, the investigator then looks up the table value for significance (in this study, the investigator had decided on the .01 level) and sees that a chi square of 9.21 is needed. The obtained chi square of 79.29 is clearly significant. This result tells us that the null hypothesis is rejected: Athletes and nonathletes respond to the statement in a significantly different way. After inspecting the table, we could conclude that a significantly greater proportion of nonathletes agreed that the player should tell the umpire about trapping the ball and, conversely, that a higher proportion of athletes thought that the player should not.

Restrictions in Using Chi Square

Although we indicated that nonparametric statistics do not require the same assumptions concerning the population as do parametric statistics, some restrictions apply in using this technique. The observations must be independent, and the categories must be mutually exclusive. By this we mean that an observation in any category should not be related to or dependent on other observations in other categories. You could, for example, ask 50 people about their activity preferences. If each gave three preferences, you would not be justified in using a total (N) of 150 because the preferences of any participant would likely be related and chi square thus inflated. Moreover, an observation can be placed in only one category. The observed frequencies are exactly that: numbers of occurrences. Ratios and percentages are not appropriate. Another related point is that the total of the expected frequencies and the total of the observed frequencies for any classification must be the same. In example 10.3, notice that the total of the expected frequencies for athletes is the same as the total of the observed frequencies. The same is seen for nonathletes and for the column totals.

Chi square is usually not applicable for small samples. The expected frequency for any cell should not be less than 1.0. Furthermore, some statisticians claim that no more than 20% of the cells can have expected values of less than 5, although opinions vary on this point. Some say that no cell should have a frequency of less than 5, whereas others allow as many as 40% of the cells to have a frequency of less than 5. A common tactic in cases in which several cells have expected frequencies less than 5 is to combine adjacent categories, thereby increasing the expected values.

Researchers generally agree that a 2×2 contingency table should have a so-called correction for continuity. This correction, usually referred to as **Yates' correction for continuity**, is to subtract 0.5 from the difference between the observed and the expected frequencies for each cell before it is squared:

Corrected $\chi^2 = \Sigma[(O - E - 0.5)^2 / E]$ (10.2)

Another limitation imposed on the 2×2 contingency table is that the total number (N) should be at least 20.

Finally, the expected distribution should be logical and established before the data are collected. In other words, the hypothesis (probability, equal occurrence, census data, and so on) precedes the analysis. Researchers are not allowed to look over the distribution and then conjure up an expected distribution that fits their hypothesis.

Yates' correction for continuity—A method of correcting a 2 × 2 contingency table by subtracting 0.5 from the difference between the observed and expected frequencies for each cell before it is squared.

Contingency Coefficient

Several correlational techniques can be used in situations in which data are discrete (i.e., not continuous). You can compute the relationship between dichotomous variables such as sex and race by using a **contingency coefficient**. The test of significance is chi square. Recall the contingency table for use in detecting differences between groups or sets of data, described in the previous section. The contingency table can also be used to determine relationships. You can have any number of rows and columns in a contingency table. After chi square is computed, the contingency coefficient C can be computed:

contingency coefficient—A method of computing the relationship between dichotomous variables such as sex and race.

$$C = \sqrt{\chi^2 / (N + \chi^2)}$$ (10.3)

If χ^2 is significant, C is also significant. The direction of the relationship is established by examining the data. Several limitations affect the ability of the contingency coefficient to estimate correlation. In general, a researcher needs several categories and many observations to obtain a reasonable estimate.

Multivariate Contingency Tables: The Loglinear Model

Categorical data can be analyzed in combination with other variables. In other words, contingency tables can be studied in more than two dimensions. The researcher can therefore identify associations among many variables, such as the interrelationships among age, sex, skill level, and teaching method. This approach is similar to parametric multivariate analysis. With continuous quantitative data, however, variables are expressed as linear composites, whereas with categorical variables, the researcher deals with contributions to the expected frequencies within each cell of the multivariate contingency table. Any given cell represents the intersection of many marginal proportions.

Loglinear models are used to analyze multivariate contingency tables. Relative frequencies are transformed to logarithms, which are additive and similar to the sum of squares in ANOVA. Main effects and interactions can be tested for significance. The probability of membership in a particular category can be predicted as a function of membership in other categories using a logistic regression equation based on the logarithmic odds of membership (called a **logit**).

The loglinear model has considerable potential application in research with categorical data (Schutz, 1989). This type of analysis is attracting much attention from researchers and theoreticians. The loglinear analysis provides a means for sophisticated study of the interrelationships among categorical variables. Readers who are interested in pursuing this topic are directed to the text by Eye and Mun (2012).

loglinear models— A system that analyzes multivariate contingency tables by transforming relative frequencies into logarithms.

logit—The probability of membership in a particular category as a function of membership in other categories in multivariate contingency tables.

PROCEDURES FOR RANK-ORDER DATA

Most books that report nonparametric procedures (including the first four editions of this book) propose a series of rank-order techniques, including these:

- Mann-Whitney U test—analogous to the parametric independent t test
- Wilcoxon matched-pairs signed-ranks test—analogous to the parametric dependent
- t test
- Kruskal-Wallis ANOVA by ranks—analogous to the parametric one-way ANOVA
- Friedman two-way ANOVA by ranks—analogous to the parametric repeated-measures ANOVA
- Spearman rank-difference correlation—analogous to the parametric Pearson r

You will still see these techniques used in the research literature, so you need to understand what each one does; our list shows you how they compare with parametric procedures. Eventually, as better methods become available, most of these techniques will pass from use, but that change will take time. Researchers who have been trained to use these procedures often do not keep up with current statistical methods; we acknowledge that staying current is difficult because researchers have their own research areas to keep up with. One jokester was heard to say that it takes 30 to 40 years for a new statistical approach to come into widespread use. This slow pace of adoption occurs because researchers rarely change the statistical procedures they use, so we have to wait for them to retire and for a more recently trained group to begin using the new procedures.

We can replace all the rank-order techniques previously listed with a standardized approach to rank-order data in which the distribution of data is not normal (for greater detail, see

Thomas, Nelson, & Thomas, 1999). The techniques parallel the parametric ones described in chapters 8 and 9. The procedures for all techniques involve these steps:

1. Change all the data values to ranks (most statistical software can do this).
2. Run the standard parametric statistical software (e.g., correlation, multiple regression ANOVA) on the ranked data.
3. Calculate the L statistic as the significance test, and compare it to the chi-square table (appendix table 7).

The nonparametric test statistic (L) used to evaluate significance for all these procedures was developed by Puri and Sen (1969, 1985) and represents the test of the null hypothesis between X and Y where X can be groups or variables and Y is variables. This method can be used in place of all the techniques previously discussed and assumes that the data fit a straight line (as all the techniques in chapters 8 and 9 did) *but does not assume normal distribution of data.* L is easy to calculate using the following formula:

$$L = (N - 1)r^2 \tag{10.4}$$

where N = the number of participants and r^2 = the proportion of true variance = $SS_{regression}$ / SS_{total}. The proportion of true variance in correlation is r^2 and in multiple regression is R^2. In a t test, $r^2 = t^2 / (t^2 + df)$, where $df = (n^1 + n^2) - 2$. In ANOVA, it is the sum of squares (SS) for the effect of interest (e.g., between groups in one-way ANOVA or for each factor in factorial ANOVA; each of these is the true variance) divided by the total sum of squares (these values are shown in the ANOVA printout from standard statistics programs).

When the L statistic is calculated for each significance test (either for correlation or ANOVA), it is compared to the chi-square table (table 7 in the appendix with degrees of freedom (df) = pq, where p = the number of df for the independent (or predictor) variable or variables and q = the number of df for the dependent (or criterion) variable or variables. Thus, for a correlation (r) between one predictor (e.g., abdomen skinfold) and one criterion (e.g., body weight), the pq df is $1 \times 1 = 1$. In a t test, $df = p$ (number of groups − 1) $\times q$ (number of dependent variables); thus, again, the pq $df = 1 \times 1 = 1$. Additional examples of how to calculate L and df are provided throughout the rest of this chapter.

> The nonparametric test statistic (L) can replace all previously discussed techniques, assumes that the data fit a straight line, and is easy to calculate.

CORRELATION

In this section, we provide sample ranked data and show how to calculate correlation and multiple regression using standard correlational techniques. We then calculate the L and df for each example. In the simple correlation example, we also show how to decide whether to use parametric or nonparametric analysis, but we do not repeat this process for subsequent procedures (you can obtain this logic for each procedure by reading Thomas, Nelson, & Thomas, 1999).

Simple Correlation

We are interested in calculating the correlation between two skinfold measurements, biceps and forearm, that were collected on 157 participants. Descriptive data are given in figure 10.1, along with a histogram showing the data distribution with the normal curve overlaid. Recall from chapter 6 that skewness and kurtosis provide information about the nature of the distribution. Here they are given in the form of a z score, where 0.0 indicates a normal distribution and positive or negative values indicate specific variations from normality in the distribution. For the biceps skinfold, the skewness is above +1.0, indicating that the hump of the curve is shifted to the left. The kurtosis is nearly +2.0, indicating that the distribution is very peaked. For the forearm skinfold, the hump of the curve is also shifted left, and the curve is extremely peaked. Neither data set seems to be normal in its distribution. Although no set rule defines how normal or nonnormal data must be, we do know that parametric techniques are not as resistant to nonnormality as once thought. In our view, these data are nonnormal enough to suggest the use of rank-order procedures rather than parametric ones.

Figure 10.1   *(a)* Biceps, *(b)* forearm.

If we use a computer program for parametric correlation to determine the relationship between these two variables (using the original data), $r = .26$, and the test of significance is $F(1, 155) = 11.46, p < .001$. But we are likely violating the assumption that data are normally distributed. If we rank subjects from the smallest biceps skinfold measurement (number 1) to the largest (number 157) and do the same for the forearm skinfold, we can then use the same computer program to correlate the ranked data; the correlation is actually larger, $r = .28$. If we square this value, we get .0784. If we then use equation 10.4 to calculate the significance test to replace F, we find that

$$L = (N - 1)r^2$$

$$= (157 - 1)(.28)^2 = (156)(.0784) = 12.23$$

This L has $df = 1$ because there is one X variable (biceps skinfold) and one Y variable (forearm skinfold), so $pq = 1 \times 1 = 1$. If we look up this L (appendix table 7) as chi square with $df = 1$, we find a value of 10.83 for $p = .001$. Our value exceeds that, so it is significant at $p < .001$. Thus, we have now used a rank-order procedure that does not violate the assumption for normality, and our r and test of significance for the r, the L statistic, were both larger than the ones we would have obtained if we had used the parametric statistics. Although this circumstance will not always occur, this approach to nonparametric analysis does have good power compared with parametric procedures when data are not normal.

Multiple Regression

We can extend the same procedures from the previous section to multiple regression. In this example we use four skinfolds—abdomen, calf, subscapular, and thigh—to predict the percentage of fat determined from hydrostatic weighting. (The data in this example are from Thomas, Keller, & Holbert, 1997. We thank these authors for allowing us to use their data.) All four skinfold measurements are changed to ranks, as are the fat percentage measurements. We then run the regular multiple-regression program from SPSS using forward-stepping procedures on the ranked data. The correlations for all ranked pairs of variables (e.g., abdomen with calf skinfold, abdomen with subscapular skinfold, and so on) range from .45 to .74. These values are about the same as for the original (unranked) data, for which the correlations range from .41 to .78. Table 10.1 summarizes the multiple-regression results for the ranked data (note that all test statistics have been changed to L). The overall multiple regression, R, was higher for the ranked data (.82) than for the original unranked data (.80).

TABLE 10.1

Summary of Forward Multiple Regression for the Ranked Data

Step	Variable	R	R^2	β	df	L
1	Subscapular skinfold	.68	.46	.332	1	35.83*
2	Calf skinfold	.77	.60	.602	2	20.29*
3	Abdomen skinfold	.80	.64	.321	3	7.85*
4	Thigh skinfold	.82	.68	−.327	4	9.56*

*$p < .05$ for all.

$L(4) = 53.01$, $p < .001$, for the linear composite of predictors.

"A generalized rank-order method for nonparametric analysis of data from exercise science: A tutorial," J.R. Thomas, J.K. Nelson, and K.T. Thomas, *Research Quarterly for Exercise and Sport*, 70(1): 11-23, 1999, Taylor and Francis, reprinted by permission of the publisher (Taylor & Francis Ltd, www.tandfonline.com).

DIFFERENCES AMONG GROUPS

We can apply the same logic and procedures from the previous two sections when we want to test differences among groups. Here we provide examples of a t test, one-way ANOVA, and factorial ANOVA. If you want to extend these procedures to repeated-measures ANOVA and multivariate ANOVA, see Thomas, Nelson, and Thomas (1999). In these examples we use data from Nelson, Yoon, and Nelson (1991); we appreciate their permission to use their data.

t Test

Data are modified push-up scores for 90 boys and 90 girls in the upper elementary school grades. The distribution of data is positively skewed and somewhat peaked. Following are the means and standard deviations for the girls and boys and the t test comparing them:

Boys $M = 18.6$ $s = 9.9$

Girls $M = 12.7$ $s = 9.8$

$t(178) = 4.00$, $p < .001$

We changed the data to ranks from 1 to 180, disregarding whether the scores were for boys or girls. We then calculated t on the ranked data; it was 3.77. But we need to change the t to L using equation 10.4. First we have to calculate r^2. Following are those calculations:

$r^2 = t^2 / (t^2 + df) = 3.77^2 / (3.77^2 + 178) = 14.21 / (14.21 + 178) = .0739$

$L = (N − 1)r^2 = (179)(.0739) = 13.23$

L is tested as a χ^2 with $df = pq$, where $p = $ the number of groups − 1 (2 − 1 = 1) and $q = $ the number of dependent variables (1), so 1 × 1 = 1 df. If we look in the chi-square table (appendix table 7), we see that our calculated value of 13.23 exceeds the table value of 10.83 at the .001 level; thus, our L is significant at $p < .001$.

You can practice this procedure by applying it to the data in example 9.8 in chapter 9. Use the data for groups 1 and 2, with 5 participants in each group ($N = 10$). Rank the scores from lowest to highest, disregarding whether the participant is in group 1 or 2. Then calculate the mean and standard deviation for each group and apply equation 9.3 to calculate t. The t value that you find will be 2.79. Then obtain r from t:

$$r^2 = t^2 / (t^2 + df) = 2.79^2 / (2.79^2 + 8) = .49$$

We then calculate L using formula 10.4:

$$L = (N - 1)r^2 = (9)(.49) = 4.41$$

This L has 1 df [(groups − 1) × number of dependent variables]. With 1 df, the critical χ^2 at the .05 level from appendix table 7 is 3.84. Because our value is larger than that, the two groups are significantly different.

One-Way ANOVA

The data from modified push-ups (Nelson, Yoon, & Nelson, 1991) was also evaluated by grade level: grades 4, 5, and 6. We use this as the example for one-way ANOVA. Grade level is the independent variable with three levels (this is really a categorical variable, but for ANOVA we treat it like an independent variable), and the modified push-up score is the dependent variable. Here are the descriptive data for each group:

Grade 4 $M = 12.1$ $s = 10.0$

Grade 5 $M = 16.3$ $s = 9.6$

Grade 6 $M = 18.5$ $s = 10.2$

We then change the data to ranks, disregarding the grade level of the children, so we have 180 scores (number of children in all three grades) ranked from the highest modified push-up score (ranked number 1) to the lowest (ranked number 180). We then run the one-way ANOVA program on the ranked data. The results are shown in table 10.2. But the printout shows F where we show L in table 10.2. To calculate the L, first calculate R^2 (the proportion of true variance in ANOVA) using

$$R^2 = SS_{between} / SS_{total}$$

$$= 31,646.61 / 485,412.00 = .065$$

Then, calculate L as

$$L = (N - 1)R^2 = (179)(.065) = 11.64$$

TABLE 10.2

Summary of ANOVA Data for the Ranked Scores

Source	df	Sum of squares	Mean squares	L	p
Between groups	2	31,646.61	15,823.30	11.64	.01
Within groups (error)	177	453,765.39	2,563.65		
Total	179	485,412.00			

"A generalized rank-order method for nonparametric analysis of data from exercise science: A tutorial," J.R. Thomas, J.K. Nelson, and K.T. Thomas, *Research Quarterly for Exercise and Sport*, 70(1): 11-23, 1999, Taylor and Francis, reprinted by permission of the publisher (Taylor & Francis Ltd, www.tandfonline.com).

The L has 2 df [(number of groups – 1) × number of dependent variables]. The value needed for significance at the .01 level with 2 df is 9.21 in appendix table 7, and our value exceeds that, so it is significant.

You can also do an example by hand if you return to example 9.7 in chapter 9 and use all three groups. As before, rank the scores from 1 to 15, disregarding which group the participants are in. Then work the one-way ANOVA as was done in example 9.8, except replace the original scores with the ranks. Calculate R^2 and L as previously described. You will find $R^2 = .657$ and $L(2) = 9.20$. The value for significance at the .02 level is 7.82 in appendix table 7, and the calculated value exceeds that, so it is significant.

Factorial ANOVA

Continuing with the same data set, we have a modified push-up score for girls and boys at each of three grade levels. Thus, we can rank the modified push-up scores from 1 to 180 as before and calculate a factorial ANOVA where one factor is sex (girls and boys) and the other is grade level (4, 5, and 6). This becomes a 2 (sex) × 3 (grade level) factorial ANOVA with the modified push-up score as the dependent variable.

These ranked data were analyzed using regular parametric statistical software. We found the outcome shown in table 10.3. We calculated R^2 using the same formula for each factor and for the interaction:

$$R^2_{grade} = SS_{grade} / SS_{total} = 31,646.61 / 485,412.00 = .065;$$

$$R^2_{sex} = SS_{sex} / SS_{total} = 35,814.01 / 485,412.00 = .074;$$

$$\text{and } R^2_{grade \times sex} = SS_{grade \times sex} / SS_{total} = 10,865.20 / 485,412.00 = .022.$$

We then used the R^2 to calculate L for each factor and for the interaction. In appendix table 7, the grade effect (with 2 df) is significant at the .005 level, the sex effect (with 1 df) is significant at the .001 level, but the interaction (with 2 df) is not significant.

TABLE 10.3

Summary of Factorial ANOVA Data for the Ranked Scores

Source	df	Sum of squares	Mean squares	L	p
Grade	2	31,646.61	15,823.30	11.64	.001
Sex	1	35,814.01	35,814.01	13.21	.001
Grade × sex	2	10,865.20	5,432.60	4.01	
Residual	174	407,086.18	2,339.58		
Total	179	485,412.00			

"A generalized rank-order method for nonparametric analysis of data from exercise science: A tutorial," J.R. Thomas, J.K. Nelson, and K.T. Thomas, *Research Quarterly for Exercise and Sport*, 70(1): 11-23, 1999, Taylor and Francis, reprinted by permission of the publisher (Taylor & Francis Ltd, www.tandfonline.com).

Extension to Repeated-Measures ANOVA and Multivariate ANOVA

Although we do not provide examples here, the use of ranked data and the L statistic can be extended to ANOVA models with repeated measures and multivariate ANOVA. Conceptually, the ideas are identical, and you follow procedures similar to the ones used here, including ranking the data, using standard computer software, and changing the test statistic to L

instead of *F*. If you need one of these techniques, see Thomas, Nelson, and Thomas (1999) for the steps to follow and ways to interpret the results.

SUMMARY

In this chapter, we provided statistical tests for data that do not meet the assumptions of parametric data (chapters 8 and 9). Included are chi-square tests for frequency data and a set of procedures for ranked data that are parallel to parametric procedures. If distributions are not normal, the data are changed to ranks. The ranked data can be analyzed by standard SPSS and SAS statistical packages. One test, the L statistic, is calculated and compared with a chi-square table for significance. These procedures can be applied to all parametric linear models.

Check Your Understanding

1. Compute chi square for the following contingency table of activity preferences of men (*n* = 110) and women (*n* = 90). Determine the significance of chi square at the .01 level using appendix table 7. When calculating the expected frequencies, round off to whole numbers.

Group	Racquetball	Weight training	Aerobic dance
Men	35	45	30
Women	28	13	49

   a. Write a brief interpretation of your results.
   b. What would be the critical value for significance at the .05 level if you had five rows and four columns?

2. Critique the statistical part of a study that uses chi square in the analysis.

11

Measuring Research Variables

A basic step in the scientific method of problem solving is the collection of data; therefore, an understanding of basic measurement theory is necessary. (We should point out that although measurement is discussed here as a research tool, measurement itself is an area of research.) In this chapter, we discuss fundamental criteria for judging the quality of measures used in collecting research data: validity and reliability. We explain types of validity and ways by which evidence of validity and reliability may be established. (The trustworthiness of qualitative research is discussed in chapter 19.) We conclude with some issues concerning the measurement of movement, of written responses used in paper-and-pencil instruments, of affective behavior, of knowledge, and of item response theory.

VALIDITY

In gathering the data on which results are based, we are also greatly concerned with the validity of the measurements we are using. If, for example, a study seeks to compare training methods for producing strength gains, the researcher must have scores that produce a valid measure of strength to evaluate the effects of the training methods. **Validity** of measurement indicates the degree to which the test, or instrument, measures what it is supposed to measure. Thus, validity refers to the soundness of the interpretation of scores from a test, the most important consideration in measurement.

There are different purposes for different measures. Consequently, there are different kinds of validity. We consider four basic types of validity: **logical**, **content**, **criterion**, and **construct**.

Although we list logical validity as a separate type of validity, the American Psychological Association and the American Educational Research Association consider logical validity a special case of content validity.

Logical Validity

Logical validity is sometimes referred to as face validity, although measurement experts dislike that term. Logical validity is claimed when the measure obviously involves the performance being measured. In other words, it means that the test is valid by definition. A static

validity—The degree to which a test or instrument measures what it purports to measure; can be categorized as *logical, content, criterion, or construct validity*.

logical validity—The degree to which a measure obviously involves the performance being measured; also known as *face validity*.

content validity—The degree to which a test (usually in educational settings) adequately samples what was covered in the course.

criterion validity—The degree to which scores on a test are related to some recognized standard or criterion.

construct validity—The degree to which a test measures a hypothetical construct; usually established by relating the test results to some behavior.

balance test that consists of balancing on one foot has logical validity. A speed-of-movement test, in which the person is timed while running a specified distance, must be considered to have logical validity. Occasionally, logical validity is used in research studies, but researchers prefer to have more objective evidence of the validity of measurements.

Content Validity

Content validity pertains largely to learning in educational settings. A test has content validity if it adequately samples what was covered in the course. As with logical validity, no statistical evidence can be supplied for content validity. The test maker should prepare a table of specifications (sometimes called a test blueprint) before making up the test. The topics and course objectives, as well as the relative degree of emphasis accorded to each, can then be keyed to a corresponding number of questions in each area.

A second form of content validity occurs with attitude instruments. Often, a researcher wants evidence of independent verification that the items represent the categories for which they were written. When this happens, experts (in many cases 20 or more) are asked to assign each statement to one of the instrument categories. These categorizations are tallied across all experts and the percentage of experts who agreed with the original categorization is reported. Typically, an agreement of 80 to 85% would indicate that the statement represents the content category.

Criterion Validity

Measurements used in research studies are frequently validated against some criterion. The two main types of criterion validity are concurrent validity and predictive validity.

Concurrent Validity

concurrent validity—A type of criterion validity in which a measuring instrument is correlated with some criterion that is administered concurrently, or at about the same time.

Concurrent validity is a type of criterion validity that involves correlating an instrument with some criterion that is administered at about the same time (i.e., concurrently). The scores from many physical performance measures are validated in this manner. Popular criterion measures include already validated or accepted measures, such as judges' ratings and laboratory techniques. Concurrent validity is usually employed when the researcher wishes to substitute a shorter, more easily administered test for a criterion that is difficult to measure.

As an illustration, maximal oxygen consumption is regarded as the most valid measure of cardiorespiratory fitness. Measurement, however, requires a laboratory, expensive equipment, and considerable time for testing; furthermore, only one person can be tested at a time. Let us assume that a researcher, Douglas Bag, wishes to screen individuals by their fitness levels before assigning them to experimental treatments. Rather than using such an elaborate test as maximal oxygen consumption, Douglas determines that it would be advantageous to use a stair-walking test that he has devised. To determine whether it is a valid measure of cardiorespiratory fitness, he could administer both the maximal oxygen consumption test and the stair-walking test to a group of participants (from the same population that will be used in the study) and correlate the results of the two tests. If a satisfactory relationship exists, Doug can conclude that his stair-walking test is valid.

Written tests can also be validated in this way. For example, a researcher might wish to use a 10-item test of anxiety that can be given in a short period rather than a lengthier version. Judges' ratings serve as criterion measures for some tests (sport skills and behavioral inventories are sometimes validated this way). A great amount of time and effort is required to secure competent judges, teach them to use the rating scale, test for agreement among judges, arrange for a sufficient number of trials, and so on. Consequently, judges' ratings cannot be used routinely to evaluate performances. The use of some skills tests would be more economical. Furthermore, the skills tests usually provide knowledge of results and measures of progress for the students. The skills tests could be initially validated, however, by giving the test and having judges rate the individuals on those skills. A validity coefficient can be obtained by correlating the scores on the skills tests with the judges' ratings.

Choosing the criterion is critical in the concurrent validity method. All the correlation can tell you is the degree of relationship between a measure and the criterion. If the criterion is inadequate, then the concurrent validity coefficient is of little consequence.

Predictive Validity

When the criterion is some later behavior, such as entrance examinations used to predict later success, **predictive validity** is the major concern. Suppose a physical education instructor wished to develop a test that could be given in beginning gymnastics classes to predict success in advanced classes. Students would take this test while they were in the beginner course. At the end of the advanced course, those test results would be correlated with the criterion of success (grades, ratings, and so on). In trying to predict a certain behavior, a researcher should try to ascertain whether there is a known base rate for that behavior. For example, someone might attempt to construct a test that would predict which female students might develop bulimia at a university. Suppose the incidence of bulimia is 10% of the female population at that school. Knowing this, one could be correct 90% of the time in predicting that no one in the sample is bulimic. If the base rate is very low or very high, a predictive measure may have little practical value because the increase in predictability is negligible.

> **predictive validity**—The degree to which scores of predictor variables can accurately predict criterion scores.

Chapters 8 and 16 discuss aspects of prediction in correlational research. Multiple regression is often used because several predictors are likely to have a greater validity coefficient than the correlation between any one test and the criterion. An example of this is the prediction of fat percentage from skinfold measurements. The criterion— fat percentage—can be measured by the underwater (hydrostatic) weighing technique. Several skinfold measures are then taken, and multiple regression is used to determine the best prediction equation. The researcher hopes to use the skinfold measures in the future if the prediction formula demonstrates an acceptable validity coefficient.

One limitation of such studies is that the validity tends to decrease when the prediction formula is used with a new sample. This tendency is called **shrinkage**. Common sense suggests that shrinkage is more likely when a small sample is used in the original study, particularly when the number of predictors is large. In fact, if the number of predictor variables is the same as N, you can achieve perfect prediction. The problem is that the correlations are unique to the sample, and when the prediction formula is applied to another sample (even similar to the first one), the relationship does not hold. Consequently, the validity coefficient decreases substantially (i.e., shrinkage occurs).

> **shrinkage**—The loss of predictive power when statistics are calculated on a different sample.

A technique recommended to help estimate shrinkage is **cross-validation**. In this technique, the same tests are given to a new sample from the same population to check whether the formula is accurate. For example, a researcher might administer the criterion measure and predictor tests to a sample of 200 people. Then, she would use multiple regression on the results from 100 of them to develop a prediction formula. This formula is then applied to the other 100 people to see how accurately it predicts the criterion for them. Because the researcher has the actual criterion measures for these people, she can ascertain the amount of shrinkage by correlating (Pearson r) the predicted scores with the actual scores. A comparison of the R^2 from the multiple prediction with the r^2 between the actual and the predicted criteria yields an estimate of shrinkage.

> **cross-validation**—A technique to assess the accuracy of a prediction formula in which the formula is applied to a sample not used in developing the formula.

Expectancy Tables

A problem of interpretation often arises when using a formula for prediction. A question frequently asked is How large must a predictive validity coefficient be before it provides useful information? For example, correlation between the attainment of advanced degrees and scores on the Graduate Record Examination (GRE) is typically quite small ($r < .40$). When we examine r^2 for meaningfulness, we see that there is only 16% or less common variance, which is discouraging. But when we look at the relationship from a different perspective, such as what percentage of students who score highly on the GRE attain a PhD and what percentage of people who score poorly attain the degree, we see that the GRE does have good validity. We can get this type of perspective by using an **expectancy table**.

An expectancy table can be used to predict the probability of some performance (e.g., academic, job-related). An expectancy tables is easily constructed; it consists of a two-way grid containing the probability of a person with a particular assessment score attaining some criterion score. As an example, assume that we have developed a test that purportedly measures good sporting behavior, called the Jolly Good Show Inventory. We wish to see how well our inventory relates to ratings of fair play by judges who have observed a group of 60 students at play (for hours and hours under various forms of competitive situations).

Our first step is to simply tally the number of students who fall in each cell. For example, three students scored between 50 and 59 on the inventory. One of these was rated average, one good, and one excellent. Table 11.1*a* shows the results of this step. Next, we convert the cell frequencies to percentages of the total number of students in each row. For example, one of the three students scoring between 50 and 59 was rated average; thus the percentage is 1 / 3 = 33%. The percentages are shown in table 11.1*b*. That is all there is to it. We can see, for example, that of the 9 students who scored in the 40 to 49 interval on the inventory, 22% were rated as excellent in sporting behavior. Further, none of the students with the top scores on the inventory were rated below average, and conversely, none of the students with the lowest inventory scores (0–9) were given ratings higher than poor. In this example, we are using an expectancy table to provide evidence of the concurrent validity of our inventory. Sometimes, expectancy table information can be used to predict success, such as grade point average in college from an aptitude test given in high school. As with all situations involving criterion validity, the availability and relevance of the criterion and its susceptibility to reliable measurement are key issues.

expectancy table—A two-way grid that predicts whether individuals with a particular assessment score will attain some criterion score.

TABLE 11.1

Expectancy Table for Jolly Good Show Inventory Scores to Ratings of Good Sporting Behavior

11.1*a* Frequency of Ratings for Each Predictor Score Level

Test score	Lousy	Poor	Average	Good	Excellent	Totals
50–59			1	1	1	3
40–49		1	2	4	2	9
30–39		4	8	2		14
20–29	2	5	7	5		19
10–19	3	6	2			11
0–9	3	1				4
						60

11.1b Expectancy Table After Converting Frequencies to Percentages

Test score	Lousy	Poor	Average	Good	Excellent
50–59			33	33	33
40–49		11	22	44	22
30–39		29	57	14	
20–29	11	26	37	26	
10–19	27	55	18		
0–9	75	25			

Construct Validity

Many human characteristics are not directly observable. Rather, they are hypothetical constructs that carry a number of associated meanings concerning how a person who possesses them to a high degree would behave differently from someone who possesses them to a low degree. Anxiety, intelligence, sporting behavior, creativity, and attitude are a few such hypothetical constructs. Because these traits are not directly observable, measurement poses a problem. Construct validity is the degree to which scores from a test measure a hypothetical construct and is usually established by relating the test results to some behavior. For example, certain behaviors are expected of someone with a high degree of good sporting behavior. Such a person might be expected to compliment the opponent on shots made during a tennis match. For an indication of construct validity, a test maker could compare the number of times a person scoring high on a test of sporting behavior complimented the opponent with the number of times a person scoring lower on the test did so.

The **known group difference method** is sometimes used in establishing construct validity. For example, construct validity of a test of anaerobic power could be demonstrated by comparing the test scores of sprinters and jumpers with those of distance runners. Sprinting and jumping require greater anaerobic power than distance running does. Therefore, the tester could determine whether the test differentiates between the two kinds of track performers. If the sprinters and jumpers score significantly better than the distance runners do, this finding would provide some evidence that the test measures anaerobic power.

An experimental approach is occasionally used in demonstrating construct validity. For example, a test of cardiorespiratory fitness might be assumed to have construct validity if it reflected gains in fitness following a conditioning program. Similarly, the originator of a motor skills test could demonstrate construct validity through its sensitivity in differentiating between groups of instructed and noninstructed children.

Correlation can also be used in establishing construct validity. Hypothesized structures or dimensions of the trait being tested are sometimes formulated and verified with factor analysis. The tester also uses correlation to examine relationships between constructs; for example, when it is hypothesized that someone with high scores on the test being developed (e.g., cardiorespiratory fitness) should also do well on some total physical fitness scale. Conversely, those with low scores on the cardiorespiratory test would be expected to do poorly on the fitness test.

Another example of using correlational techniques to assess construct validity is in the development of an attitude instrument. A discussion of the procedures for attitude research can be found in the paper by Silverman and Subramaniam (1999). Confirmatory factor analysis is used to see whether the data from a pilot study fit the proposed model of constructs. If the data do not fit the model, additional pilot testing occurs until the model fits established

known group difference method—A method used for establishing construct validity in which the test scores of groups that should differ on a trait or ability are compared.

standards. Examples of this technique with attitude measurement can be found in physical education and other areas (e.g., Keating & Silverman, 2004; Kulinna & Silverman, 1999; Subramaniam & Silverman, 2000).

Actually, all the other forms of validity we have discussed are used for evidence of construct-related validity. Indeed, using evidence from all the other forms is usually necessary to provide strong support for the validity of scores for a particular instrument and the use of its results.

RELIABILITY

> A valid test is reliable: It yields the same results on successive trials.

An integral part of validity is reliability, which pertains to the consistency, or repeatability, of a measure. A test cannot be considered valid if it is not reliable. In other words, if the test is not consistent—if you cannot depend on successive trials to yield the same results—then the test cannot be trusted. Of course, scores from a test can be reliable yet not valid, but it can never be valid if it is not reliable. For example, weighing yourself repeatedly on a broken scale would give reliable results but not valid ones. Test reliability is sometimes discussed in terms of **observed score**, **true score**, and **error score**.

> **observed score**—In classical test theory, an obtained score that comprises a person's true score and error score.

> **true score**—In classical test theory, the part of the observed score that represents the person's real score and does not contain measurement error.

> **error score**—In classical test theory, the part of an observed score that is attributed to measurement error.

A test score obtained by a person is the observed score. It is not known whether this is a true assessment of this person's ability or performance. Measurement error may occur because of the test directions, the instrumentation, the scoring, or the person's emotional or physical state. Thus, an observed score theoretically consists of the person's true score and error score. Expressed in terms of score variance, the observed score variance consists of true score variance plus error variance. The goal of the tester is to remove error to yield the true score. The coefficient of reliability is the ratio of true score variance to observed score variance. Because true score variance is never known, it is estimated by subtracting error variance from observed score variance. Thus, the reliability coefficient reflects the degree to which the measurement is free of error variance.

Sources of Measurement Error

Measurement error can come from four sources: the participant, the testing, the scoring, and the instrumentation. Measurement error associated with the participant includes many factors, including mood, motivation, fatigue, health, fluctuations in memory and performance, previous practice, specific knowledge, and familiarity with the test items.

Errors in testing are related to how clear and complete the directions are, how rigidly the instructions are followed, whether supplementary directions or motivation is applied, and so forth. Errors in scoring relate to the competence, experience, and dedication of the scorers and to the nature of the scoring itself. The extent to which the scorer is familiar with the behavior being tested and the test items can greatly affect scoring accuracy. Carelessness and inattention to detail produce measurement error. Measurement error because of instrumentation includes such obvious examples as inaccuracy and the lack of calibration of mechanical and electronic equipment. It also refers to the inadequacy of a test to discriminate between abilities and to the difficulty of scoring some tests.

Expressing Reliability Through Correlation

> **interclass correlation**—The most commonly used method of computing correlation between two variables; also called *Pearson r* or *Pearson product moment coefficient of correlation*.

The degree of reliability is expressed by a correlation coefficient, ranging from 0.00 to 1.00. The closer the coefficient is to 1.00, the less error variance it reflects and the more the true score is assessed. Reliability is established in several ways, which are summarized later in this chapter. The correlation technique used for computing the reliability coefficient differs from that used for establishing validity. Pearson r is often called **interclass correlation**. This coefficient is a bivariate statistic, meaning that it is used to correlate two variables,

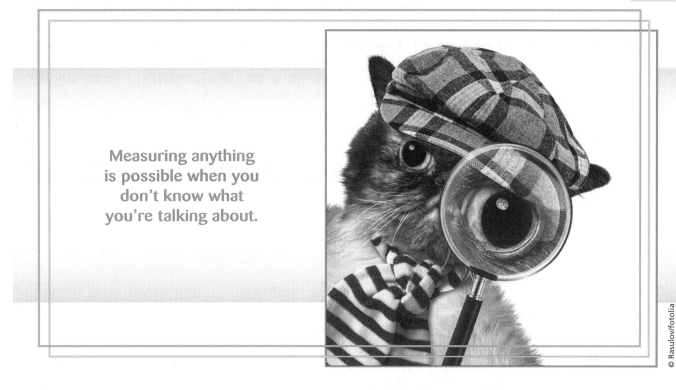

Measuring anything is possible when you don't know what you're talking about.

as when determining validity by correlating judges' ratings with scores on a skill test. But interclass correlation is not appropriate for establishing reliability because two values for the same variable are being correlated. When a test is given twice, the scores on the first test are correlated with the scores on the second test to determine their degree of consistency. Here, the two test scores are for the same variable, so interclass correlation should not be used. Rather, **intraclass correlation** is the appropriate statistical technique. This method uses ANOVA to obtain the reliability coefficient.

intraclass correlation— An ANOVA technique used for estimating the reliability of a measure.

Interclass Correlation

There are three main weaknesses of Pearson r (interclass correlation) for reliability determination. The first is that, as mentioned previously, the Pearson r is a bivariate statistic, whereas reliability involves univariate measures. Second, the computations of Pearson r are limited to only two scores, X and Y. Often, however, more than two trials are given, and the tester is concerned with the reliability of multiple trials. For example, if a test specifies three trials, the researcher must either give three more trials and use the average or best score of each set of trials for the correlations, or correlate the first trial with the second, the first with the third, and the second with the third. In the first case, extra trials must be given solely for reliability purposes; in the second case, meaningfulness suffers when several correlations among trials are shown. Finally, the interclass correlation does not provide a thorough examination of different sources of variability on multiple trials. For example, changes in means and standard deviations from trial to trial cannot be assessed with the Pearson r method but can be analyzed with intraclass correlation.

Intraclass Correlation

Intraclass correlation provides estimates of systematic and error variance. For example, systematic differences among trials can be examined. The last trials may differ significantly from the first trials because of a learning phenomenon or fatigue effect (or both). If the tester recognizes this, then perhaps initial (or final) trials can be excluded or the point at which performance levels off can be used as the score. In other words, through ANOVA the tester can truly examine test performance from trial to trial and then select the most reliable testing schedule.

The procedures leading to the calculation of intraclass correlation (R) are the same as those of simple ANOVA with repeated measures, discussed in chapter 9. The components are subjects, trials, and residual sums of squares. An example of a summary ANOVA is shown in table 11.2. We use these data to compute R.

TABLE 11.2

Summary of ANOVA for Reliability Estimation (Three Trials)

Source	SS	df	MS	F
Subjects	14.9	4	3.73	
Trials	7.6	2	3.80	5.94*
Residual	5.1	8	0.64	

*$p < .05$.

M (trial 1) = 2.4; M (trial 2) = 3.8; M (trial 3) = 4.0.

An F for trials determines whether any significant differences exist among the three trials. Refer to table 6 in the appendix and read down the 2 df column to the 8 df row. Our F of 5.94 is found to be greater than the table F of 4.46 at the .05 level of probability. Thus, significant differences exist among trials. At this point we should mention that opinions vary about what should be done with trial differences (Baumgartner & Jackson, 1991; Johnson & Nelson, 1986; Safrit, 1976). Some test authorities argue that the test performance should be consistent from one trial to the next and that any trial-to-trial variance should be attributed simply to measurement error. If we decide to do this, the formula for R is

$$R = (MS_S - MS_E) / MS_S \tag{11.1}$$

in which MS_S is the mean squares for subjects (from table 11.2) and MS_E is the mean squares for error, which is computed as follows:

$$MS_E = \frac{SS \text{ for trials} + SS \text{ for residual}}{df \text{ for trials} + df \text{ for residual}}$$
$$= \frac{7.6 + 5.1}{2 + 8} \tag{11.2}$$
$$= 1.27$$

Thus, $R = (3.73 - 1.27) / 3.73 = .66$.

Another way of dealing with significant trial differences is to discard the trials that are noticeably different from the others (Baumgartner & Jackson, 1991). Then a second ANOVA is calculated on the remaining trials, and another F test is computed. If F is not significant, R is calculated using equation 11.1, in which trial variance is considered a measurement error. If F is still significant, additional trials are discarded, and another ANOVA is conducted. The purpose of this method is to find a measurement schedule that is free of trial differences (i.e., to find a nonsignificant F that yields the largest possible criterion score) and that is most reliable. This method is especially appealing when there is an apparent trend in trial differences, as when a learning phenomenon (or release of inhibitions) is evident from increased scores on a later trial or trials. For example, if five trials on a performance test yield mean scores of 15, 18, 23, 25, and 24, you might discard the first two trials and compute another analysis on the last three trials. Similarly, a fatigue effect may be evidenced by a decrease in scores on the final trials in some types of tests.

At the bottom of table 11.2, note that the mean of trial 1 is considerably lower than the means of trials 2 and 3. Therefore, we discard the first trial and compute another ANOVA on trials 2 and 3 only (results are shown in table 11.3). The F for trials in the table is nonsignificant, so first we compute MS_E by combining the sums of squares for trials and residual and dividing by their respective degrees of freedom: $MS_E = (0.1 + 3.4) / (1 + 4) = 0.7$. Then we compute R with equation 11.1:

$$R = (2.85 - 0.7) / 2.85 = .75$$

We see that R is considerably higher when we discard the first trial.

TABLE 11.3

Summary of ANOVA for Reliability With Trial 1 Discarded

Source	SS	df	MS	F
Subjects	11.4	4	2.85	
Trials	0.1	1	0.10	0.11*
Residual	3.4	4	0.85	

*$p > .05$.

A third approach is simply to ignore trial-to-trial variance as a measurement error. In this approach, trial-to-trial variance is not considered to be true score variance or error score variance. Consequently, R is notably higher than in the previous approaches because all trial-to-trial variance is removed. Although some measurement authorities advocate this approach, we do not recommend it because it does not seem to follow the theory that observed score variance equals true score variance plus error score variance. We believe that the tester should ascertain whether there are trial-to-trial differences. If significant differences are found, the tester can decide whether to eliminate some trials (as with a learning trend) or simply to consider trial-to-trial differences as measurement error.

Intertester reliability is determined by the same procedures we just outlined. Thus, the objectivity of judges or different testers is analyzed by intraclass R, and judge-to-judge variance is calculated in the same way as trial-to-trial variance. Of course, more complex ANOVA designs can be used in which trial-to-trial, day-to-day, and judge-to-judge sources of variance all can be identified. Baumgartner (1989) and Safrit (1976) discussed some models that can be used for establishing reliability.

intertester (interrater) reliability—The degree to which different testers can obtain the same scores on the same participants; also called *objectivity*.

METHODS OF ESTABLISHING RELIABILITY

Establishing reliability is easier than establishing validity. We first look at three types of coefficients of reliability: stability, alternate forms, and internal consistency.

Stability

The coefficient of **stability** is determined by the test–retest method on separate days. This method is used frequently with fitness and motor performance measures but less often with pencil-and-paper tests. This is one of the most severe tests of consistency because the errors associated with measurement are likely to be more pronounced when the two test administrations are separated by a day or more.

In the **test–retest method**, the test is given one day and then repeated a day or so later. The interval may be governed to some extent by how strenuous the test is and whether more than a day's rest is needed. Of course, the interval cannot be so long that actual changes in ability, maturation, or learning occur between the two test administrations.

Intraclass correlation should be used to compute the coefficient of stability of the scores on the two tests. Through ANOVA procedures, the tester can determine the amount of variance accounted for by the separate days of testing, test trial differences, participant differences, and error variance.

stability—A coefficient of reliability measured by the test–retest method on different days.

test–retest method—A method of determining stability in which a test is given one day and then administered exactly as before a day or so later.

Alternate Forms

The **alternate-forms method** of establishing reliability involves the construction of two tests that supposedly sample the same material. This method is sometimes referred to as the *parallel-form method* or the *equivalence method*. The two tests are given to the same people. Ordinarily, some time elapses between the two administrations. The scores on the two tests are then correlated to obtain a reliability coefficient. The alternate-forms method is widely used with standardized tests, such as those of achievement and scholastic aptitude. The method is rarely used with physical performance tests, probably because constructing two sets of good physical test items is more difficult than writing two sets of questions.

Some test experts maintain that, theoretically, the alternate-forms method is the preferred method for determining reliability. Any test is only a sample of test items from a universe of possible test items. Thus, the degree of relationship between two such samples should yield the best estimate of reliability.

Internal Consistency

Reliability coefficients can be obtained by several methods that are classified as **internal consistency** techniques. Some common methods are the same-day test–retest, the split-half method, the Kuder-Richardson method of rational equivalence, and the coefficient alpha technique.

The **same-day test–retest method** is used almost exclusively with physical performance tests because practice effects and recall tend to produce spuriously high correlations when this technique is used with written tests. The test–retest on the same day results in a higher reliability coefficient than does a test–retest on separate days. We would certainly expect more consistency of performance within the same day than on different days. Intraclass correlation is used to analyze trial-to-trial (internal) consistency.

The **split-half technique** has been widely used for written tests and occasionally in performance tests that require numerous trials. The test is divided in two, and the two halves are then correlated. A test could be divided into first and second halves, but this approach is usually not deemed satisfactory. People may tire near the end of the test, or easier questions may be placed in the first half. Usually, the odd-numbered questions are compared with the even-numbered ones. That is, the number of odd-numbered questions a person got correct is correlated with the number of even-numbered questions the person got correct.

Because the correlation is between the two halves of the test, the reliability coefficient represents only half of the total test; that is, behavior is sampled only half as thoroughly. Thus, a step-up procedure, the **Spearman-Brown prophecy formula**, is used to estimate the reliability for the entire test because the total test is based on twice the sample of behavior (twice the number of items). The formula is

$$\text{Corrected reliability coefficient} = \frac{2 \times \text{reliability for } 1/2 \text{ test}}{1.00 + \text{reliability for } 1/2 \text{ test}} \qquad (11.3)$$

If, for example, the correlation between the even-numbered items and the odd-numbered items was .85, the corrected reliability coefficient would be

$$\frac{2 \times .85}{1.00 + .85} = \frac{1.70}{1.85} = .92$$

Another split-half method is the **Flanagan method**, which analyzes the variances of the halves of the test in relation to the total variance. No correlation or Spearman-Brown step-up procedure is involved.

In the **Kuder-Richardson (KR) method of rational equivalence**, one of two formulas, known as KR-20 and KR-21, is used for items scored dichotomously (i.e., right or wrong).

alternate-forms method—A method of establishing reliability involving the construction of two tests that both supposedly sample the same material; also called *parallel-form method* or *equivalence method*.

internal consistency—An estimate of the reliability that represents the consistency of scores within a test.

same-day test–retest method—A method of establishing reliability in which a test is given twice to the same participants on the same day.

split-half technique—A method of testing reliability in which the test is divided in two, usually by making the odd-numbered items one half and the even-numbered items the other half. The two halves are then correlated.

Spearman-Brown prophecy formula—An equation developed to estimate the reliability for the entire test when the split-half technique is used to test reliability.

Flanagan method—A process for estimating reliability in which the test is split into two halves, and the variances of the halves of the test are analyzed in relation to the total variance of the test.

Kuder-Richardson (KR) method of rational equivalence—Formulas developed for estimating the reliability of a test from a single test administration.

Only one test administration is required, and no correlation is calculated. The resulting coefficient represents an average of all possible split-half reliability coefficients.

Highly regarded by many test experts, the KR-20 involves the proportions of students who answered each item correctly and incorrectly in relation to the total score variance. The KR-21 is a simplified, less accurate version of the KR-20.

The **coefficient alpha** technique, sometimes referred to as the *Cronbach alpha coefficient* (see Cronbach, 1951), is a generalized reliability coefficient that is more versatile than other methods. One particularly desirable feature of coefficient alpha is that it can be used with items that have various point values, such as essay tests and attitude scales that have as possible answers *strongly agree, agree,* and so on. The method involves calculating variances of the parts of a test. The parts can be items, test halves, trials, or a series of tests such as quizzes. When the items are dichotomous (i.e., either right or wrong), the coefficient alpha results in the same reliability estimate as KR-20 (in fact, KR-20 is just a special case of coefficient alpha). When the parts are halves of the test, the results are the same as the Flanagan split-halves method. And when the parts are trials or tests, the results are the same as intraclass correlation. Coefficient alpha is probably the most commonly used method of estimating reliability for standardized tests.

> **coefficient alpha**—A technique used for estimating the reliability of multiple-trial tests; also called *Cronbach alpha coefficient.*

INTERTESTER RELIABILITY (OBJECTIVITY)

A form of reliability that pertains to the testers is called intertester (or interrater) reliability, or, often, **objectivity.** This facet of reliability is the degree to which different testers can achieve the same scores on the same subjects.

> **objectivity**—The degree to which different testers can achieve the same scores on the same subjects, also known as *intertester reliability.*

Overall, most teachers and students prefer objective to subjective measures because so much depends on how valid and reliable the measures are. Objective measures are not automatically better than subjective ones, but they do yield quantitative scores that are more understandable and that can be statistically handled more easily. In most research techniques, objective measurements are essential.

The degree of objectivity (intertester reliability) can be established by having more than one tester gather data. Then the scores are analyzed with intraclass correlation techniques to obtain an intertester reliability coefficient. It is possible to assess a number of sources of variance in one analysis, such as variance caused by testers, trials, days, participants, and error (for a discussion of the calculations involved, see Safrit, 1976).

Some forms of research in physical activity involve the observation of certain behaviors in real-world settings, such as during physical education classes or sport participation. This approach involves the use or development of some sort of coding instrument. Most frequently, the instrument has a series of categories into which motor behaviors may be coded. The behaviors are observed using techniques described in chapter 16, such as event recording, time sampling, and duration recording.

In these types of scales, as with any measure, validity and reliability are important. But it is usually much more difficult to obtain consistency in recording activities in a physical education class or sport than in recording error measurements from a laboratory task such as the linear slide. Consequently, observational researchers are concerned about coder consistency. Typically, coders are trained to a criterion level of reliability, and reliability is checked regularly throughout the project. A common way of estimating reliability among coders, often used in event recording and time sampling, is called **interobserver agreement (IOA)**, which uses the following formula:

> **interobserver agreement (IOA)**—A common way of estimating reliability among coders by using a formula that divides the number of agreements in behavior coding by the sum of the agreements and disagreements.

$$IOA = \frac{\text{agreements}}{\left(\text{agreements} + \text{disagreements}\right)} \tag{11.4}$$

Agreements are behaviors that are coded the same, whereas disagreements are behaviors that are coded differently. IOA is typically reported as the percentage of agreement.

STANDARD ERROR OF MEASUREMENT

Previous chapters touched on the idea of standard error several times with regard to confidence intervals, the t and F tests, and interpreting significance levels. Chapter 8 also discussed standard error of prediction in correlational research. The standard error of measurement is an important concept in interpreting the results of measurement. Sometimes we get too carried away with the aura of scientific data collection and fail to realize that the possibility of measurement error always exists. For example, maximal oxygen consumption ($\dot{V}O_2max$) has been mentioned several times in this book as the most valid measure of cardiorespiratory fitness. Field tests are frequently validated through correlation with $\dot{V}O_2max$. We must be careful, however, not to consider $\dot{V}O_2max$ as a perfect test that is error free. Every test yields only observed scores; we can obtain only estimates of a person's true score. It is much better to think of test scores as falling within a range that contains the true score. The formula for the standard error of measurement ($S_{Y \cdot X}$) is

$$S_{Y \cdot X} = s\sqrt{1.00 - r}$$

where s is the standard deviation of the scores and r is the reliability coefficient for the test. In the measurement of percentage of body fat for female adults, assume that the standard deviation is 5.6% and the test–retest correlation is .83. The standard error of measurement would be

$$S_{Y \cdot X} = 5.6\sqrt{1.00 - .83} = 2.3\%$$

Assume further that a particular woman's measured fat percentage is 22.4%. We can use the standard error of measurement to estimate a range within which her true fat percentage probably falls.

Standard errors are assumed to be normally distributed and are interpreted in the same way as standard deviations. About two-thirds (68.26%) of all test scores fall within ±1 standard error of measurement of their true scores. In other words, a person's true score has about a 68% chance of being found within a range of the obtained score plus or minus the standard error of measurement. In the example of the woman who had an obtained score of 22.4% fat, chances are 2 in 3 that her true fat percentage is 22.4% ± 2.3% (standard error of measurement), or somewhere between 20.1% and 24.7%.

We can be more confident if we multiply the standard error of measurement by 2 because about 95% (95.44%) of the time the true score can be found within a range of the observed score plus or minus two times the standard error of measurement. Thus, in the present example, we can be about 95% sure that the woman's true fat percentage is 22.4% ± 4.6% (±2.3% × 2); that is, between 17.8 and 27.0%.

From the formula we see that the standard error of measurement is governed by the variability of the test scores and the reliability of the test. If we had a higher reliability coefficient, the error of measurement would obviously decrease. In the present example, if the reliability coefficient was .95, the standard error of measurement would be only 1.3%. With the same reliability (.83) but with a smaller standard deviation, such as 4%, the standard error of measurement would be 1.6%.

Remember the idea of the standard error of measurement when you are interpreting test scores. As indicated earlier, people sometimes have blind faith in measurements, particularly if the measurements seem scientific. With fat percentage estimation from skinfold thicknesses, for example, we need to keep in mind that error is connected not only with

Although a test never yields a person's true score, test scores should be considered as being in the range containing the true score.

the skinfold measurements but also with the criterion that these measurements are predicting—that is, the body density values obtained from underwater weighing and the fat percentage determined from them. Yet we have observed people accepting as gospel that they have a certain amount of fat because someone measured a few skinfolds. Newspapers have reported that some athletes have only 1% fat, an impossibility from a physiological standpoint. Moreover, obtaining a predicted negative fat percentage from regression equations is possible. Please do not misunderstand. We are not condemning skinfold measurements. We are simply trying to emphasize that all measurements are susceptible to errors and that common sense, coupled with knowledge of the concept of standard error of measurement, can help us better understand and interpret the results of measurements. We revisit standard error in chapter 15 when we discuss errors associated with surveys.

USING STANDARD SCORES TO COMPARE PERFORMANCE

Direct comparisons of scores are not meaningful without some point of reference. Is a score of 46 cm on the vertical jump as good as a score of 25 push-ups? How can you compare centimeters and repetitions? If we know that the class mean for the vertical jump is 40 cm and that 20 is the mean for push-ups, we know that the performances of 46 cm and 25 push-ups are better than average, but how much better? Is one performance better than the other?

One way to compare the performances is to convert each score to a standard score. A standard score is a score expressed in terms of standard deviations from the mean. We now discuss how to determine standard scores by using z scores and T scales.

z Scores

The basic standard score is the **z score**. The z scale converts raw scores to units of standard deviation in which the mean is zero and a standard deviation is 1.0. The formula is

$$z = (X - M) / s \qquad (11.6)$$

Suppose the mean and standard deviation for vertical-jump scores are 40 cm and 6 cm, respectively, and for push-ups they are 20 and 5, respectively. Thus, a score of 46 cm for the vertical jump is a z score of +1.00: $z = (46 - 40) / 6 = 6 / 6 = 1.00$. A score of 25 push-ups is also a z score of +1.00: $z = (25 - 20) / 5 = 5 / 5 = 1.00$.

We see that the person performed exactly the same on the two tests. Both performances were 1 standard deviation above the mean. Similarly, scores of different students on the same test can be compared by z scores. A student who jumps 37 cm has a z score of –0.5, a student who jumps 44 cm has a z score of 0.67, and so on. All standard scores are based on the z score. But because z scores are expressed in decimals and can be positive or negative, they are not as easy to work with as some other scales.

T Scales

The **T scale** sets the mean at 50 and the standard deviation at 10. Hence, the formula is $T = 50 + 10z$. This removes the decimal and makes all the scores positive. A score 1 standard deviation above the mean ($z = 1.0$) is a T score of 60. A score 1 standard deviation below the mean ($z = -1.0$) is a T score of 40. Because more than 99% (99.73%) of the scores fall between ±3s, it is rare to have T scores below 20 ($z = -3.0$) or above 80 ($z = +3.0$). Some standardized tests that use different transformations of means and standard deviations using the z-score distribution are shown in table 11.4.

z score—The basic standard score that converts raw scores to units of standard deviation in which the mean is zero and a standard deviation is 1.0.

T scale—A type of standard score that sets the mean at 50 and standard deviation at 10 to remove the decimal found in z scores and to make all scores positive.

TABLE 11.4

Standardized Means and Standard Deviations of Well-Known Tests

Scale	M	s
Graduate Record Examination Quantitative	151.3	8.7
Stanford-Binet IQ	100	16
College Entrance Examination	500	100
National Teachers Examination	500	100
Wechsler IQ	100	15

The decision of which standard score to use depends on the nature of the research study and the extent of interpretation required for the test takers. In essence, then, it is a matter of choice in light of the use of the measures.

MEASURING MOVEMENT

Much of the research in physical activity obviously involves movement. The measurement of physical fitness has fascinated exercise physiologists and physical educators for years. The concept of health-related physical fitness is widely accepted as being represented by the components of cardiorespiratory endurance, muscular strength, muscular endurance, flexibility, and body composition. In addition, many research studies have examined other fitness parameters that are needed primarily for skilled performances, such as in athletics and dance. These components include power, speed, reaction time, agility, balance, kinesthetic perception, and coordination. Research in motor behavior generally deals with the acquisition and control of motor skills. Types of measurements include tests of basic movement patterns, sport skills, and controlled (often novel) laboratory tasks.

Research in biomechanics involves such measurements as high-speed cinematography, force transduction, and electromyography. Research in pedagogy typically involves observations of behavior in real-world settings, such as during physical education classes or sport participation. Video-recorded observations are often used to allow more precise analyses.

The validity and reliability of all forms of measuring movement must be defended by the researchers. In some respects, measuring movement is simpler and more straightforward than measuring cognitive or affective behaviors. For example, the amount of force a person can apply in a given movement can be measured accurately and directly. The distance a person can jump can be assessed with a measuring tape, and everyone accepts the result as a valid and reliable measure. On the other hand, cognitive and affective behaviors must usually be inferred from marks made on a sheet of paper. Nevertheless, the measurement of movement is rarely a simple manner of measuring distance jumped. It is usually complex and often difficult to standardize. Each type of measurement has its own methodological difficulties, which pose problems for the researcher with regard to validity and reliability. We do not attempt a thorough description of measuring movement because much of this material is specific to your area of specialization (e.g., motor behavior, exercise, wellness). The focus of your graduate program is on many of these issues.

MEASURING WRITTEN RESPONSES

The measurement of written (and oral) responses is part of the methodology of numerous research studies in kinesiology. Research questions frequently deal with affective behavior, which includes attitudes, interests, emotional states, personality, and psychological traits.

MEASURING AFFECTIVE BEHAVIOR

Affective behavior includes attitudes, personality, anxiety, self-concept, social behavior, and sporting behavior. Many general measures of these constructs have been used over the years (e.g., *Cattell's 16 Personality Factor [PF] Questionnaire,* Spielberger's *State-Trait Anxiety Inventory*). Nelson (1989) presented an extensive discussion of the construction, strengths, and weaknesses of affective tests and scaling.

Attitude Inventories

A large number of attitude inventories have been developed. Undoubtedly, many test developers see a direct link between attitude and behavior. For example, if a person has a favorable attitude toward physical activities, that person will participate in such activities. Research, however, has seldom substantiated this link between attitude and behavior, although such a link seems logical.

Researchers usually try to locate an instrument that has already been validated and that is an accepted measure of attitude rather than construct a new one. Finding a published test that closely pertains to the research topic and that has scores validated with a similar sample is a problem. Another problem is that often the researcher wants to determine whether some treatment will bring about an attitude change. A source of invalidity discussed in chapter 18 is the reactive effects of testing. The pretest sensitizes the participant to the attitudes in question, and this sensitization, rather than the treatment, may promote a change.

Another problem (mentioned in chapter 15) inherent in any self-report inventory is whether the person is truthful. It is usually evident what a given response to an item in an attitude inventory indicates. For example, a person may be asked to show his or her degree of agreement or disagreement with the statement "Regular exercise is an important part of my daily life." The person may perceive that the socially desirable response is to agree with the statement regardless of his true feelings. Some respondents deliberately distort their answers to appear good (or bad). Tests sometimes use so-called filler items to make the true purpose of the instrument less visible. For example, a test designer might include several unrelated items, such as "Going to the opera is a desirable social activity," to disguise the fact that the instrument is designed to measure attitude toward exercise. This issue is especially important when social desirability considerations may be biasing factors.

When a researcher seeks an attitude instrument to use in a study, validity and reliability must, of course, be primary considerations. Unfortunately, published attitude scales have not always been constructed scientifically, and limited information is provided about their validity and reliability. Reliability of methodology can be established easily. (We are not saying that attitude scales are easily made reliable.) Validity is usually the problem because of the failure to develop a satisfactory theoretical model for the attitude construct.

Personality Research

Many research studies in physical education and sport have attempted to explore the relationship between personality traits and aspects of athletic performance. Interest in this topic can be attributed to several factors, including the great deal of public attention focused on athletics. Athletes are obviously special people with regard to physical characteristics. Beyond that, however, is the hypothesis that athletes have certain personality traits that distinguish them from nonathletes.

One area of investigation is to identify personality traits that might be uniquely characteristic of athletes in different sports. For example, do people who gravitate toward vigorous contact sports differ in personality from those who prefer noncontact sports? That is, is there an American football type or a bowler type? Are superior athletes different from average athletes in certain personality traits? Can participation in competitive sport modify a person's personality structure? Moreover, do athletes within a sport differ on some traits that, if known, could point to different coaching strategies or perhaps be used to screen and predict which athletes will be difficult to coach or will lack certain qualities associated with success?

People with a strong interest in sports, such as former coaches and players, have been greatly attracted to the study of personality and athletics. But having a strong interest in athletics does not compensate for a lack of preparation and experience in psychological evaluation. Unfortunately, in some cases, personality-measuring instruments have been misused. Recently, sport psychologists have recognized the value of sport-specific measures of various behaviors and perceptions. Much of the impetus for this approach can be traced to Martens' *Sport Competition Anxiety Test* (Martens, 1977).

Some sport-specific measures relate to group cohesiveness (Carron, Widmeyer, & Brawley, 1985), intrinsic and extrinsic motivation (Weiss, Bredemeier, & Shewchuck, 1985), confidence (Vealey, 1986), and sport achievement orientation (Gill & Deeter, 1988). The rationale for the sport-specific approach is that general measures of achievement, anxiety, motivation, self-concept, social behavior, and fair play do not have high validity for sport situations. Recent studies in test development indicate a commitment by sport psychologists to develop measures of affective behavior that are multidimensional and specific to the competitive environment.

SCALES FOR MEASURING AFFECTIVE BEHAVIOR

In measuring affective behavior, a variety of scales are used to quantify the responses. Two of the most commonly used are the Likert-type scale and the semantic differential scale.

Likert-Type Scale

Likert-type scale—A type of closed question that requires the participant to respond by choosing one of several scaled responses; the intervals between items are assumed to be equal.

The **Likert-type scale** is referred to in chapter 15 in connection with survey research techniques. It is usually a five- or seven-point scale with assumed equal intervals between points. The Likert-type scale is used to assess the degree of agreement or disagreement with statements and is widely used in attitude inventories. An example of a Likert-type item follows:

> **I prefer quiet recreational activities such as chess, cards, or checkers rather than activities such as running, tennis, or basketball.**

> **Strongly agree   Agree   Undecided   Disagree   Strongly disagree**

A principal advantage of scaled responses such as the Likert type is that they permit a wider choice of expression than responses such as *always* or *never,* or *yes* or *no.* The five, seven, or more intervals may help increase the reliability of the instrument. (For more comprehensive information concerning the Likert-type scale, semantic differential scale, and other scales, see Edwards, 1957, and Nunnaly, 1978.)

Semantic Differential Scale

semantic differential scalee—A scale used to measure affective behavior in which the respondent is asked to make judgments about certain concepts by choosing one of seven intervals between bipolar adjectives.

The **semantic differential scale** employs bipolar adjectives at each end of a seven-point scale. The respondent is asked to make judgments about certain concepts. The scale is based on the importance of language in reflecting a person's feelings. A sample of a semantic differential item is shown here:

The 1–7 scale between adjectives is scored, with 7 being the most positive judgment. Factor analysis studies have consistently identified the same three dimensions that are assessed by the semantic differential technique: evaluation (e.g., fair–unfair), potency (e.g., powerful–feeble), and activity (e.g., dynamic–static).

Rating Scales

Rating scales are sometimes used in research to evaluate performance. For example, in a study that compares strategies in teaching diving, the dependent variable (diving skill) would most likely be derived from expert ratings because diving does not lend itself to objective skill tests. Thus, after the experimental treatments have been applied, people knowledgeable in diving rate all the participants on their diving skills. To do this in a systematic and structured manner, the raters need to have some kind of scale with which to assess skill levels in different parts or phases of the performance.

A self-rating scale concerning a person's perceived efforts during exercise that has been widely used in research is the **rating of perceived exertion (RPE)** scale by Borg (1962). The underlying rationale for the scale is to combine and integrate the many physiological indicators of exertion into a whole, or gestalt, of subjective feeling of physical effort. This feeling of perceived exertion was quantified by Borg into a scale with numbers ranging from 6 to 20, reflecting a range of exertion from "very, very light" to "very, very hard."

Types of Rating Scales

There are different kinds of rating scales. Some use numerical ratings, some use checklists, some have verbal cues associated with numerical ratings, some require forced choices, and still others use rankings. Some scales are simple, whereas others are complex. Whatever the degree of complexity, however, practice in using the scale is imperative.

When more than one judge is asked to rate performances, some common standards must be set. Training sessions with video-recorded performances of people of different ability levels are helpful in establishing standard frames of reference before judging the actual performances. Intertester agreement and reliability were discussed earlier in this chapter.

Rating Errors

Despite efforts to make ratings as objective as possible, pitfalls are inherent in the process. Recognized errors in rating include leniency, central tendency, the halo effect, proximity, observer bias, and observer expectation.

Leniency is the tendency for observers to be overgenerous in their ratings. This error is less likely to occur in research than in evaluating peers (e.g., coworkers). Thorough training of raters is the best means of reducing leniency.

rating scale—A measure of behavior that involves a subjective evaluation based on a checklist of criteria.

rating of perceived exertion (RPE)—A self-rating scale developed by Borg (1962) to measure a person's perceived effort during exercise.

leniency—The tendency for observers to be overly generous in rating.

Central tendency errors result from the inclination of the rater to give an inordinate number of ratings in the middle of the scale, avoiding the extremes of the scale. Several reasons are suggested as the cause. Ego needs or status may play a part. For example, the judge is acting in the role of an expert and, perhaps unconsciously, may grade good performers as average to suggest that she is accustomed to seeing better performances. Sometimes, errors of central tendency are due to the observer's wanting to leave room for better future performances. A common complaint in large gymnastics, diving, and skating competitions is that the performers scheduled early in the meet are scored lower for comparable performances than athletes scheduled later in the competition. Another central tendency error results when the judge avoids assigning very low scores, likely because he is reluctant to be too harsh. This kind of rating is, of course, a form of leniency.

The **halo effect** is the commonly observed tendency for a rater to allow previous impressions or knowledge about a certain person to influence all ratings of that person's behaviors. For example, knowing that a person excels in one or more activities, a judge may rate that person highly on all other activities. The halo effect perhaps is not the most appropriate term because negative impressions of a person tend to lead to lower ratings in subsequent performances.

Proximity errors are often the result of overdetailed rating scales, insufficient familiarity with the rating criteria, or both. Proximity errors are manifested when the rater tends to rate behaviors that are listed close together as more nearly the same than when the behaviors are separated on the scale. For example, if the qualities "active" and "friendly" are listed side by side on the scale, proximity errors result if raters evaluate performers as more similar on those characteristics than if the two qualities were listed several lines apart on the rating scale. Of course, if the rater does not have adequate knowledge about all facets of the behavior, she may not be able to distinguish between behaviors that logically should be placed close together on the scale. Thus, the different phases of behavior are rated the same.

Observer bias errors vary with the judge's own characteristics and prejudices. For example, a person who has a low regard for movement education may tend to rate students from such a program too low. Racial, sexual, and philosophical biases are potential sources of rating errors. Observer bias errors are directional because they are consistently too high or too low.

Observer expectation errors can operate in various ways, often stemming from other sources of errors such as the halo effect and observer bias. Observer expectations can contaminate the ratings because a person who expects certain behaviors is inclined to see evidence of those behaviors and interpret observations in the expected direction.

Research has demonstrated the powerful phenomenon of expectation in classroom situations in which teachers are told that some children are gifted or slow learners. The teachers tend to treat the pupils accordingly, giving more attention and patience to the "gifted learners" and less time and attention to the "slow learners."

In the research setting, observer expectation errors are likely when the observer knows what the experimental hypotheses are and is thus inclined to watch for these outcomes more closely than if the observer were unaware of the expected outcomes. The double-blind experimental technique described in chapter 18 is useful for controlling expectation errors. In the double-blind technique, the observers do not know which participants received which treatments. They also should not know which performances occurred during the pretest and which occurred during the posttest.

In summary, rating errors are always potentially present. The researcher must recognize and strive to eliminate or reduce them. One way to minimize rating errors is to define the behavior to be rated as objectively as possible. In other words, avoid having the observer make many value judgments. Another suggestion is to keep observers ignorant of the hypotheses and of which participants received which treatments. Bias and expectation can be reduced if the observer is given no information about the participants' achievements, intelligence, social status, and other characteristics. To achieve high levels of accuracy and interrater reliability, the most important precaution the researcher can take is to train observers adequately.

MEASURING KNOWLEDGE

Obviously, the measurement of knowledge is a fundamental part of the educational aspects of physical education, exercise science, and sport science. But the construction of knowledge tests is relevant for research purposes as well. The procedures for establishing the validity and reliability of most pencil-and-paper measuring instruments used in research are similar to the procedures previously discussed. But we must also determine whether the difficulty of individual items and their ability to discriminate between levels of ability are functioning in the desired manner; that is, we must perform item analysis. (For a more thorough discussion of measurement methodology for knowledge tests, see Mood, 1989.)

The purpose of **item analysis** is to determine which test items are suitable and which need to be rewritten or discarded. Two important facets of item analysis are determining the difficulty of the test items and determining their power to discriminate among levels of achievement.

item analysis—The process in analyzing knowledge tests in which the suitability of test items and their ability to discriminate are evaluated.

item difficulty—The analysis of the difficulty of each test item in a knowledge test, determined by dividing the number of people who correctly answered the item by the number of people who responded to the item.

Analyzing Item Difficulty

The analysis of **item difficulty** is usually accomplished easily. You simply divide the number of people who correctly answered the item by the number of people who responded to the item. For example, if 80 people answered an item and 60 answered it correctly, the item would have a difficulty index of .75 (60 / 80). The more difficult the item is, the lower its difficulty index is. For example, if only 8 of 80 answered an item correctly, the difficulty index is 8 / 80, or .10. Most test authorities recommend eliminating questions with difficulty indices below .10 or above .90. The best questions are those that have difficulty indices around .50. Occasionally, a test maker may wish to set a specific difficulty index for screening purposes. For example, if only the top 30% of a group of applicants are to be chosen, this could be accomplished by using questions with difficulty indices of .30. Questions that everyone answers correctly or that everyone misses provide no information about people differences in norm-referenced measurement scales.

index of discrimination—The degree to which a test item discriminates between people who did well on the entire test and those who did poorly.

Analyzing Item Discrimination

Item discrimination (also known as **index of discrimination**), or the degree to which test items discriminate between people who did well on the entire test and those who did poorly, is an important consideration in analyzing norm-referenced test items. There are many ways to compute an index of discrimination. The simplest way is to divide the completed tests into a high-scoring group and a low-scoring group and then use the following formula:

Index of discrimination = $(n_H - n_L) / n$ (11.7)

where n_H is the number of high scorers who answered the item correctly, n_L is the number of low scorers who answered the item correctly, and n is the total number in either the high or the low group. To illustrate, if we have 30 in the high group and 30 in the low group, and if 20 of the high scorers answered an item correctly and 10 of the low scorers answered it correctly, the index of discrimination would be $(20 - 10) / 30 = 10 / 30 = .33$.

Various percentages of high and low scorers, such as the upper and lower 25, 30, or 33%, are used in determining discrimination indices. The Flanagan method uses the upper and lower 27%. The proportion of each group answering each item correctly is calculated; then a table of normalized biserial coefficients is consulted to obtain the item reliability coefficient. Thus, item reliability is the relationship between responses to each item and total performance on the test.

If approximately the same proportion of high scorers and low scorers answer an item correctly, the item is not discriminating. Most test makers strive for an index of discrimination of .20 or higher for each item. Obviously, a negative index of discrimination would be unacceptable. In fact, when this happens, the question needs to be examined closely to see whether something in the wording is throwing off the high scorers.

ITEM RESPONSE THEORY

Most of the information concerning the validity, reliability, and item analysis presented thus far pertains to what is called **classical test theory (CTT)**. Some radical changes have occurred recently in the study of the measurement of cognitive and affective behaviors. The advance that has received the most attention in the educational and psychological literature is **item response theory (IRT)**.

Characteristics of Item Response Theory

In CTT, inferences are made about items and student abilities from *total* test score information, whereas item response theory (IRT), as the name implies, attempts to estimate examinees' abilities based on their responses to test items. CTT requires only a few assumptions about the observed scores and the true scores of people on a test. Group statistics regarding the total score on a particular test for the total group being examined are used to make generalizations to an equivalent test and population. The estimate of error is assumed to be the same for all participants.

IRT is based on stronger assumptions than CTT is. The two major assumptions are unidimensionality and local independence. Unidimensionality means that a single ability or trait is being measured. This ability is not directly measurable, so IRT is sometimes referred to as *latent trait theory*. For an introduction to IRT and some of its applications, see Spray (1989). According to Spray, the real advantage of IRT is that the measurement of an examinee's ability from responses to test items is not limited to a particular test. Rather, it can be measured by any collection of test items that are considered to be measuring the same trait.

In CTT, item difficulty is measured as a function of the total group. In IRT, item difficulty is fixed and can be assessed relative to an examinee's ability level. Thus, the probability that an examinee with a particular ability level will make a correct response to an item can be mathematically described by an **item characteristic curve (ICC)**. The ICC is a nonlinear regression for any item that increases from left to right, indicating an increase in the probability of a correct response with increased ability, or latent trait. **Parameter invariance** means that the item difficulty remains constant regardless of the group of examinees. The discriminating power of the item is indicated by the steepness of the curve. The ICC can be used to analyze the difficulty of the item, the discriminating power of the item, and a so-called guessing parameter.

Application of Item Response Theory

Space limitations (and lack of knowledge by the authors) prohibit a detailed description and discussion of IRT. The concept is not simple, and complex computer programs are required. Large sample sizes are needed for item calibration and ability estimates. The IRT model has been the subject of intense research in psychology and education for several years. It definitely has potential application for assessment problems in physical education, exercise science, and sport science.

classical test theory (CTT)—A measurement theory built on the concept that observed scores are composed of a true score and an error score.

item response theory (IRT)—A theory that focuses on the characteristics of the test item and the examinee's response to the item as a means of determining the examinee's ability; also called *latent trait theory*.

IRT does not limit measurements to one test. Multiple tests that measure the same trait can be used.

item characteristic curve (ICC)—A nonlinear regression curve for a test item that increases from left to right, indicating an increase in the probability of a correct response with increased ability, or latent trait.

parameter invariance—A postulate in item response theory that the item difficulty remains constant regardless of the populations of examinees and that examinees' abilities should not change when a different set of test items is administered.

Spray (1989) described several ways that IRT can be used: item banking, adaptive testing, mastery testing, attitude assessment, and psychomotor assessment. **Item banking** is the creation of large pools of test items that can be used for constructing tests that have certain characteristics concerning the precision of estimating latent ability. **Adaptive testing**, sometimes called tailored testing, refers to selecting items that best fit (are neither too difficult nor too easy for) the ability level of each person. This function must be done on a computer by using items drawn from an item bank. Some of the widely used standardized tests such as the Graduate Record Examination use adaptive testing. IRT works in a way similar to computer games that vary the level of difficulty based on the player's performance.

In criterion-referenced measurement, tests are constructed that use a cutoff score to show the proportion of items that should be answered correctly to represent mastery of the subject matter. Item response theory can be used to select the optimal number of items that yield the most precise indication of mastery for an examinee. In other words, people of different ability levels would require different numbers of items.

Item response theory has considerable potential application for assessing attitudes and other affective behaviors. Models of IRT have been proposed that estimate the attitude or trait parameter of each respondent on an interval scale regardless of the ordinal nature of the scale. A score for each trait level is available for each category of each item. Changes in attitude or other traits over time can also be assessed with IRT models. To date, few affective measuring instruments have been constructed using IRT procedures.

The potential use of IRT for psychomotor assessment has been postulated (Spray, 1987) but has not yet been implemented to any extent. Motor performance tests are different from written tests in terms of the number of items and trials. Also, some assumptions (particularly local independence) of IRT are not easily accommodated in psychomotor testing. Preliminary research on the application of IRT to motor performance has been conducted by Safrit, Cohen, and Costa (1989). More research will undoubtedly be done on the application of IRT models to our field.

item banking—The creation of large pools of test items that can be used for constructing tests that have certain characteristics concerning the precision of estimating latent ability.

adaptive testing— Selecting test items that best fit the ability level of each person; also called *tailored testing*.

SUMMARY

In this chapter, we discussed the concepts of validity and reliability of measurements and their application to research. Criterion validity (which includes both concurrent and predictive validity) and construct validity are two of the most popular methods of validating measures used in research studies. One problem often identified regarding predictive measures is population or situation specificity. The value of using expectancy tables was discussed.

The topic of test reliability has prompted hundreds of studies and innumerable discussions among test theorists and researchers. The reason for this interest is that a measure that does not yield consistent results cannot be valid. Classical test theory views reliability in terms of observed scores, true scores, and error scores, and the coefficient of reliability reflects the degree to which the measure is free of error variance. The rationale for using intraclass R instead of r for reliability and the computational procedures for intraclass R were presented. Various methods of estimating reliability were mentioned, such as stability, alternate forms, and internal consistency. An understanding of the concept of standard error of measurement is vital. Any test score should be viewed as only an estimate of a person's true score that probably falls within a range of scores.

Standard scores allow direct comparisons of scores on different tests using different types of scoring. The basic standard score is the z score, which interprets any score in terms of standard deviations from the mean.

Research studies in physical activity frequently use instruments involving written responses to measure knowledge or affective attributes such as attitudes, interests, emotional states, and psychological characteristics. More recently, sport- and exercise-specific affective measures have been developed.

Responses to items on these scales often use the Likert-type scale (a five- or seven-point scale ranging from *strongly agree* to *strongly disagree*) and the semantic differential scale (a seven-point scale anchored at the extremes by bipolar adjectives). When using scales like these, researchers must be aware of problems such as rating errors, leniency, central tendency, the halo effect, proximity errors, observer bias errors, and observer expectation errors.

Knowledge testing requires item analysis techniques to ensure objective responses. The difficulty and discrimination of each test item must be established to determine the quality of the knowledge test.

Item response theory (IRT, also called latent trait theory) differs from classical test theory. It allows item difficulty to be fixed so that the participant's ability level can be determined. Item response theory is useful in tailoring tests for adaptive testing, mastery testing, attitude assessment, and psychomotor assessment.

Check Your Understanding

1. Briefly describe two ways that evidence of construct validity could be shown for a motor performance test (such as throwing), a test of power, or a test of manipulative skill. How could criterion validity be shown in your example?

2. Find a study in a research journal that used a questionnaire (e.g., an attitude inventory). Describe how the author reported the reliability of the instrument scores. What other technique could have been used to establish reliability?

3. A girl received a score of 78 on a test for which the reliability is .85 and the standard deviation is 8. Interpret her score in terms of the range in which her true score would be expected to fall with 95% confidence.

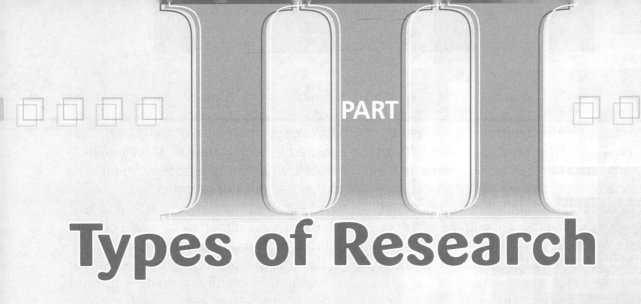

Types of Research

Plan—Any scheme that would have gone off without a hitch except for the intervention of reality.

—Anonymous

Research may be divided into five basic categories: analytical, descriptive, experimental, qualitative, and mixed. In particular, researchers are beginning to plan studies across these basic categories because better and more complete descriptions and explanations of human behavior can be developed. These approaches are often called mixed methods, and chapter 20 addresses that topic.

Chapter 12 discusses sociohistorical research, which is a type of analytical research that answers questions through the use of past knowledge and events. Substantial changes have occurred in recent years in the reporting of sociohistorical research, particularly regarding the integration of the events of interest with other related events. In this chapter, David Wiggins and Daniel Mason provide an excellent overview of sociohistorical research methods.

Chapter 13 describes another type of analytical research that can be applied to physical activity: philosophical research. R. Scott Kretchmar and Tim Elcombe do a fine job of explaining and grouping philosophical research methods and use examples to identify the strengths and weaknesses of each.

Chapter 14 presents research synthesis (specifically, meta-analysis), another type of analytical research that focuses on the shortcomings of the typical literature analysis. Meta-analysis is a useful method for analyzing a large body of literature. Although several other types of research are considered analytical, historical research, philosophical research, and meta-analysis are the most common and useful in the study of physical activity.

Chapter 15 discusses descriptive research and focuses on survey techniques. Then, other descriptive techniques, such as case studies, correlational studies, developmental studies, and observational studies, are presented in chapter 16. In general, descriptive research shows relationships among people, events, and performances as they currently exist.

Chapter 17, written by Barbara Ainsworth and Charles Matthews, provides methods in a relatively new type of descriptive research in our field: exercise epidemiology. Often, this approach uses large databases on exercise and health behaviors from various sources.

Chapter 18 introduces experimental research, which deals with future events or the establishment of cause and effect. Which independent variables can be manipulated to create change in the future in a certain dependent variable? After reviewing how difficult cause and effect are to establish, we divide the chapter according to the strengths of various designs: preexperimental, true experimental, and quasi-experimental. Our purpose is to demonstrate which designs and principles are best suited for controlling the sources of invalidity that threaten experimental research.

Chapter 19 presents qualitative research techniques, which are increasingly used in the study of physical activity. The assumptions underlying qualitative research differ from those used in research that adheres to the traditional scientific method. This difference does not mean that qualitative research is not science (i.e., systematic inquiry), but the techniques for acquiring and analyzing knowledge differ from the typical steps in the scientific method. This chapter discusses the differences in the quantitative and qualitative paradigms, procedures used in qualitative research, the interpretation of qualitative data, and theory construction.

Finally, in chapter 20, we present the mixed-methods approach to research. This approach combines one of the quantitative approaches (correlational, survey, experimental) with the qualitative model. The idea is that both approaches and types of data represent human behavior, and that achieving a full understanding of behavior is difficult without using both methods.

If you are a producer or a consumer of research in physical activity, you must understand accepted techniques for solving problems systematically. The nine chapters in part III provide the basic underpinnings of research planning. Many research types reported here are closely associated with the appropriate statistical analyses previously presented. We refer to the appropriate statistics as we discuss the types of research. Use the knowledge gained from the previous section to understand this application. You should learn about the relationship between the correct type of design and the appropriate statistical analysis as soon as possible.

12

Sociohistorical Process in Sport Studies

David K. Wiggins
George Mason University

Daniel S. Mason
University of Alberta in Edmonton

David K. Wiggins
George Mason University

Daniel S. Mason
University of Alberta in Edmonton

> **Often it does seem a pity that Noah and his party did not miss the boat.**
>
> —Mark Twain

In 1974, Marvin H. Eyler, one of the founders of the academic subdiscipline of sport history, contended that "the past can never be precisely replicated," but must be "reconstructed on the basis of evidence which has been selected from pre-suppositions" (Eyler, 1974, p. 45). Furthermore, in contrast to science, he noted that history uses a "different standard of objectivity," and is concerned with knowledge that is "inferential and indirect" (Eyler, 1974, p. 45). "Historical research attempts to systematically recapture the complex nuances, the people, meanings, events and even ideas of the past that have influenced and shaped the present" (Berg, 2001, pp. 210-211). Another definition offered by Burke (1992) is that history is "the study of human societies in the plural, placing the emphasis on the differences between them and also on the changes which have taken place in each one over time" (p. 2). Quite simply, a history is an account of some event or series of events that has taken place in the past (Berg, 2001). To put historical research within the broader context of this book, perhaps the best way to justify the use of historical methods is through the following statement by Burke (1992): "If we want to understand *why* social change takes place, it may be a good strategy to begin by examining *how* it takes place" (p. 139). Thus, the role of the historian is to uncover the evidence and analyze it in a way that provides new insights into social change over time.

This chapter examines the status of sport history as an academic subdiscipline and the major issues and methodological approaches associated with sport history, identifies exemplary research studies in sport history, and discusses future avenues of inquiry in the field. It is hoped that this review will provide a greater awareness of sport history as an academic subdiscipline and, more specifically, present the strengths and weaknesses of the research process in this field.

DEVELOPMENT OF THE DISCIPLINE

At the beginning of the 1970s, there was unbridled optimism about the future of sport history as an academic subdiscipline; the first four years of that decade alone would witness the creation of two organizations, two journals, and three symposiums devoted to sport history (Berryman, 1973; Pope, 1997b; Struna, 1997). Early work such as that of Eyler (1974) provided an initial blueprint for sport historians regarding the meaning of their work, methods for recovering information from the past and seeking historical truth, and a greater understanding of the development of sport.

Much of the momentum established in sport history during the early 1970s would be maintained over the next 25 years. Courses in various aspects of sport history at both the undergraduate and graduate levels were added to university curricula, and a number of research studies on sport history proliferated. Using a variety of methodologies and approaches, scholars examined what seemed to be every conceivable topic (Adelman, 1983; Baker, 1983; Guttmann, 1983; Hill, 1996; Kruger, 1990; Morrow, 1983; Park, 1983; Pope, 1997b; Struna, 1985, 1997, 2000; Walvin, 1984; Wiggins, 1986, 2000). Perhaps most important, the number and quality of research outlets increased significantly. Besides *Sport History Review*, the *Journal of Sport History*, and *International Journal of the History of Sport,* scholars could choose to publish their work in one of the journals from the parent discipline or in such sport studies–related journals as *Sport, Education and Society; Sporting Traditions;* and *Olympika*. In addition to academic journals, both university and commercial presses added sport history titles to their publication lists.

As Berryman, Struna, Pope, and others have pointed out, many important articles and books on aspects of sport history were published during the first half of the 20th century, including Frederic L. Paxson's "The Rise of Sport" in the *Mississippi Valley Historical Review* (1917), John A. Krout's *Annals of American Sport* (1929), Jennie Holliman's *American Sports, 1783-1835* (1931), Herbert Manchester's *Four Centuries of Sport in America, 1490-1890* (1931), and John R. Betts' "Organized Sport in Industrial America" (1951). The watershed for sport history, however, in regards to organizational structure and institutional support, occurred during the 1960s and early 1970s. Buoyed by the era's newfound interest in the history of the common person and everyday life (often called histories from below), a group of people, largely from departments of physical education, held symposiums, founded organizations, and began publishing academic journals devoted to sport history.

Unfortunately, much of the enthusiasm for sport history has seemingly abated in recent years (Booth, 1997, 1999, 2005; Nauright, 1999; Phillips, 1999, 2001; Ward, 2013). Although research studies in sport history continue to be produced and both individual and institutional memberships remain steady in sport history organizations, the field has gradually been marginalized as an academic subdiscipline. As a group, historians generally view sport as frivolous and unimportant in the grand scheme of things. The reasons for this attitude are impossible to explain with any certainty. Gorn and Oriard, in their appropriately titled and frequently cited essay "Taking Sports Seriously," noted that "athletes' bodies remain curiously off-limits" in cultural studies, history, and other disciplines and speculate that intellectuals are simply "uncomfortable with physicality" or are "playing out the long-standing faculty antagonism to the distorted priorities of universities with multimillion-dollar athletics programs" (Gorn & Oriard, 1995, p. A52).

The status of sport history is not much different in departments of physical education, human movement, and kinesiology, where academicians in biomechanics or exercise physiology, or even what is termed the pedagogical sciences, do not see the relevance of sport history (Booth, 1997, 1999). An interesting, but not all that surprising, response by some sport historians to this transformation in these fields has been to focus more directly on the history of the body and exercise rather than the institution of sport. Currently, a group of sport historians, as well as a group of sport sociologists, are more interested in examining the development of physical culture; representations of the human body; the interconnections among health, exercise, and medicine; and other such topics (Berryman & Park, 1992; Todd, 1998; Vertinsky, 1990; Whorton, 1982).

However, the marginalization of sport history in departments of physical education, human movement, and kinesiology is not simply a result of the kinds of topics and phenomena explored. In fact, the relatively low status granted sport history in these departments is as much a result of the perceived limitations of its research techniques and methodological approaches as anything else (Booth, 1997). Scholars from various subdisciplines, either out of ignorance or deep philosophical differences, have been highly critical of the approaches taken by many sport historians. Some sport sociologists have railed against sport historians for their refusal to use models and adequately conceptualize their research data. In response, sport historians have criticized sport sociologists for their technical jargon, overemphasis on theory, and inability to grasp the complexities of the past (Phillips, 2001). In reality, there are elements of truth in the criticisms of both groups. Unfortunately, both groups, but particularly sport historians, generally pay little attention to these criticisms and have been reluctant to reflect on methodological approaches and the meaning of their work. Although debatable, several scholars have surmised that historians avoid self-reflection because they are simply too busy collecting data and adding to the body of knowledge. It is hoped that this chapter provides a model for conducting sport history research that can at least partially reconcile some of the divisiveness within the field of sport studies.

THEORY AND SPORT HISTORY

Despite the issues identified in the preceding section, the field of sport history has much to offer to the realm of sport studies. In many respects, the problems within the discipline largely reflect a marginalization in other areas of sport studies and kinesiology or physical education programs. However, a number of very competent scholars have used a variety of approaches to study the history of sport and the role sport has had within a broader social context. As noted earlier, one of the roles that sport historians play is in discovering and uncovering evidence that gives us a better understanding of how our world works today.

Burke (1992) identified two basic approaches to historical knowledge, which seem to work as ends of a continuum. He called them *particularizing* and *generalizing*. Others have used the terms *historical* and *theoretical,* and *splitting* and *lumping*. The difference between the approaches lies in the level and type of analysis used by the historian. For example, particularizers are interested in minute details related to the past, whereas generalizers are more interested in how things fit into a bigger picture. Similarly, many traditional historians have been interested in simply uncovering the events of the past, whereas others are keener on discovering the broader theoretical implications of their research. Splitting can be described as chipping away at a theory like a sculptor, by examining how theories hold against empirical data uncovered by the historian. In contrast, lumping consists of a complementary process of lumping together seemingly disparate facts, "working outwards from the particular to the general" (Burke, 1992, p. 130).

These two basic approaches shed some light on a movement within the history literature away from traditional histories that sought to examine the deeds of elite individuals (typically male), and toward understanding the social structures and processes of change that affected a given time period (Iggers, 1997). This can be described simply as a movement away from studying the limited and myopic realm of the great deeds of great men to more inclusive studies of what can be called histories from below. An example of this can be seen in the histories of Muhammad Ali. It could be argued that his athletic accomplishments (the great deeds of a great man) were not as important as the impact he had on millions of

© Rohit-Seth/Fotolia

Looks like a potential historian—fitting pieces of the puzzle together.

people in his generation (how his deeds affected those "from below"). In other words, how society was affected by his outspokenness and exploits during the turbulent period of the late 1960s could be considered of greater interest and importance than the number of knockouts he had in his career.

RELATIONSHIP BETWEEN THEORY AND METHOD

It is important to note that Muhammad Ali would not have had such an impact were it not for his boxing ability. Thus, it is critical for historians to be aware of both splitting and lumping. A sport historian needs to be able to obtain evidence, critically ascertain its trustworthiness, and then understand how it adds to our greater understanding of the phenomenon being investigated. As this section will reveal, the way historians have sought to gain a greater understanding is directly related to the theoretical perspectives taken by researchers.

paradigm—An assumption about the nature of reality and the appropriate methodological approach in gathering and analyzing data.

Sport historians have chosen a number of ways to undertake this scholarly process, and some have criticized sport historians for their reliance on several popular approaches. In this section, we review several analytical **paradigms**: structural functionalism and modernization, Marxism, and postmodernism. However, before we examine them in more detail, it is important to note that theory can never be applied to the past. What theory can do, on the other hand, is suggest new questions for historians to ask about their period, or new answers to familiar questions (Burke, 1992, p. 165). The questions sport historians ask, and the conclusions they come to, are to a large degree influenced by the theories they use to draw broader inferences about the past. The more dominant approaches that have emerged in the field of sport history are outlined next.

Structural Functionalism and Modernization

One of the earliest and most enduring approaches used by sport historians relates to social structures. Functionalism is characterized by an interest in social systems, institutionalization, specific periodization, and urbanization. Structural functionalism has been an especially popular approach among sport historians, particularly in the United States. Many historians during World War II and immediately thereafter adopted this approach assuming that sport reflected America's superior way of life, democratic institutions, and unabated progress. Works of this genre include Foster Rhea Dulles' *A History of Recreation: America Learns to Play* (1965), Frederick W. Cozens and Florence Scovil Stumpf's *Sports in American Life* (1953), and John Rickards Betts' *America's Sporting Heritage, 1850-1950* (1974). Of these three works, Betts' survey is the most well-known and influential. Published posthumously, the book was reconstructed from Betts' 1951 dissertation "Organized Sport in Industrial America" (1951) by noted sport sociologist John Loy with assistance from Betts' son Richard and prominent sport historian Guy Lewis. The book, wonderfully researched and characterized by a combination of social and business history, provides a framework and organizational structure that has been used in subsequent monographs and sport history surveys. Betts discussed such topics as the technological revolution and the rise of sport, the Golden Age of Sport in the 1920s, the influence of the Great Depression and World War II on organized sport, and the interrelationships between sport and institutions in the fields of education, religion, and art. The structural functionalist approach to the study of sport continues to dominate much of the work in the subdiscipline.

According to Burke (1992), the appeal of functionalism to historians lies in the fact that it has compensated for the long-standing tendency of historians to explain too much of the past in terms of the actions and intentions of specific individuals (p. 106). Instead, according to structural functionalists, the past is best explained by the broader social structures and forces exerted during a given time period. The structural functionalist approach does have its critics, however. Sport studies scholar Doug Booth, for one, believes the system has serious flaws. "Functionalist explanations in sport history," noted Booth, "typically gloss over disparities in sporting interests between different social and economic groups, exaggerates social stability and harmony, and ignores the inevitable losers and outsiders in local (community, regional, national) consensus building activities" (Booth, 2003, p. 11).

Certainly one of the most influential analytical designs in sport history is the modernization paradigm first adopted by Allen Guttmann in his classic study *From Ritual to Record: The Nature of Modern Sports* (1978). According to Pope, modernization theory has developed as a way of investigating how society has evolved from a rural agrarian society to an urban and industrial society, at different times and in different places throughout the world (Pope, 1997b, p. 4). *From Ritual to Record* charts the formal-structural characteristics of modern sports and the extent to which these characteristics were evident in sport during earlier periods in history. It was considered so influential that an entire issue of *Sport History Review* was recently devoted to a retrospective critique of the book, (Booth, 2001; Brownell, 2001; Guttmann, 2001; Howell, 2001; Lippe, 2001). With its impressive list of diverse sources, astute historical and statistical comparisons, and cogent analysis and thoughtful questions, Guttmann's book is essentially a structural functional analysis that outlines seven characteristics of modern sport: secularism, equality, rationalization, specialization, bureaucratization, quantification, and an obsession with records.

Melvin Adelman also used the modernization paradigm with much success in his path-breaking study *A Sporting Time: New York City and the Rise of Modern Athletics, 1820-70* (1986). The first book in the University of Illinois Press' sport and society series, *A Sporting Time* is one of the most frequently cited and important studies in sport history. Adelman, through painstaking research and perceptive analysis, examined the transformation of American sport from its premodern to modern forms by exploring early baseball, horse and harness racing, boxing, and such traditionally neglected activities as cricket, rowing, yachting, and billiards. Adelman challenged previous research by arguing that sport, particularly in New York City, evolved from an informal, unorganized activity to a more highly structured and organized phenomenon between 1820 and 1870 rather than during the late stages of the 19th century. The 1820 to 1870 time period witnessed the growth of commercialized athletics and evidence of a fledgling sport information system, the standardization of records and equipment, specialized player roles, and philosophical rationales for sport participation.

The modernization paradigm adopted by Guttmann and Adelman has received some criticism. In fact, in the aforementioned retrospective critique of *From Ritual to Record* in *Sport History Review*, several prominent scholars took turns poking holes in Guttmann's approach. Howell was critical of the book for its sweeping approach to history, arguing that it helped identify the process of change in sport over time, but was less successful in explaining the transformation of sport. "While many processes are identifiable in history," wrote Howell, "it is another thing altogether to understand history itself as a process. *From Ritual to Record* was written at a time of grand theorizing, when the grand narratives of development that postmodernist scholars now disdain were clearly in vogue" (2001, p. 13). Booth generally had high praise for the book, but claimed that Guttmann's work provides a "limited conceptualization of the process of modernization and the nature of modernity" and "lacks adequate political and dialectical analyses" (2001, p. 22). Pope also weighed in on the debate, noting that critics of modernization theory have "exposed modernization as a historical construct that distorted the past by emphasizing structure at the expense of human agency" (Pope, 1997b, p. 5). In some ways, this criticism is similar to that leveled against supporters of structuralism.

Guttmann countered these criticisms and others with his usual aplomb and erudition. He acknowledged that his modernization paradigm has limitations, including the fact that it can be "too abstract to account for every detail in history's vast panorama," "can easily be misinterpreted to mean that the observed changes occurred as part of some uniform and inevitable process," and "can be misused as a facile instrument of ethical judgment—as if modern ways were somehow a moral advance as well as a technological advance beyond traditional customs" (2001, p. 5). He argued, however, that in spite of the obvious limitations of the modernization paradigm, no alternative model has been postulated that does a better job in chronicling the changes in sport over time and marking the differences between sport of yesteryear and that of today. "Despite the manifest possibility that the concept of modernization can be misunderstood and misrepresented" wrote Guttmann, "I continue to find it more useful than the postmodernism [to be described in more detail later in this chapter, and throughout this book] that is currently fashionable in what once were called the social sciences. An imperfect and admittedly selective account of the historical facts seems preferable to the startled discovery that there is no such thing as absolute truth" (2001, p. 6).

The last three decades, however, have witnessed the adoption of new and varied theoretical approaches and paradigms in sport history. Taking their cues from the vast changes taking place in historical scholarship during the 1970s and 1980s, sport historians increasingly turned to the analytical frameworks used in such fields as anthropology, cultural studies, and sociology to examine the development of sport. Although it was not always easy to discern specific patterns of inquiry, there was now a seeming emphasis in the subdiscipline on such concepts as community formation and power and how socially constructed notions of race, ethnicity, gender, class, and religion affected sport and vice versa. Increasingly, sport historians adopted the works of Max Weber, Norbert Elias, Clifford Geertz, and the like, to frame their analyses and interpretations (Pope, 1997b).

Marxism

Many sport historians have employed versions of Marxism to analyze the development of sport. Interested in the mode of production and the use of political and social power, Marxist sport historians have focused on sport in regards to social class aspirations and the way sport has been redefined by blacks, women, workers, and other groups in society. Richard Gruneau, the noted Canadian sociologist, has had a profound effect on the work of sport historians through his many publications on sport and Marxist cultural studies (Gruneau, 1983). Equally significant to sport historians have been the theoretical insights provided by such cultural Marxists as E.P. Thompson, Raymond Williams, Eric Hobsbawm, and Antonio Gramsci. Historian Colin Howell correctly observed that "over the past few years it has been difficult to pick up an issue of one of the major sport history journals and not find any article employing Gramsci's notion of **hegemony** or referring to notions of human agency' (1998, p. 101).

hegemony—The cultural or political dominance or authority over others.

Howell might also have added that the theoretical approaches of Gramsci as well as other cultural Marxists have found their way in one form or another into some of the most important sport history monographs. One example of this is Pope's *Patriotic Games: Sporting Traditions in the American Imagination, 1876-1926* (1997a). Using Eric Hobsbawm's concept of invented traditions, Pope examined the national sporting culture that was established in the United States between the 1876 centennial and the 1926 sesquicentennial. He illustrated how sport becomes a cultural transmitter of patriotic meaning, intertwined with Thanksgiving and Fourth of July holiday celebrations and military preparedness. Perhaps most important, Pope made clear that sport served as both a metaphorical activity and class drama and helped define public discourse on American hopes and dreams.

Pope's book is significant in that it is reflective of what Hill (1996) and Holt (1998) referred to as the subtle shift in sport history from causation to meaning. Over the last two decades, scholars have been less concerned about the determinants of sport participation and more interested in the multiple and complex meanings sport has had for athletes, spectators, and other individuals, groups, and communities. Examples of this scholarship are numerous. MacAloon focused on rituals and cultural symbols in his book *This Great Symbol: Pierre de Coubertin and the Origins of the Modern Olympic Games* (1981). Peter Levine argued in *Ellis Island to Ebbetts Field: Sport and the American Jewish Experience* (1992) that sport served as a middle ground for American Jews who were striving to become full participants in American society while at the same time attempting to maintain their sense of ethnic pride and identity. Using an impressive list of source materials and wonderful stories of athletes and teams, Levine made clear that for East European Jewish immigrants and their children during the first half of the 20th century, sport was seen as an avenue of assimilation and a means to pursue the American dream. Elliott Gorn provided a fascinating analysis of the meaning of prizefighting in 19th-century America in his *The Manly Art: Bare-Knuckle Prize Fighting in America* (1986). Integrating the work of anthropologists, sociologists, folklorists, labor and social historians, and American and gender studies specialists, Gorn examined how prizefighting was transformed from a working-class ritual that symbolized physical prowess and toughness to a middle-class sport that illustrated dominant ideas of social mobility and meritocracy.

Postmodernism

Finally, it is important to recognize the presence of postmodern approaches to sport history. It is probably safe to say that most sport historians reject many of the basic tenets of the postmodernist approach, believing that it only results in a "dazzling kaleidoscope of impressions where anything can be construed to mean anything" and ultimately anarchic nihilism (Holt, 1998, p. 18). As historian Georg Iggers explained, "The basic idea of postmodern theory of historiography is the denial that historical writing actually refers to an actual historical past" (Iggers, 1997, p. 118).

However, a few brave and adventurous souls in the field have employed versions of this approach in their research. Two studies that depend to some extent on postmodernist theory are Patricia Vertinsky's *The Eternally Wounded Woman: Women, Exercise, and Doctors in the Late Nineteenth Century* (1990) and Susan Cahn's *Coming on Strong: Gender and Sexuality in Twentieth-Century Women's Sport* (1994). Tina Parratt (1998) was absolutely correct when she noted that both of these works are indebted to postmodernism, particularly the notion that power and knowledge are the result of discursive practices.

The first and perhaps only genuine postmodernist study in sport history, however, is Synthia Sydnor's "A History of Synchronized Swimming" (1998). Published in a special issue of the *Journal of Sport History* titled "Sport History: Into the 21st Century," Sydnor's essay follows in the path of Marxist theorist Walter Benjamin in that it avoids such standard historical practices as recording facts and periodization to ponder synchronized swimming through an assemblage of symbols, images, and spectacles. Sydnor, designating her "essay as a modern text with postmodern concerns" did not analyze or explain or contextualize; rather she provided a number of evocative ideas through an "exhibition or performance" presented in a poetic and literary form (1998, p. 260). However, works such as Sydnor's remain very much in the minority in the field of sport history (Phillips, 2006).

An approach that is, in the words of Genevieve Rail, "an amalgam of often purposely ambiguous and fluid ideas," postmodernism is variously thought of as a revolt against modernism and structuralism, a radical break from the past that leans toward a deconstruction of linguistic systems, a distaste for disciplinary boundaries, and a belief that the world is fragmented into many isolated and autonomous discourses that cannot be explained by any grand theory (Rail, 1998, p. X). Michel Foucault, one of the most prominent postmodernist thinkers, argues that truth is only a partial, localized version of reality and that power relations and meaning are the function of language and discourse (Rail, 1998).

All of the approaches described in this section have been shown to be viable ways to study the history of sport. However, each has shortcomings in helping us better understand the past. In a recent article, sport historian Colin Howell critiqued the use of the concept of hegemony in sport history. Although his criticisms were directed at one specific lens for historical research, in some respects similar criticisms could also be leveled at the other approaches we have discussed:

> **Unfortunately, the idea of hegemony is often employed by the unimaginative as a formula to lend a patina of academic respectability to an academic paper or journal article. In turn, its overuse has made the concept appear hackneyed, especially when it is misconstrued as a more academically credible form of social-control analysis. One should keep in mind that the end of historical inquiry is not to confirm Gramsci, but rather to use his theoretical insights to better understand what happened. (Howell, 1998, p. 101)**

Perhaps the lesson for aspiring sport historians is to make sure they use an approach that will help the researcher more capably investigate the question at hand. In the meantime, as Howell warned, researchers must not get too caught up in an approach and use it simply to justify their own research.

Perhaps the most important issue at hand with regard to the theoretical approaches to the study of sport history is the fact that sporting activities, like social practices in general, have always tended to indicate, and at times been reinforced by, class distinctions (Booth & Loy, 1999, p. 6). Within these class distinctions, it is also critical to identify issues and differences concerning gender, race, and other social concerns. For this reason, sport historians must try to gain a greater understanding of the broader social context of the periods they study, particularly in terms of the relationships among classes. Thus, sport history must be informed by social theories that seek to draw inferences from a number of contexts and time periods to better explain the role of sport in society.

RESEARCH SOURCES

Sport historians generally adhere to a specific and recognizable historical method. The first, and in some ways most important, task of the sport historian is to become conversant with the secondary literature. According to historians, there are two basic types of sources—primary and secondary. Primary sources are those that were created or witnessed by someone or something directly associated with the topic that is being investigated. For example, a primary source for a paper on an Olympic boxing match that took place in 1988 would be a person who was involved in the match itself, or a firsthand witness to the event. Primary source information can be obtained in a number of ways, such as interviewing a spectator, participant, or judge, or reading a newspaper account written by someone who was at the event (e.g., a sports reporter who covered the event and wrote an article that appeared in the next day's paper). Another way would be to watch video footage of the event. Primary sources are critical for sport historians because they bring them as close as possible to the event.

Secondary sources are sources such as books and articles written by people who were not directly associated with the event or topic. In many cases, they are written by researchers who have done their own primary source–based research. However, the degree of filtering resulting from having been examined by others makes secondary sources less valuable to historians. This is because someone else has analyzed the primary source material and come to his or her own independent conclusions.

At the outset of doing research, secondary sources are extremely important for the sport historian because they provide background information. Thus, when starting out, the historian usually explores the secondary literature. To locate secondary literature, sport historians traditionally turn first to a library catalog to find entries for book titles, subjects, and names of authors. They also depend on such resource aids as historical dictionaries, large general histories, specialized academic journals, biographical dictionaries, periodical indexes, bibliography of bibliographies, and current national bibliographies. Times have changed, however. In addition to the aforementioned tools, sport historians now conduct computer searches to locate secondary sources or contact colleagues on e-mail lists to find secondary material (Atkinson, 1978; Gottschalk, 1969; D.H. Porter, 1981; Shafer, 1980). Databases such as *SPORTDiscus* also provide opportunities to search online for topics using keywords. In addition, other athletic bodies, such as the LA84 Foundation (la84.org), have made journal and other sport sources available in full text through their websites.

RESEARCH TOPICS

According to Berg (2001), "The major impetus in historical research, as with other data-collection strategies, is the collection of information and the interpretation or analysis of the data" (p. 213). Sport historians are fortunate that the number of improved search tools has coincided with a dramatic increase in both the quality and quantity of secondary works dealing with various aspects of sport history. Thought-provoking, creative, and imaginatively designed studies have been completed on such topics as the pattern of sport in both rural and urban settings (Adelman, 1986; Berryman, 1982; Borish, 1996; Gems, 1997; Hardy, 1981, 1982; Riess, 1989; R. Roberts, 2000; Somers, 1972; Uminowicz, 1984); the history of baseball and other team and individual sports (Goldstein, 1989; Kirsch, 1989; Oriard, 1993, 2001; Riess, 1980; Sammons, 1988); the sport participation patterns of African Americans and those in other racial and ethnic groups (Cahn 1994; Festle, 1996; Guttmann, 1991; Mangan & Park, 1987); the interconnections among organized sport, physical culture, and the medical profession (Berryman & Park, 1992; Fair, 1999; H. Green, 1986; Todd, 1998; Verbrugge, 1988; Vertinsky, 1990; Whorton, 1982); the origin and evolution of the modern Olympic movement (Dyreson, 1998; Guttmann, 1984, 1992; MacAloon, 1981; Senn, 1999; Young, 1996); the role of women in sport at the national and international levels of competition (Bass, 2002; Bloom, 2000; Eisen & Wiggins, 1994; Hoberman, 1997; P. Levine, 1992; Regalado, 1998; Riess, 1998; Shropshire, 1996; Wiggins, 1997; Wiggins & Miller, 2003); and the conception of leisure and transformation of public recreation and organized sport at different times in history and among different classes and geographical locations (Cavello, 1981; Hardy, 1990; Rosenzweig, 1983; Rosenzweig & Blackmar, 1992).

The responsibility of sport historians is not simply to familiarize themselves with this literature, but to assess and analyze it in regard to the types of evidence used, generalizations drawn, and historical frameworks employed. A thorough and detailed understanding of the secondary literature assists scholars in formulating paradigms and asking good questions, which are at the heart of the historical process. It is precisely for this reason that the more experienced scholar has little difficulty selecting a topic and delimiting that topic to manageable proportions, whereas the fledgling scholar usually struggles to find a topic that will contribute to the body of knowledge and add to the research literature. Armed with a firm grasp of the secondary literature, the more experienced sport historian can usually establish appropriate themes and set parameters conducive to consistent and logical arguments. On the other hand, the less experienced members of the subdiscipline, equipped with a limited understanding of the secondary literature, often pick topics that are too broad, underestimating rather than overestimating the amount of source material available.

Novice sport historians may tend to choose topics that are too broad because they want to make their mark and are worried about choosing a topic that is too narrow and insignificant to be worthy of study. Peter Burke called historical examinations on a small, local scale micro histories. One thing to keep in mind when formulating a topic is that good primary sources that address small, obscure topics can shed light on broader social phenomena and should not be trivialized. As Burke explained:

> **One might begin with the charge that the micro historians trivialize history by studying the biographies of unimportant people or the difficulties of small communities. . . . However, the aim of the micro historian is generally more intellectually ambitious than that. If they do not aspire to show the world in a grain of sand, these historians do claim to draw general conclusions from local data. (Burke, 1992, p. 41)**

For sport historians, then, a thorough grounding in the secondary literature is essential for formulating good questions, and these questions need not be on a grand scale. Scholars in the subdiscipline know full well, however, that good questions are also those of interest to the researcher, those that are answerable, and those that are tied to the primary evidence. Sport historians would be wise to listen to the advice of advisors and colleagues, but ultimately they must address those questions that are of interest to them and not particularly to anyone else. A traditional but still very valuable strategy is to consider some basic questions. The first have to do with the phenomenon to be studied. Is the researcher most interested in examining sport? Athletics? Exercise? Leisure? Recreation? The next questions have to do with chronology. Is the researcher interested in a particular year? The Renaissance period? The late 19th century? The 1920s? The post–World War II period? Other questions have to do with geography. Is the researcher interested in Great Britain? Canada? The United States? Latin America? China? Finally, the researcher needs to ask questions about the individuals or groups to be studied. Is the researcher interested in elite athletes? Particular racial and ethnic groups? Ancient Greeks? Physical culturists? Health reformers?

In reality, of course, topic questions are not mutually exclusive. The sport historian must arrive at a final set of manageable questions that beg answers gleaned from primary sources.

RESEARCH DESIGN

Once their topic questions have been finalized, sport historians put together research designs. Although they do not construct experiments or manipulate and control data, sport historians are similar to all scholars in that they map strategies and develop research designs characterized by a systematic arrangement of questions and the approach necessary for answering those questions.

The two most commonly cited designs in sport history are descriptive history and analytical history. Descriptive history primarily addresses what happened in sport, and analytical history addresses the complexity of relationships and draws connections to determine how and why sport developed the way it did and the role it served in the lives of participants and spectators alike. Although the differences between the two basic research designs are not always discernible, most sport historians strive to complete analytical history, which has been most evident from the beginning (although some scholars in the subdiscipline would certainly disagree). As our earlier discussion of the need to link theory and method in historical research suggests, a reliance on descriptive histories may reinforce and rationalize the lack of respectability the field has within both sport studies and the parent discipline of history. As a result, we encourage aspiring sport historians to employ more analytical approaches to their research.

Regardless of the design chosen, all sport historians must be concerned with the collection of evidence. To answer their questions, sport historians seek as many primary sources or firsthand accounts as possible. Although recognizing the impossibility of locating all documents, and the inevitable incompleteness of the historical process, sport historians search archives, libraries, private collections, and a host of other repositories to find the primary sources necessary for gaining an understanding of the sporting past. Primary materials may be obtained through sources that are not necessarily considered first. For example, Iacovetta and Mitchinson (1998) suggested exploring case files, which could include employment records, court proceedings, patient records, and psychiatrists' case histories. They explain that historians have turned to case files because they offer us a rare window on human interaction and conflict. . . . These records can illuminate the ways in which dominant class, gender, and racial ideologies shaped official discourses and actions and relations between experts and clients. (p. 6)

Thus, the aspiring sport historian in many ways has an opportunity to start doing historical research without many of the traditional restrictions to primary sources faced by sport historians of earlier years.

One consideration that bodes well for the sport historian is the greater ease of access to primary sources created by new technologies. As Cox and Salter (1998) explained: "Technology will release the sport historian from unskilled, repetitive, time-consuming tasks, allow him or her to concentrate on the core activities of a sport historian—analysis, synthesizing, theorizing, and communicating information" (p. 294). For example, many newspapers, which historians typically research in microform format in libraries, are now available in pdf format through online databases. In many cases, these databases can be searched using keywords. So, rather than performing the time-consuming task of reading pages and pages of old newspapers on a microform reader looking for articles concerned with a topic, people can now have a database pull up the appropriate newspaper pages on a given issue. An example is the *Toronto Star* newspaper's database at www.thestar.com/about/archive-search.html.

Databases are just part of a host of digitized sources, tools, and social media platforms available to historians, including Wikipedia, blogs, Facebook, Twitter, YouTube, and Flickr. These sources are easy to access and provide information that is potentially valuable to historians in an assortment of ways. They do, however, pose methodological and philosophical questions for historians interested in assessing past events and analyzing changes over time.

One question involves the accuracy and superficial nature of these materials. Another has to do with context, whether the selective digitalization process provides the relevant background information necessary for charting key events and providing historical narratives. In spite of these questions, historians, whether they embrace the digitized world or not, would certainly benefit from exploring it combined with standard print sources and archival material (Cohen & Rosenzweig, 2006; Evans & Rees, 2012; O'Malley & Rosenzweig, 1997; Rosenzweig, 2003).

DATA ANALYSIS AND INTERPRETATION

Having identified and located primary sources, the sport historian then exposes the sources to a rigorous two-step evaluative process known as external and internal criticism (Atkinson, 1978; Gottschalk, 1969; D.H. Porter, 1981; Shafer, 1980).

External Criticism

authenticity—The state of being true and legitimate.

External criticism involves establishing the **authenticity** of the primary source. The sport historian, using tests similar to those employed by lawyers or detectives, must determine whether the chosen documents either are forged or have been tampered with since the time of the event. To differentiate between a forgery or misrepresentation and an authentic document, the sport historian must determine whether events, objects, people, or customs are correctly placed chronologically. Was a particular type of paper being used at the time the document was supposedly written? Could the alleged author of the document possibly have been at the location where it was supposedly written? Were the punctuation, spelling, proper names, and signatures in the document consistent with other contemporary writings? Does the alleged author of the document refer to events about which he or she could not possibly have known? Another common issue relates to authorship. For example, a book or article may be credited to a person who might be considered a primary source (such as an autobiography). However, these types of books are often ghostwritten and therefore less reliable.

Internal Criticism

Once authenticity has been established, the sport historian then assesses the credibility of documents through a process known as internal criticism (Atkinson, 1978; Gottschalk, 1969; D.H. Porter, 1981; Shafer, 1980). A time-consuming and extremely important part of the historical method, internal criticism is concerned with determining who left documents and the relationship of the source of those documents to the event or events. In other words, what interests or biases might the author of the document have regarding the event? Internal criticism is also concerned with the wording of documents, particularly how words were defined and the way they were used during the historical period being studied.

Internal criticism is also concerned with the information included and omitted from historical documents. Historians must be cognizant of the fact that all historical documents are only partial glimpses into the past and that some information is often either intentionally or unintentionally left out of written and oral accounts. This is precisely why scholars in sport history always seek corroboration by searching for the independent testimony of two or more trustworthy witnesses. If only one witness of an event is known to exist, scholars must seek other forms of corroboration, including the reputation of the author of the document, the absence of contradictions within the document, and the lack of anachronisms.

A good example of when internal criticism is critical is when relying on newspapers for data. According to Reah (1998), newspaper reports are simply accounts of selected information on recent events. Reporters, subconsciously or not, put ideological spins on events, providing a skewed viewpoint (p. 9). McNeil (2001) addressed this issue in a study of a worker's riot in Britain, in which he found that newspaper coverage was biased toward management. Thus, newspaper reports must be viewed with some skepticism because they tend to present the reader with aspects of news offered in a way that will influence the reader in a certain direction (Reah, 1998, p. 50). Baker (1994) wrote:

> **Journalists are actors and creators as well as reporters. They collect, select, and color the information they dispense to their customers; they frame the issues and assumptions from which cultural debates arise. Reporters thus shape opinion just as surely as do newspaper editors and columnists. (p. 530)**

With this in mind, the sport historian must try to gain an understanding of the potential biases that might influence the reporter. For example, it is likely that early coverage of professional sports teams in local newspapers favored the home team. This was due to civic boosterism and also to the very close ties early newspaper reporters had with their local sporting clubs. Thus, the historian cannot rely solely on information from such a source, because it represents only one viewpoint. The sport historian needs to seek other sources to corroborate facts or identify biases in the coverage. Perhaps the easiest way of doing this is to compare coverage in other newspapers (such as that of the hometown of the visiting team), or to find other accounts from other sources (e.g., interviews with spectators).

SOURCE CREDIBILITY

Every sport historian can probably relate personal stories about establishing the credibility of documents. David Wiggins, one of the authors of this chapter, found that his most challenging struggle with the problem of internal criticism took place while he was doing research on the role of leisure and recreation in the lives of slaves on plantations in the antebellum south (Wiggins, 1983a, 1983b). A very sensitive and controversial subject, he was faced with the problem of how to reconcile the differing conceptions of slave leisure and recreation expressed by an enormously diverse group of witnesses who had their own biases and vantage points. He had access to British travel accounts located in the U.S. Library of Congress. British travelers, who visited the South in fairly large numbers between 1780 and 1860, varied greatly in their observational skills and represented such diverse professions and occupations as land scouts, missionaries, scientists, politicians, educators, and actors. White plantation owners provided another vantage point from which to view the institution. Wiggins used letters and plantation diaries from such repositories as the Southern Historical Collection (Chapel Hill, North Carolina) and the Library of Congress to analyze a group of people who obviously had a strong allegiance to the institution of slavery and a reluctance to point out its many cruelties and horrors. The final group of witnesses to the cruel institution constituted the slaves themselves, who typically did not know how to read or write and frequently could recount their lives on southern plantations only after realizing their freedom. For their view of the institution, Wiggins turned to both published slave narratives and interviews conducted by the Works Progress Administration (WPA) during the Great Depression of the 1930s.

To assess the credibility of these divergent sources, Wiggins was forced to call on everything he had ever learned about the historical method. First of all, he had to constantly remind himself that what was most important was the credibility of statements made by authors rather than their overall credibility as witnesses. For example, plantation owners and their families, although generally providing a rather romantic view of slavery and refusing to acknowledge the insidiousness of

(continued)

Source Credibility (*continued*)

the institution, were able to detail with some accuracy the play activities and games participated in by slaves because of their frequent contact and close proximity to them. Second, Wiggins had to constantly remind himself of the great variability among witnesses in regard to the degree of attention they paid to the phenomenon being studied and the literary style they employed. For example, British travelers, many of whom were searching for large audiences and obviously seeking to appear omniscient, often recounted the leisure patterns and recreation of slaves based on nothing more than hearsay or tradition. Last, Wiggins had to remind himself that many witnesses were far removed in time from the events they were describing. For example, many former slaves were at an advanced age when they recounted their days in slavery, a fact that obviously tested their memory and begged the question of whether their longevity was due to better treatment than their fellow slaves. In addition, former slaves were often describing their childhood days, a time when their lives were typically more carefree and devoid of the intense labor and horrors more characteristic of the institution of slavery as a whole.

In addition to establishing the credibility of documents, sport historians must carefully read the evidence to ascertain the information while keeping in mind the close interrelationship between empirical particulars and theory formation. Wiggins certainly discovered this in a study he conducted on Muhammad Ali's relationship to the Nation of Islam (Wiggins, 1995). *Muhammad Speaks*, the official organ of the Nation of Islam, provided important details of Ali's membership in the organization, but not a comprehensive view of the champion's religious beliefs. Ali's autobiography, *The Greatest: My Own Story*, provides insights into the champion's involvement with the Nation of Islam, but virtually nothing about his strained and sometimes acrimonious relationship with both Elijah Muhammad and Malcolm X. Black newspapers covered only Ali's relationship to the Nation of Islam in a cursory fashion, seemingly either disinterested or unwilling to discuss the champion's involvement with the politically active and, some would say, racist organization. To piece together these seemingly incomplete sources and craft as accurate a portrayal of Ali's relationship to the Nation of Islam as possible, Wiggins had to pay particular attention to content, meaning, and historical context. This necessitated a close examination and comparison of written documents and a constant search for the biases that inevitably pervaded any discussion and analysis of Ali's involvement in and influence on the Nation of Islam. Wiggins also had to be speculative on occasion and draw historical generalizations that helped explain not only Ali's connection to the Nation of Islam but how the great heavyweight champion fit into the larger black power movement and came to symbolize the tumultuous Vietnam Era.

Clearly, the sport historian's job is not as simple as having a good research topic grounded in the secondary literature.

> **The historian is still bound by his or her sources, and the critical apparatus with which he or she approaches them remains in many ways the same. Nevertheless we view these sources more cautiously. We have become more aware of the extent to which they do not directly convey reality but are themselves narrative constructs that reconstruct these realities, not willy-nilly, but guided by scholarly findings and by a scholarly discourse. (Iggers, 1997, p. 10)**

In other words, today's sport historians should not be so bold as to believe that they can know exactly what happened in the past. Instead, they should attempt to understand what happened and how it fits into our broader understanding of the conditions that shaped both that time and the present. This understanding will certainly be influenced by the researcher's own experiences, the availability of sources, any biases of the resources, the underlying social dynamics of the time period studied (and the present as it affects the researcher), and the general epistemological and theoretical frameworks that guide the researcher and his or her subsequent research questions.

RESEARCH FINDINGS

The limited information provided by each source is why all sport historians seek as much evidence as possible about their chosen topics. Once all the available evidence is collected and the sport historian has attempted to determine what it all means, the next job is to disseminate the findings to colleagues and hopefully a larger public. There is no specific formula for accomplishing these two tasks, but there are certain principles to which all good sport historians adhere when writing their final products. They present their evidence in as logical a fashion as possible, being careful to discuss as well as analyze and interpret the primary material within an overall theme. They present their evidence within the proper historical context, providing a meaningful framework for the data they have assembled. They must also consider the arrangement of historical data, recognizing the importance of establishing at least some form of framework that both reveals and clarifies the progression of events. And last, but certainly not least, they pay special attention to writing style and composition. Sport historians recognize, as do other scholars involved in qualitative approaches to the study of sport, that how one says something is just as crucial as the information being imparted. Thus, the narrative, or the way the researcher pieces together the data and results into a story, is absolutely critical. Words must convey exact meanings to avoid generalities and typification when possible, and to refrain from using artificial rhetoric. In many respects, because sport historians have adhered to this practice, writing in the field of sport history has remained some of the most accessible in the field of sport studies.

EXEMPLARY STUDIES IN SPORT HISTORY

Now that we have reviewed the field of sport history and given an overview of how to go about doing research in this area, we now turn to reviewing several works that we believe exemplify the quality research that is being done in this subdiscipline. Two pathbreaking studies that address the complex meaning of sport and stand as outstanding examples are Michael Oriard's *Reading Football: How the Popular Press Created an American Spectacle* (1993) and *King Football: Sport & Spectacle in the Golden Age of Radio & Newsreels, Movies & Magazines, the Weekly & the Daily Press* (2001). Oriard's books are significant in that they are two of only a handful of works in the subdiscipline of sport history to employ the narrative form so popular in some recent historical studies. According to Phillips (2001, p. 336), "Oriard's book is pioneering in the context of sport history because it typifies a wider trend, identified by Peter Burke as the inventive use of alternative forms of narrative to portray the past." Thus, as explained by Phillips (2001), in *Reading Football*, Oriard uses Clifford Geertz' classic work on the Balinese cockfight to read early American football as a cultural text.

Although it provides empirical data on the history of the game, *Reading Football* is more concerned with how American football at the turn of the 20th century is best understood as a textual narrative, an open-ended story of mediated interpretations. Oriard melded sport history with narrative theory and popular cultural studies to illustrate that football emerged as a multivalent story through a process of reading, and that process implicated such key cultural themes as violence, immigration, race, masculinity, and power. Thus, "his book is a remarkable attempt to critically engage with narrative by exploring football as a cultural text through multiple voices and perspectives" (Phillips, 2001, p. 336).

In *King Football*, Oriard continued his story by exploring American football through the media from the 1920s through the 1950s. Through an analysis of "conventional, formulaic narratives and images" portrayed in newspapers, magazines, movies, and other powerful communications media, Oriard traced what football meant to members of the American middle class who were as different from each other as they were similar (p. 16). Most important, Oriard did something remarkable for a historian. He moved "freely between fact and fiction: between the local newspaper's account of last Saturday's game, a magazine profile of a prominent coach, the magazine's cover painting of a lovely lady ministering to her wounded football hero, and a short story or film about the son of a coal miner subsidized to play football at Ivy College or State U" (p. 17). This approach was important, said Oriard, because an understanding of the "fantasies concocted by magazine covers and short stories" provide insights into "how people thought about football" (p. 17).

In all his works, including his more recent *Brand NFL: Making and Selling America's Favorite Sport* (2007a) and *Bowled Over: Big Time College Football From the Sixties to the BCS Era* (2007b), Oriard also employed what Booth referred to as a more reflexivity methodological approach (Booth, 2005). Combining a more personal context (an advantage for Oriard, who played college football at Notre Dame and professional football with the Kansas City Chiefs) and traditional historiographical approach to his writing about past events, Oriard provided interpretations that are marked by his own perspectives and a recognition of the importance the present has on the past.

Other studies that take a more reflexivity approach are Dan Nathan's, *Saying It's So: A Cultural History of the Black Sox Scandal* (2003); Peter Levine's *Ellis Island to Ebbetts Field: Sport and the American Jewish Experience* (1992); Douglas Hartmann's, *Race, Culture, and the Revolt of the Black Athlete: The 1968 Olympic Protests and Their Aftermath* (2003); John Hoberman's, *Darwin's Athletics: How Sport Has Damaged Black America and Preserved the Myth of Race* (1997); and Patricia Vertinsky and Sherry McKay's, *Disciplining Bodies in the Gymnasium: Memory, Movement, Modernism* (2004). Of these studies, Levine's is the oldest, but it fits squarely into the reflexivity approach and indicates how important the present is to the study of the past. In the acknowledgments in the book, Levine pointed out that historians "regardless of what they write about and however much they may seek to avoid it, always bring to their work their own values, interests, and beliefs. I am not uncomfortable with this fact" (p. X). He then proceeded, through interviews, personal conversations and questions, and autobiographies of Jewish men between the ages of 60 and 80, to trace the involvement of Jews in American sport. Levine ultimately concluded that participation in sport "both confirmed a meaningful Jewish identity while promoting assimilation and American acceptance" (p. 7). Perhaps even more important to Levine, however, was that the writing of the book "has been a chance for my father and myself to 'talk' in ways we never did when he was alive and for me to appreciate and accept what we had and didn't have together" and that hopefully the "story of the Jewish experience in American sport also has meaning for Jews today who debate whether or not Jewish life is still possible in the United States" (pp. 275–276).

SUMMARY

This chapter provided a brief overview of the subdiscipline of sport history, the theories underpinning research in this field, and a basic explanation of the process sport historians undergo to explore their research questions. In doing so, we have also brought to light some of the issues facing sport history as a field of study within the broader context of sport studies. We hope that we have identified the issues in a way that allows aspiring sport historians to continue undertaking interesting and worthwhile research projects that can contribute to this relevant and rewarding area of study.

Although it could be argued that the field of sport history has been marginalized to a certain degree, we would like to present several key points that we believe exemplify why sport history can and should continue to be an integral component of sport studies. As explained by Briesach,

> **Once we accept that human life is marked both by change as that which makes past, present, and future different from each other and continuity as that which links them together, we begin to understand why historians have played so central a role in Western civilization. (p. 3)**

We can only understand our present (and begin to ponder our future) with a greater understanding and appreciation of what occurred in the past. The process whereby sport historians have gone about achieving this understanding is, to a large degree, affected by a number of influences, including their epistemological and theoretical approaches, the context in which past events occurred, and the context in which they investigate their research questions. Thus, it can be argued that "historians are practitioners of an imperfect craft" (Iacovetta & Mitchinson, 1998, p. 13), and an awareness of this fact can only help them with their research endeavors.

We end this chapter with comments from two prominent sport historians, Colin Howell and Steven Pope. According to Howell (2001):

> **Any of you who know my work will be aware of my attempt to nudge sport historians away from their preoccupations with elite, urban, or metropolitan sporting experiences, away from sport at the national and international levels, to a study of hinterland localities and borderland regions. (p. 16)**

This comment in some ways reflects a desire to move away from exploring the traditional great deeds of great men and toward a more inclusive historical paradigm. Similarly, Steven Pope (1998) noted:

> **Sport historians can no longer conceive of their works in a specialized, narrowly-focused fashion and expect to receive wide institutional approval. Historical case studies cannot be written in a scholarly vacuum; they must be relevant to contemporary issues and problems. (p. vii)**

With this in mind, we argue that the field of sport history remains an interesting and rewarding field of study. By adhering to Howell and Pope's advice and following some of the basic research tools outlined in this chapter, aspiring sport historians may carve a significant niche in this foundational area of sport studies and produce work that continues to show the relevance of this field within the academy.

Check Your Understanding

Find and read a historical study on some aspect of sport and physical activity.

1. Describe the research design.

2. Identify the primary sources.

13

Philosophical Research in Physical Activity

R. Scott Kretchmar
Pennsylvania State University

Tim Elcombe
Wilfrid Laurier University

Think about this . . . No one ever says "It's only a game" when his team is winning.

—From Lachlan's Laws (Scottish Highland philosopher)

Philosophy has two fundamental purposes—descriptive and prescriptive. Philosophers reflect on things; they try to describe existence as accurately as possible. Reflection is critical; problems might exist with the way things are, with what we believe, or with how we act. When these issues arise, philosophers should not merely describe them; they should also recommend, counsel, urge, cajole, convince, and otherwise promote more culturally attuned values or more ethically justifiable behaviors. In other words, philosophers need to go beyond describing existence and initiate inquiry into how we ought to live and act.

In this chapter, we invite you to engage in both the descriptive and prescriptive aspects of philosophical research. We suggest that, in many ways, prescriptive activities are both more interesting and more important than descriptive ones. In fact, you will be encouraged to enter an ongoing and still lively debate—at least within sport philosophy circles—over the justification for banning the use of performance-enhancing substances in sporting contexts. We will also present the difficulty, if not impossibility, of separating ethical questions like this one from more foundational questions about the nature of sport and competition. Thus, you will be involved in some descriptive exercises as well. First, however, a few words of introduction are needed.

Unfortunately, some consider *philosophical research* a contradiction in terms. This viewpoint rests on the rise of empirical science, historical relationships of physical activity professions with medicine, and widespread contemporary doubts about the validity of reflective, reason-based procedures (Kretchmar, 1997, 2005). Some believe that philosophy involves little more than establishing personal philosophies or sharing opinions, albeit while using impressively long sentences and, sometimes, incomprehensible words. Critics point out that philosophers themselves do not agree

fallibilist—A person who espouses a theory about limitations in seeking knowledge. This theory suggest that final and complete evidence is rarely if ever available for any truth claims. Because of this, the quest for knowledge is an ongoing process and any specific claims are open to revision.

on the utility of their methods and the validity of their results. They suggest that nearly all philosophers today are necessarily **fallibilists**. That is, even the best and sharpest minds in philosophy admit (to varying degrees) that their evidence is often incomplete and that their conclusions may be limited.

These criticisms and admissions have not proven fatal for philosophy or philosophers. In many ways, all researchers studying physical activity sail on the same boat. Most scientists, if honest, would acknowledge the fallibility of their own work. As we know all too well, many scientific discoveries or findings related to health, diet, safety, training, and sporting performance are regularly revisited and, just as regularly, revised. As you have read (or will read) in other chapters in this book, scientific paradigms change. (Newtonian physics opens certain doors to understanding, but quantum and chaos theory provides a very different platform from which to view the world.) New technologies change how and what can be measured. (The genome project was hardly conceivable 50 years ago.) Research protocols evolve. Standards for verifiability, or significance, are contested and change from time to time. (For some research, a p value of .01 is needed. For other research, it is .05. For yet other research, scientists disagree over what p values should be required.)

In short, and at least in the main, the unavoidable limits of inquiry affect scholars across the subdisciplines involved in physical activity research. To be sure, philosophy is unique because it does not literally measure anything, gather data, or use p values. Relying heavily on logic and other reflective techniques, philosophical inquiry faces distinctive challenges in justifying its methodologies and asking others to trust its conclusions. But once again, we invite you to try them out rather than assume that they do not work or, take it on faith from us that they can be effective.

Historical evidence suggests that philosophical inquiry does real work in the real world, despite its intangible subject matter and fallibilistic nature. Virtually all colleges and universities around the world house philosophy departments and offer philosophy classes. Among the best-selling books in the East and West are those dealing with ethics, philosophy, wisdom, religion, and tips on how to live the good life. Casual conversations on the street corner and serious discussions in academic halls invariably turn to ethics—that is, to claims about how things should be, not how they are. Exercise scientists are well prepared to analyze the mechanisms and particularities of performance enhancement in classes and within their research. But students (and colleagues, friends, and community members) regularly want to talk about whether these techniques are fair or ethical. Engaging informed and critical arguments against and in favor of doping bans requires a different kind of expertise and form of inquiry. Even books like this one that attempt to describe techniques for good research in a wholly objective and dispassionate way cannot avoid ethical issues. The authors, in fact, devote an entire chapter to ethical issues in research (chapter 5).

At one time, virtually every physical education, exercise science, and kinesiology program taught philosophy and ethics. Although physical activity–related curricula in some parts of the world still include philosophical content, this no longer is the case at a majority of North American institutions of higher education. Many schools, for economic, pedagogical, or other reasons, consider their many science-based courses as not only necessary for physical activity professionals, but also sufficient. We think this is unfortunate. Although we agree that a full complement of science-based coursework is necessary, it is far from sufficient. We hope your journey through this chapter allows you to draw a similar conclusion, as have many others (cf. Anderson, 2001; Fahlberg & Fahlberg, 1994; Glassford, 1987; *ICSSPE Bulletin*, no. 27, Fall, 1999; Kretchmar, 2005, 2007; Lawson, 1993; Sage, Dyreson, & Kretchmar, 2005; Sheets-Johnstone, 1999).

This chapter includes an examination of the purposes of philosophical research and draws distinctions among objective, subjective, and normative inquiry. You will have the opportunity to try your hand at one objective analysis in an area called metaphysics (the description of the nature of reality) and one combined objective–subjective analysis in the domain of axiology (the description of values). We also draw attention to the similarities and differences among philosophical inquiry, qualitative research, and empirical science. This discussion is followed by descriptions of four general approaches to philosophical research and brief analyses of the strengths and weaknesses of each. The material is designed to help you read an article in normative inquiry by Tamburrini (2000) that outlines a provocative position in the contemporary sport ethics debate over doping. You will be invited to enter that discussion.

IDENTIFYING THE PURPOSES OF PHILOSOPHICAL RESEARCH

The purposes of philosophy have changed over time. In its earliest forms—for instance, in the work of Plato and Aristotle—philosophy's role was to provide answers to virtually all of life's questions and offer solutions to most of its problems. Indeed, before the rise of science in the 15th and 16th centuries, most scholarship in the West was produced by philosophers working independently as secular thinkers or, as was often the case, under the influence and support of the Church. Philosophy's purposes reflected its expansive ambitions. These aspirations were seen to be both synoptic (dedicated to answering the so-called big questions, pulling things together, and trying to make sense of the whole) and sufficient (considered independently capable of generating all important truths). However, after the appearance of such thinkers as Copernicus, Kepler, Galileo, and Newton, the world changed. It was reconceptualized as a complex mechanism, a "self-governing machine," an entity to eventually be fully understood in geometrical and arithmetic terms (Tarnas, 1991, p. 263). Astronomy, physics, and mathematics pushed philosophy and religion off center stage.

In the 17th century, matters got worse for traditional philosophy—understood as the love of wisdom—when Francis Bacon articulated the principles of science: the explanatory utility of careful observation, objective measurement, inductive reasoning, and other foundation stones of empirical research. Put on the defensive, philosophy needed to explain how and why reflective methodologies still capably provided avenues to important insights—that is, truths and plausible hypotheses inaccessible to physicists, chemists, and other scientists using their own methods. Although its purposes continued to be synoptic, and although philosophy still attempted to answer big questions related to the nature and purpose of human existence, it now shared the stage with the many branches of natural science, the social sciences, and the other humanities such as history and literature.

The revised purposes of philosophical research acknowledged a fundamental fact. Research today in physical activity is necessarily multidisciplinary or interdisciplinary. This text, in effect, argues that no single discipline and no single methodology can go it alone. Complex human problems will be solved by combining the efforts and insights of researchers from any number of research silos.

This can happen in one of two ways. Scholars can work largely independently and later come together to attempt to fit their research puzzle pieces together, or they can collaborate before research projects are designed and possibly develop more effective studies. The former has been identified as interdisciplinary collaboration; the latter, as cross-disciplinary study (Kretchmar, 2005, 2008).

Today, the purposes of philosophy are still described as descriptive and prescriptive. Descriptive inquiry can be further divided into objective and subjective inquiry (W. Morgan, 2007). Prescriptions recommending values and behavior go under the heading of normative inquiry.

Objective Inquiry

objective inquiry—A form of inquiry aimed at producing universal knowledge that accurately describes reality. It works to eliminate biases from language, history, socialization, or any other influences that would relativize truth claims to a time, a place, or personal perspective.

Objective inquiry describes the way things are, independent of idiosyncratic experience, personal preference, or individual perspective. It is the most accepted and recognized form of inquiry and is shared by science and philosophy alike. It typically analyzes phenomena such as health, fitness, and physical activity using empirical methodologies such as looking through microscopes at muscle tissue, collecting respiratory or blood pressure data, and employing statistical procedures to determine the strengths of possible causal relationships between group-based participation and rates of exercise adherence. Objective inquiry ideally elicits facts or laws that describe existence as accurately and universally as possible. For example, it is accepted as fact that water freezes when it reaches a temperature of 0° C. In philosophy, the claim that competitors must play the same game to compare results and determine a winner is generally regarded as logically required or, as philosophers would say, true *a priori*. It too might be identified as a fact about the world—in this case, about how sport works.

All objective inquiry, as the theory of fallibilism suggests, seeks but rarely achieves impartial factual accuracy and macro-level generalizability. Most facts, particularly when related to human life, undergo revision and generate debate as ongoing empirical and reflective analyses reveal new evidence, challenge previous interpretations of the results, and lead to the emergence of alternate conceptual frameworks.

Subjective Inquiry

subjective inquiry—A form of inquiry that takes personal experience and individual preferences seriously. It speaks to the fact that people are not entirely alike, that individual perspectives and personal backgrounds matter.

Subjective inquiry, like objective inquiry, emphasizes description. Unlike objective inquiries, however, subjective inquiries seek to explain individual (or small group) preferences by examining nongeneralizable micro-level variation and personal context. For instance, subjective inquiry may explore whether and why certain people prefer running to cycling, or chocolate ice cream to vanilla. The validity and reliability of subjective inquiry depends entirely on the authenticity of beliefs articulated by the person or people under study rather than the repeatability of the methods and generalizability of the results. A result is true in subjective inquiry if the individual (or small group) genuinely accepts it as such. It makes sense, in other words, that some people might prefer running to cycling or chocolate ice cream to vanilla even though it makes equal sense that others might prefer the opposite. Individual preference claims need to be taken at face value. Consequently, subjective inquiry that explores the realm of opinion and preference is regularly considered less scientific than objective inquiry, which trades in facts and neutrality.

Normative Inquiry

normative inquiry—A form of inquiry that seeks to challenge, reinforce, or apply the norms or ideal standards of a culture, which recommend behavior, promote action, indicate how life ought to be lived.

Normative inquiry is the form of philosophical inquiry that moves description to prescription, is to ought, life as we find it to life as it should be lived. It responds to the widely held view that things could be different . . . and better! The process of establishing norms requires constant analysis to ensure that the ideals of a culture or social phenomenon (such as sport) are best reflected in its practices. When norms are in place, they function as ideal standards to use when making evaluations about good or bad, right or wrong, or better or worse.

A crucial distinction between normative and subjective inquiry rests on the degree to which the implications of beliefs or behaviors extend beyond mostly innocuous individual preference. Choosing to ride in a bike race wearing a favored blue jersey instead of a red one hardly matters to anyone besides the cyclist. On the other hand, deciding to covertly use a banned substance in the same race deeply affects others. In other words, purely subjective inquiry describes idiosyncratic preferences; normative inquiry seeks to prescribe justifiable positions.

Normative inquiry also differs from objective inquiry because its recommendations are not likely to generate agreement from all parties and for all time. Objective inquiry in science may "factually" conclude that a completely sedentary lifestyle increases the risk of type 2 diabetes, and metaphysicians in philosophy may conclude with similar levels of consensus that dance is not the same thing as sport. But it is improbable that everyone would agree with normative claims that nonmovers ought to be denied the same level of health care provided to physically active people as a consequence of their lifestyle choices; or, in the realm of physical activity, that dancers (as artists) should be treated with more cultural reverence than athletes.

Although philosophers may engage in both objective and subjective inquiry, normative inquiry is arguably the most important contribution of philosophical research to human life. For one, data about a phenomenon derived from the descriptive methodologies of objective inquiry stop short of prescribing what participants in a culture ought to do. A physiologist can provide objective facts about the effects of blood doping on a marathoner's red blood cell production, but these facts on their own do not provide an answer to the question Should marathoners be permitted to inject their own blood back into their bodies? Furthermore, the response to this question, no matter what that answer is specifically, has wide-ranging implications. Thus, it goes beyond the description of preferences that subjective inquiry provides. This is not to say that objective and subjective descriptive analyses play no role in normative inquiry. Facts emerging from objective inquiry offer important evidence to consider when formulating normative positions (e.g., What are the physiological effects of blood doping?); subjective preferences inform normative inquiry of the host of values people hold relative to the issue (e.g., Do people watching the New York City Marathon want to see world record times challenged?). Philosophical research is needed not because empirical methodologies are ineffective but because they are, in isolation, insufficient.

IDENTIFYING THE RANGE OF PHILOSOPHICAL RESEARCH

As noted, philosophers can engage in both objective and subjective forms of descriptive inquiry. However, they do so by employing reflective rather than empirical methods to explore their topics of interest. The range of that reflective activity is expansive.

Metaphysics

A philosopher may reflect on the conceptual features of sport in relation to games, or compare sporting activities to the creation of artwork to offer clearer and more generalizable descriptions of the differences between sport and art. This form of objective inquiry in philosophy is traditionally referred to as **metaphysics**.

In metaphysics, philosophers attempt to analyze the nature of things. The scope of this work runs the gamut from simple distinctions (e.g., distinguishing sport from dance) to more complex and controversial theses (e.g., describing the nature of excellence in competitive events). It may also portray categories as clearly defined with sharp edges, or alternately as less distinct, perhaps in terms of family resemblances with fuzzy or blurry lines of demarcation. Like the empirical scientist, philosophers doing metaphysics carefully examine a phenomenon and try to describe it as accurately as possible. Unlike empirical scientists, however, philosophers cannot use tools to measure, examine a response to a stimulus, or collect surveys to explain something such as sport. Metaphysics—translated as "beyond the physical"—relies solely on the results of a philosopher's reflections.

metaphysics—A branch of philosophy that addresses the nature of things.

PHILOSOPHICAL EXERCISE 1

Here is your first chance to engage in some philosophical reflection—in this case, some metaphysics. Sport philosophers have given a great deal of attention to the differences and the relationships among sport, games, and play. One way of depicting such differences and relationships is by using a Venn diagram, such as figure 13.1 showing phenomenon B (say, an apple) as a species of phenomenon A (fruit). That is, all instances of B are also instances of A. But the reverse is not true: All instances of A do not necessarily include B—there are many kinds of fruit, and apples are only one of them. How then would you describe the relationships among sport, games, and play?

You should begin with a description of each phenomenon. What is it to be at play? What is a game? What is sport? Once you have some descriptions down on paper, draw your three-circle Venn diagram (one circle for each phenomenon). If used as a class exercise, it would be interesting for everyone to work alone and then compare diagrams. How much agreement is there? Can those with a high degree of confidence in their diagrams provide good reasons for that assuredness and convince classmates of the superiority of their thinking?

Use your conclusions to challenge (or confirm) the Venn diagram and logic used by a sport philosopher found at the end of this chapter. Although this kind of metaphysical exercise cannot produce universally accepted truths or facts, it can result in better and worse outcomes—that is, more plausible results on the one hand, or less credible conclusions on the other.

This metaphysical task can then be used prescriptively. Your description of sport, for example, can be used to determine what activities should or should not be included in the Olympic Games. Similar examples could include evaluating whether someone qualifies as exhibiting athletic excellence or fair play based on a metaphysical conception of athletic excellence or fair play.

Figure 13.1 Logical relationship between a genus (fruit) and a species (apple).

Axiology

Philosophers also engage in a form of inquiry that combines subjective and objective approaches. In traditional philosophical language, this second branch of philosophy is called **axiology** and focuses on inquiry related to values. Axiological work explores the values and meanings attached to achievements, acquisitions, or states of affairs such as fitness, health, knowledge, and excellence (theories of nonmoral value), the display of appealing human qualities such as good sporting behavior (moral value) and respect for the social tenets of fair play (ethical value), as well as the dramatic or sensuous qualities of movement (aesthetic value). All four brands of axiology take personal opinion and preference seriously. Thus, they are necessarily subjective. However, all four types of reflection on value also aim at a degree of objectivity. For instance, although many behaviors might promote fair play, not every behavior qualifies. Although many elements of sporting experiences are beautiful, not everything can be described this way. Thus, axiology involves an interesting blend of objective and subjective description.

This unique objective–subjective reflective approach in axiology allows philosophers to pose and answer questions that empirical methodologies cannot (or do not) address. For instance, after chemists, physiologists, sociologists, psychologists, and historians gather and analyze data on exercise, questions still remain about the human meanings and values associated with it. Why should we exercise? To live longer? To live better? Both? If both, which is more important, more life per se or a certain quality of life? On what criteria would we determine quality of life? Would it be freedom, power, wealth, novelty, love, knowledge, security, adventure? Something else? Each person might respond differently to these questions with no correct answer in sight. Can people be considered wrong for exercising to extend their lives more so than to improve the daily quality of their existence? Is it correct to prefer cycling to running for the goal of developing aerobic capacities? Ambiguous responses to these questions show that we need to respect personal perspectives on the one hand and acknowledge the limited range of lifestyles and ethical choices that genuinely merit respect on the other.

PHILOSOPHICAL EXERCISE 2

Here is your second chance to exercise and improve your philosophical research skills. Currently, a number of sport philosophers are debating the purpose and value of sport—that is, attempting to conceptualize sport in its best light. What is sport about? What do sporting activities look like when they are going well? Alternatively, what harms the sporting project? What prevents it from reaching its potential?

Begin by identifying five "goods," or values, the sporting experience produces or includes by asking yourself the following questions: What satisfies sport participants? What would persuade them to describe a particular game as a good game? Why are people interested in sport? What do they get out of it? See if you can articulate to others your responses by writing a one- or two-sentence description of each of the goods, or values, identified. Complete the exercise by circling the one or two you believe are most important.

If you did a good job, you should notice three things. First, all five of your values are found in sport and are (or can be) considered noncontroversial goods. (This is the objective side to your axiological reflections.) Second, the two goods, or values, circled are true for you, assuming you honestly and conscientiously made the selections. Third, even though your choices are true for you, other choices can be equally true for other people. These latter two points constitute the subjective side of your axiological work.

Upon completion of this exercise, check the end of the chapter to see if your five goods, or values, overlap any of the six goods, or values, one of the authors (Kretchmar, 2015) described. Do you agree with Kretchmar that these six values are, more or less, universal goods experienced across cultures and across time?

I'm working out my philosophy of number theory scoring for chasing and catching cats.

© Zhershudio/fotolia

Much philosophical ink has been spilled on logical problems of moving from *is* to *ought,* but most modern philosophers believe that the mandate for action is found in the nature of the value or behavior described in careful axiological analyses. For instance, if our axiological reflections showed that cheating makes it difficult to determine how well one played, and if this knowledge of how well one played is on the list of noncontroversial sport values, it is clear that athletes should not cheat. The move from description to prescription, in other words, is not hard to negotiate.

Normative recommendations parallel the axiology areas for descriptive analysis listed earlier. Thus, *should* statements might accompany the following four value-related topics. Here is how the conversion from description to prescription would look:

1. Because nonmoral values such as excellence, knowledge, and happiness are good, people *should* seek to realize nonmoral values in physical activity.

2. Because moral values such as prudence, determination, honesty, and respect are good, people *should* seek to realize moral values in physical activity.

3. Because ethical values of playing fairly and attempting to avoid injuring an opponent are good, people *should* seek to employ and model ethical values in physical activity.

4. Because close, dramatic finishes are exciting and because sports such as diving and figure skating include any number of beautiful physical movements, people *should* promote aesthetic values in physical activity.

Epistemology

The distinction between objective and subjective inquiry is evident in yet another branch of philosophy. It has to do with how we know things, with what assurance we can claim that our findings are factual (or warrantable assertions). In some ways, this purpose undergirds all the others—those of metaphysics, axiology, and normative inquiry. It is foundational because extreme positions make a huge difference in the efficacy of philosophy itself. Radically subjective-oriented philosophers (skeptics or relativists) believe that no knowledge or truth exists and thus regard the role of philosophy to reinforce the validity of group preferences and cultural perspectives. Objective-oriented philosophers, at the other extreme (absolutists or realists), believe that reason is powerful and capable of uncovering enduring knowledge and truths about human existence and how we should live. This battle is ongoing in an area called **epistemology** or, in common terminology, the theory of knowledge.

epistemology—A branch of philosophy that addresses securing knowledge and understanding its status.

When you begin your reflections on a difficult normative problem in the next section, you will have to decide where you stand on this matter of epistemology. Figure 13.2 shows three broad positions along a continuum: from minimal to maximal confidence. We begin with those who place minimal faith in the power of philosophical reasoning, those who believe that objectivity is not available and that truth finding is not philosophy's purpose.

Figure 13.2 Epistemological confidence: from minimal to maximal.

Low Expectation Philosophy: Skepticism, Nominalism, and Relativism

Philosophers at this end of the continuum are subjectively oriented and thus pessimistic about the validity of objective knowledge, or truth claims, for two fundamental reasons. The first has to do with the world, with what is to be reflected on. The second has to do with oneself—that is, the power of reflection itself. First, people called skeptics or nominalists doubt that the world is composed of neat and tidy objects simply out there waiting to be discovered. They are skeptical because they do not believe the objects exist or, what amounts to the same thing, that what appear to be objects have no firm nature and are actually shaped and constructed by us. They are nominalists because the identification of things in the world amounts to nothing more than an arbitrary process of naming chunks of reality. Thus, for skeptics and nominalists, a phenomenon such as sport is not simply one distinct thing among others, but is built and rebuilt by us, by those who perceive it. In other words, if we wanted sport to look different, we could change it immediately. Sport, in this view, is socially constructed, not found. Because such constructions are subjectively controlled, objective truth about sport is not available.

Second, philosophers with minimal confidence in the power of reason doubt the validity of objectively oriented inquiry by pointing to themselves rather than the world. These people are typically called relativists. Although a stable, objective world awaiting discovery might exist, relativists claim that we lack reliable access to it. Because of language, socialization, learning, idiosyncratic perspectives, and all other manner of reflective contamination, we cannot trust our powers of reasoning. Anyone's claims about sport, for instance, would be relative to his or her own perspective or upbringing. Once again, no truths about sport can be found.

High Expectation Philosophy: Absolutism, Realism, and Rationalism

The epistemological position opposite to low expectation philosophy is often identified as absolutism, realism, or rationalism. Realists and rationalists, unlike skeptics and relativists, assume that both physical and nonphysical entities exist independently of the perceiver. Valid truth claims thus correspond to the way things really are in the world. In other words, existence goes beyond what a person thinks about some phenomenon. There is a truth about the phenomenon that is the same for everyone for all time. Sport, from this perspective, has an essence in need of discovery and clarification—and that essence cannot be altered for an activity to be considered sport. Consequently, realists and rationalists place a great deal of confidence in the power of logic or reasoning to solve philosophical problems. Rationalists perceive the world as a giant math problem: just as mathematicians can be sure that $1 + 1 = 2$, more complex problems will be solved in the future with equal certainty. They believe there are right answers and, like many empirical scientists, argue that eventually these truths will be uncovered.

Moderate Expectation Philosophy: Fallibilistic Realism

Many philosophers today adopt a middle position, believing that the world is reasonably durable even if it is not fixed. Pure objectivity and pure thoughtfulness are considered unhelpful abstractions. Because of this, philosophers need to return to their objects of inquiry, study them again and again, and thereby acknowledge both their dynamic qualities and the inherent difficulty in knowing them completely. To be sure, something like sport is a more or less durable thing, but it may evolve in some ways along with the people who play it, and it is extremely complex. Likewise, reason is powerful, but there is no Archimedean or external vantage point from which to view something such as sport. As noted in the introduction, reason is acknowledged by most philosophers today to be inherently and significantly limited. And because of this, philosophers are inherently fallible. Although philosophers make progress, there will always be mistakes requiring correction, and there will always be more to be seen, described, and recommended.

Where do you stand along this continuum? You will be asked in the next section to determine whether current rules against steroid and other performance enhancers in sport are morally justified. Professor Tamburrini believes that bans should be lifted and provided a number of arguments to support his position. In suggesting how sport *should* be conducted, he was engaging, as you now know, in normative inquiry. As you enter into this debate yourself, will you be a skeptic and think that no good, all-encompassing "true" reasons can be given for or against the bans, or a realist and argue that a final, best answer is available on a foundation of reason? Or will you have moderate expectations and believe that persuasive but not conclusive evidence can be placed on the debate table?

If you have not already done so, read Tamburrini's article "What's Wrong with Doping?" (2000). You will need this background to work through the next two sections of this chapter on locating and analyzing a research problem.

LOCATING A RESEARCH PROBLEM

Philosophical problems are not difficult to find. They are literally everywhere. You cannot wake up in the morning without being confronted with potential philosophical issues. What is worth doing today? What is the good life? How should I go about my work or play? Why should I be ethical—particularly if everyone else is cutting corners and there is little chance of getting caught? Or, for the purposes of this text, what counts as valuable research? How does one conduct research in an ethical manner? Metaphysical, axiological, and normative questions lie beneath every rock in the landscape.

In the domain of physical activity, the situation is the same. One cannot deal with health, disease, fitness, movement, sport, games, and the like, without any number of philosophical issues presenting themselves. Tamburrini selected an enduring issue challenging humans throughout the history of game playing—namely, what methods should be allowed and disallowed in our pursuit of success and victory?

Athletes have always attempted to find ways to improve their performances. The quest for higher, faster, and stronger seems built into sport itself. One cannot study competitive activity historically without finding athletes pushing the performance envelope—whether ethically, unethically, or by means difficult to categorize as clearly one or the other.

A first step in locating a research problem is ensuring that the question at hand is, in fact, worthy of examination. If answers are obvious, if no good arguments can be rallied on one side or the other of the issue, then the problem is probably a pseudo-problem and not worth examining. This would be the case with the following two ethical questions in the area of performance enhancement:

1. Should 6-year-old athletes be allowed to experiment on their own with dangerous steroid use?

2. Should ambitious parents be free to force their underage children to use steroids to promote sporting success?

Neither of these questions is particularly interesting because normative answers appear obvious. It would be difficult to rally an argument on the affirmative side for either of these issues. These questions, in other words, constitute an unsuccessful attempt to locate a research problem. Tamburrini was aware of this and thus limited his discussion to the issue of performance enhancement among informed adult athletes. The question becomes more complex in this context because we do, in fact, allow adults to make decisions about their lives as long as those choices do not negatively affect others. We permit informed adults to use dangerous tobacco and alcohol products, for instance. Thus, a potentially interesting normative question arises: Why allow smoking and alcohol consumption and not steroids?

On closer inspection, we see that the issue of deregulating bans on certain performance enhancers actually taps into a durable dilemma in the history of modern philosophy. It pits libertarians wanting to maximize human freedoms, on the one hand, against conservatives or communitarian philosophers who, for a variety of reasons, believe that limitations are warranted. You should be able to feel the tension here. You value your freedoms and undoubtedly resist the heavy hand of paternalism whether from your parents, the government, the church, or some other source. You resist the intrusion of others telling you how to live your life. On the other hand, you benefit from many social arrangements that put restrictions on yourself and others—for example, the requirement to pay taxes, the need (in some countries) to serve in the military or the importance of obeying noise ordinances. Thus, on reflection, you can see that this debate goes beyond sport. It is about broader issues related to the limits of human freedom—particularly among informed, consenting adults. This further suggests that Tamburrini had hold of a good research problem.

A second step in locating a philosophical research problem is more technical. Analogous to the limitations and delimitations accompanying some empirical study, philosophers need to clarify the issue under their reflective microscope. Tamburrini did this early in his article by identifying three kinds of doping: harmful but legal, harmful and illegal, and both harmless and legal. Anabolic steroids, the kind of enhancement that lies at the center of much debate over the bans, fall into the first category. Steroids are legal in the sense that professionals can prescribe steroids for medical purposes, yet they cause harm when taken inappropriately or in large doses. Tamburrini also drew an important distinction between banned substances (such as steroids) and banned techniques (such as blood doping, in which one's own blood is withdrawn and then injected back into the body prior to competition). The three categories of doping substances and the distinction between substances and techniques all come into play in his argument.

Tamburrini took two additional steps to locate and establish his research problem. He identified three reasons for eliminating the ban (with a fourth—the preservation and promotion of human freedom—underlying all three) and considered four prominent arguments used by supporters of the ban. This set the stage, as it were, for the philosophical debate to begin—after all, there would be little rationale for Tamburrini's essay if no good or harm came from eliminating the ban.

The final preliminary step was to identify his thesis. Tamburrini indicated that he would focus on the four arguments supporting the current prohibition and attempt to show that they lack validity. He wrote, "I will scrutinize these [four, pro-ban] arguments to support the claim that the ban on doping is not justified and should therefore be lifted" (p. 202). He chose this strategy because of the "widespread intuition" that doping is "inherently wrong." That is, many take the four arguments used to support the ban—whether intuitively or reflectively—as conclusive. Tamburrini wanted to challenge that assumption.

> Good philosophical research identifies a good problem, clarifies the problem, and establishes a clear purpose or thesis.

ANALYZING A RESEARCH PROBLEM

This is where much of the hard work requiring good logic, clear thinking, unbiased description, attention to detail, and other factors related to solid philosophical practice begins. We call this work hard because the criticisms of extreme skeptics and relativists—those arguing that persuasive justifications for certain perspectives do not exist—are difficult to overcome. But once again, we want you to walk through these methods to see their utility. Even if they cannot provide final answers, they may point us in the right direction.

This section provides examples of four kinds of reasoning that philosophers use—inductive, deductive, descriptive, and speculative/critical. All four can be applied to Tamburrini's project and are implicit in his reasoning. Even though Tamburrini did not do so, we highlight them to show how they work.

Inductive Reasoning

Inductive reasoning is reasoning that moves from a limited number of specific observations to general conclusions about the phenomenon (or thing) under consideration. It relies on intelligent discernment to identify common elements or similarities in the samples.

A major issue underlying doping bans is the question of paternalism. When and under what conditions can second parties, such as sport governing bodies, justifiably limit an athlete's individual freedom? One obvious answer is that externally imposed prohibitions are appropriate when a behavior may bring serious harm to that athlete and fellow competitors. Safety rules in sport are common . . . and commonly accepted! This logic seems to provide a very strong counterargument to Tamburrini's case. However, he can use inductive logic to provide a rejoinder. He can identify a limited number of activities considered dangerous yet permitted, such as skydiving, deep sea diving, climbing very high mountains, playing tackle (American) football, chewing tobacco, riding motorcycles, and eating high-fat meats or high-calorie desserts.

With this list in hand, a philosopher can look for commonalities or general conclusions regarding their similarities, such as the following:

- They are all dangerous to the point of causing debilitating injury or illness or even premature death.

- All of them are legal. (No paternalistic agency prohibits participation by informed adults.)

- All of them are (or can be) enjoyable. Many people engage in them simply because they serve as a source of pleasure.

These inductively generated insights led Tamburrini to a couple of interesting questions and a potentially persuasive conclusion. The questions are these: What if athletes found that the athletic effects of taking steroids enhance their enjoyment? And what if the risks were no greater than those incurred by others involved in this list of dangerous but permissible activities? The conclusion would point to a potentially indefensible inconsistency. Society tolerates any number of high-risk behaviors in the name of individual freedom and the pursuit of personal pleasure or gratification. So, too, should it be with steroids. Consistency requires it.

You would be right to question the force of this argument because any number of additional reasons may support the prohibition. Moreover, these other reasons may trump the "tolerance for risk" and the "pursuit of pleasure" arguments. But inductive reasoning proves to be a reliable tool in pointing out an apparent inconsistency in how we deal with risk. It raises important questions about double standards and, possibly, a kind of irrationality underlying doping prohibitions. Even the use of the term *doping* suggests something illegal, surreptitious, and otherwise distasteful. Why not call it a technology or simply a performance enhancement so that more objective and dispassionate reviews of its status can take place?

Inductive reasoning has both strengths and weaknesses. For one, the development of the original list is crucial. If items are not representative of the phenomenon at hand (here it was "dangerous but permitted activities"), then the conclusions will likely be wrong or misleading. Also, the process of identifying generalities can prove challenging. If too obvious (e.g., all of these activities involve people), the inductive generalization may be true but uninteresting and lacking helpful insight. A generality might also be wrong. A philosopher thinking only of skydiving and riding motorcycles might consider these risks external to the person. But of course, chewing tobacco and eating high-calorie desserts are internal. Although this error is obvious, others might be more subtle and go unnoticed.

Philosophers in general, but particularly those taking the cautious middle epistemological position we identified as fallibilistic realism, test their conclusions derived from inductive reasoning repeatedly. This is the case because neither scientists nor philosophers can be certain that the items scrutinized the first time around represent those that were not examined. The previous sample, in other words, might be flawed or contaminated and lead to unwarranted conclusions. Or the way the sample was tested might have introduced errors. Scientists therefore test and retest. Philosophers reflect inductively on a set of items; then reconsider the items and reflect some more. Progress is made, but certainty is rarely achieved.

Deductive Reasoning

Deductive reasoning is a companion technique of inductive reasoning, and many philosophers adroitly and spontaneously intermix the two. Deduction requires intellectual movement in the opposite direction from induction. Whereas inductive reasoning starts working from particulars to construct abstractions or generalities, deductive reasoning starts with general claims to see what particulars follow.

The initial general claims or premises used in deduction are of two sorts. The first are statements of fact often phrased as *Because such and such is true, then it follows that . . . .* But premises can also be hypothetical: *If such and such were true, then it follows that . . . .* We will take a look at a deductive line of reasoning using hypothetically stated premises.

One standard justification offered by those supporting the current bans is the so-called coercion argument. Those employing this argument might agree with the libertarian contention that society should err on the side of freedom even when a person chooses to accept a high degree of risk. However, they would disagree that people should be allowed to coerce others into similar actions. Because steroids confer considerable competitive advantages (so the argument goes), athletes otherwise inclined to avoid such risks would feel coerced to accept them. This coerciveness is what some doping ban defenders deem unacceptable and consequently what justifies paternalistic prohibition. Tamburrini, however, advanced a series of counterarguments that can be put in deductive form:

> Premise 1: If it is true that high-risk behaviors occur in business (e.g., working all night, taking pep pills to stay awake), and
>
> Premise 2: If it is true that such behaviors confer a clear competitive advantage on those willing to take such risks and are, for this reason, potentially coercive for fellow workers, and
>
> Premise 3: If it is true that many fellow workers still choose to resist coercive pressures and do not adopt such behaviors, then
>
> Conclusion: It follows that high-risk, advantage-conferring behaviors in business are not (strongly) coercive.

Would lawyers, accountants, or small business owners accept a mandatory 40-hour week with fines imposed for working overtime? Unlikely. So Tamburrini can then ask, Why should it be any different in sport? Just like people working in offices, athletes can say no and, in their own case, simply choose not to dope. Moreover, it could be argued that those willing to take the risks (in business or in sport) should reap the benefits and suffer the consequences. Justice is being served.

Tamburrini was aware of another argument, one that hinges on a difference by degree. Working all night on the job may confer *some* advantage, but this may also lead to burnout, inefficiency, increased mistakes due to fatigue, and the like. The advantage, in other words, may not be significant. And besides, relatively few workers might risk compromising their own health and family life (if they have one), thus making those working inordinately long days statistical outliers. Prudent workers therefore will be compared with a large number of other prudent workers (apples compared to apples). They will not be at a disadvantage. Thus, for both reasons (justice is served and unfair advantages are mitigated), prohibiting overtime in the workplace is not needed.

deductive reasoning—Reasoning that moves from general claims or premises to particular conclusions. (See figure 2.2 in chapter 2.)

Sport is different because advantages are clear and decisive. In the case of the Tour de France, ban supporters claimed that clean athletes simply could not compete with those using illegal enhancements (justice is absent). Also, in this environment, it was a common belief that almost everyone doped. The coercive pressures were enormous. If one wanted to compete, one had to dope. A few prudent cyclists would be compared to an overwhelmingly large number of imprudent cyclists (apples compared to oranges). This is patently unfair.

Tamburrini presented the following rejoinder. Again, this could be put in a deductive format.

Premise 1: If athletes choosing not to dope are less successful competitively, but

Premise 2: If less successful athletes still have access to a number of athletic benefits, then

Conclusion: Athletes choosing not to dope still have access to a number of athletic benefits

Tamburrini speculated on two scenarios in which nondoping athletes plausibly could receive benefits even if bans were lifted. The first suggests that the national and international level of interest in athletics is so great that even second-tier athletes (those choosing not to dope) can have successful professional careers. The second argument points to the perceived superior moral position maintained by those who do not dope. Many fans may prefer a doping-free version of the activity, and thus support the athletic aspirations of individuals and groups choosing on moral grounds not to partake in certain sanctioned performance-enhancement practices.

These arguments may strike you as a little far-fetched. In a world in which winning seems valued over all else and fans regularly ignore poor behavior exhibited by favorite players or teams, Tamburrini's scenarios may seem unrealistic. But the deductive logic, as far as it goes, is sound. The two premises necessarily generate the conclusion, and this is perhaps the greatest strength of deduction. As much as any other method, it aims at certainty. St. Thomas Aquinas used it to good effect during the Middle Ages to argue that if certain premises about God's nature, or God's will, or claims of Holy Scripture were true, then important theological conclusions necessarily followed. This is much like a geometric proof and thus favored by those with high degrees of confidence in the powers of reason and strong tendencies toward objectively oriented epistemological realism.

However, two important weaknesses must be addressed. The first relates to the validity of the premises. If any of the premises are false, the deductive conclusion is likely false as well. You might question, for instance, Tamburrini's second premise that athletic benefits will remain available to elite athletes choosing not to dope. If that premise is wrong, the argument fails. Thus, it is commonly known in philosophy that the best way to attack a deductive conclusion is to raise questions about one or more of the premises resulting in that conclusion.

Of course, the deduction itself can be faulty. The conclusion can introduce terms or concepts not included in any of the premises. See if you can pick out the missing premise in the following line of reasoning:

Premise 1: If it is true that Armstrong doped, and

Premise 2: If it is true that doping is both illegal and unethical, then

Conclusion: It follows that Armstrong should be stripped of all victories gained while he was doping.

The conclusion, although it may seem reasonable enough, is not deductively valid. It introduces concepts not included in either of the premises—concepts about grounds for stripping medals or victories. Thus, a third premise is required, one that states, "If it is true that all illegal and unethical conduct warrants the vacation of victories," then the conclusion follows. The logical chain of reasoning from Armstrong's doping to the stripping of medals has been connected.

Descriptive Reasoning

Descriptive reasoning provides a direct and disarmingly simple way to conduct philosophical research. Philosophers using this method describe phenomena in the context of lived or hypothetical experience. To be sure, they manipulate that experience adroitly, as will be described shortly, but descriptive techniques rely on the fidelity of observations and resultant descriptions.

Descriptive philosophers prefer to conduct their research in less roundabout ways than those provided by inductive and deductive logic. They do not line up a series of particulars and ask themselves what they all have in common (induction), nor do they begin with a set of givens or premises to see what follows (deduction). They simply describe what they see reflectively when examining a phenomenon.

The issue of fairness, one central to the doping debate, can be addressed descriptively. We can reflect, for instance, on elite downhill skiers from Norway, who happen to be identical twins separated at birth, now vying for top world ranking. We wonder if their competition is fair. It would seem that it is . . . or at least largely so. They both play by the rules, so neither gains an advantage by cheating. Moreover, raised in nearby towns, they had (let us say) equal amounts of winter weather as well as access to the same equipment, the same coaches, and the same mountainous slopes during their formative years while learning to ski. And, perhaps most important of all, they share the same genes. On this particular afternoon, weather conditions are the same for both. Multiple runs lessen the possible effects of chance or luck, and the order in which they ski is reversed each time to neutralize advantages from slope conditions. The competition, as much as humanly possible, seems to pit the merits of one sister against the merits of the other. We intuit this as a level playing field and describe it as a fair competition.

But then we vary the situation. We picture two unrelated downhill competitors who grew up under very different conditions—one in Norway as before, but the second one in a desert environment, from a relatively poor family, in possession of only inferior equipment, with poor coaching, and finally, with very limited access to good downhill training facilities. In spite of this, she has become reasonably accomplished and now competes against the Norwegian woman. Do we still intuit this as a fair competition?

In a way, we do. Both are following the rules. Nobody is gaining an advantage by cheating. Both are skiing down the same slope. However, in another important sense, it is very unfair. The merits of the two skiers are not the only things being tested. Their backgrounds are very much in play. The Norwegian woman will probably win, but she can hardly take credit for her many advantages. We might even speculate that sport is not designed to measure the kinds of things that produced the very different skills of these women—benefits of birth, geography, weather, wealth, and other such factors that seem beside the point. Sport is supposed to show who deserves top ranking, who earned the victory. Thus, we describe this competition as unfair or invalid.

These kinds of descriptive or intuitive realizations led Tamburrini to draw a surprising conclusion. He suggested that if they were genuinely interested in promoting fair play, international sporting organizations would allow selective doping to level the playing field and better test the true merits of those competing. It would amount to a kind of high-tech handicapping designed to nullify unearned advantage. In our scenario, only the second skier from the desert climate would be permitted to dope to equalize the competition. Once again, this may strike you as far-fetched, but Tamburrini, like many good philosophers, followed the logic where it led him. He would suggest we not dismiss this idea out of hand.

The processes of looking carefully and describing accurately require considerable skill, and like all skills, they can be performed well or poorly. Because human beings can be inattentive, often jump to conclusions, regularly fail to notice details, take things for granted, and often confuse parts and wholes, producing accurate, insightful, and careful description

descriptive reasoning— Reasoning that describes lived experiences and relies on descriptions of observations.

is not particularly easy. When done well, however, descriptive reasoning can be extremely useful and enlightening.

Descriptive reasoning highlights at least one advantage philosophers enjoy over their scientific colleagues. We can vary our subject matter at will. In the preceding example, we did not have to find a skier from the desert, and we did not have to conduct an interview. We could simply create the new scenario and describe it. If we did our work well, our conclusions would map onto the real world in effective ways. In this case, we could say with a high degree of assurance that they would. We see many sport organizations trying, as best they can, to neutralize irrelevant factors that could influence performance. They implement weight classes in sports in which crude physical mass can create advantages. They standardize equipment to ensure that the contest is less likely to turn on the merits of the equipment manufacturer. And we learned that, at least in principle, selective doping could promote rather than detract from fair play.

Descriptive methodologies place a great deal of confidence in the capacity of reason to portray existence accurately. It is important to recognize that this confidence resides in traits and skills of human intelligence depending on the level of description attempted. If description is intended to depict an actual lived experience—with its real perceptions, feelings, ups, and downs—then there must be a strong reliance on good memory, honesty, attention to detail, an unwillingness to embellish facts, and so on. This subjective level of description (similar to that used in qualitative methodologies) in effect involves making claims about what someone really experienced, and such portrayals can vary between fact and fiction.

A second, more objective level of descriptive work, such as the one modeled earlier, is disinterested in the actual subjective experience itself. At this level, the concern is with the nature of things in principle. This level of description places confidence in the power of reason to notice important differences when varying the subject matter. For instance, we noticed differences in two species of fairness, the kinds of factors that influence fairness, as well as the varying roles steroids can play in the pursuit of fair play.

> Descriptive methodologies rely on reason's capacity to reliably portray reality, whether depicting an experience or working on distinctions at a more abstract level.

The danger with this second type of descriptive methodology is not poor memory, dishonesty, or tendencies to exaggerate events, but rather people's internal biases that contaminate the ability to see clearly. For instance, your religious upbringing might affect the way you observe the world. You may have been raised in a family that was significantly harmed by drugs. If so, you might not be able to reflect on the issue of drug taking in sport objectively and dispassionately.

Speculative and Critical Forms of Reasoning

speculation—Claims about subjects, such as values, that are significant to humans but cannot be forcefully defended.

critical reasoning—An activity intended to show reasoning errors, faulty claims, and misplaced beliefs.

Speculation can be thought of as an extension of both descriptive and prescriptive philosophy. It is a philosophical method that raises possibilities that cannot be substantiated or otherwise forcefully defended. **Critical reasoning** is the other side of the coin. It is a philosophical method that deconstructs or debunks philosophical analyses of many types, particularly those that are speculative. Critical reasoning attempts to uncover the biases shaping philosophical conclusions. Some critical philosophy regards all attempts to find the truth as futile or otherwise misguided.

Speculative philosophy attempts to answer questions without applying simple logic or using clear description. In this sense, speculative conclusions lack supporting evidence. It does not follow, however, that speculation requires no argumentation whatsoever. Again, speculating can be done well or poorly. Even though speculative conclusions cannot be proven or demonstrated, they can vary in their degree of plausibility.

Tamburrini addressed one speculative claim about sport and took a more or less neutral position on its validity. It has been speculated that much of the value and charm of sport lies in its uncertainty, the attempt in sport to pull something off without any assurances of success. In other words, if sport became predictable, it would lose much of its appeal, and quite likely, fewer people would want to play or watch it. It could be further speculated that this uncertainty has two roots—the test and the contest. The test is the game problem—such as running 4 miles in a short period of time. This produces the uncertainty of passing or not passing the test, of getting an A, a B, a C, or a worse grade as a result of our efforts. We also evaluate against ourselves. Can I do better than I did yesterday? Can I set a new personal record? This delicious uncertainty draws us to the track, the golf course, the bowling alleys—wherever we can face this tension or uncertainty.

A second source of uncertainty lies in the contest. Regardless of how well I do on the test today, will it be good enough to beat you? This produces a dichotomous contest result of victory or defeat, but in actuality it is a graduated comparison running from no difference (a tie game) to an extreme difference in two directions (a lopsided victory or defeat). Regardless, there is also some delightful drama and uncertainty here too (Kretchmar & Elcombe, 2007). Will I win or lose? If I lost to you the last time by a wide margin, can I keep it closer today?

Other philosophers speculate that different factors account for much of the value of sport. Once again, this relates to the metaphysical exercise you encountered earlier in this chapter. Different plausible answers present themselves seemingly without any way to adjudicate among and between them. Thus, speculative philosophy leaves room for subjective preference. Some may play sport for the dramatic tensions it includes. Others may play sport for the knowledge they can gain about their personal abilities. This is like preferences for vanilla and chocolate ice cream. We have to respect these unique tastes or opinions.

However, Tamburrini was not willing to give too much credence to subjective preference. He believed that there is at least a degree of truth to claims about the importance of uncertainty. Thus, he took pains to show how doping does not threaten this quality of sport. He provided several reasons for claiming that doping and uncertainty are compatible. You will have to read these arguments to see if you agree.

This raises a point made earlier in the chapter. Descriptive metaphysics and prescriptive normative inquiry are often interrelated. To know how we should act in X, we need to know what X is, how it works, what its values are. Tamburrini acknowledged this relationship by doing some speculative metaphysics first (determining that uncertainty is part of the nature of sport) and then moving on to an axiological description (determining that it is valuable), and finally making a normative recommendation (bans should be lifted because doping does not threaten uncertainty).

The advantage of speculative philosophy is that it broaches subjects with deep human significance and the potential to markedly change human lives. Wars, for instance, have been fought over a way of life grounded in the perceived superiority of some values in relation to others—say, democracy over some form of totalitarianism. Accordingly, given that people mold their lives around values, it may well be important to do research on intangibles, even if knockdown arguments are not available.

The research principles here would be twofold. First, it is better to know a little (even if it is inconclusive or incomplete) than to know nothing at all. And second, one can be unable to prove a claim and still be right about it. Some speculate that a spiritual reality of some sort exists, and they provide evidence, they believe, pointing in that direction. Others disagree and present their own arguments to the effect that nothing is real but atoms and void. Neither side can present knockdown arguments. Yet one side is still right. Or perhaps, in some hard-to-imagine way, both of them are right.

Disadvantages also exist. Narrow, selfish, manipulative, and even mean philosophies can masquerade as legitimate speculative statements about reality. Either purposely or unintentionally, some people spew groundless opinions, biases, and self-serving and status-quo-preserving visions of life in the name of philosophy. Even arguments intended to show the plausibility and attractiveness of certain claims might only play to the biased predispositions of the audience. For instance, claims about the superiority of sport when conceptualized as a mutual quest for excellence may seem persuasive in a capitalistic, achievement-oriented culture in which most people are raised to be all they can be. In another society, such arguments for excellence might seem odd or misplaced.

In some ways, Tamburrini's unusual defense of doping can be read as critical philosophy. It debunks the metaphysical superstructures built up around sport. Although it does not move significantly in a political direction, it raises questions related to power and politics. Who benefits by maintaining the status quo? Why would sporting organizations make a show of enforcing doping regulations? When some folks argue that doping violates the nature of sport, why should their version of sport be accepted as definitive or absolute?

At one time in history, and still to a certain extent today, amateur competition was regarded as the only true form of sport. Critical sociologists and philosophers debunked that myth and exposed it as a very limited and self-serving ideological view of sport. Tamburrini may have done much the same thing by exposing the preeminence of drug-free competition as a myth. Or at least he tried to do so.

Critical reasoning is based on skeptical attitudes toward the power of reason and therefore seems unduly negative. Some critical philosophy reads like an exercise in tearing things down, not building them back up. However, exposing myths and errant thinking is no trivial activity—such as showing supposed truths about women, races, religious organizations, sexual orientation, or certain privileged versions of sport as biased and harmfully inaccurate. It is far better to question what is held to be true than to perpetuate a lie or tacitly support potentially injurious beliefs.

This is the dilemma that Tamburrini's challenging article lays out. Are bans a mistake? Or are they grounded in defensible concerns? Hopefully, you will be able to interact with the position he developed and add further arguments to either support or dispute his conclusions. A bibliography is provided at the end of the chapter for those who wish to look at what a number of philosophers have written about this issue.

SUMMARY

In some ways, doing philosophy well is like performing a motor action skillfully. And like many acts that require expertise, confidence grows as skill improves. Within broad boundaries of orthodoxy, different strategies and styles work for different people. They allow us to get in touch with our preferences. This is the important subjective side of philosophy, the part that respects a range of likes and dislikes that vary from one person to another, and from one group of people to another. They also allow us to appreciate the descriptive and prescriptive purposes of philosophy—the metaphysics and axiology of description, and the normative recommendations of prescription. As many philosophers have noted, the constant subtext of philosophy is the reminder that the world can be different . . . and better.

Check Your Understanding

1. Have the purposes of philosophy remained constant across history? If not, how have they changed?

2. What is the difference between objective philosophy and subjective philosophy? Why are both important? Why does much traditional philosophy favor objective analyses?

3. What is the difference between metaphysics and axiology? Why are both identified as descriptive aspects of philosophy? Can you identify the four subcategories of axiology and describe their differences?

4. What is the difference between metaphysics and axiology, on the one hand, and normative philosophy on the other? See if you can identify three normative claims, one that is highly subjective, one that falls between the subjective and objective ends of the continuum, and one that is highly objective—that is, one that would be difficult to refute.

5. Why is epistemology such a foundational aspect of philosophy? Which of the three epistemological positions did you find most comfortable, and why?

6. What counts as a good philosophical problem? Why would Tamburrini's question count as a good problem? Why are definitions important? What role does a thesis play in philosophical research?

7. How do deduction, induction, description, speculation, and critical philosophy work? What are their strengths and weaknesses?

Appendix: Comparing Sport Philosopher's Logic

Philosophical Exercise 1

You were asked to do some metaphysics—specifically, to identify the nature of sport, games, and play and the relationships among them. The diagram in figure 13.3 was proposed by Klaus Meier. It suggests that games and play are not the same, yet compatible or overlapping, phenomena. Play is a way of participating in which the end lies in the doing. Play is not necessitated or done out of duty. It is intrinsically interesting, joyful, or meaningful. Games, on the other hand, are artificial problems.

Given these brief definitions, relationships between play and games can be identified. Some play is not a game (relationship 1). An example would be reading a good book for intrinsic reasons—simply for the pleasure of reading. However, some play is compatible with games (relationship 2). A person could be intrinsically engaged in solving an artificial problem. An example would be a person participating in a baseball game for intrinsic reasons—simply for the joy of facing the challenges of the activity. Finally, some game experiences are not also play (relationship 3). Examples: A father forces his child to play against his will, or a professional ballplayer plays simply to make money (i.e., for extrinsic reasons).

Sport is described as a kind of game; that is, a species of artificial problems. It is a kind of game that requires physical prowess to solve the game problem. Thus, crossword puzzles, sudoku, and chess are not sports because physical skills are not essential for solving their respective problems. However, baseball, basketball, and other physically demanding games deserve that title. Consequently, sport is entirely within the game circle (all sports are also games). But sport can be engaged in within the spirit of play (relationship 4) or not (relationship 5).

How did your Venn diagram compare to Meier's?

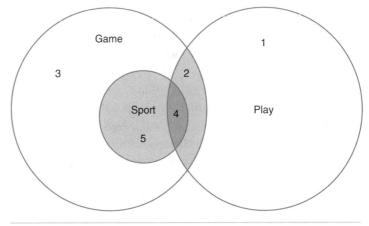

Figure 13.3   The interrelationships among sports, games, and play.

"Triad trickery," K. Meier, *Journal of the Philosophy of Sport* xv: 26, 1988, Taylor and Francis, reprinted by permission of the publisher (Taylor & Francis Ltd, www.tandfonline.com).

Philosophical Exercise 2

You were asked to describe values that can be found in sporting activities. This exercise was designed to indicate that subjective preferences are important, but that objective constraints still operate in most subjective domains. In other words, there may be no single version of sport that deserves top billing. On the other hand, not every preference has to be respected. In other words, the range of valid preferences is limited. In a recent article, Kretchmar (2015) suggested that there may be at least six versions of sport that deserve support and that people or cultures might legitimately choose any one over the others. Before initiating speculations on ways of shaping sport, Kretchmar accepted a common definition of sport as an artificial physical test or contest (or both). Briefly, the six versions are as follows:

1. Achievement model—shaped by the ethic of work, aimed at progress and excellence, based on merit, and generally opposed to luck playing a role in the outcome.

2. Serendipitous model—shaped by the ethic of play, revitalization, hope, having a chance, and gaining relief from the normalcy of work and meritocratic arrangements.

3. Epistemological model—shaped by human curiosity and the need to know, and influenced by the values of education and research.

4. Aesthetic model—shaped by the values of art, theater, and literature; aimed at the axiological values of drama, beauty, and other aesthetic qualities; influenced by the requirements of coherence and living a connected or meaningful story line.

5. Existential–individualist model—shaped by existential values of authenticity, realizing one's freedom, and taking responsibility for one's own choices; aimed at personal expression and creativity; fears conformity or other forms of socially controlled sporting behavior.

6. Communitarian model—shaped by the values of the polis, the family, the group; aimed at belonging, cooperating, and solving problems together; formed around sports and competitive skill levels.

Selected Bibliography on the Ethics of Performance Enhancement (Chronologically Arranged)

Brown, M. (1980). Drugs, ethics, and sport. *Journal of the Philosophy of Sport, 7,* 15–23.

Simon, R. (1985). Good competition and drug-enhanced performance. *Journal of the Philosophy of Sport, 11,* 6–13.

Gardner, R. (1989). On performance-enhancing substances and the unfair advantage argument. *Journal of the Philosophy of Sport, 16,* 59–73.

Schneider, A.J., & Butcher, R.B. (1993). Why Olympic athletes should avoid the use and seek the elimination of performance-enhancing substances and practices for the Olympic Games. *Journal of the Philosophy of Sport, 20*(1), 64–81.

Burke, M., & T. Roberts. (1997). Drugs in sport: An issue of morality or sentimentality? *Journal of the Philosophy of Sport, 24,* 99–113.

Kretchmar, S. (1999, Fall). The ethics of performance-enhancing substances in sport. *Bulletin: International Council of Sport Science and Physical Education, 27,* 19–21.

Tamburrini, C.M. (2000). What's wrong with doping? In T. Tansjo & C. Tamburrini (Eds.), *Values in sport: Elitism, nationalism, and gender equality and the scientific manufacture of winners* (pp. 200–216). London: Spon Press.

Holowchak, A. (2002). Ergogenic aids and the limits of human performance in sport: Ethical issues and aesthetic considerations. *Journal of the Philosophy of Sport, 29,* 75–86.

Simon, R. (2005). *Fair play: The ethics of sport* (2nd ed., chapter 4). Boulder, CO: Westview.

Petersen, T.S., & Kristensen, J.K. (2009). Should athletes be allowed to use all kinds of performance-enhancing drugs? A critical note on Claudio M. Tamburrini. *Journal of the Philosophy of Sport, 36,* 88–98.

14

Research Synthesis (Meta-Analysis)

> Within the last few chapters we have used ten puns hoping to make you laugh. No pun in ten did.
>
> —Jerry R. Thomas

An analysis of the literature is a part of all types of research. The scholar is always aware of past events and how they influence current research. Sometimes, however, the literature review stands by itself as a research paper that analyzes, evaluates, and integrates the published literature. A term used to describe this is *research synthesis*. As mentioned in chapter 2, many journals consist entirely of literature review papers, and nearly all research journals publish review papers occasionally.

All the procedures discussed in detail in chapter 2 apply to research synthesis. This chapter addresses using the literature to draw empirical and theoretical conclusions rather than to document the need for a particular research problem. A good research synthesis results in several tangible conclusions and should spark interest in future directions for research. Sometimes, a research synthesis leads to a revision to or the proposal of a theory. The point is that a research synthesis is not simply a summary of the related literature; it is a logical type of research that leads to valid conclusions, hypothesis evaluations, and the revision and proposal of theory.

The approach to a research synthesis is like the approach to any other type of research. The researcher must clearly specify the procedures that are to be followed. Unfortunately, the literature review paper seldom specifies the procedures that the author used. Thus, the basis for the many decisions made about individual papers is usually unknown to the reader. Of course, an objective evaluation of a literature review then becomes nearly impossible. Questions that are important yet usually unanswered in the typical review of literature include the following:

- How thorough was the literature search? Did it include a computer search and hand search? In a computer search, what descriptors were used? Which journals were searched? Were theses and dissertations searched and included?

> A research synthesis leads to valid conclusions, to hypothesis evaluations, and possibly to the proposal of a theory or of future directions for research. It is more than a summary of related literature.

■ On what basis were studies included or excluded from the written review? Were theses and dissertations arbitrarily excluded? Did the author make decisions about inclusion or exclusion based on the perceived internal validity of the research, on sample size, on research design, or on appropriate statistical analysis?

■ How did the author arrive at a particular conclusion? Was it based on the number of studies supporting or refuting the conclusion (called vote counting)? Were these studies weighted differentially according to sample size, the meaningfulness of the results, the quality of the journal, and the internal validity of the study?

Many more questions could be asked about the decisions made in the typical literature review paper, because good research involves a systematic method of problem solving. But in most literature reviews, the author's systematic method remains unknown to the reader, thus prohibiting an objective evaluation of these decisions.

In recent years, several attempts have been made to solve the problems associated with literature reviews. The most notable of these attempts was made in a paper by Glass (1976) and followed up with a book by Glass, McGaw, and Smith (1981), in which they proposed a technique called **meta-analysis**. The major purpose of this chapter is to present an overview of meta-analysis and the procedures developed for use in meta-analysis (for a more general approach to research synthesis, see Cooper & Hedges, 1994).

meta-analysis—A technique of literature review that contains a definitive methodology and quantifies the results of various studies to a standard metric that allows the use of statistical techniques as a means of analysis.

USING META-ANALYSIS TO SYNTHESIZE RESEARCH

Since Glass introduced meta-analysis in 1976, thousands of meta-analyses have been published, especially in the social, behavioral, and medical sciences. At a national conference in 1986 sponsored by the U.S. National Institutes of Health, Workshop on Methodological Issues in Overviews of Randomized Clinical Trials, participants focused considerable attention on meta-analysis for health-related and medical research. *The Handbook of Research Synthesis,* edited by Cooper and Hedges (1994), is still one of the most useful sources of information for planning and conducting meta-analytical reviews.

Purpose of Meta-Analysis

Like any research procedure, meta-analysis involves the selection of an important problem to address. However, meta-analysis involves two steps lacking in the typical literature review paper. First, a definitive methodology is reported concerning the decisions in a literature analysis. Second, the results of various studies are quantified to a standard metric called effect size (ES) that allows the use of statistical techniques as a means of analysis. Here are the steps in a meta-analysis:

1. Identify a problem.
2. Perform a literature search by specified means.
3. Review identified studies to determine inclusion or exclusion.
4. Carefully read and evaluate to identify and code important study characteristics.
5. Calculate effect size.
6. Apply appropriate statistical techniques.
7. Report all these steps and the outcomes in a review paper.

Of course, one of the major problems in a literature review paper is the number of studies that must be considered. To some extent, analyzing all these studies is like trying to make sense of all the data points collected on a group of participants. However, within a study, statistical techniques are used to reduce the data to make them understandable. The procedures of meta-analysis are similar. The findings within individual studies are considered the data points to use in a statistical analysis of the findings of many studies.

How, then, can findings based on different designs, data collection techniques, dependent variables, and statistical analyses be compared? Glass (1976) addressed this issue by using the estimate effect size (ES, or the symbol Δ). Note that we discussed this general concept in chapters 7 and 9 as a way of judging the meaningfulness of group differences. ES is determined by the following formula:

$$ES = (M_E - M_C) / s_C \tag{14.1}$$

where M_E is the mean of the experimental group, M_C is the mean of the control group, and s_C is the standard deviation of the control group. Note that this formula places the difference between the experimental and control groups in control group standard deviation units. For example, if $M_E = 15$, $M_C = 12$, and $s_C = 5$, then $ES = (15 - 12) / 5 = 0.60$. The experimental group's performance exceeded the control group's performance by 0.60 standard deviations. If this were done across several studies addressing a common problem, the findings of the studies would be in a common metric, ES (Δ), which could be compared. The mean and standard deviations of ES can be calculated from several studies. This allows a statement about the average ES of a particular type of treatment.

Suppose we want to know whether a particular treatment affects males and females differently. In searching the literature, we find 15 studies on males comparing the treatment effects and 12 studies on females. We calculate an ES for each of the 27 studies and the mean (and standard deviation) of the ES for males ($n = 15$) and females ($n = 12$). If the ESs were distributed normally, an independent t test could then be used to see whether the average ES differed for males and females. If the t values were significant and the average ES for the females were greater, we could conclude that the treatment had more effect on females than on males. Cooper and Hedges (1994) provide considerable detail on the methods of meta-analysis, including literature search strategies, ways to calculate ES from the statistics reported in various studies, suggestions for how and what to code from studies, and examples of the use of meta-analysis.

Certainly, meta-analysis is not the answer to all the problems associated with research synthesis. But Glass provided an objective way to evaluate the literature. Advances (Hedges, 1981, 1982a, 1982b; Hedges & Olkin, 1980) in studying the statistical properties of ES have contributed considerably to the appropriate use of meta-analysis. The text by Hedges and Olkin (1985) provides a complete accounting of the procedures and statistical analyses appropriate for meta-analysis. (Other books have appeared on meta-analysis advocating slightly different procedures, e.g., Hunter & Schmidt, 2004.) A tutorial by Thomas and French (1986) is an overview of Hedges and colleagues' techniques and includes examples from the study of physical activity. Some examples of meta-analyses that have appeared in the physical activity literature follow.

Examples of Meta-Analysis

Meta-analyses have been published since 1976, including many about physical activity. A brief overview of some of the more recent studies shows the value of meta-analysis.

In studying sex differences in motor performance across childhood and adolescence, Thomas and French (1985) reported findings from 64 studies based on 31,444 participants. They found motor performance differences to be related to age in 12 of the 20 motor tasks. These 12 tasks followed four general types of curves across age. Figure 14.1 shows one typical type of curve for three tasks (long jump, shuttle run, and grip strength). The differences are moderate (about 0.5 to 0.75 standard deviation units) before puberty but then increase dramatically during and following puberty (over 1.5 standard deviation units). They concluded that the differences before puberty were likely to be induced by environmental factors (differential treatment by parents, teachers, coaches, and peers), but that they represented an interaction of biology and environment beginning at puberty. Note the difference between the curves in figures 14.1 and 14.2. The effect size for throwing performance is 1.5 standard deviation units at ages 3 to 4 and increases constantly across childhood and adolescence until the differences are 3.5 standard deviation units by age 18. Thomas and French (1985) suggested that differences that are so large so early in life might have some basis in biology and in the influence of cultural treatment and expectations for girls and boys.

Payne and Morrow (1993) reported a meta-analysis of the effects of exercise on children's $\dot{V}O_2$max. From 28 studies yielding 70 effect sizes, they compared cross-sectional (trained vs. untrained children) and pre- to posttraining designs. Effect sizes were large in cross-sectional studies (0.94 ± 1.00) but small in training studies (0.35 ± 0.82). In training studies, the average improvement in $\dot{V}O_2$max was only 2 ml · kg · min^{-1}. Effect size for the training studies was not influenced by sex, training protocol, or the way participants were tested on $\dot{V}O_2$max.

Carron, Hausenblas, and Mack (1996) reported a meta-analysis of social influence on exercise. They found 87 studies containing 224 ESs based on 49,948 participants. The outcomes indicated that social influence has a positive but generally small to moderate relationship (ESs ranged from 0.2 to 0.5) to exercise. Four variables, however, had a moderate to large relationship (0.5 to 0.8): "family support and attitudes about exercise, task cohesion and adherence behavior, important others and attitudes about exercise, and family support and compliance behavior" (p. 1).

Figure 14.1 ES for three motor performance tasks. *Note:* Dotted lines are confidence intervals.

Figure 14.2 ES by age and sex for throwing for distance. *Note:* Dotted lines are confidence intervals.

Schieffer and Thomas (2012) evaluated the findings from 12 school-based intervention programs in which data were reported on both intervention and control schools using components of the coordinated school health model. The benefits from interventions were generally small except for health knowledge. Outcomes were better when more components of the model were included and when interventions were longer. However, they reported problems with some of the designs and analyses.

Gordon, Tucker, Burke and Carron (2013) examined the effectiveness of physical activity interventions on preschool children. Using effect sizes from 15 studies, they found that interventions had small to moderate effects on general physical activity (ES = 0.44). They reported that most children were not meeting general physical activity guidelines, but that the interventions did increase general physical activity.

We use these examples to illustrate the value of the meta-analysis approach. Meta-analysis, when applied appropriately and interpreted carefully, offers a means of reducing a large quantity of studies to underlying principles. These principles can become the bases for program development, future research, and theory testing, as well as practical applications, such as practice and training.

> A properly applied and interpreted meta-analysis will reveal the underlying principles for a large number of studies.

Methodological Considerations

Although meta-analysis has been a widely used technique in the last 30 years, early on there was considerable controversy about its validity (e.g., Carlberg et al., 1984; Slavin, 1984a, 1984b). Among the most severe criticisms of meta-analysis is that it combines findings from studies representing different measurement scales, methodologies, and experimental designs. This mixing of apples and oranges was compounded by the results of early meta-analyses using the methods of Glass (1977), which tended to reveal few differences in ESs, even between studies in which internal validity and methodological control clearly varied. Hedges (1981, 1982a, 1982b) and Hedges and Olkin (1983, 1985) extended the original work of Glass (1977) and proposed a new set of techniques and statistical tests to address the following questions and criticisms of meta-analysis:

- How should the process of deciding the variables to code and the systematic coding be organized?
- What should be used as the standard deviation when calculating an ES?
- Because sample ESs are biased estimators of the population of ESs, how can this bias be corrected?
- Should ESs be weighted for their sample size?
- Are all ESs in a sample from the same population of ESs? This is the apples and oranges issue: Is the sample of ESs homogeneous?
- What are appropriate statistical tests for analyzing ESs?
- If a sample of ESs includes outliers, how can they be identified?

In the following sections, we address each of these questions, summarize the theoretical basis for the statistical procedures introduced by Hedges (1981, 1982a, 1982b) and Hedges and Olkin (1983, 1985), and suggest applications of these techniques in the study of physical activity. Rosenthal (1994) has an excellent chapter (16) that provides all the formulas for conducting a meta-analysis when data are normally distributed.

Deciding What to Code

One of the most difficult yet most important tasks in doing a meta-analysis is choosing variables to code and developing a scheme for coding them. Stock (1994) provided a useful discussion of this topic. In particular, the meta-analyst must realize the trade-off between

Being familiar with the relevant literature and proficient with meta-analysis procedures help in the planning of coding.

the number and importance of items to code and the time required to code. Of course, the more information that is coded about a study, the more time is required to develop a coding scheme and do the coding. On the other hand, omitting potentially important items to code also creates major problems because the scholar then has to go back through all of the studies to pick up these omitted items.

The best way to be successful in selecting items to code and a coding scheme is to know the theoretical and empirical literature on which you are doing the meta-analysis and to understand meta-analysis procedures. Stock (1994) suggested the following two important considerations for planning the coding for a meta-analysis: selecting the items to code and organizing them.

The meta-analyst is looking for items to code that are likely to influence (or be related to) the effect sizes that will be calculated. For example, when Thomas and French (1985) calculated effect sizes for sex differences in motor performance, they thought that children's ages would be related to sex differences. Specifically, they hypothesized that sex differences would become greater as children grew. After important items to code are selected, the meta-analyst needs an organizational scheme for the items. At the end of this chapter is a partial sample of a coding form and code book developed by Sheri Parks from the School of Human Movement Studies at the University of Queensland, Brisbane, Australia, for her dissertation review on sport expertise. We appreciate her permission to use the materials. Parks' coding scheme used these categories (with several subparts of each): descriptive information about the paper, type of sport or area of expertise, information about the task, type of measures used for each task, and information about the participants.

Stock (1994) provided a generalized example of a coding form. An important issue here is how to code variables, or how to assign numbers to characteristics to be coded. Sometimes specific information is maintained; for example, if the average age of participants from each study is important, this average age can be entered as the code. In other instances, variables are grouped into categories for coding; for example, categories of level of expertise could be coded as 1 = internationally ranked athletic performers, 2 = nationally ranked athletic performers, 3 = college or university athletic performers, 4 = high school athletic performers. Other important issues include the training of coders and establishing and maintaining intra- and intercoder reliability.

Choosing the Standard Deviation for the Effect Size

Originally, Glass and colleagues (1981) proposed the use of the control group's standard deviation as the most appropriate measure of group variability. From an intuitive perspective, the control group's variability represents the "normal" variation in an untreated population. The control group's standard deviation also has the advantage of assigning equal ESs to equal treatment means when a study contains two or more experimental groups that have heterogeneous variances. Therefore, the control group's standard deviation serves as a common standard metric from which treatment differences can be compared.

In most cases, estimates of group variance are homogeneous. Hedges (1981) argued that a pooled estimate of the variance provides a more precise estimate of the population variance. (The square root of the pooled estimate of the variance is the pooled estimate of the standard deviation.) One advantage of pooling variances is an increase in the degrees of freedom associated with the estimate of variance. The pooled estimate Hedges (1981) suggested is given in equation 14.2. Note that the variance of each group is weighted by the sample size of that group in a way that is similar to the procedure used in t tests with unequal ns. The formula for the pooled estimate of standard deviation (s_P) is

$$s_P = \sqrt{\frac{\left(N_E - 1\right)s_E^2 + \left(N_C - 1\right)s_C^2}{N_E + N_C - 2}} \qquad (14.2)$$

where N_E is the sample size of the experimental group (group 1), N_C is the sample size of the control group (group 2), s_E^2 is the variance of the experimental group (group 1), and s_C^2 is the variance of the control group (group 2).

Many studies involve tests of effects between or among categorical variables (e.g., race and sex) where there is no control condition. Early meta-analyses used an average of the standard deviations of the groups compared (Glass & Smith, 1979; Hyde, 1981; Smith, 1980). Hedges and Olkin (1985) suggested using the weighted pooled standard deviation (s_P, equation 14.2) as the estimate of the standard deviation (for an example using sex differences in motor performance, see Thomas & French, 1985).

In some cases the variance for the groups is heterogeneous. Glass and colleagues (1981) showed that the ESs are biased when the group variances are unequal. Because parametric tests are based on the assumption of equal variances, ESs calculated from t and ANOVA may be biased if the group variances are unequal.

Researchers should evaluate whether heterogeneous variances are a common phenomenon in the area of research for the meta-analysis. We have encountered unequal variances when comparing the motor performance of children of different age levels (French & Thomas, 1987). We offer these suggestions: When the ES compares an experimental group and a control group and the group variances are unequal, use the standard deviation of the control group to calculate the ES in all studies. If the ES compares two groups (such as different ages or sexes) in which no clear control condition is present, we believe that the weighted pooled estimate (equation 14.2) suggested by Hedges (1981) is the best choice.

Calculating Effect Sizes for Within-Subjects Designs

Often, a researcher wants to calculate an effect size for a within-subjects design (or repeated-measures effect), usually between a pretest and posttest for a treatment effect (i.e., experimental group). The appropriate formula for this effect size is to use the pretest standard deviation in the denominator of equation 14.1 (Looney, Feltz, & VanVleet, 1994). This represents the best source of untreated variance against which to standardize differences between the pre- and posttest means.

Using Effect Size as an Estimator of Treatment Effects

Hedges (1981) provided a theoretical and structural model for the use of ES as an estimator of treatment effects. An individual ES may be viewed as a sample statistic that estimates the population of possible treatment effects within a given experiment.

Hedges (1981) also demonstrated that ESs are positively biased in small samples; the bias is 20% or less, however, when the sample size exceeds 20. A virtually unbiased estimate of ES can be obtained by multiplying the ES by the correction factor given in the following formula:

$$c = 1 - \frac{3}{4m-9} \tag{14.3}$$

where $m = N_E + N_C - 2$ when a pooled estimate is used as the standard deviation, $m = N_1 + N_2 - 2$ if a pooled estimate is used in a categorical model, or $m = N_C - 1$ when the control-group standard deviation (or pretest standard deviation in the case of a within-subjects ES) is used.

Each ES should be corrected before averaging or further analysis. If ESs are not corrected before averaging, the average of even a large number of ESs remains biased and simply estimates an incorrect value more precisely (Hedges, 1981). Early meta-analyses in exercise and sport (Feltz & Landers, 1983; Sparling, 1980) did not correct each ES for small sample bias. The ESs reported in these studies are most likely slight overestimates.

Correct ESs before averaging to help avoid a biased and incorrect result.

Although the individual ES estimate can be corrected for bias, the variability associated with the estimate remains a function of the sample size. Hedges (1981) showed that the variance of an individual ES can be directly calculated from the following formula:

$$\text{var}\left(\text{ES}_i\right) = \frac{N_E + N_C}{N_E N_C} + \frac{\text{ES}_i^2}{2\left(N_E + N_C\right)} \qquad (14.4)$$

where N_E is the sample size of the experimental group (group 1), N_C is the sample size of the control group (group 2), and ES_i is the estimate of ES. Note that the variability associated with the sample statistic, or ES, is a function of the value of the ES and the sample size. An ES based on a large sample has a smaller variance than an ES based on a small sample. Therefore, ESs based on large samples are better estimates of the population parameter of treatment effects. Hedges (1981) and Hedges and Olkin (1985) suggested that each ES should be weighted by the reciprocal of its variance, that is, 1 / variance (as estimated from equation 14.4). Therefore, ESs that are more precise receive more weight in each analysis.

Testing Homogeneity

Meta-analysis has been criticized for mixing apples and oranges, or combining studies with various measurement scales, designs, and methodologies. No appropriate test existed to determine whether all the ESs were estimating the same population treatment effect until Hedges (1982b) introduced his test for homogeneity. The homogeneity statistic, H, is specifically designed to test the null hypothesis, H_0: $\text{ES}_1 = \text{ES}_2 = \ldots = \text{ES}_i$. This is equivalent to saying that all ESs tested come from the same population of ESs.

The statistic H is the weighted sum of squared deviations of ESs from the overall weighted mean. The contribution of each ES to the overall mean is weighted by the reciprocal of its variance (equation 14.4). Effect sizes with smaller variances receive more weight in the calculation of the overall mean. Under the null hypothesis, H has a chi-square distribution with $N - 1$ degrees of freedom, where N equals the number of ESs.

When the null hypothesis is not rejected, all ESs are similar and represent a similar measure of treatment effectiveness. In this case, the researcher should report that the homogeneity statistic indicates that ESs are homogeneous and use the weighted mean ES with confidence intervals for interpretation. If the null hypothesis is rejected, ESs are not homogeneous and do not represent a similar measure of treatment effectiveness, or grouping. Two methods have been proposed by Hedges (1982b) and Hedges and Olkin (1983) to examine explanatory models for ESs. We look at these models next.

Analyzing Variance and Weighted Regression

The first method to fit an explanatory model to ES data is analogous to ANOVA, in which the sum of squares for the total statistic is partitioned into a sum of squares between (or among groups of) ESs (H_B) and a sum of squares within groups of ESs (H_W). Each sum of squares can be tested as a chi square with $k - 1$ df for H_B and $N - k - 1$ df for H_W (where N is the number of ESs and k is the number of groups). Therefore, a test can be conducted for between- or among-group differences (H_B), and a test can be conducted to determine whether all ESs within a group are homogeneous (H_W). Further discussion of the categorical model is presented in Hedges (1982b) and Hedges and Olkin (1985).

The second method proposed by Hedges and Olkin (1983) to fit an explanatory model to ES data is a weighted regression technique. We have chosen to present a more detailed discussion of the regression techniques for several reasons. First, one of the meta-analyses discussed previously in exercise and sport used regression techniques to analyze ES (Thomas & French, 1985). Second, it is common for a continuous variable to influence ES. Third, often, more than one study characteristic influences ES, especially when there are many ESs. The regression procedures can accommodate a larger number of variables in the analysis without inflating the alpha level of the statistical test. Conducting many tests using the categorical or ANOVA-like procedure results in alpha inflation, so the researcher would need to report the experimentwise error rate or adjust the alpha level using the Bonferroni technique (see chapter 9). Fourth, categorical variables can easily be dummy- or effect-coded and entered into the regression procedures. For example, published versus unpublished papers can be dummy-coded (1 or 0) and entered into the regression. Thus, conducting a separate analysis for categorical variables is not necessary.

Each ES is weighted by the reciprocal of its variance in the weighted regression technique. Most standard statistical packages (e.g., SAS, SPSS, BIOMED) have an option to perform weighted regression. Thus, the computations can be done easily on most computers.

The sum of squares total for the regression is equivalent to the homogeneity statistic, H. It is partitioned into a sum of squares regression and a sum of squares error. The **sum of squares** for regression yields a test of the variance due to the predictor variables (H_R). The H_R is tested as a chi square with $df = p$, where p equals the number of predictor variables. The sum of squares for error (H_E) provides a test of model specifications, or whether the ESs are homogeneous when the variance caused by the predictors is removed. The H_E is tested as a chi square with $df = N - p - 1$, where N equals the number of ESs and p equals the number of predictor variables. The test of model specification evaluates the deviation of the ESs from the regression model. A nonsignificant test of model specification indicates that the ESs do not deviate substantially from the regression model, whereas a significant test for model specification shows that one or more ESs deviate substantially from the regression model. Ideally, the researcher wants the H_R for regression to be significant and the H_E for model specification to be nonsignificant.

When the test for model specifications is significant, one or more ESs do not follow the specified regression. Often, the ESs that do not follow the same pattern may suggest other characteristics that can be added to the model (e.g., published vs. unpublished studies). Moreover, some ESs may represent outliers (unrepresentative scores). In either case, the use of outlier techniques is helpful in identifying these ESs.

sum of squares—A measure of the variability of scores; the sum of the squared deviations from the mean of the scores.

Testing for Outliers

Hedges and Olkin (1985) outlined procedures to identify outliers in categorical and regression models. We use an example from a regression model (for more information concerning the categorical model, see chapter 7 in Hedges & Olkin, 1985).

Outliers in regression models can be identified by examining the residuals of the regression equation. The absolute values of the residuals are standardized to z scores by subtracting the mean and dividing by the standard deviation. Effect sizes with standardized residuals larger than 2 are often examined as potential outliers, because they fall outside 95% of the distribution.

According to Hedges and Olkin (1985), each ES makes a contribution to the regression model. The outliers identified by standardized residuals may vary, depending on which ESs are in the model at the time. Hedges and Olkin (1985) suggested computing the standardized residuals multiple times with a different ES deleted from the model each time. For example, if you have 10 ESs, standardized residuals would be computed 10 times with a different ES deleted from the model each time. But some of our preliminary work suggested that if more than 20 ESs are included in the regression model, going through all possible combinations of deleting ESs from the model may be of limited value. Simply calculate the residuals with all the ESs in the model, standardize the residuals, and evaluate whether ESs with a z score larger than 2 might be outliers.

Alternatively, if the ESs are not normally distributed (as Hedges & Olkin, 1985, reported is often the case), the rank-order procedures described in chapter 10 (and developed in greater detail by Thomas, Nelson, & Thomas, 1999) are effective if ESs are changed to ranks before statistical analysis.

Accounting for Publication Bias

Journals have a tendency to accept papers that report significant findings. Thus, there may be some number of studies on any topic (including one on which a meta-analysis is being

Just because things are on the same path does not mean they fit together in one analysis.

done) that have not been submitted to or published in scholarly journals. This problem has been labeled the file-drawer effect.

An important question in a meta-analysis is how many of these studies are sitting around in file drawers. The meta-analyst should find every study possible, but how many studies have been done that have not been reported? Hedges and Olkin (1985) suggested a technique for estimating how many unpublished studies with no significant effect on the variable of interest would be needed to reduce the mean ES to nonsignificance.

$$K_0 = [K(d_{ES\ mean} - d_{ES\ trivial})] / d_{ES\ trivial} \tag{14.5}$$

where K_0 is number of studies needed to produce a trivial ES, K is the number of studies in the meta-analysis, $d_{ES\ mean}$ is mean of all the ESs in the meta-analysis, and $d_{ES\ trivial}$ is an estimate of a trivial ES.

For example, if there were 75 studies in the meta-analysis (K), if the mean ES ($d_{ES\ mean}$) was 0.73, and if a trivial effect size ($d_{ES\ trivial}$) was established as 0.15, to estimate the number of file-drawer studies needed to produce a trivial ES (K_0), we solve equation 14.5:

$$K_0 = [75(0.73 - 0.15)] / 0.15 = 290$$

Thus, there would need to be an inordinate number (290) of unpublished studies on this topic to reduce the average ES of 0.73 in the study to a trivial ES of 0.15.

Other Considerations

Sometimes all the information needed to calculate an ES is not available in a research report. Rosenthal (1994) provided an excellent discussion, formulas, and examples of how to estimate effect sizes from other statistics that are often provided in individual studies. On occasion, however, only the statement that two groups do not differ significantly is reported. What should the meta-analyst do? To drop the study is to bias the results of the meta-analysis toward significant differences. Thomas and French (1985) inserted a zero ES, using the rationale that no significant differences means that the differences did not reliably differ from zero. But Thomas and French (1985) had few zero ESs. Using many zero ESs may cause problems in the analysis of homogeneity by making the studies look overconsistent. Hedges (1984) suggested a correction factor that adjusts for this problem. Regardless of the solution selected, the meta-analyst must cope with this issue and select a logical solution.

PRESENTING EFFECT SIZE DATA

Graphic displays are extremely useful when presenting effect size data. According to Light, Singer, and Willett (1994), graphs enable the reader to see important characteristics of relationships in the effect size data. In particular, graphic displays of effect size data are useful for showing sources of variation in effect size displaying the center, spread, and shape of the distribution; such displays include stem-and-leaf plots, schematic plots (box and whiskers), and funnel graphs (effect sizes plotted against sample sizes). Graphic displays of effect size data are also useful for sorting effect sizes by important study characteristics, such as treatment intensity, age, or sex. These displays should also reflect the variability of effect sizes for sorted characteristics (e.g., 95% confidence intervals; see figure 14.2 for an example).

Light, Singer, and Willett (1994, pp. 451–452) listed five valuable concepts for graphic displays of meta-analytic data (of course, these concepts apply equally well to studies reporting original data):

1. *The spirit of good graphics can even be captured by a table of raw effect sizes, provided that they are presented in a purposeful way* (e.g., by magnitude or important study characteristics).

2. *Graphic displays should be used to emphasize the big picture* (e.g., make the main points clear).

3. *Graphics are especially valuable for communicating findings from a meta-analysis when each individual study outcome is accompanied by a confidence interval* (this demonstrates not only the variability of the effect sizes of interest but also whether they are different from the null value).

4. *Graphic displays should encourage the eye to compare different pieces of information.*

5. *Stem-and-leaf plots and schematic plots are especially efficient for presenting the big picture that emerges in a research synthesis.*

SUMMARY

A research synthesis is not just a summary of studies arranged in some kind of sequence. It should be structured, analytical, and critical and lead to specific conclusions. The meta-analysis is a definitive methodology that quantifies results and allows statistical techniques to be used as a means of analysis.

The meta-analysis explicitly details how the search was done, the sources, the choices made regarding inclusion or exclusion, the coding of study characteristics, and the analytical procedures used. The basis of a meta-analysis is the effect size, which transforms differences between experimental and control groups' performances to a common metric, which is expressed in standard deviation units.

The meta-analysis technique has been widely used in recent years in many fields, including the study of physical activity. Meta-analysis offers a means of reducing large quantities of studies to underlying principles. Refinements in technique continue to be made regarding the choice of the standard deviation to be used in calculating an effect size, the weighting of effect sizes for sample bias, the test for outliers, and the statistical procedures that can be used for the analysis.

Check Your Understanding

Select a possible problem (e.g., sex differences in running speed, the effects of weight training on males and females, or the influence of training at 60% vs. 80% of ($\dot{V}O_2$max). Find five studies that compare the characteristics that you select (e.g., males and females). Make sure each study has the means and standard deviations for the variables of interest. Calculate the ES for each study and correct it for bias. Calculate the average ES for the five studies and their pooled standard deviations. Interpret this finding.

SAMPLE CODING FORM FOR A META-ANALYSIS OF SPORT EXPERTISE

This section provides a detailed account of the procedures used to code the 92 characteristics outlined for each paper in Sheri Parks' meta-analysis in the area of expertise in sport.

STEP 1 CHOOSING APPROPRIATE PAPERS

This review is interested in any and all research that investigates expertise in sport using some form of between-group (expert–novice) comparison and addressing it using a motor behavior approach.

Each paper must provide the means and standard deviations for each of the performance measures listed so that comparisons on the basis of effect sizes can be calculated and compared with other like papers. In other words, all papers and research in this review must be of a database format.

STEP 2 PROVIDING DESCRIPTIVE INFORMATION ABOUT EACH PAPER

Column 1: Article number

Each paper chosen for inclusion in analysis should be given a unique number code.

Column 2: Year completed

Column 3: Is the research (1) published or (0) unpublished?

Column 4: Source information

A. If published, is it from a

    (2) refereed journal?

    (1) nonrefereed journal?

    (0) conference proceedings?

B. If unpublished, is it from a

    (2) PhD dissertation?

    (1) master's thesis?

    (0) unpublished manuscript?

Column 5: If published, where did the research originate?

    (4) Sport science journal

    (3) Movement science journal

    (2) Mainstream psychology journal

    (1) Other source

    (0) Not applicable

Column 6: If published, name of journal

Keep a list of the journals that have been coded (1 to 1,000), where 0 denotes "not applicable," as in the case of unpublished work.

Column 7: Affiliation

This column includes a coding for where the specific research was done (i.e., which university, which lab, which particular group of researchers).

Column 8: Internal validity ranked on a scale of

(2) Strong

(1) Moderate

(0) Weak

This item is developed from the following five-point checklist.

1. Did this experiment include a control group? Y/N

2. Did the experiment use randomization techniques to control for the following threats to internal validity?

(2a) History and maturation—by random placement or matched pairing of participants into each of a control and experimental group? Y/N

(2b) Order effects and learning—by random ordering or presentation of the tasks? Y/N

3. Did the experiment establish appropriate instrumentation (or test reliability or psychometrics)? Y/N

4. Was the participant retention or dropout rate limited? Y/N

5. Were the participants unaware of the purposes and intentions of this experiment? Y/N

Total yes 5 = strong

Total yes 2–4 = moderate

Total yes 0–1 = weak

STEP 3 PROVIDING INFORMATION ABOUT THE TYPE OF SPORT OR AREA OF EXPERTISE

Column 9: Type of sport (area of expertise) studied

Column 10: Classification of sport or activity

Which cell from Higgin's eight-cell model does each sport or activity fit?

Nature of environment (open vs. closed) × body stability × body transport

Column 11: Is it (1) a team sport or (0) an individual or dual sport?

Column 12: Type of motor skills involved in sport or activity

(2) Gross motor skills

(1) Fine motor skills

(0) A combination of both gross and fine motor skills

STEP 4 PROVIDING INFORMATION ABOUT THE TASK OR TASKS PERFORMED

Column 13: How many tasks are involved in the study?

This item involves counting the number of experimental tasks identified in the experiment. Many experiments keep the same participants and have them undergo a series of different tasks. This method is acceptable, as long as the same participants are not "reused," so to speak, in completely different research sources, as previously stated in notes for column 4.

Column 14: Type of task

Itemize all tasks in studies, and try to group like tasks within a similar range of numbers; for example, all perceptual tasks are identified by numbers 1–10, all decision-making tasks should be identified by numbers 15–25, and so on.

(continued)

Column 15: Levels or conditions of the independent measure for each task

The independent measure can be defined as the variable of interest, or the part of the experiment that the researcher is manipulating (e.g., blocks of practice, occlusion condition, and so on). This can also be referred to as the experimental or treatment variable.

Column 16: Levels or conditions of the dependent measure for each task

The dependent measure can be defined as the particular performance measures used in the experiment (e.g., reaction time, error scores, and so on). This can also be considered the effect of the independent variable or the yield of the experiment.

Column 17: Scoring direction for task, where

(1) Higher score is better, lower is worse (i.e., performance scores)

(0) Lower score is better, higher score is worse (i.e., error scores)

STEP 5 PROVIDING INFORMATION ABOUT THE TYPE OF MEASURES USED IN EACH TASK

Column 18: Ecological validity ranked on a scale of

(2) Strong (total ≥7)

(1) Moderate (total 4–6)

(0) Weak (total ≤3)

This item is developed from the following nine-point checklist. Refer to column 8 for further explanation of the rating scale for ecological validity used in this analysis.

1. Is the *action* involved in performing the task . . . that of the real-world sporting situation?

(a) Exactly the same, +3

(b) Somewhat the same, +2

(c) Nothing like, +1

(d) Not applicable, 0

2. Is the *perceptual display* offered in performing the task . . . that of the real-world sporting situation?

(a) Exactly the same, +3

(b) Somewhat the same, +2

(c) Nothing like, +1

(d) Not applicable, 0

3. Are the *time constraints* involved in performing the task (e.g., preparation time, decision time, performance time, and so on) . . . that of the real-world sporting situation?

(a) Exactly the same, +3

(b) Somewhat the same, +2

(c) Nothing like, +1

(d) Not applicable, 0

Column 19: Validity of measures ranked on a scale of

(4) Strong

(3) Moderate

(2) Weak

(1) Not mentioned or cannot be interpreted on the basis of the information provided in the source

(0) If column 20 reliability of measures is rated as weak or not mentioned in study (0), rate validity as 0 in this column.

Column 20: Reliability of measures ranked on a scale of

(2) Strong

(1) Moderate

(0) Weak

Column 21: Were the statistics and statistical procedures and analyses used appropriate, as rated on the following scale:

(2) Appropriate

(1) All right or adequate

(0) Not appropriate

Column 22: Is it a (1) lab-based or (0) field-based assessment or report?

Column 23: Is the action (1) natural or (0) contrived?

Column 24: Does the task duplicate normal time constraints? (1) Yes or (0) No

Column 25: If a visual display is used, is it

(2) Static

(1) Dynamic

(0) Not applicable

Column 26: If a visual display is used, are the stimuli

(2) Task specific

(1) Alphanumeric

(0) Not applicable

Column 27: If a visual display is used, how well does it mimic the natural situation across all modalities?

(3) Strong

(2) Moderate

(1) Weak

(0) Not applicable

STEP 6 PROVIDING INFORMATION ABOUT THE EXPERIMENTAL GROUP OF PARTICIPANTS

Column 28: Group number code, where groups can be one of the following:

(6) Competitive at an international or Olympic level

(5) Competitive at a national level

(4) Competitive at a state or regional level

(3) Participating in the sport at a recreational intramural or local level

(2) Competitive in sports other than the one of interest

(1) Nonathletes

Column 29: Quality of definition of the group (expert, elite, skilled, experienced, and so on)

(2) Well defined

(1) Not enough information provided

(0) Not defined

(continued)

Column 30: Confidence rating in performance selection

    (2) High

    (1) Average

    (0) Low

This column attempts to address the following questions: Are these groups the best at this level? Are they only moderately competitive at this level, or do they lose first round at the Olympics, for example?

Column 31: How was this level determined?

    (3) Previous performance record

    (2) Coach's determination

    (1) Other criteria

    (0) Not mentioned

Column 32: Total number of participants in group

Column 33: Sex of participants

    (3) All males

    (2) All females

    (1) Mixed

    (0) Not stated

Column 34: Mean age of sample used in group

    (5) ≥30

    (4) 20–29

    (3) 16–19

    (2) 13–15

    (1) ≤12

    (0) Not stated

Column 35: Mean number of years' experience with sport

    (5) >10

    (4) 7–10

    (3) 4–6

    (2) 2–3

    (1) <2

    (0) Not stated

Column 36: Status of participants

    (2) Paid

    (1) Unpaid

    (0) Not stated

Column 37: Previous experience with task

    (2) Familiar

    (1) New or novel

    (0) Not stated

Column 38: Mean group performance score on task

Column 39: Group standard deviation on task

STEP 7 PROVIDING INFORMATION ABOUT THE CONTROL GROUP OF PARTICIPANTS

Column 40: Group number code, where groups can be one of the following:

(6) Competitive at an international or Olympic level

(5) Competitive at a national level

(4) Competitive at a state or regional level

(3) Participating in the sport at a recreational intramural or local level

(2) Competitive in sports other than the one of interest

(1) Nonathletes

Column 41: Quality of definition of the group (expert, elite, skilled, experienced, and so on)

(2) Well defined

(1) Not enough information provided

(0) Not defined

Column 42: Confidence rating in performance selection

(2) High

(1) Average

(0) Low

Column 43: How was this level determined?

(3) Previous performance record

(2) Coach's determination

(1) Other criteria

(0) Not mentioned

Column 44: Total number of participants in group

Column 45: Sex of participants

(3) All males

(2) All females

(1) Mixed

(0) Not stated

Column 46: Mean age of sample used in group

(5) ≥30

(4) 20–29

(3) 16–19

(2) 13–15

(1) ≤12

(0) Not stated

Column 47: Mean number of years' experience with sport

(5) >10

(4) 7–10

(continued)

(3) 4–6

(2) 2–3

(1) <2

(0) Not stated

Column 48: Status of participants

(2) Paid

(1) Unpaid

(0) Not stated

Column 49: Previous experience with task

(2) Familiar

(1) New or novel

(0) Not stated

Column 50: Mean group performance score on task

Column 51: Group standard deviation on task

STEP 8 PROVIDING INFORMATION ABOUT THE INTERMEDIATE GROUP OF PARTICIPANTS

Column 52: Group number code, where groups can be one of the following:

(6) Competitive at an international or Olympic level

(5) Competitive at a national level

(4) Competitive at a state or regional level

(3) Participating in the sport at a recreational intramural or local level

(2) Competitive in sports other than the one of interest

(1) Nonathletes

Column 53: Quality of definition of the group (expert, elite, skilled, experienced, and so on)

(2) Well defined

(1) Not enough information provided

(0) Not defined

Column 54: Confidence rating in performance selection

(2) High

(1) Average

(0) Low

Column 55: How was this level determined?

(3) Previous performance record

(2) Coach's determination

(1) Other criteria

(0) Not mentioned

Column 56: Total number of participants in group

Column 57: Sex of participants

(3) All males

(2) All females

(1) Mixed

(0) Not stated

Column 58: Mean age of sample used in group

(5) ≥30

(4) 20–29

(3) 16–19

(2) 13–15

(1) ≤12

(0) Not stated

Column 59: Mean number of years' experience with sport

(5) >10

(4) 7–10

(3) 4–6

(2) 2–3

(1) <2

(0) Not stated

Column 60: Status of participants

(2) Paid

(1) Unpaid

(0) Not stated

Column 61: Previous experience with task

(2) Familiar

(1) New or novel

(0) Not stated

Column 62: Mean group performance score on task

Column 63: Group standard deviation on task

STEP 9 PROVIDING INFORMATION ABOUT ANY OTHER GROUP OF PARTICIPANTS INCLUDED IN STUDY

Column 64: Group number code, where groups can be one of the following:

(6) Competitive at an international or Olympic level

(5) Competitive at a national level

(4) Competitive at a state or regional level

(3) Participating in the sport at a recreational intramural or local level

(2) Competitive in sports other than the one of interest

(1) Nonathletes

Column 65: Quality of definition of the group (expert, elite, skilled, experienced, and so on)

(2) Well defined

(1) Not enough information provided

(0) Not defined

(continued)

Column 66: Confidence rating in performance selection

(2) High

(1) Average

(0) Low

Column 67: How was this level determined?

(3) Previous performance record

(2) Coach's determination

(1) Other criteria

(0) Not mentioned

Column 68: Total number of participants in group

Column 69: Sex of participants

(3) All males

(2) All females

(1) Mixed

(0) Not stated

Column 70: Mean age of sample used in group

(5) ≥30

(4) 20–29

(3) 16–19

(2) 13–15

(1) ≤12

(0) Not stated

Column 71: Mean number of years' experience with sport

(5) >10

(4) 7–10

(3) 4–6

(2) 2–3

(1) <2

(0) Not stated

Column 72: Status of participants

(2) Paid

(1) Unpaid

(0) Not stated

Column 73: Previous experience with task

(2) Familiar

(1) New or novel

(0) Not stated

Column 74: Mean group performance score on task

Column 75: Group standard deviation on task

Our thanks to Sheri Parks from the School of Human Movement Studies, University of Queensland, Brisbane, Australia, for the use of her coding booklet.

Surveys

> **Most surveys represent the average opinion of people who don't know.**
>
> —Unknown

escriptive research is a study of status and is widely used in education and the behavioral sciences. Its value is based on the premise that problems can be solved and practices improved through objective and thorough description. The most common descriptive research method is the **survey**. The survey is generally broad in scope. The researcher usually seeks to determine present practices (or opinions) of a population. The survey is used in the fields of education, psychology, sociology, and physical activity. The questionnaire, the Delphi method, the personal interview, and the normative survey are the main types of surveys.

QUESTIONNAIRES

The **questionnaire** and the **interview** are essentially the same except for the method of questioning. Questionnaires are usually answered in writing, whereas interviews are usually conducted orally. As noted later in this chapter and in chapter 19, interviews for qualitative research are generally open-ended and may have different goals than the interview-based survey. The procedures for developing questionnaire and interview items are similar. Consequently, much of the discussion regarding the steps in the construction of the questionnaire also pertains to the interview. Similarly, much of this advice is the same for either a paper survey or one conducted electronically on the Internet. The method of delivering the survey may depend on the sample—electronic if participants are geographically dispersed and paper if it is easier to gain direct access to the participants in person.

Researchers use the questionnaire to obtain information by asking participants to respond to questions rather than by observing their behavior. The obvious limitation of the questionnaire is that the results consist simply of what people say they do or what they say they believe or like or dislike. But certain information can be obtained only in this manner, so the questionnaire must be planned and prepared carefully to ensure the most valid results. The survey research process includes eight steps, and we discuss each in this section, which focuses on paper-based surveys and using the postal service for sending and returning the survey. The sections that follow address Internet surveys.

survey—A technique of descriptive research that seeks to determine present practices or opinions of a population; it can take the form of a questionnaire, interview, or normative survey.

questionnaire—A paper-and-pencil or electronic survey used in descriptive research in which information is obtained by asking participants to respond to questions rather than by observing their behavior.

interview—A survey technique similar to the questionnaire except that participants are questioned and respond orally rather than in writing.

Although a questionnaire is limited by what respondents say they do, believe, or prefer, it is the only technique that can gather certain information.

Determining the Objectives

This step may seem too obvious to mention, yet countless questionnaires have been prepared without clearly defined objectives. In fact, poor planning may account for the low esteem in which survey research is sometimes held. The investigator must have a clear understanding of what information is needed and how each item will be analyzed. As with any research, the analysis is determined in the planning phase of the study, not after the data have been gathered.

The researcher must decide on the specific purposes of the questionnaire: What information is wanted? How will the responses be analyzed? Will they be described by merely listing the percentages of participants who responded in certain ways, or will the responses of one group be compared with those of another?

One of the most common mistakes in constructing a questionnaire is not specifying the variables to be analyzed. In some cases, when investigators fail to list the variables, they ask questions unrelated to the objectives. In other cases, investigators forget to ask pertinent questions.

Delimiting the Sample

Most researchers who use questionnaires have in mind a specific population to be sampled. Obviously, the participants selected must be the ones who have the answers to the questions. In other words, the investigator must know who can supply what information. If information about policy decisions is desired, the respondents should be those involved in making such decisions.

Sometimes the source used in selecting the sample is inadequate. For example, some professional associations are made up of teachers, administrators, professors, and other allied professional workers. Thus, the membership of such an association is not a good choice for participants for a study that is geared only for public school teachers. Unless some screening mechanism is used regarding place of employment, many incomplete questionnaires will be returned because the questions were not applicable.

The representativeness of the sample is an important consideration. Stratified random sampling, as discussed in chapter 6, is sometimes used. In a questionnaire surveying the recreational preferences of university students, the sample should reflect the proportion of students at the various class levels. Thus, if 35% of the students are freshmen, 30% are sophomores, 20% are juniors, and 15% are seniors, then the sample should be selected using those percentages. Similarly, if a researcher is studying school program offerings and 60% of the schools in the state have fewer than 200 students enrolled, then 60% of the sample should be from such schools.

The selection of the sample should be based on the variables specified to be studied. This affects the generalizability of the results. If an investigator specifies that the study deals with just one sex, one educational level, or one institution, then the population is narrowly defined, and it may be easy to select a representative sample of that population. But the generalizations that can be made from the results are also restricted to that population. On the other hand, if the researcher is aiming the questionnaire at everyone in a population (e.g., all athletic directors or all fitness instructors), then the generalizability is enhanced, but the sampling procedures become more difficult. The representativeness of the sample is more important than its size.

In some instances, surveys address populations that are difficult to sample in the ways noted earlier. For example, identifying all the possible participants for a national study of physical education teachers may be impossible. Obtaining a listing from every state or school district is impractical, and each may have records that are more or less current. After identifying the type of participant, in these types of studies the researcher works to get the largest and most diverse sample size possible. This procedure occurs in a number of ways and is discussed further in the section on electronic surveys.

Sampling Error

Frequently, you read newspaper reports of national polls about some national issue or about whether a public figure should resign or whether a defendant received a fair trial. Generally, the reports include a statement about the sample size and sampling error, such as "The poll involved 1,022 Americans, and the results were subject to a sampling error of plus or minus 3.1 percentage points." Do you ever wonder how the pollsters arrive at that error range? We discussed standard error in earlier chapters. Standard error of sampling can be expressed as

$$SE = \sqrt{variance / n} \qquad (15.1)$$

where SE is the standard error of sampling and n is the size of the sample. We know that variance can be defined as the sum of the squared deviations from the sample mean divided by n. In this case, however, variance is calculated as a function of proportions; that is, the percentage of a sample that has a certain characteristic or gives a certain response. Variance of a proportion is calculated as $p(1 - p)$, where p is the proportion that has a characteristic or that responds a certain way and $1 - p$ is the proportion that does not have the characteristic or that responds differently. These proportions could be of many things, such as males and females, voters and nonvoters, smokers and nonsmokers, or those who believe one way versus those who believe differently, as is common in surveys. So the formula can be written

$$SE = \sqrt{p(1-p)/n} \qquad (15.2)$$

If we do not know the proportions (or expected proportions), $p = .50$ is used. This proportion results in the largest error. Let us go back to our example of the survey of 1,022 Americans with the 3.1% sampling error. Using equation 15.2, we have

$$SE = \sqrt{.50(1-.50)/1022} = .0156, \text{ or } 1.6\% \qquad (15.3)$$

But remember that standard errors are interpreted as standard deviations in that ±1 SE includes roughly 68% of the population; to say it another way, we are only 68% confident that our sample is representative. So most polls use a 95% confidence interval, which we know is ±1.96 SE. Hence, our finalized formula should be

$$\begin{aligned} SE &= 1.96\sqrt{p(1-p)/n} \\ &= 1.96\sqrt{.50(1-.50)/1022} \\ &= 3.1\% \end{aligned} \qquad (15.4)$$

Sample Size

The size of the sample needed is an important consideration from two standpoints: (a) for adequately representing a population and (b) for practical considerations of time and cost. Certain formulas can be used to determine the adequacy of sample size (Vockell, 1983, pp. 111–118). These formulas involve probability levels and the amount of sampling error deemed acceptable. The practical considerations of time and cost need attention in the planning phase of the study. Surprisingly, students often ignore these considerations until they are forced to sit down with a calculator and tabulate the costs of printing, initial mailing (which includes self-addressed, stamped return envelopes), follow-ups, scoring, and data analysis. Sometimes the costs are so substantial that a sponsoring agency or grant must be found to subsidize the study; otherwise, the project must be narrowed or abandoned entirely. Time is also important with regard to the availability of participants, possible seasonal influences, and deadlines. See Fink (2003) and Fowler (2009) for discussions on survey sampling, including the calculation of sample size.

Constructing the Questionnaire

The notion that constructing a questionnaire is easy is a fallacy. Questions are not simply "thought up." Anyone who prepares a questionnaire and asks someone to read it soon discovers that it is not an easy task after all. The questions that were clear and concise to the writer may be confusing and ambiguous to the respondent.

One of the most valuable guidelines for writing questions is to ask yourself continually what specific objective each question is measuring. Then, ask how you are going to analyze the response. While you are writing questions, it is a good idea to prepare a blank table that includes the categories of responses, comparisons, and other breakdowns of data analysis so that you can readily determine exactly how each item will be handled and how each will contribute to the objectives of the study. Next, you must select the format for the questions, some examples of which follow.

Open-Ended Questions

open-ended question— A category of question in questionnaires and interviews that allows the respondent considerable latitude to express feelings and expand on ideas.

closed question— A category of question in questionnaires and interviews that requires a specific response and that often takes the form of rankings, scaled items, or categorical responses.

Open-ended questions such as "How do you like your job?" or "What aspects of your job do you like?" may be the easiest to write. Such questions allow the respondent considerable latitude to express feelings and expand on ideas. But several drawbacks to open-ended questions usually make them less desirable than **closed questions**. For example, most respondents do not like open-ended questions. For that matter, most people do not like questionnaires because they think that surveys encroach on their time. Also, open-ended questions require more time to answer than closed questions do. Another drawback is limited control over the nature of the response: The respondent often rambles and strays from the question. In addition, such responses are difficult to synthesize and to group into categories for interpretation. Although open-ended items can yield valuable information, they are hard to analyze by any means other than simple description.

Sometimes open-ended questions are used to construct closed questions. Student course evaluations are often developed by having students list all the things they like and dislike about a course. From such lists, closed questions are constructed by categorizing the open-ended responses.

Closed Questions

Closed questions come in a variety of forms. A few of the more commonly used closed questions are rankings, scaled items (some measurement scales were covered in chapter 11), and categorical responses.

ranking— A type of closed question that forces the respondent to place responses in a rank order according to some criterion.

A **ranking** forces the respondent to place responses in a rank order according to some criterion. As a result, value judgments are made, and the rankings can be summed and analyzed quantitatively. Here is an example of a rank-order response question:

> **Rank the following activities with regard to how you like to spend leisure time. Use numbers 1 through 5, with 1 being the most preferred and 5 the least preferred.**
>
> ___ **Reading**
>
> ___ **Watching television**
>
> ___ **Arts and crafts**
>
> ___ **Vigorous sports such as tennis and racquetball**
>
> ___ **Mild exercise activity such as walking**

Scaled items are one of the most commonly used types of closed questions. Participants are asked to indicate the strength of their agreement or disagreement with some statement or to cite the relative frequency of some behavior, as in the following example:

Indicate the frequency with which you are involved in committee meetings and assignments during the academic year.

Rarely   **Sometimes**   **Often**   **Frequently**

A Likert-type scale is a scale with three to nine responses, in which the intervals between responses are assumed to be equal:

In a required physical education program, students should be required to take at least one dance class.

Strongly agree   **Agree**   **Undecided**   **Disagree**   **Strongly disagree**

The difference between *strongly agree* and *agree* is considered equivalent to the difference between *disagree* and *strongly disagree*. Following are examples of words and phrases that can be used in scaled responses: *excellent*, *good*, *fair*, *poor*, and *very poor*; *very important*, *important*, *not very important*, and *of no importance*.

Categorical responses offer the respondent only two choices. Usually, the responses are *yes* and *no* or *agree* and *disagree*. An obvious limitation of categorical responses is the lack of other options such as *sometimes* or *undecided*. Categorical responses do not require as much time to administer as scaled responses do but do not provide as much information about the respondent's degree of agreement or the frequency of the behavior.

Sometimes, questions in questionnaires are keyed to the responses of other items. For example, a question might ask whether the respondent's institution offers a doctoral degree. Respondents who answer *yes* are directed to answer subsequent questions about the doctoral program. Those who answer *no* are directed either to stop or to move on to the next section.

Gall, Gall, and Borg (2006) offered the following rules for the construction of questionnaire items:

- The items must be clearly worded so that they have the same meaning for all respondents. Avoid words that have no precise meaning, such as *usually, most,* and *generally.*

- Use short questions rather than long questions because short questions are easier to understand.

- Do not use items that have two or more separate ideas in the same question, as in this example: "Although everyone should learn how to swim, passing a swimming test should not be a requirement for graduation from college." This item cannot really be answered by a person who agrees with the first part of the sentence but not the second, or vice versa. Another example is "Does your department require an entrance examination for master's and doctoral students?" This question is confusing because the department may have such an examination for doctoral students but not for master's students, or vice versa. If the response choices are only *yes* and *no,* a *no* response would indicate no examination for either program, and a *yes* response would mean exams for both. This subject should be divided into two questions.

- Avoid using negative items, such as "Physical education should not be taught by coaches." Negative items are often confusing, and the negative word is sometimes overlooked, causing respondents to answer in exactly the opposite way from what they intended.

- Avoid technical language and jargon. Attempt to achieve clarity and the same meaning for everyone.

- Be careful not to bias the answer or lead the respondent to answer in a certain way.

Sometimes questions are stated in such a way that the person knows what the "right" response is. For example, you would know what response was expected if you encountered this question: "Because teachers work so hard, shouldn't they receive higher salaries?" The same advice applies to the problem of threatening questions. A respondent who perceives certain items to be threatening will probably not return the questionnaire. A questionnaire for teachers on grading practices, for example, may be viewed as threatening because poor grading practices would indicate that the teacher is not doing a good job. Thus, a teacher who feels threatened either may not return the questionnaire or may give responses that seem to be "right" answers.

Results of experiments on surveys show that minor changes in wording, or even in ordering the alternatives, can cause differences in responses. At a seminar at Washington State University, Don Dillman, a recognized authority on the survey method, reported results of experiments on the effects of question form and survey mode on survey results. As an illustration of some of the variations studied, when students were asked how many hours they studied per day, 70% said more than 2.5 hours, but when the same question was asked another way, only 23% indicated more than 2.5 hours.

Appearance and Design

Finally, the entire appearance and format of the questionnaire can have a significant bearing on the return rate. Questionnaires that appear to be poorly organized and poorly prepared are likely to be "filed" in the wastebasket or clicked out of if accessed online. Remember, many people have negative attitudes toward questionnaires, so anything the researcher can do to overcome this negative attitude enhances the likelihood that the questionnaire will be answered. Some suggestions are merely cosmetic, such as the use of colored paper or an artistic design. Even small enhancements, such as having dotted lines from the questionnaire item to the response options or grouping related items, may pay big dividends.

The questionnaire should provide the name and address of the investigator. The instructions for answering the questions must be clear and complete, and examples should be provided for any items that are expected to be difficult to understand.

The first few questions should be easy to answer; respondents are more likely to start answering easy questions and are more apt to complete the questionnaire after they are committed. A poor strategy is to begin with questions that require considerable thought or time to gather information. Every effort should be made to make difficult questions as convenient to answer as possible. For example, questions that ask for enrollment figures, the size of a faculty, or the number of graduate assistants can often ask for responses in ranges (e.g., 1–10, 11–20, 21–30). First, you will probably group the responses for analysis purposes anyway. Second, the respondent can often answer range-type questions without having to consult the records or at least can supply the answers more quickly. An even more basic question to ask yourself is Do I really need that information? or Does it pertain to my objectives? Unfortunately, some investigators simply ask for information off the top of their heads with no consideration for the time required to supply the answer or the relevance of the information.

Generally, shorter questionnaires have higher response rates and more validity than longer ones do. The issue is not more data, but better data (Punch, 2003). An analysis of 98 questionnaire studies by Gall, Gall, and Borg (2006) showed that, on average, each page added to a questionnaire reduced the number of returns by 0.5%. Because many people are reluctant to complete questionnaires, the cover letter (discussed later) and the size and appearance of the questionnaire are crucial. A lengthy questionnaire requiring voluminous information will likely be put aside until later if not discarded immediately.

Conducting the Pilot Study

A pilot study is recommended for any type of research but is imperative for a survey. In fact, the designer of a questionnaire may be well advised to do two pilot studies. The first trial run consists simply of asking a few colleagues or acquaintances to read over the questionnaire. These people can provide valuable critiques about the format, content, expression and importance of items, and feasibility of adding or deleting questions.

After revising the questionnaire in accordance with the criticisms obtained in the first trial run, respondents who are a part of the intended population are selected for the second pilot study. The questionnaire is administered, and the results are subjected to item analysis (discussed in chapter 11). In some questionnaires, correlations may be determined between scores on each item and the whole test to see whether the items are measuring what they are intended to measure. Responses are always examined to determine whether the items seem clear and appropriate. First, questions that are answered the same way by all respondents need to be evaluated; they probably lack discrimination. Unexpected responses may indicate that the questions are poorly worded. Some rewording and other changes might also be necessary if the participants, who might be sensitive to some questions, do not respond to them. Furthermore, the pilot study determines whether the instructions are adequate. Another value of the pilot study relates to determining the length of the survey. The researcher should record how long it takes the average participant to complete the survey.

A trial run of the analysis of results should always be accomplished in the pilot study. The researcher can see whether the items can be analyzed in a meaningful way and then ascertain whether some changes are warranted for easier analysis. This is one of the most profitable outcomes of the pilot study. Of course, if substantial changes are mandated by the results of the pilot study, another pilot study is recommended to determine whether the questionnaire is ready for initial mailing.

Writing the Cover Letter

Unquestionably, the success of the initial mailing depends largely on the effectiveness of the **cover letter** that accompanies the questionnaire. If the cover letter explains the purposes and importance of the survey in a succinct and professional manner (and if the purposes are worthy of study), the respondent will likely become interested in the problem and will be inclined to cooperate.

cover letter—The letter attached to a survey that explains the purposes and importance of the survey.

An effective cover letter should also assure respondents that their privacy and anonymity will be maintained. Furthermore, the cover letter should make an appeal for the respondent's cooperation. Some subtle flattery about the respondent's professional status and the importance of his or her response may be desirable. This effort must be done tactfully, however, and only when appropriate. In this situation, the person's name and address should appear on the cover letter. The letters should look as if they were individually typed, even when a word processor is used. People will be insulted if it is suggested that they have been handpicked for their expertise and valued opinions when the letters are addressed "Dear Occupant."

If the survey is endorsed by recognized agencies, associations, or institutions, specify this in the cover letter. If possible, use the organization's or institution's stationery. Respondents will be much more cooperative if some respected person or organization is supporting the study. In addition, acknowledge whether financial support is being given and by whom. Identify yourself by name and position. If the study is part of your thesis or dissertation, give your advisor's name. Having the department chair or the dean of the college sign the letter may be advantageous. To increase the response rate, contact individuals by e-mail, letter, card, or telephone, asking for their participation in the survey. If the purposes of the study are worthwhile, offering the respondent a summary of the results is usually effective at this time. Before making this offer, give ample consideration to how much work and money such a summary will entail. If you do offer to provide a summary, be sure to follow through on your promise.

Many universities require informed consent by those completing surveys. The survey topic and university procedures may determine the type of consent required (i.e., passive or active) and the method by which participants provide consent (e.g., by checking a box on an electronic survey or sending a signed consent in a separate envelope when a sensitive topic is being addressed). As the study is being designed and institutional review board documents are being prepared, determine what type of consent is required and how to make this process efficient for the participants.

Because a questionnaire imposes on a person's time, strategies involving rewards and incentives are sometimes used in an effort to amuse or involve the respondent. Some questionnaires include money (perhaps a dollar) as a token of appreciation. This incentive may appeal to a person's integrity and evoke cooperation. Then again, you may just be out a dollar. The disadvantage is that inflation plays havoc with the effects of such rewards. A quarter may have been effective years ago, whereas one or more dollars may now be required to elicit the same sense of guilt for not responding. Research results concerning the effectiveness of monetary incentives have been mixed. Denton and Tsai (1991) compared several sizes of monetary incentives and a raffle for a subscription to a professional journal and failed to find significant benefits.

The cover letter should request that the questionnaire be returned by a certain date. (This information should also be specified on the questionnaire.) When establishing the date for the return of data, consider such things as the respondents' schedules, responsibilities, and vacations. Be reasonable about the time that you allow the person to respond, but do not give the respondent too much time because of the tendency to put a questionnaire aside and later forget it. One week is ample time for the respondent to answer (in addition to the mailing time).

The appearance of the cover letter is just as important as the appearance of the questionnaire. Grammatical errors, misspelled words, and improper spacing and format give the respondent the impression that the author does not attach much importance to details and that the study will probably be poorly done.

Researchers have tried a number of subtle and tactful approaches to establish rapport with respondents. Some have tried a solemn appeal to the monumental importance of the survey, some have attempted casual humor, and others have tried a folksy approach. Obviously, the success of any approach depends on the skill of the writer and the receptiveness of the reader. Any of these attempts can backfire.

A number of years ago, one of this book's authors (Nelson) received from his department head a note to which a letter like this was attached:

Dear Dr. _____:

It has been said: "There are two kinds of information, knowing it or knowing where to find it." I asked three men you know, Doctors Eeney, Meeney, and Moe, a certain question, and each said this: "Sorry. I don't know his name." But they came back with this helpful suggestion: write to you.

Now for the question: What is the name of the person in your department or school who is in charge of your graduate program in physical education? . . .

Cordially yours,

Harry Homespun

The department head answered the letter, naming Nelson as head of the graduate program; shortly thereafter, this letter arrived:

Dear Dr. Nelson:

Your good work in administering a graduate program is well known throughout our profession. Even though there are many knotty problems, everyone wants to upgrade the resources for graduate studies. But how? Because of your scholarly approach, my associates Doctors Eeney, Meeney, and Moe suggested that I write to you.

It has been said, "If you want to get something important done, ask a busy person." My associates tell me you are busier than a bird dog in tall grass. They also say that you have a high regard for excellence. . . . If you could find it convenient to return your checked copy on or before May 1, you might help reduce the cultural lag.

Cordially yours,

Harry Homespun

We have, of course, omitted the main parts of the letters concerning what was being studied and what the instructions specified, but those parts were generally well done.

About a year later, another questionnaire arrived from the same university with essentially the same folksy approach, including the analogy of a bird dog in tall grass. It is safe to assume that these two graduate students had the same research methods course at the same institution. Most likely, in the discussion of the survey method, some examples of cover letters were given with different approaches. Professors Eeney, Meeney, and Moe would undoubtedly be embarrassed if they knew that the two students had written such nearly identical letters.

In most respects, the cover letters contained the essential information, and the topics were worthwhile. The authors were simply too heavy handed in their attempts at a down-home approach, and their efforts to flatter the respondent regarding his "well-known" expertise lacked subtlety.

The sidebar Sample E-Mail Cover Letter shows a cover letter for a questionnaire that deals with a potentially threatening topic. The letter effectively explains to the respondent how confidentiality will be ensured. It also focuses on the importance of the topic and explains how the process will work. Bourque and Fielder (2003) offered helpful suggestions regarding what should be included in cover letters.

SAMPLE E-MAIL COVER LETTER

Dear _____:

I am a physical education doctoral student conducting a study for my dissertation titled *Teachers' Perceptions of School District Teacher Assessment Procedures.* As a part of this study I will compare the attitudes of more- and less-experienced teachers and (a) how they view the teacher assessment process, (b) how the assessment information is used by their supervisors, and (c) how they use assessment information. My experience as a public school physical education teacher suggests that this information is important—a great deal of time is spent evaluating teachers and how teachers view assessment often is ignored. I hope you will agree to participate in this study. Your director of physical education has agreed that your district will participate, and your name was randomly selected from a group of physical educators who teach in the district.

Your participation will require approximately 20 minutes to complete an online survey. If you would like to participate, all you need to do is click on the link below and it will take you to a site where you can begin. Before starting the survey, you will be informed of your rights and asked to give consent to participate. Included in the introductory material is a description of how the information you provide will be kept completely confidential. You will be anonymous, and at no time will your answers be identified with your name or any other information that could identify you. In addition, the only data that will be reported are means and other summary scores. You and your director of physical education will receive a summary of the results. Neither you nor your school district will be identified in this report, the dissertation, or any papers that are published once the dissertation is complete. Over 20 school districts have agreed to participate, and data will be aggregated by the sizes of the school districts.

I hope you will agree to participate in this study. I would be happy to answer any questions. You can respond to this e-mail or call me at 212-555-5555.

Thank you for your time and for considering this request. Your help will enable me to conduct a strong study that will provide valuable results for our field.

Sincerely,
Ronald Smith
Graduate Student

Click Here to Participate in this Study

Sending the Questionnaire

You must take care in choosing the time for the initial mailing. Such considerations include holidays, vacations, and especially busy times of the year for the respondents. A self-addressed, stamped envelope should be included. It is almost an insult to expect respondents not only to answer your questionnaire but also to provide an envelope and a stamp to mail it back to you.

Other matters regarding the mailing of the questionnaire, such as establishing the date to be returned, have been covered in previous sections. The initial mailing represents a substantial cost to the sender. Securing a sponsor to underwrite or defray expenses and using bulk-mailing services are important considerations.

Many surveys are sent via e-mail or provided on the Internet. Specific information on these forms of communication is provided later in this chapter.

Following Up

This should not come as a big shock, but it is unlikely that you will get 100% return on the initial mailing. A follow-up letter is nearly always needed, and this task can be done in many ways. One is to wait about 10 days after the initial mailing and then send a card to all people in the sample, stressing the importance of their participation (and apologizing if they have already returned the questionnaires). Approximately 10 days after the card, another letter with another copy of the questionnaire and another self-addressed, stamped envelope should be sent to those who have not responded. This approach is expensive, of course, but follow-ups are effective. In a number of cases, the person has forgotten to respond, and a mere reminder will prompt a return. Other nonrespondents who had not initially planned to return the questionnaire may be influenced by the researcher's efforts in reiterating the significance of the study and the importance of the person's input.

The follow-up letter should be tactful. The person should not be chastised for failing to respond. The best approach is to write as though the person would have responded had it not been for some oversight or mistake on your part (see the sidebar).

SAMPLE FOLLOW-UP LETTER

Dear _____:

I sent a questionnaire regarding criteria for planning physical fitness work in physical education classes to you a few weeks ago and have not heard from you. As you can appreciate, it is important that we obtain response from everyone possible inasmuch as only a few select individuals were contacted. Our school district is planning an immediate study and updating of its physical education program based on the results of this study so it is of vital concern.

The questionnaire was sent during the summer when you may have been away from your office. I have included another copy, however, and it would be most helpful if you could take 30 to 45 minutes to give your opinions on the information requested.

Thank you so much for your cooperation.

Sincerely yours,
Physical Education Specialist
San Diego Unified School District, Reprinted by permission from Ash Hayes.

A lack of response is especially likely when the questionnaire deals with a sensitive area. For example, surveys about program offerings and grading practices are often not returned by schools with inadequate programs and poor grading practices. Thus, the obtained responses are apt to be biased in favor of the better programs and better teachers. Regardless of the nature of the survey, the results from a small return rate (e.g., 10 to 20%) cannot be given much credibility. People who have a particular interest in the topic being surveyed are more likely to respond than those who are less interested. These respondents are self-selected, and the responses are almost invariably biased in ways that are directly related to the purposes of the research (Fowler, 2009). K.E. Green (1991) reported that people who responded to the initial mailing had more favorable attitudes toward the topic and had more positive attitudes about themselves than reluctant respondents did.

In fact, in most studies in which more than 20% of the questionnaires are not returned, it is recommended that a sample of the nonrespondents be surveyed. Of course, this task is not easy and may be impossible if you are recruiting participants through professional organizations or government agencies that will not provide direct contact information. If people have ignored the initial mailing and one or two follow-ups, the chances are not good that they will respond to another request, but it is worth a try. The preferred technique is to select a

> Survey results with a small return rate have less credibility because respondents are more likely to be those who are interested in the topic. This pattern almost always produces a bias directly related to the purposes of the research.

small number (e.g., 5 to 10%) of the nonrespondents randomly by using the table of random numbers (this technique is sometimes called double-dipping). Then, make contact either by telephone or by a special cover letter (perhaps even by registered mail). After the responses have been obtained, comparisons are made between the answers of the nonrespondents and the answers of the people who responded initially. If the responses are similar, you can assume that the nonrespondents are not different from those who replied. If there are differences, you should either try to get a greater percentage of the nonrespondents or be sure these differences are noted and discussed in the research report.

To follow up on nonrespondents, you have to know who has not responded. Keeping records of those who have and have not responded may seem to fly in the face of guaranteed anonymity. However, this is not a difficult task. An identifying number written on the questionnaire or return envelope can be used. Of course, it is recommended that you inform the respondents about the use of the number in your cover letter. You might offer an explanation such as the following:

> Your responses will be strictly confidential. The questionnaire has an identification number for mailing purposes only. With this numbering system, I can check your name off the mailing list when your questionnaire is returned. Your name will never be placed on the questionnaire.

Another approach, which works well, is to send a separate postcard (self-addressed and stamped) with the questionnaire. The card contains an identifying number and is to be mailed back separately when the person returns the questionnaire. The card simply states that the person is sending the questionnaire and therefore that you do not need to send any more reminders. This procedure ensures anonymity and yet informs you that the person has responded (Fowler, 2002).

A good return rate is essential. We frequently read journal and newspaper reports of surveys in which something like 2,000 surveys were mailed and 600 were returned. The number of returns sounds respectable, but these 600 people represent only a 30% return rate. They are not a random sample. They are self-selected, and their responses may be quite different from those of the 1,400 who did not respond for whatever reason.

Analyzing the Results and Preparing the Report

These last two steps are discussed in chapter 21, which deals with the results and discussion sections of the research report. The main consideration is that the method of analysis should be chosen in the planning phase of the study. Many questionnaires are analyzed merely by tallying the responses and reporting the percentage of the respondents who answered one way and the percentage who answered another way. Often, not much in the way of meaningful interpretation can be gained if only a simple tally is given. For example, when the researcher states simply that 18% of the respondents strongly agreed with some statement, 29% agreed, 26% disagreed, 17% strongly disagreed, and 10% had no opinion, the reader's reaction may be So what? Questionnaires, like all surveys, must be designed and analyzed with the same care and scientific insight as experimental studies.

ELECTRONIC SURVEYS

As noted throughout the previous section, much of the information pertaining to surveys prepared and distributed on paper is applicable to electronic surveys. The care in determining the objectives, constructing and designing the questionnaire, writing a cover letter, and communicating with participants is the same. And, of course, conducting a pilot study is equally important. The use of electronic surveys, however, has some different demands, which are discussed here.

After you have finalized the objectives and are in the process of writing questions, you should determine which survey software is available on your campus. There are both Internet-based survey programs and programs that reside on a local computer. In the second type, the survey is placed on the Internet after it is constructed. These programs are relatively easy to use. Most have fees associated with their use, but your university is likely to have a site license that allows you to get a user name for Internet surveys or use it on a department computer without cost. If you have more than one choice about which program to use, talk with those at the computer help desk who specialize in this software. The best choice is the software that is easiest to use and allows you to construct the survey, post it, and get the results in the most efficient way possible.

After you determine which survey software to use, we highly recommend attending an orientation session on campus or completing the online tutorial. Either approach will alert you to the capabilities that the software has to offer and help you learn how to insert questions, change appearance and design, and prepare data for analysis. A few hours invested here will likely save many hours as you develop and pilot-test the survey.

After you have written the questions and had them reviewed by advisors and colleagues, you can construct the online survey. If you have become familiar with the software, this step should be relatively straightforward. Most electronic surveys, as with paper surveys, begin with an introduction of the purpose and provide directions for completing the survey (e.g., "Click *next page* after you have competed the questions" or "You cannot go back and complete questions if you already have submitted a page"). Consent for participating in the study appears early in the survey and it should be worded according to the stipulations of the institutional review board.

Care is needed in designing a survey that is visually appealing and easy to use (Fink, 2009). Most survey software permits an infinite array of color and design options. Choose these carefully because people may react differently to similar color schemes and layouts. One of us was recently asked to complete an online survey that alternated between a light blue and a dark blue color with each question. Although this design was appealing from an artistic viewpoint, it was nearly impossible to read the questions in dark blue because the text was posted in black. All decisions about design should be made to enhance the number of participants who complete the survey.

As you design the cover letter, use the information from the previous section. This letter will direct the participant to the link to complete the survey. Check and double-check that the link in the e-mail letter is correct. If the link is not correct, participants may be lost and may not respond to a second request if they wasted their time trying to connect the first time.

The pilot study should be conducted in the same way that the final survey will be conducted. A good approach is to have colleagues and your major professor respond to the e-mail request and complete the survey before you use a larger group. This step allows you to correct formatting and design issues, as well as make sure the link to the survey works and the survey is accessible from a variety of locations. This step, which often can be conducted in a few days, may identify problems that would create havoc to the study time frame if a second, larger pilot study has to be conducted. After the initial pilot study with colleagues, and then again with the larger pilot study, modify the questions and adjust the design and layout of the survey as needed.

If you are using a sampling procedure, confirm that the e-mail addresses for participants are correct before sending them anything. As noted earlier, for many studies it is not possible to use sampling techniques, so requests for participation are made in cooperation with professional groups and government agencies and through requests on often-visited websites. Arranging to get help with requesting participation and deciding how to conduct follow-ups should be started early. Determine who in the organization will be responsible for giving permission and who will send out e-mail requests. Work with that person to make the process efficient and as pleasant as possible. The final sample size will be determined by who is solicited to participate, and this may be contingent on how much others want to help you.

After all follow-ups have been sent and you are ready to begin the analysis, download the data file with all the completed surveys. Most survey software programs have a number of format and file type options for saving the data. As we noted earlier, the pilot is an ideal time to determine which formats and file types will enhance subsequent analysis. You generally will have no need to reconfigure anything to use SPSS or another statistical program—if you plan ahead.

DELPHI METHOD

Delphi survey method— A survey method that uses a series of question-naires in such a way that the respondents (usually experts) reach a consensus about the subject.

round—A stage of the Delphi survey method in which respondents are asked for their opinions and evaluations of various issues, goals, and so on.

The **Delphi survey method** uses questionnaires but in a different manner than the typical survey. The Delphi method uses a series of questionnaires in such a way that the respondents finally reach a consensus about the topic. It is basically a method of using expert opinion to help make decisions about practices, needs, and goals.

The procedures include the selection of experts, or informed people, who are to respond to the series of questionnaires. A set of statements or questions is prepared for consideration. Each stage in the Delphi method is called a **round**. The first round is mostly exploratory. The respondents are asked for their opinions on various issues, goals, and so on. Open-ended questions may be included to allow the participants to express their views and opinions.

The questionnaire is then revised as a result of the first round and sent to the respondents, asking them to reconsider their answers in light of the analysis of all respondents to the first questionnaire. In subsequent rounds, the respondents are given summaries of previous results and asked to revise their responses, if appropriate. Consensus about the issue is finally achieved through the series of rounds of analysis and subsequent considered judgments. Anonymity is a prominent feature of the Delphi method, and the consensus of recognized experts in the field provides a viable means of confronting important issues. For example, the Delphi method is used sometimes to determine curricular content for programs, to decide on the most important objectives of a program, or to choose a solution to a problem.

PERSONAL INTERVIEWS

As mentioned earlier, the steps for the interview and the questionnaire are basically the same. This section focuses only on the differences.

Preparing for the Interview

The most obvious difference between the questionnaire and the interview is in the gathering of the data. In this respect, the interview is more valid because the responses are apt to be more reliable. In addition, the percentage of returns is much greater.

Participants should be selected using the same sampling techniques used for a questionnaire. Generally, the interview uses smaller samples, especially when a graduate student is doing the survey. Cooperation must be secured by contacting the participants selected for interviewing. If some refuse to be interviewed, you must consider possible bias to the results, as with nonrespondents in a questionnaire study.

Conducting an effective interview takes a great deal of preparation. Graduate students sometimes have the impression that anyone can do it. Because the procedures used for the questionnaire are followed in preparing the items, you must be very familiar with them. Carefully rehearse the interview techniques. One source of invalidity is that interviewers tend to improve with experience, and thus the results of earlier interviews may differ from those of interviews conducted later in the study. A pilot study is extremely important. Make sure the vocabulary level is appropriate and the questions are equally meaningful given the ages and educational backgrounds of the participants. Use careful planning in organizing the questions and visual aids to provide a comfortable flow of presentation and transitions from question to question.

Training is required in making initial contact and presenting the oral "cover letter" by phone. At the meeting, establish rapport and help the person feel at ease. If you are audio-recording the interview, obtain permission. If you are not audio-recording, have an efficient system of coding the responses without consuming too much time and appearing to be taking dictation. You must not inject your own bias into the conversation and certainly should not argue with the respondent. Although you hold many advantages over the questionnaire with regard to the flexibility of the questioning, it introduces the risk that the respondent will stray from the questions and get off the subject. You must tactfully keep the respondent from rambling, and doing this requires skill.

The key to getting good information is to ask good questions. Some surveys that employ in-person interviews are quantitative, and the questions may be similar to those in a mailed questionnaire. Other surveys deal with qualitative data. In these studies, standardization is less of a concern, and more emphasis is given to description. Good interviewers in qualitative surveys do not ask yes-or-no questions, they do not ask multiple questions disguised as a single question, and they try to avoid inserting their own points of view. Merriam (2001) defined four major categories of questions: hypothetical, devil's advocate, ideal position, and interpretive. Here are some examples of each:

- Hypothetical: "Let's suppose that this is my first day of student teaching. What would it be like?"

- Devil's advocate: "Some people say that professional educational courses are of little value for the student-teaching experience. What would you say to them?"

- Ideal position: "What do you think the ideal student-teaching preparation program would be like?"

- Interpretive: "Would you say that the student-teaching experience is different from what you expected?"

You may recognize these approaches from interviews that you have seen on television. The interview has the following advantages over the mailed or e-mailed questionnaire:

- The interview is more adaptable. Questions can be rephrased, and clarification can be sought through follow-up questions.

- The interview is more versatile with regard to the personality and receptiveness of the respondent.

- The interviewer can observe how the person responds and can thus achieve greater insight into the sensitivity of the topic and the intensity of the respondent's feelings. This feature can add considerably to the validity of the results, because respondents' tendency to avoid sensitive topics is one of the greatest threats to validity in questionnaire studies.

- Because each person is contacted before the interview, interviews have a greater rate of return. Moreover, people tend to be more willing to talk than to fill out a questionnaire. A certain amount of ego is involved because a person who is interviewed feels more flattered than one who receives only a list of questions.

- An advantage of in-person interviews is that visual aids such as flash cards can be used to simplify long questions and explain the response lists.

A variety of resources provide information to help prepare interview questions and perfect techniques for qualitative interviews. Books by Patton (2002) and Merriam (2001) provide general advice. More specific advice is found in books by Rubin and Rubin (2012) and Seidman (2006). The references you select and use will be determined by the research questions and the type of interview that will yield the best responses to them.

A potential problem in interviews is losing the data. If you audio-record your interviews and back them up appropriately, this problem may be less consequential because you will have a transcript and can go back to the recording if questions arise. Because of confidentiality restrictions in some studies, recordings will not be used and names may not be placed on the interview data instrument. Consequently, the researcher may later become confused about which data belong to which respondents and what each respondent said. Use several identifiers for each interview, and practice taking notes during pilot studies. Oishi (2003) provided an excellent coverage of the dos and don'ts in planning and conducting the personal interview.

Conducting a Telephone Interview

Interviewing by telephone is becoming more prevalent. Telephone interviewing has a number of advantages over face-to-face interviewing:

- Telephone interviewing is less expensive than traveling to visit respondents. Bourque and Fielder (2003) reported that telephone interviews cost half as much as personal interviews and about twice as much as mail surveys.
- Researchers can conduct telephone interviews in significantly less time than in-person interviews.
- The interviewer can work from a central location, which facilitates the monitoring and the quality control of the interviews and provides a better opportunity for using computer-assisted interviewing techniques (Gall, Gall, & Borg, 2006).
- Many people are more easily reached by telephone than by personal visit.
- The telephone interview enables the researcher to reach a wide geographical area; geography is a limitation of the personal interview. This advantage also can increase the validity of the sampling.
- Some evidence indicates that people respond more candidly to sensitive questions over the telephone than they do in personal interviews, in which the presence of the interviewer may inhibit some responses.

Using a computer in telephone interviewing eases data collection and analysis. For example, a computer can display the questions for the interviewer to ask, to which the interviewer types the responses obtained. Each response triggers the next question, so the interviewer does not have to turn pages and is less likely to ask inappropriate questions. Responses are stored for analysis (thus reducing scoring errors), and the stored responses can then be accessed for statistical analysis.

Among the disadvantages of telephone surveys is that obtaining a representative sample is increasingly difficult because of cell phones, answering machines, caller ID, and call-blocking devices. Telemarketing has greatly soured the public on telephone interviews.

A precall letter can be effective in securing cooperation, thereby increasing the response rate. The drawback is that you must have the names, addresses, and telephone numbers of the target population. In some surveys, the researcher uses random-digit dialing (RDD) to select the sample. In this method, a table of random numbers can be used to select the four numbers following the three-number exchange. Both listed and unlisted numbers can be reached using this method.

Much time is consumed in callbacks because of no answer, answering machines, unwanted people answering the phone, busy signals, and language barriers. The researcher usually must hire other people because of the many calls that must be made. An important source

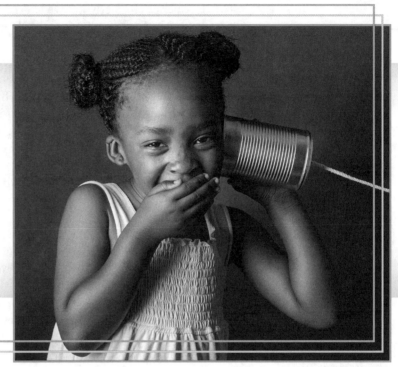

Surveys are
sometimes the
average opinion
of people who
don't know.

© Belinda Pretorius/fotolia

of bias concerns the time when telephone surveys are done. If conducted during the day, the sample is biased toward the people most likely to be home: homemakers, the unemployed, and the retired. Thus, interviewers usually must work after 5:00 p.m. In addition, it is estimated that one-third of the people contacted are not home on the initial call, which means that callbacks are necessary (sometimes as many as 10 or 12!).

A commonly cited weakness of telephone surveys has been that the sample is limited to people who have telephones, although cell phones are replacing many traditional landlines. Today, however, a vast majority of people in the United States have telephones. In 2000, 97.25% of all households had telephones (Bourque & Fielder, 2003). The percentage varies somewhat from state to state. Low-income people, particularly in inner cities and rural areas, are less likely to have access to telephones. Certainly, if the target population is low-income respondents, the telephone survey is not recommended.

Bourque and Fielder (2003) gave a thorough description of the relative advantages, disadvantages, and recommended methodology of the telephone survey.

NORMATIVE SURVEYS

The **normative survey** is not described in most research methods textbooks. As implied by the name, this method involves establishing norms for abilities, performances, beliefs, and attitudes. A cross-sectional approach is used: Samples of people of various ages, both sexes, and other classifications are selected and measured. The steps in the normative survey are generally the same as in the questionnaire, the difference being the way the data are collected. Rather than asking questions, the researcher selects the most appropriate tests to measure the desired performances or abilities, such as the components of physical fitness.

In any normative survey, the test must be administered in a rigidly standardized manner. Deviations in the way measurements are taken give meaningless results. The researcher collects and analyzes the data from the survey by some norming method, such as percentiles, T scores, or stanines, and then constructs norms for the categories of age, sex, and so on.

The timing of a telephone call can bias the results. Calling before 5:00 p.m. may skew the sample toward homemakers, the unemployed, and the retired.

normative survey—A survey method that involves establishing norms for abilities, performances, beliefs, and attitudes.

SHAPE America (formerly known as AAHPERD) has sponsored several normative surveys. Probably the most notable was the Youth Fitness Test (see American Association for Health, Physical Education and Recreation, 1958), conducted in response to the furor caused by the results of the Kraus-Weber test (Kraus & Hirschland, 1954), which had revealed that American children were inferior to European children in minimum muscular fitness. The Youth Fitness Test was originally given to 8,500 boys and girls in a nationwide sample. Follow-up testing was done in 1965 and 1975.

In the 1958 AAHPER normative survey, a committee determined a seven-item motor fitness test battery. The University of Michigan Survey Research Center selected a representative national sample of boys and girls in grades 5 through 12 and then requested the cooperation of each school. They prepared directions for giving the test items and selected and trained physical education teachers in various parts of the country to administer and supervise the testing.

In addition, AAHPERD conducted a sport skills testing project, which established norms for boys and girls of various ages in the skills of a number of sports. The National Children and Youth Fitness Study (1985, 1987) was a normative survey that established norms for a health-related physical fitness test.

Sometimes comparisons are made between the norms of different populations. In other studies, the major purpose is simply to establish norms. The primary drawbacks of any normative survey occur in test selection and the standardization of testing procedures. With many test batteries, there is a danger of generalizing based on the test results. For example, when only one test item is used to measure a component (such as strength), making unwarranted generalizations is possible.

The standardization of testing procedures is essential for establishing norms. But when a normative survey involves many testers from different parts of the country, problems with standardization become a potential source of measurement error. Published test descriptions simply cannot address all the aspects of test administration and the ways of handling the many problems of interpreting test protocol that arise. Extensive training of testers is the answer, but this is often logistically impossible.

SUMMARY

The most common descriptive research technique is the survey, which includes questionnaires and interviews. The two are largely similar except for the method of asking questions. Surveys can be conducted on paper or on the Internet, and similar steps are required for getting good results for both methods. The questionnaire is a valuable tool for obtaining information over a wide geographical area. A good cover letter is important in getting cooperation. One or two follow-ups are often necessary to secure an adequate response. Personal interviews usually yield more valid data because of the personal contact and the opportunity to make sure that respondents understand the questions. Telephone interviews are becoming increasingly popular. They have most of the advantages of personal interviews and the flexibility to involve more participants over a larger geographical area. The Delphi survey method uses a series of questionnaires in such a way that the respondents eventually reach a consensus about the topic. It is frequently used to survey expert opinion to make decisions about practices, needs, and goals. The normative survey is designed to obtain norms for abilities, performances, beliefs, and attitudes. The AAHPERD fitness tests developed over the years are examples of normative surveys. In all surveys, representative sampling is extremely important.

Check Your Understanding

1. Find a thesis or dissertation that used a questionnaire to gather data. Briefly summarize the methodology, such as the procedures used in constructing and administering the questionnaire, selecting participants, and following up.

2. Critique a study that used interviews for gathering data. Discuss the strengths and weaknesses of its methods.

16

Other Descriptive Research Methods

Analyzing humor is like dissecting a frog. Few people are interested and the frog dies of it.

—E.B. White (1899-1985)

orms of the survey method of descriptive research were discusscd in chapter 15. In this chapter, we address several other descriptive research techniques. One is developmental research. Through cross-sectional and longitudinal studies, researchers investigate the interaction of growth and maturation and of learning and performance variables. The case study, in which the researcher gathers a large amount of information about one or a few participants, is widely used in a number of fields. The job analysis is a technique used to describe the working conditions, requirements, and preparation needed for a particular job. Through observational research, the researcher obtains quantitative and qualitative data about people and situations through observing their behavior. Some research studies employ unobtrusive methods in which the participant is not aware of being studied. Correlational studies determine and analyze the relationships among variables and generate predictions.

DEVELOPMENTAL RESEARCH

Developmental research is the study of changes in behaviors across years. Although much of the developmental research has focused on infancy, childhood, and adolescence, research on senior citizens and even across the total human life span is increasingly common.

developmental research—The study of changes in behaviors across years.

Longitudinal and Cross-Sectional Designs

Developmental research focuses on cross-age comparisons. For example, researchers can compare how far children can jump when they are 6, 8, and then 10 years old, or they can compare adults who are 45, 55, and 65 years old on their knowledge of the effects of obesity on life expectancy. Both of these examples are developmental studies. The major distinction between the two basic approaches in developmental studies is whether researchers follow the same participants over time (longitudinal design) or select different participants at each age level (cross-sectional design).

Longitudinal studies are powerful because the changes in behavior across the time span of interest are seen in the same people. Longitudinal designs, however, are time-consuming. A longitudinal study of children's jumping performances at ages 6, 8, and 10 obviously requires 5 years to complete. That choice of design is probably not wise for a master's thesis. Longitudinal designs have additional problems besides the time required to complete them. First, over the several years of the study, some of the children are likely to move away when parents change jobs or change schools when school districts are rezoned. In longitudinal studies of senior citizens, some may die over the course of the study. The problem is not knowing whether the sample characteristics remain the same when participants are lost. For example, when children are lost from the sample because parents change jobs, is the sample then composed of children of lower socioeconomic levels because the more affluent parents moved? Furthermore, if obesity is related to longevity, are older people more likely to be less obese and consequently have increased knowledge because the more obese people with less knowledge have died? Thus, knowledge about obesity may not change from ages 45 to 65; rather, the sample may change.

Another problem with longitudinal designs is that participants become increasingly familiar with the test items, and the items may cause changes in behavior. The knowledge inventory on obesity may prompt the participants to seek information about obesity, thus changing their knowledge, attitudes, and behaviors. Therefore, the next time they complete the knowledge inventory, they have gained knowledge. This gain of knowledge, however, is the result of having been exposed to the test earlier and might not have occurred without that exposure.

Cross-sectional studies are usually less time-consuming to carry out than longitudinal studies. Cross-sectional studies test several age groups (e.g., 6, 8, and 10) at the same point in time. Although these studies are more time efficient than longitudinal studies, a limitation called the **cohort problem** exists: Are all the age groups really from the same population (group of cohorts)? Asked another way, Are the environmental circumstances that affect jumping performance for 6-year-olds the same today as when the 10-year-olds were 6, or have physical education programs improved over this 4-year span so that 6-year-olds receive more instruction and practice in jumping than the 10-year-olds did when they were 6? If the latter is true, then we are not looking at the development of jumping performance but rather at some uninterpretable interaction between normal development and the effects of instruction. The cohort problem exists in all cross-sectional studies.

Examples of longitudinal developmental studies are Halverson, Roberton, and Langendorfer (1982), who studied the velocity of the overarm throw in children from early elementary school through junior high school, and Nelson, Thomas, Nelson, and Abraham (1986), who investigated sex differences in throwing in children in kindergarten through third grade. Thomas and colleagues (1983) conducted a cross-sectional developmental study in which they looked at the development of memory for distance information in different age groups. They compared the effects of a practiced strategy for remembering distance at each age level to show that the appropriate use of strategy reduces age differences in remembering distance. Each of these studies suffers from the specific defects associated with the type of developmental design. Both Halverson and colleagues (1982) and Nelson and colleagues (1986) lost participants over the several years of the study. In the latter study, measurements were taken from 100 children in kindergarten at a single school, and three years later only 25 children were still at that school. In a recent longitudinal study on children's throwing performance, Robertson and Konczak (2001) reported that 73 children originally were filmed but only 39 children completed the 7-year study. On the other hand, in their cross-sectional study, Thomas and colleagues (1983) could not establish whether the younger children were more familiar with memory strategies than the older children were when they were younger.

Although both longitudinal and cross-sectional designs have some problems, they are the only means available to study development. Thus, both are necessary and important types of research. These two types of design are considered descriptive research. However, either can also be experimental research (chapter 18); that is, an independent variable can be manipulated within an age group. Thomas and colleagues' (1983) study manipulated the use of memory strategy within each of the three age groups. Thus, age was a categorical variable, whereas strategy was a true independent variable. This point is addressed here because developmental research is not covered in chapter 18 on experimental research.

Methodological Problems of Developmental Research

Whether the developmental research is longitudinal or cross-sectional, several methodological problems exist (for a more detailed discussion, see Thomas, 1984). Some examples are addressed in this section.

Unrepresentative Scores

One of the most common problems in developmental research is unrepresentative scores, also called outliers. Mentioned in previous chapters, they occur in all research but are particularly problematic at the extremes of developmental research (children and senior citizens). Outliers frequently result from shorter attention spans, distraction, and a lack of motivation to perform the task. The best way to handle unrepresentative scores is to do the following:

- Conduct the testing situation within a reasonable time frame to account for attention span.
- Set up the testing situation so that distractions do not occur.
- Be aware of what an unrepresentative score is and retest when one occurs.

The last thing a researcher wants is to use unrepresentative scores. Therefore, outliers not detected at testing should be found when the distribution of the data is studied. There are several ways to test for these extreme and unrepresentative scores (e.g., see Barnett & Lewis, 1978). The developmental researcher should expect and plan to handle outliers in the data set.

Unclear Semantics

A second problem in developmental research involves semantics. Selecting the words to use in explaining a task to children in various age groups is a formidable challenge. If the researcher is not careful, older children will perform better than younger children primarily because they grasp the idea of what to do more quickly. Although the standard rule in good research is to give identical instructions to all the participants, researchers must bend this rule for developmental studies with children. They must explain the test in a way that the participants can understand, and obtain tangible evidence that children of various ages understand the test before conducting it. To ensure that the children understand the test, the researcher could consider having them demonstrate the activity to some criterion level of performance before collecting the data.

A good example of the semantics problem involved a group of early elementary school children taking a computer course through a continuing education program at a local college. The children were doing fine until the teacher (a college instructor of computer science) began to write instructions on the blackboard. Then everyone stopped working. Finally, one of the children whispered to the teacher, "Some of us can't read cursive." After the teacher printed the instructions in block letters, all the children happily returned to their computing work (*Chronicle of Higher Education*, 1983).

Lack of Reliability

A third developmental research problem is the lack of reliability in younger children's performances. When obtaining a performance score for a child, you must ensure that it is a reliable one; that is, if the child is tested again, the performance score should be about the same. Obtaining reliable performance is frequently a problem when testing younger children for many of the same reasons that outliers occur. Of course, making sure that the child understands the task must be the first consideration, and maintaining motivation the second. A task made fun and enjoyable is more likely to elicit a consistent performance. This can be done by using cartoon figures, encouragement, and rewards (for some ideas on how to use cartoon figures to improve motivation on many gross motor tasks, see Herkowitz, 1984). Perform frequent reliability checks during testing sessions (for appropriate techniques, see chapter 11).

Statistical Problems

The final developmental research issue to mention is statistical problems. A frequent means of making cross-age comparisons is ANOVA (see chapter 9), which assumes that the groups being compared have equal variances (spread of scores about the mean). However, researchers often violate this assumption in making cross-age comparisons. Depending on the nature of the task, older children may have considerably larger or smaller variances than younger children. Be aware of this potential issue and some of the solutions. In particular, pilot work using the tasks of interest in the research should provide insight into this problem.

Protecting Participants

In chapter 5 we discussed the protection of human participants in research. Of course, this protection also pertains to children. Parents or guardians must grant permission for minors to participate in research. Researchers should obtain this permission as they do for adults, except that they give the explanation and consent forms to the parents or guardians. If minors are old enough to understand the methodology, researchers should also obtain their consent. This means explaining the purpose of the research in terms they can understand. Most public and private schools have their own requirements for approval of research studies.

The normal sequence of events for obtaining permission is as follows:

1. Planning the research
2. Acquiring the approval of your university's committee for the protection of human participants
3. Locating and gaining the approval of the school system, the school involved, and the teachers
4. Obtaining the approval of parents and, when appropriate, children

You can see that this step requires a good deal of paperwork. Thus, beginning the process well before you plan to begin data collection is essential.

Retrospective Developmental Studies

Another approach to developmental studies is to look backward in time rather than forward. Sometimes data have been collected on a group of people over many years, and the researcher can use that data to evaluate development. For example, Thomas and Thomas (1999) discovered that two professional athletes were from the same Iowa town and that the same physical education teachers had taught these two athletes from grades K through 6. They interviewed the physical education teachers about characteristics of these professional athletes as they progressed through the elementary grades.

The Thomas and Thomas (1999) study was qualitative, but collecting quantitative data post hoc is also possible. For example, the Professional Golfers Association (PGA) collects performance data each year on players in all tournaments. Vermilio, Thomas, Thomas, and Morrow (2013) used this data for a study of the performance of the same professional golfers over a 6-year period. Of course, they first had to identify players who competed over this time period, thus excluding young players just starting on the tour and older players who may have dropped off the tour. If you go to the PGA website where we obtained the data used in this study, you can find data on professional golfers over the past several years. Using this sort of data, a researcher could analyze quantitative changes in performance of individual or groups of golfers during the period for which data are available.

CASE STUDIES

In the **case study**, the researcher strives for in-depth understanding of a single situation or phenomenon. This technique is used in many fields, including anthropology, clinical psychology, sociology, medicine, political science, speech pathology, and educational areas such as disciplinary problems and reading difficulties. It has been widely used in the health sciences and to some extent in exercise science, sport science, and physical education.

> **case study**—A form of descriptive research in which a single case is studied in depth to reach a greater understanding about other similar cases.

The case study is a form of descriptive research. Whereas the survey method obtains a rather limited amount of information about many participants, the case study gathers a large amount of information about one or a few participants. Although the study consists of a rigorous, detailed examination of a single case, the underlying assumption is that the case is representative of many other such cases. Consequently, through the in-depth study of a single case, a greater understanding about similar cases is achieved. This is not to say, however, that the purpose of case studies is to make generalizations. On the contrary, drawing inferences about a population from a case study is not justifiable. On the other hand, the findings of a number of case studies may play a part in the inductive reasoning involved in the development of a theory.

The case study is not confined to the study of an individual but can be used in research involving programs, institutions, organizations, political structures, communities, and situations. The case study is used in qualitative research to deal with critical problems of practice and to extend the knowledge base of the various aspects of education, physical education, exercise science, and sport science (qualitative research is discussed in chapter 19).

Information about case study methodology is difficult to find. As Merriam (1988) observed, material on case study research strategies can be found everywhere and nowhere. Methodological material on case study research is scattered about in journal articles, conference proceedings, and research reports of the many fields that use this form of research. Frustration in trying to find substantive material about case study research in an educational setting prompted Merriam (1988) to write her interesting and informative text *Case Study Research in Education*. Yin (2003) presented a comprehensive discussion on design and methods for case study research in the social sciences.

Types of Case Studies

In many ways, case study research is similar to other forms of research. It involves identifying the problem, collecting data, and analyzing and reporting results. As in other research techniques, the approach and the analysis depend on the nature of the research problem. Case studies can be descriptive, interpretive, or evaluative. The case study approach enables a more in-depth, holistic approach to the problem than may be possible with survey studies.

Descriptive Studies

A descriptive case study presents a detailed picture of the phenomenon but does not attempt to test or build theoretical models. Sometimes, descriptive case studies are historical, and sometimes they are done for the purpose of achieving a better understanding of the present status. Descriptive case studies frequently serve as an initial step or database for subsequent comparative research and theory building (Merriam, 1988).

Interpretive Studies

Interpretive case studies also employ description, but the major focus is interpreting the data to classify and conceptualize the information and perhaps theorize about the phenomena. For example, a researcher might use the case study approach to gain a better understanding of the cognitive processes involved in sport.

Evaluative Studies

Evaluative case studies also involve description and interpretation, but the primary purpose is to use the data to evaluate the merit of some practice, program, movement, or event. The efficacy of this type of case study relies on the competence of the researcher to use the available information to make judgments (Guba & Lincoln, 1981).

Case Study Participants

The selection of participants in a case study depends, of course, on the problem being studied. The individual (or case) may be a person (e.g., student, teacher, coach), a program (e.g., Little League baseball), an institution (e.g., a one-room school), a project, or a concept (e.g., mainstreaming). In most studies, random sampling is not used because the purpose of a case study is not to estimate a population value but to select cases from which one can learn the most. Chien (1981) used the term *purposive sampling*. Goetz and LeCompte (1984) referred to this concept as *criterion-based sampling*. The researcher establishes criteria necessary to include in the study and then finds a sample that meets the criteria. Criteria can include age, years of experience, evidence of level of expertise, and situation and environment. The case may be one classroom that meets certain criteria or a state that is involved in a specific program.

Characteristics of the Case Study

The case study involves the collection and analysis of information from many sources. In some respects, the case study has some of the same features found in historical research. Although it consists of the intensive study of a single unit, the ultimate worth of a case study may be that it provides insight and knowledge of a general nature for improved practices. The generalizability of a case study is ultimately related to what the reader is trying to learn from it (Kennedy, 1979). The case study approach is probably most frequently used to understand why something has gone wrong.

Gathering and Analyzing Data

The case study is flexible with regard to the amount and type of data that are gathered and the procedures used in gathering the data. Thus, the steps in the methodology are not distinct or uniform for all case studies.

Data for case studies can be interviews, observations, or documents. It is not uncommon for a case study to employ all three types of data. A case study involving a child, for example, may include interviews with the child and the child's teachers and parents. The researcher may systematically observe the child in the class or in some other setting. Documents could include medical examination reports, physical performance test scores, achievement test results, grades, interest inventories, scholastic aptitude tests, teachers' anecdotal records, and autobiographies. As noted previously, some case studies are mainly descriptive, others focus on interpretation, and others are evaluative. Some case studies propose and test hypotheses, and some attempt to build theories through inductive processes.

According to Yin (2003), the analysis of case study data is one of the least developed and most difficult aspects of conducting case studies. Yin maintained that too often the researcher begins a case study without the slightest notion as to how the data are to be analyzed. Analyzing data in a case study is a formidable task because of the nature of the data and the massive amount of information. According to Merriam (1988), analysis continues and intensifies after the data have been collected. The data must be sorted, categorized, and interpreted. As in any research, the ultimate value of a study rests on the insight, sensitivity, and integrity of the researcher, who is the primary instrument in the collection and analysis of the data in the case study. This attribute is both a strength and a weakness. It is a weakness if the researcher fails to use the appropriate sources of information. The researcher can also be guilty of either oversimplifying the situation or exaggerating the actual state of affairs (Guba & Lincoln, 1981). On the other hand, a competent researcher can use the case study to provide a thorough, holistic account of a complex problem.

Applying Case Study Research in Physical Activity

Several case studies in physical education were directed by H. Harrison Clarke at the University of Oregon (Clarke & Clarke, 1970). The studies dealt with people of low and high fitness levels. Most of the studies sought to discover factors that may contribute to low fitness or lack of strength. A remarkable example of the case study approach and effective follow-up was seen in a demonstration project undertaken by Frederick Rand Roger and Fred E. Palmer (Clarke, 1968). The cause of low fitness in 20 junior high school boys with the lowest physical fitness scores was studied. Information used included somatotype, IQ, academic scholarship, medical history, and status. A follow-up project involved individual attention and special class meetings. Through exercises, improved health habits, medical referrals, and counseling, the students showed vast improvements in fitness, scholarship, and behavior. Two examples of case studies in physical education are Thomas and Thomas (1999), who described the practice patterns of two adult sport experts during their elementary years and Yan and Cardinal (2013), who studied physical activity in Chinese female graduate students.

One of the principal advantages of the case study approach is that it can be fruitful in formulating new ideas and hypotheses about problem areas, especially areas for which there is no clear-cut structure or model. The researcher selects the case study method because of the nature of the research questions being asked. The case study, when used effectively, can play an important role in contributing to knowledge in our field.

> Case studies can help to formulate new ideas and hypotheses, especially for areas that lack established clear-cut structures or models.

JOB ANALYSIS

A **job analysis** can be considered a type of case study. Its purpose is to determine the nature of a particular job and the types of training, preparation, skills, working conditions, and attitudes necessary for success in the job.

The procedures in conducting a job analysis vary. The objective is to obtain as much relevant information as possible about the job and the job requirements. One method is to observe someone in the occupation. Of course, this method is time-consuming and probably bothersome for the person in the job. Nevertheless, the procedure is recommended because the researcher can acquire a kind of vicarious on-the-job experience and gain valuable insight into the whole atmosphere connected with the occupation. A limitation to this method is the lack of sufficient time to observe all facets of the job, particularly seasonal duties and responsibilities.

> **job analysis**—A type of case study that determines the nature of a particular job and the types of training, preparation, skills, working conditions, and attitudes necessary for success in the job.

Questionnaires and interviews are two effective job analysis techniques. They allow people to respond to questions about the kinds of duties they perform, the types and degree of preparation required or recommended for accomplishing their tasks, and the perceived advantages and disadvantages of their jobs.

Systematic job analysis techniques can result in highly specific and quantitative descriptions of the job. For example, a job description for an administrative position reported (a) a percentage breakdown of principal accountabilities; (b) the number of hours per day spent sitting, standing, and walking; (c) the physical effort requirements for the number of pounds lifted waist high and carried alone and the distance carried; (d) the sensory requirements for vision, hearing, and speaking; (e) mental effort requirements; (f) the number of hours spent under time pressure and hours spent working rapidly; and (g) the percentage of time spent indoors at a desk and in an office. Other requirements and conditions included the minimum academic degree required, the necessity of working weekends and evenings, and the ability to use a computer.

The limitations of the job analysis as a research technique include the fallibility of both participants' memory and self-reports. Participants may tend to accent either the positive or the negative aspects of the job, depending on the time, circumstances, and subject. Also, there is a danger that the approach can be too mechanical, thus neglecting some of the more abstract aspects of the job and its requirements.

OBSERVATIONAL RESEARCH

Observation is used in a variety of research endeavors. It provides a means of collecting data and is a descriptive method of researching certain problems. In the questionnaire and interview techniques, the researcher relies on self-reports about how the participant behaves or what the participant believes. A weakness of self-reports is that people may not be candid about what they really do or feel and may give what they perceive to be socially desirable responses. An alternative descriptive research technique is for the researcher to observe people's behavior and qualitatively or quantitatively analyze the observations. Some researchers claim that this technique yields more accurate data. There are, of course, several limitations to observational research.

Basic considerations in observational research include the behaviors that will be observed, who will be observed, where the observations will be conducted, and how many observations will be made. Many other considerations are connected to these basic ones. Depending on the problem and the setting, each investigation has unique procedures. Therefore, only the basic considerations can be discussed in rather general terms.

What Behaviors Will Be Observed?

Deciding what behaviors will be observed relates to the statement of the problem and to the operational definitions. For example, a study on teacher effectiveness must have clearly defined observational measures of teacher effectiveness. Definite behaviors must be observed, such as the extent to which the teacher asks students questions. Some other aspects of teacher effectiveness include giving individual attention, demonstrating skills, dressing appropriately for activities, and starting class on time. In determining the behaviors to observe, the researcher must also limit the scope of the observations to make the study manageable.

Who Will Be Observed?

As with any study, the population from which samples will be drawn must be determined. Will the study focus only on elementary school teachers? Which grades? Will the study include only physical education specialists, or will it also include classroom teachers who teach physical education? Another question concerns the number of participants who will be observed. Will the study include observations of students in addition to teachers? In other words, the researcher must describe the participants of the study precisely.

Where Will the Observations Be Conducted?

The setting for the observations must be considered, in addition to the basic considerations of the size of the sample and the geographical area. Will the setting be unnatural or natural? Using an unnatural setting means bringing the participant to a laboratory, room, or other locale for observation.

An unnatural setting offers some advantages in terms of control and freedom from distractions. For example, a one-way mirror is advantageous for observation because it removes the influence of the observer on the behavior of the participant. That the participants' behavior is affected by the presence of an observer is also shown in classroom situations. When the observer first arrives, the students (and perhaps the teacher) are curious about the observer's presence. Consequently, they may behave differently than they would if the observer were not there. The teacher may also act differently, possibly by perceiving the observer as a threat or being aware of the purpose of the observations. In any case, the researcher should not make observations on the initial visit. Allowing the students to become gradually accustomed to the observer's presence is best.

Whether a participant will be observed alone or in a group is also related to the setting. In a natural setting, such as the playground or classroom, the child may behave more typically, but additional extraneous influences on behavior are likely to be present.

How Many Observations Will Be Made?

Many factors determine how many observations to make. First are the operationally defined behaviors in question and the time constraints of the study itself. For example, if you are studying the amount of activity or participation of students in a physical education class, you

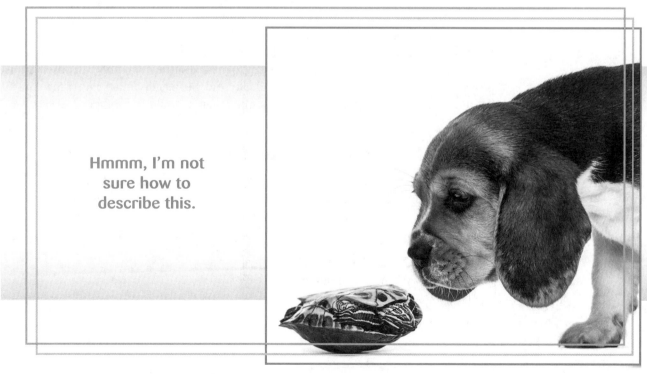

Hmmm, I'm not sure how to describe this.

must consider several things. In the planning phase of the study, you must decide the type of activity unit and the number of units encompassed in the study. The number of observations for each activity depends on the particular stage of learning in the unit (e.g., introductory stage, practice stage, playing stage). This must be specified in the operational definitions, of course, but both the length of the unit and the subsequent length of each stage within each unit play major roles in determining the number of observations that are feasible.

Another factor to consider is the number of observers. If only one person is doing the observations, either the number of persons being observed or the number of observations per person (or both) are restricted. Attempting to generalize typical behavior from observing a few individuals on a few occasions is hazardous.

Some types of behavior may not be manifested frequently. Fair play, aggression, leadership, and other traits (as operationally defined) are not readily observable because of the lack of opportunity to display such traits (among other things). The occasion must present itself, and the elements of the situation must materialize so that the participant has the opportunity to react. Consequently, the number of observations is bound to be extremely limited if left to chance occurrence. On the other hand, situations that are contrived to provoke certain behavior are often unsuccessful because of their artificiality.

We cannot say how many observations are necessary, but can only warn against too few; thus, we recommend a combination of feasibility and measurement considerations. This question is readdressed in the discussion on scoring and evaluating observations.

When Will the Observations Be Made?

You can easily understand that all the basic considerations being discussed are related and overlap. The determination of when to make the observations includes decisions about the time of day, day of the week, phase in the learning experience, season, and other time factors.

In our previous example of observing the amount of student activity in a physical education course, different results would be expected if the observations were made at the beginning of the unit than if they were made at the end. Allowing the students to become accustomed to the situation so that the observational procedures do not interfere with normal activity is another consideration. In observing student teachers, for example, differences would certainly be expected if some were observed at the beginning of their student-teaching experience and others were observed at the end.

Graduate students encounter major problems in observational research with regard to time. They find it difficult to spend the time necessary to make a sufficient number of observations to provide reliable results. Furthermore, graduate students usually must gather the data by themselves, making it much more time-consuming than it would be if other observers were available.

How Will the Observations Be Scored and Evaluated?

Researchers employ a number of techniques for recording observational data. The use of computers and other computer-assisted event-recording methods have alleviated many of the technical problems that plagued observational research in the past. Following are some of the more commonly used procedures for recording observational data:

- Narrative, or continual recording
- Tallying, or frequency counting
- Interval method
- Duration method

In the **narrative**, or *continual-recording*, **method**, researchers record in a series of sentences the occurrences that they observe as they happen. This method of recording is the slowest and least efficient. The observer must be able to select the most important information to record because not everything that occurs in a given situation can be recorded. Probably the best use of this technique is in helping to develop more efficient recording instruments. The researcher first uses the continuous method and then develops categories for future recording from the narrative.

The **tallying**, or *frequency-counting*, **method** involves recording each occurrence of a certain behavior. The researcher must clearly define the behavior and make the frequency counts within a certain period, such as the number of occurrences in a 10 min or 30 min session.

The **interval method** is used when the researcher wishes to record whether the behavior in question occurs in a certain interval of time. This method is useful when counting individual occurrences is difficult. One of the leading standardized systems for interval recording is the Flanders' Interaction Analysis System (Flanders, 1970), in which the observer records the behavior of the student according to 10 specific behavior classifications within each time interval. The Academic Learning Time in Physical Education (ALT-PE) is an observational instrument developed by Siedentop and graduate students at Ohio State University for use in physical education (Siedentop, Birdwell, & Metzler, 1979; Siedentop, Trousignant, & Parker, 1982). The instrument entails time sampling in which a child is observed for a specified period, during which the child's activities are coded. The Cheffers' Adaptation of the Flanders' Interaction Analysis System (CAFIAS) was developed by Cheffers (1973) to allow the systematic observation of physical education classes and classroom situations. The CAFIAS provides a device for coding behavior through a double-category system so that any behavior can be categorized as verbal, nonverbal, or both. The CAFIAS permits the coding of the class as a whole when the entire class is functioning as one unit, of smaller groups when the class is divided, or of individual students when the students are working individually or independently with no teacher influence. Rink and Werner (1989) developed a scoring instrument for teacher observation termed the Qualitative Measures of Teaching Performance Scale (QMTPS).

The **duration method** involves timed behavior. The researcher uses a stopwatch or other timing device to record how much time a person spends engaged in a particular behavior. A number of studies have used this method in observing student time on task and off task. In the previous example about the amount of activity in a physical education class, a researcher could simply record the amount of time a student spends in actual participation or the amount of time spent standing in line or waiting to perform. The researcher usually observes a student for a given unit of time (e.g., a class period), starting and stopping a stopwatch as the behavior starts and stops to record cumulative time on task (or off task).

Using Video for Observation

Video is a potentially valuable instrument for observational research. Its greatest advantage is that the researcher need not worry about recording observations when the behavior is occurring. Furthermore, it allows the researcher to observe a number of people at one time. For example, teachers and students can be observed simultaneously, which is difficult in normal observational techniques. In addition, the researcher can replay a video as needed for evaluating the behavior and retain a permanent record.

The use of video does have a disadvantage. The presence of a video-recording device may also alter behavior to the extent that the participants do not behave normally. But if this disadvantage can be resolved, video can be effective for observational research.

narrative method—A method of recording in observational research in which researchers describe their observations as they occur; also called *continual-recording method.*

tallying method—A method of recording in observational research in which researchers record each occurrence of a clearly defined behavior within a certain period; also called *frequency-counting method.*

interval method—A method of recording in observational research that is used when counting individual occurrences is difficult; the researcher records whether the behavior in question occurs in a certain interval of time.

duration method—A method of recording in observational research in which the researcher uses a stopwatch or other timing device to record how much time a participant spends engaged in a particular behavior.

Weaknesses of Observational Research

Problems and limitations of observational research include the following:

■ A primary danger in observational research lies in the operational definitions of the study. The behaviors must be carefully defined and observable. Consequently, the actions may be so restricted that they do not depict the critical behavior. For example, teacher effectiveness encompasses many behaviors, and to observe only the number of times the teacher asked questions or gave individual attention may be an inadequate sample of effectiveness.

■ Using observation forms effectively requires much practice. Inadequate training therefore represents a major pitfall in this form of research. Also, trying to observe many things can be difficult. Often, the observation form is too ambitious for one person to use.

■ Certain behaviors cannot be evaluated as finely as some observation forms dictate. A common mistake is to ask the observer to make discriminations that are too precise, thus reducing the reliability of the ratings.

■ The presence of the observer usually affects the behavior of the participants. The researcher must be aware of this likelihood and try to reduce the disturbance.

■ Generally, having more than one observer greatly expedites observational research. Failure to use more than one observer results in decreased efficiency and objectivity.

UNOBTRUSIVE RESEARCH TECHNIQUES

unobtrusive measures—Measures of behavior taken on people who are not aware that the researcher is gathering data.

Many methods of gathering information about people are available other than questionnaires, case studies, and direct observation. Webb, Campbell, Schwartz, and Sechrest (1966) discussed a variety of approaches that they termed **unobtrusive measures**. Examples they mentioned include the replacement rate for floor tiles around museum exhibits as a measure of the relative popularity of exhibits, and the shrinking diameter of a circle of seated children as a measure of the degree of fear caused by telling ghost stories. Researchers have measured boredom by the amount of fidgeting movements in an audience. The rate of library withdrawals of fiction and nonfiction books has been studied to determine the influence of television in communities. Researchers have also noted that children demonstrate their interest in Christmas by the size of their drawings of Santa Claus and the amount of distortion in the figures.

In some of the preceding examples, the experimenter was not present when the data were being produced. In other situations, the researcher is present but acts in a nonreactive manner (i.e., the subjects are not aware that the researcher is gathering data). For example, a researcher in psychology measured the degree of acceptance of strangers among delinquent boys by measuring the distance maintained between a delinquent boy and a new boy to whom the researcher introduced the delinquent participant. Sometimes, the researcher intervenes to speed up the action or force the data but in a manner that does not attract attention to the method. In studying the cathartic effect of activity on aggression, Ryan (1970) had an accomplice behave in an obnoxious manner and then measured the amount of electric shock the participants administered to that accomplice and to innocent bystanders. Other researchers have intervened by causing participants to fail or succeed so that they could observe responses to winning and losing in competitive situations.

Levine (1990) described an interesting study on the pace of life using unobtrusive measures involving people in various cities and countries. Among the measurements used were (a) the walking speed of randomly chosen pedestrians over a distance of 100 ft (30 m), (b) the accuracy of outdoor bank clocks in a downtown area, and (c) the speed with which postal clerks fulfilled a standard request for stamps.

Levine (1990) did a similar study on the pace of U.S. cities. Four unobtrusive measures were used: (a) walking speed for 60 ft (18 m); (b) the time required for bank clerks to complete a simple request; (c) the talking speed of postal clerks explaining the differences among regular, certified, and insured mail; and (d) the proportion of people wearing wristwatches.

Unobtrusive measures also include records such as birth certificates, political and judicial records, actuarial records, magazines, newspapers, archives, and inscriptions on tombstones. An interesting use of city records (Webb et al., 1966) was the analysis of city water pressure as an index of television viewing interest. Immediately after a television show, the water pressure dropped as drinks were obtained and toilets flushed. Mabley (1963) presented data on Chicago's water pressure on the day of an exciting Rose Bowl game that showed a drastic drop in pressure just after the game ended.

The ethical issue of invasion of privacy arises in some forms of unobtrusive measures. Informed-consent compliance has placed considerable restraints on certain research practices, such as those that involve entrapment and experiments that aim to induce heightened anxiety.

> Unobtrusive measures can raise ethical concerns regarding privacy and informed consent.

CORRELATIONAL RESEARCH

Correlational research is descriptive in that it explores relationships that exist among variables. Sometimes predictions are made on the basis of the relationships, but correlation cannot determine cause and effect. The basic difference between experimental research and correlational research is that the latter does not cause something to happen. Correlational research involves no manipulation of variables or no administration of experimental treatments. The basic design of correlational research is to collect data on two or more variables on the same people and determine the relationships among the variables. But, of course, the researcher should have a sound rationale for exploring the relationships.

> **correlational research—** Research that explores relationships among variables and that sometimes involves the prediction of a criterion variable.

Correlational techniques were discussed in chapter 8 through examples of situations that lend themselves to correlational research. The two main purposes for doing a correlational study are to analyze the relationships among variables and to predict.

Steps in Correlational Research

The steps in a correlational study are similar to those used in other research methods. The problem is first defined and delimited. The selection of the variables to be correlated is of critical importance. Many studies have failed in this regard. Regardless of how sophisticated the statistical analysis may be, the statistical technique can deal only with the variables that are entered; as the saying goes, Garbage in, garbage out. The validity of a study that seeks to identify basic components or factors of fitness hinges on the identification of the variables to be analyzed. A researcher who wishes to discriminate between starters and substitutes in a sport is faced with the crucial task of determining which physiological or psychological variables are the important determinants of success. The researcher must lean heavily on past research in defining and delimiting the problem.

Participants are selected from the pertinent population using recommended sampling procedures. The magnitude and even the direction of a correlation coefficient can vary greatly, depending on the sample used. Remember that correlations show only the degree of relationship between variables, not the cause of the relationship. Consequently, because of other contributing factors, a researcher might obtain a correlation of .90 between two variables in a sample of young children but a correlation of .10 between the same variables in adults, or vice versa. Chapter 8 presented examples of how factors such as age can influence certain relationships.

Another aspect of correlation is that the size of the correlation coefficient depends to a considerable extent on the spread of the scores. A sample that is fairly homogeneous in certain traits seldom yields a high correlation between variables associated with those traits. For example, the correlation between a distance run and maximal oxygen consumption with a sample of elite track athletes is almost invariably low because the athletes are so similar; the variability is insufficient to permit a high correlation because the scores on the two measures are too uniform. If some less-trained runners were included in the sample, the size of the correlation coefficient would increase dramatically.

In prediction studies, the sample must be representative of the population to whom the study is directed. One of the major drawbacks to prediction studies is that the prediction formulas are often sample specific, which means that the accuracy of a formula is greatest (or maybe only acceptable) when it is applied to the particular sample on whom it was developed. Chapter 11 discussed this shrinkage phenomenon as well as cross-validation, which is used to counteract shrinkage.

The collection of data requires the same careful attention to detail and standardization that all research designs do. A variety of methods for collecting data can be used, such as physical performance tests, anthropometric measurements, pencil-and-paper inventories, questionnaires, and observational techniques. The scores must be quantified, however, to be correlated.

Analysis of the data can be performed by a number of statistical techniques. Sometimes the researcher wishes to use simple correlation or multiple correlation to study how variables, either by themselves or in a linear composite of variables, are associated with some criterion performance or behavior. Factor analysis is a data-reduction method that helps determine whether relationships among a number of variables can be reduced to small combinations of factors or common components. Structural modeling is a technique used to test some theoretical model about causal relationships among three or more variables. (Simple correlation, factor analysis, and structural modeling are discussed in chapter 8.)

Prediction studies usually employ multiple regression because the accuracy of predicting some criterion behavior is nearly always improved by using more than one predictor variable. Discriminant analysis is a technique used to predict group membership, and canonical correlation is a method for predicting a combination of several criterion variables from several predictor variables. (Multiple regression and canonical correlation are discussed in chapter 8, and discriminant analysis in chapter 9.)

Limitations of Correlational Research

> The easier a criterion is to define operationally and measure reliably, the more effective correlational research will be.

Limitations of correlational research include those of both planning and analysis. We have already pointed out the importance of identifying pertinent variables and selecting proper tests to measure those variables. Hypotheses should be based on previous research and theoretical considerations; simply correlating a set of measurements to see what happens is not an effective way to approach research. Selecting an inadequate measure as a criterion in a prediction study is a common weakness in correlational research. For example, a criterion of success in some endeavor is often difficult to define operationally and may be more difficult to measure reliably.

SUMMARY

Descriptive research encompasses many techniques. In this chapter, we described the basic procedures and strengths and weaknesses of six techniques in the category of descriptive research.

Developmental research seeks to study growth measures and changes in behavior over a period of years. In a longitudinal design, the same people are followed over time. When participants of different ages are sampled, the design is cross-sectional. Of paramount importance in developmental research is whether the participants represent their populations and their performances. A developmental study may involve experimental treatment in which the researcher attempts to determine the interaction of some treatment with age. Developmental studies may also be done in a post hoc manner in which the researcher evaluates data that have been collected over a number of years.

In a case study, the researcher attempts to gather a lot of information about one or a few cases. An in-depth study of a single case can result in a greater understanding of other similar cases. Case studies can be descriptive, interpretive, or evaluative. One of the principal advantages of the case study approach is that it can lead to the formulation of ideas and hypotheses about problem areas.

The job analysis is a type of case study in which extensive information is gathered and evaluated about job conditions, job duties, and the types of training, preparation, skills, and attitudes that are necessary for success in certain working conditions.

Observational research is a descriptive technique that involves the qualitative and quantitative analysis of observed behaviors. Unlike the survey method, which relies on self-reports about how a person behaves, observational research attempts to study what a person actually does. Behavior is usually coded by what occurs and when, how often, and how long. Standardized observation instruments such as the ALT-PE and the CAFIAS are frequently used. Researchers often video-record their participants to store the observations for later analysis.

Some studies use unobtrusive measures in which the participants are unaware that the researcher is gathering data. For example, instead of (or in addition to) asking a person how much he smokes, the researcher might count the cigarette butts in an ashtray after a certain period. Although the methods can be interesting and innovative, they may be constrained by ethical considerations such as invasion of privacy and entrapment.

Correlational research examines relationships among variables. Sometimes the relationships are used for prediction. Although correlations are often used with experimental research, the study of relationships is descriptive in that it does not involve the manipulation of variables. A major pitfall in correlational research is assuming that because variables are related, one causes another.

Check Your Understanding

1. Locate a cross-sectional study and a longitudinal study in the literature, and answer the following questions about each:

   a. Is the study descriptive or experimental?

   b. What age levels are studied?

   c. What are the independent and the dependent variables?

   d. What statistics are used to make cross-age comparisons?

2. Write an abstract of a case study found in the literature. Indicate the problem, the sources of information used, and the findings.

3. Find an observational study in the literature and write a critique of the article, concentrating on the methodology.

4. Write a brief abstract of a correlational research study.

17

Physical Activity Epidemiology Research

Barbara E. Ainsworth
Arizona State University

Charles E. Matthews
National Institutes of Health

> The older you get, the tougher it is to lose weight because by then your body and your fat have gotten to be really good friends.
>
> —Thinkexist.com

The emergence of the mid-20th-century heart disease epidemic fueled a number of large-scale observational epidemiological studies aimed at identifying the determinants of heart disease so that preventive measures could be undertaken to improve the public's health. A series of observational studies was initiated between the late 1940s and early 1960s. A number of them were particularly important to the development of the field of physical activity epidemiology because they were the first to develop methods for measuring physical activity and to conduct systematic studies of the link between physical activity and a life-threatening disease. These studies, to name a few (and the important people associated with each), include the Framingham Heart Study, London busmen/British civil servants study (Jeremy Morris), Tecumseh Community Health Study (Henry Montoye), the College Alumni Health Study (Ralph Paffenbarger), and early studies from Minnesota (Henry Taylor).

By the early 1980s, researchers had begun to amass solid evidence demonstrating that low levels of physical activity were associated with increased risk of heart disease and overall mortality. Because heart disease is and was the leading cause of death in the United States and the countries of the European Union, physical activity emerged as an important public health concern. Accordingly, the U.S. Public Health Service initiated surveillance programs to quantify the leisure-time physical activity patterns in the U.S. population in the mid-1980s. These programs demonstrated that more than 60% of U.S. adults were not physically active on a regular basis. In the 1990s, epidemiological evidence continued to accumulate, showing that low levels of physical activity were associated with increased risk for a number of important health conditions, including all-cause and cause-specific mortality, cardiovascular disease, osteoporosis, some forms of cancer, and mental health and quality of life problems (U.S. Department of Health and Human Services, 1996). Major initiatives are currently underway to understand the best way to change the behavior of the population at both the individual and community level.

One such initiative is Healthy People 2020, a plan developed by the U.S. Department of Health and Human Services and updated every 10 years. (U.S. Department of Health and Human Services, 2013). The physical activity objectives for 2020 focus on reducing inactivity and increasing physical activity to levels recommended in the U.S. Physical Activity Guidelines, increasing physical activity promoting structural environments, and developing policies to provide sustainable opportunities for physical activity.

The 11 objectives established for Healthy People 2020 that have ongoing data collection systems with which to evaluate progress are as follows:

1. Reduce the proportion of adults who engage in no leisure-time activity.

2. Increase the proportion of adults who meet federal physical activity guidelines for aerobic activity and for muscle-strengthening activity.

3. Increase the proportion of adolescents who meet current federal physical activity guidelines for aerobic physical activity and for muscle-strengthening activity.

4. Increase the proportion of the nation's public and private schools that require daily physical education for all students.

5. Increase the proportion of adolescents who participate in daily school physical education.

6. Increase regularly scheduled elementary school recess in the United States.

7. Increase the proportion of school districts that require regularly scheduled elementary school recess.

8. Increase the proportion of children and adolescents who do not exceed recommended limits for screen time.

9. Increase the number of states with licensing regulations for physical activity provided in child care.

10. Increase the proportion of the nation's public and private schools that provide access to their physical activity spaces and facilities for all persons outside of normal school hours (that is, before and after the school day, on weekends, and during summer and other vacations).

11. Increase the proportion of physician office visits that include counseling or education related to physical activity.

The following four developmental objectives do not have ongoing data collection systems but are deemed important to increase physical activity and reduce sedentary behaviors:

1. Increase the proportion of employed adults who have access to and participate in employer-based exercise facilities and exercise programs.

2. Increase the proportion of trips made by walking.

3. Increase the proportion of trips made by bicycling.

4. Increase legislative policies for the built environment that enhance access to and availability of physical activity opportunities.

The 2020 objectives for aerobic activity reflect levels recommended in the 2008 physical activity guidelines for Americans (U.S. Department of Health and Human Services, 2008). Specifically, adults ages 18 and older should accumulate at least 150 min per week of moderate physical activity, or 75 min per week of vigorous physical activity, or an equivalent combination. Adults should also engage in 2 days of muscle-strengthening activities per week. Figure 17.1 shows the prevalence of Americans meeting the physical activity guidelines for aerobic activity in 2010 by sex, race/ethnicity, education, and activity limitation. To reach the 2020 target of 47.9% of adults meeting the guidelines (a 10% increase from 2010 levels), targeted strategies are needed for helping specific groups increase their activity levels.

This minihistory of physical activity epidemiology mirrors the major goals of epidemiological inquiry that will be described in the remainder of this chapter. Epidemiological methods are used to provide the scientific backbone for public health endeavors, including (a) quantifying the magnitude of health problems, (b) identifying factors that cause disease (i.e., **risk factors**), (c) providing quantitative guidance for the allocation of public health resources, and (d) monitoring the effectiveness of prevention strategies using population-wide surveillance programs (Caspersen, 1989). In this chapter, we define epidemiology, describe methods of measuring physical activity in epidemiological studies, describe epidemiological study designs, consider threats to the validity to these designs, and provide an outline for reading and interpreting an epidemiological study.

risk factor—An exposure that has been found to be a determinant of a disease outcome or health behavior.

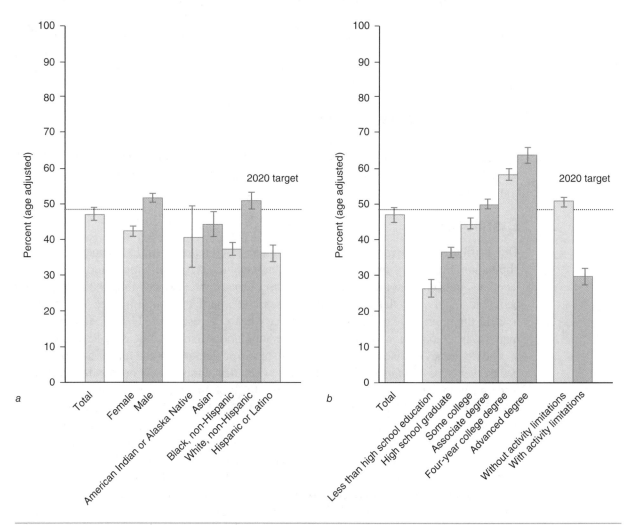

Figure 17.1   Percentage of adults who meet physical activity objective 2.0 for moderate and vigorous activity in 2010, and 2020 targets for *(a)* all adults combined and for adults by sex and race and ethnicity, and *(b)* by education and activity limitations. I = 95% confidence interval.

National Health Interview Survey, CDC/NCHS.

OBSERVATIONAL VERSUS EXPERIMENTAL RESEARCH

A key to understanding epidemiological methods lies in considering the difference between observational and experimental research. It has been said that an epidemiologist is most happy when conditions conspire to produce circumstances much like those of a true experiment (Rothman, 1986).

Observational research uses existing differences in factors that may cause disease within a population, such as physical activity, dietary habits, or smoking. Some people in a population choose to be physically active, whereas others do not. Epidemiologists use these naturally occurring differences in a population to observe, and therefore understand, the effect of these differences on specific disease **outcomes**.

A primary reason epidemiological research has become such an important discipline is that it is virtually the only way to obtain a quantitative understanding of the health risks of many behaviors. This circumstance arises from the fact that it would be unethical to conduct true experimental research on health behaviors such as physical inactivity. For example, researchers could not randomize a group of people to either a completely sedentary lifestyle or an active lifestyle, wait 5 to 10 years, and then see how much death and heart disease resulted from physical inactivity. Although an experimental study such as this would unequivocally demonstrate that physical inactivity is a causal factor in the development of heart disease, the ethical problems of the experiment are obvious.

> Observational research considers how existing differences (e.g., in physical activity, diets) affect disease outcomes.

> **outcome**—The dependent variable in an analysis.

WHAT IS EPIDEMIOLOGY?

Epidemiology has been defined as "the study of the distribution and determinants of health related states or events in specified populations, and the application of this study to the control of health problems" (Last, 1988, p. 141).

Distribution

The distribution of disease relates to the *frequency* and *patterns* of disease occurrence in a population. Frequency, or how often the disease occurs, is typically measured as the prevalence, incidence, or mortality rate of a disease. Disease prevalence refers to the number of people in a given population that have a disease at a particular point in time. For example, in 2011 the prevalence of adults with coronary heart disease in the United States was estimated to be 26.5 million, or about 9% of the population (Schiller, Lucas, & Peregoy, 2012).

The frequency of disease occurrence may also be calculated as the rate of new disease or health events, as *incidence* or *mortality rates* (i.e., new cases or new deaths from a specific disease within a specified period). For example, the mortality rates for coronary heart disease were 193.6 per 100,000 adults in the United States in 2010 (Murphy, Xu, & Kochanek, 2013).

A presentation of the frequency of disease occurrence relative to the number in the population of interest enables a comparison of prevalence, incidence, or mortality rates across populations. For example, the mortality rates clearly indicate that cardiovascular disease mortality is about 50% higher for men than it is for women in the United States.

Evaluating the basic *patterns* of disease occurrence within a specified population is often useful for developing hypotheses about risk factors for the disease. Patterns of disease occurrence refers to characteristics related to *person, place,* and *time.* Personal characteristics include demographic factors such as age, sex, and socioeconomic status. Characteristics of place include geographic differences, urban–rural variation, and, particularly important in the history of physical activity epidemiology, differences in occupation. Historically, differences in occupational classifications enabled researchers to make crude comparisons

---

COMPONENTS OF EPIDEMIOLOGICAL RESEARCH

Distribution
Frequency—prevalence, incidence, mortality rate
Patterns—person, place, time
Determinants
Defined characteristics—associated with change in health
Application
Translation—knowledge to practice

---

of occupational physical activity levels and were used in some of the first epidemiological studies of the relationship between physical activity and heart disease.

Time of disease occurrence refers to annual, seasonal, or daily patterns of occurrence. Quantification of temporal changes in disease rates often lead to hypotheses that generate a more detailed examination of the factors that caused this change. A good example of this was the observation that the incidence of upper-respiratory infections (i.e., common colds) in runners increased in the 14-day period after running an ultramarathon (56K) (Peters & Bateman, 1983). Of the 141 marathoners studied, 47 (33%) came down with a cold after the race, whereas only 19 (15%) of the 124 people who did not run the marathon had a cold during the same period. Thus, the incidence of colds was more than twice as high among the runners, apparently because of the **exposure** of the marathon event. This observation led to intensive research in the area of exercise immunology and ultimately resulted in a greater understanding of the relationship between exercise and the immune system (Nieman, 1994).

exposure—Factors (variables) in epidemiological studies that are tested for their relationship with the outcome of interest.

determinant—A factor that changes a characteristic.

Determinants

A **determinant** is "any factor, whether event, characteristic, or other definable entity, that brings about change in a health condition, or other defined characteristic" (Last, 1988, p. 500). In physical activity research, the goals are usually to test the hypothesis that activity is or is not a determinant for a particular disease outcome, or to identify the determinant of physical activity behaviors. Determinants of disease are often called risk factors because they increase a person's risk for disease. Epidemiological studies have been instrumental in identifying risk factors for heart disease, including obesity, high blood pressure, high LDL cholesterol, low HDL cholesterol, and physical inactivity. The identification and surveillance of specific determinants of a particular disease allow for targeted health promotion campaigns that present this new health knowledge to the public.

Application

The application of the established understanding of the causal factors related to disease is a major goal of public health. Thus, after epidemiologists have identified the cause of disease, health educators interact with communities to make them healthier places to live.

Other names for the application of research in a community setting are *translation* and *dissemination*. Successful public health dissemination strategies are those that engage the community using a variety of methods, from motivating people to change their behaviors to affecting public policy. This is referred to as the ecological model for health promotion

(McLeroy, Bibeau, Steckler, & Glanz, 1988). Public health disseminators translate knowledge from epidemiological studies to help increase physical activity among individuals and within social groups and community organizations. Strategies are also used to affect the community environment (e.g., by building walking trails) and encourage policy makers to enact legislation or appropriate funds to enable people to lead physically active lives.

PHYSICAL ACTIVITY MEASUREMENT DEFINITIONS

Because physical activity can be described in many ways, definitions are necessary to increase measurement consistency and reduce variability across studies. *Physical activity* is a global term that is defined as all movement produced by the contraction of skeletal muscle that substantially increases energy expenditure. In this context, physical activity may occur on the job, in the home, during leisure and recreational periods, and in getting from place to place (Caspersen, 1989). Physical activity includes all forms of movement done in occupational, exercise, home and family care, transportation, and leisure settings. *Exercise* is planned, structured, and repetitive bodily movement done to improve or maintain one or more components of physical fitness (Caspersen, 1989). *Physical fitness* is a multidimensional concept associated with a set of attributes that people have or achieve that relates to the ability to perform physical activity. Types of physical fitness are cardiorespiratory, muscular, flexibility, metabolic, and motor fitness. *Leisure* is a concept that includes the elements of free choice, freedom from constraints, intrinsic motivation, enjoyment, relaxation, personal involvement, and opportunity for self-expression (Henderson, Bialeschki, Shaw, & Freysinger, 1996). *Volitional* physical activity refers to activities done for a purpose in either structured or unstructured settings. *Spontaneous* physical activity refers to brief periods of movement that results in energy expenditure through fidgeting, gesticulation, or unintentional short accumulated periods of movement.

When physical activity is measured in epidemiological studies, the frequency, duration, and intensity of volitional physical activities performed at some period in the past are studied in relation to their effect on health.

- *Frequency* relates to the number of times per week or year that one is physically active. For example, in an updated 1995 consensus report from the U.S. Centers for Disease Control and Prevention (CDC) and the American College of Sports Medicine (ACSM), physical activity obtained on five days of the week, preferably all days, was recommended to reduce some risks for chronic disease and enhance health-related quality of life (Haskell et al., 2007). According to the ACSM, the minimum frequency of vigorous exercise needed to increase cardiorespiratory fitness is 3 days per week (American College of Sports Medicine, 1994).

- *Duration* refers to the time spent in a specific activity as hours or minutes per session. In the Multiple Risk Factor Intervention Trial (MRFIT) of 12,138 men at high risk for cardiovascular disease, analyses were performed to determine the relationship between physical activity and mortality (Leon, Connett, Jacobs, & Rauramaa, 1987). Physical activity was measured in minutes per day and categorized into three periods described as short (15 min/day), middle (47 min/day), and long (133 min/day) duration. Results showed lower mortality rates among men who performed leisure-time activity of middle and long duration compared with men in the short-duration group. Based on this study and other epidemiological studies reviewed by Pate and colleagues (1995), at least 30 min per day is a recommended duration of activity to reduce the risks for premature morbidity and mortality.

- *Intensity* refers to the difficulty of an activity and is generally classified as light, moderate, or vigorous in epidemiological settings. Intensity may be expressed in *absolute* or *relative* terms. The recommended unit for intensity in absolute terms is the metabolic equivalent (MET), which is defined as the ratio of the activity's metabolic rate to the resting metabolic rate. One MET is approximately equal to 3.5 ml · kg^{-1} · min^{-1} of oxygen consumption or about 1 kcal· kg^{-1}· hr^{-1} of energy expenditure for a 60 kg (132 lb) person. To provide consistency in assigning intensity levels to activities, the Compendium of Physical Activities, which provides MET intensities for over 500 activities, was developed (Ainsworth et al., 2011). Expressing physical activity intensity in relative terms allows for the adjustment of difficulty based on individual differences. Relative intensity measures include the percentage of maximal oxygen uptake (% $\dot{V}O_2$max) or maximal heart rate (% max heart rate) and rating of perceived exertion (RPE). A difficulty in using relative intensities in epidemiological studies is that comparison of intensity levels for similar activities is unequal between studies. Thus, it is preferable to express physical activity intensity in relative terms when measuring activity for a given person, such as in prescribing exercise programs on an individual basis.

■ *Dose* refers to the combination of the frequency, intensity, and duration of physical activity and is expressed as kcal/day, MET-hours/day, minutes on a treadmill graded exercise test, or other units. In the Aerobics Center Longitudinal Study, Blair and colleagues (1989) studied the association between the time to exhaustion on a treadmill graded exercise test and all-cause mortality. They showed a strong inverse dose–response relation between maximal cardiorespiratory fitness levels (represented as maximal MET level) and all-cause mortality. The greatest reduction in mortality was seen between men in the lowest and next-to-lowest levels of fitness.

PHYSICAL ACTIVITY RECOMMENDATIONS

2008 U.S. Physical Activity Guidelines
Avoid inactivity

Frequency = activities spread throughout the week

Moderate intensity and duration = 150 to 300 min (5 hours) a week and strength training 2 days a week

Vigorous intensity and duration = 75 to 150 min a week and strength training 2 days a week

Energy expenditure = 500 to 1,000 metabolic equivalent (MET) min a week

CDC-ACSM Moderate Recommendation
Frequency = 5 or more days a week

Duration = 30 or more min a day

Intensity = 3.0-5.9 METs

ACSM Vigorous Recommendation
Frequency = 3 or more days a week

Duration = 20 to 45 min per event

Intensity = 60 to 90% maximal heart rate

ASSESSMENT OF PHYSICAL ACTIVITY

Physical activity can be measured using a variety of methods ranging from direct measurement of the amount of heat a body produces during activity to asking people to rate how active they recall being during the past week or year. From the 1950s to the 1980s, job titles were used to classify physical activity patterns in epidemiological studies involving occupational physical activity. But with the changing profile of the labor market, occupational titles no longer reflect the physical requirements of a job, eliminating the use of job titles to classify occupational energy expenditure (Montoye, Kemper, Saris, & Washburn, 1996).

Because of the exceptionally large number of people in many epidemiological studies, self-administered questionnaires or brief interviews are often used to capture activity spent at work, in exercise, at home, in transportation, and in leisure settings. Questionnaires are classified as global, short recall, and quantitative histories depending on the length and complexity of the items.

- *Global* questionnaires are instruments of 1 to 4 items that present a general classification of a person's habitual activity patterns. Global surveys are most accurate for classifying people according to their level of vigorous-intensity physical activity. Because these surveys may take little time to administer (less than 2 min), they are preferred for use in epidemiological studies.

- *Short recall* questionnaires generally have 5 to 15 items and reflect recent physical activity patterns (during the past week or month). They are effective for classifying people into categories of activity, such as inactive, insufficiently active, or regularly active based on health guidelines and recommendations for minimal levels of activity. Short recall questionnaires may take from 5 to 15 minutes to complete and are recommended for surveillance activities and descriptive epidemiological studies designed to assess the prevalence of adults and children who obtain national recommendations for physical activity and health.

- *Quantitative history* questionnaires are detailed instruments that have from 15 to 60 items and reflect the intensity, frequency, and duration of activity patterns in various categories, such as occupation, household, sports and conditioning, transportation, family care, and leisure activities. This type of survey was used by Leon and colleagues (1987) in the MRFIT study. Quantitative history questionnaires allow researchers to obtain detailed information about physical activity energy expenditure and patterns of activity from the

© germanskydive110/fotolia

Understanding the "risk factors" in research is important.

past month to year. Because of their length and complexity, however, they may take from 15 to 30 minutes to complete and are usually interviewer administered. Quantitative history questionnaires are appropriate for studies designed to examine issues of dose–response in heterogeneous populations with a wide variety of physical activity patterns.

Since their introduction as an objective measure of free-living physical activity in the early 1980s, waist-mounted activity monitors (accelerometers and pedometers) have become a staple of the physical activity assessment repertoire. They have been used extensively in the validation of self-reported physical activity surveys, as outcome measures of physical activity in intervention studies, and in research designed to identify the psychosocial and environmental correlates of physical activity behaviors. The great advantage of objective physical activity measures is that they overcome some of the limitations inherent in self-report methods that necessarily rely on information reported by participants. With activity monitors, there are no reporting errors or errors introduced by interviewers, and real-time data collection and automated data reduction can provide a rich description of the activity profiles of people and populations. After the logistics of monitor delivery and retrieval are overcome, activity monitors offer a relatively simple and efficient method of measurement that is suitable for small clinical studies and epidemiological studies of intermediate size (e.g., fewer than 5,000).

The ActiGraph is a small battery-operated accelerometer-based motion sensor commonly worn on the waist. It provides a computerized record of the intensity and duration of ambulatory movement, presented as movement counts, for up to 28 days. The ActiGraph accelerometer is used in the U.S. National Health and Nutrition Examination Surveys to augment physical activity questionnaires in a subsample of participants. Minute-by-minute activity count data for the ActiGraph accelerometer on 2 days of monitoring for a middle-aged woman participating in a physical activity intervention study are presented in figure 17.2. Activity count levels above 1,950 counts per minute reflect purposeful moderate-intensity walking in the range of 3 to 4 mph (4.8 to 6.4 km/h). Panel *a* depicts a relatively sedentary day with few extended bouts of walking. Data summary for this day indicated that only 6 min of walking and 3,691 steps were accumulated. In contrast, panel *b* represents a more active day in which there was a walking bout (22 min) in the middle of the day and several

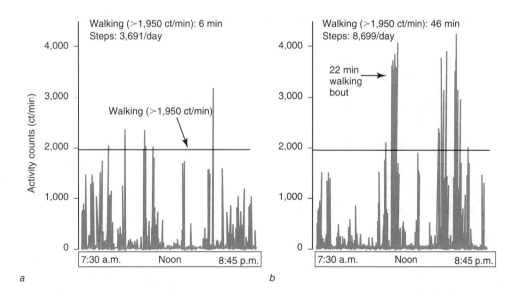

Figure 17.2 ActiGraph output and interpretation: *(a)* sedentary day, *(b)* active day.

additional shorter episodes of activity later in the day, eventually accumulating 46 walking min and 8,699 steps over the course of the day. Objective information such as this can be used to track changes in activity patterns in response to an intervention or to describe differences in the activity profiles of different populations using the same objective standard. A number of other accelerometer-based activity monitors (e.g., GENEActiv, Tritrac, SenseWear) that are comparable to the ActiGraph devices are commercially available.

Pedometers also have been used in epidemiological studies to measure the amount of accumulated steps in free-living settings. Among the less expensive pedometers (i.e., less than $50), the Yamax Digiwalker (Model CW-701) has demonstrated the highest accuracy in controlled field tests for walking (within 1% of counted steps), distance traveled (within 10% at 80 meters per minute), and gross kilocalories expended (Crouter, Schneider, Karabulut, & Bassett, 2003). Pedometers are used extensively in intervention and health promotion programs to provide feedback to participants about their activity levels, in community trials to measure the effect of interventions designed to increase walking behaviors, and to characterize the walking behaviors of community residents. For example, to quantify levels of walking among residents in a small southern community, Tudor-Locke, Williams, Reis, and Pluto (2004) mailed pedometers to 209 adults residing at randomly selected households who agreed to wear the pedometer for 7 consecutive days and record their accumulated steps in a logbook each day. The mean step counts were 5,932 ± 3,664 steps per day. Steps per day were higher in men (7,192 ± 3,596) than in women (5,210 ± 3,518), higher in whites (6,628 ± 3,375) than in nonwhites (4,792 ± 3,874), and inversely related to body mass index (normal weight = 7,029 ± 2,857; overweight = 5,813 ± 3,441; obese = 4,618 ± 3,359).

To learn more about the technological aspects and validity of pedometers, accelerometers, and fitness measures, see Tudor-Locke and colleagues (2004) for a review of studies.

EPIDEMIOLOGICAL STUDY DESIGNS

Initial hypotheses are developed and crudely tested using simple descriptive studies that use cross-sectional and ecological study designs. More refined study designs can then be used to test specific hypotheses in analytical studies. Analytical studies use case-control and **cohort** study designs, and mortality or disease incidence is often the dependent variable for the investigations. These two designs, the workhorses of epidemiological research, have been used for the last 50 years, and a solid body of evidence has been amassed to support their basic methods.

cohort— A group of individuals who are followed over a period of time.

After analytical epidemiological studies have consistently demonstrated an exposure–disease link, in conjunction with supportive laboratory evidence, experimental studies can then be initiated to test, in a rigorous experimental design, the validity of the observational findings. The design of choice for experimental epidemiological studies is the randomized controlled trial. Outcomes may be mortality, disease incidence, or an intermediate end point such as blood cholesterol levels or blood pressure. In the case of physical activity intervention research, the dependent outcome variable could be the physical activity level of either a person or a community.

Descriptive Epidemiology

Descriptive epidemiology has been defined as the relationship of basic characteristics such as age, sex, race, occupation, social class, or geographic location to disease states (Last, 1988). The major objective of descriptive studies is to quantify the magnitude of specific health problems, identify population subgroups that may have higher rates of disease, and develop hypotheses about specific factors that may be determinants of disease. Because descriptive studies use information sources that are readily available or easily obtained, they are inexpensive to conduct, in terms of both the cost of acquiring the data and the time required to complete the investigation.

Cross-Sectional Designs

The cross-sectional design is perhaps the most frequently conducted type of study examining the relationship between physical activity and health outcomes. Frequently, rather than using a disease end point as the main outcome (or dependent variable), known risk factors for a disease are used as intermediate end points. For example, a great deal of the initial work examining the effect of physical activity on cardiovascular disease risk factors (e.g., high blood cholesterol or blood pressure levels) used simple cross-sectional designs. Researchers simply assembled two groups of people, highly active people (usually athletes) and sedentary control participants, and then measured the cardiovascular risk factors of interest. A simple test of the mean differences between groups formed the analysis. Because these studies typically measure the outcome variables at a single point in time, they are relatively simple and easy to conduct.

> Descriptive studies are useful in identifying the magnitude of a health problem, adversely affected population subgroups, and risk factors (or determinants).

An example of a cross-sectional study shows the effect of physical activity on estrogen metabolism in postmenopausal women. The dependent variable is the ratio score of two estrogen metabolites, 2-hydroxyestrone to 16a-hydroxyestrone (2/16 ratio). Women with low 2/16 ratios were hypothesized to have an increased risk for breast cancer compared with women with high 2/16 ratios. Physical activity was measured using a self-administered questionnaire with leisure-time physical activity dose calculated as MET-hours per day. Results showed an interaction between physical activity and adiposity (measured as body mass index [BMI, kg/m^2]), as shown in figure 17.3. Physically active women had higher 2/16 ratios in each of three categories of BMI ranging from low to high. In contrast, inactive women had lower 2/16 levels as their BMI increased. The researchers concluded that a reduction in 2/16 ratios among inactive women with higher BMI levels may be predictive of increased risk for breast cancer.

An advantage of cross-sectional studies is that the physiological differences of individuals (rather than groups) can be compared, and with planning, investigators can control for factors that could potentially confound the relationship of interest. Confounding factors can be controlled using statistical methods when the size of the study is adequate, or by matching the two comparison groups on an important factor. For example, to control for the confounding effect of body fat on the physical activity–blood cholesterol relationship, one could recruit highly active and sedentary people with similar levels of body fat (e.g., ± 2%). Comparisons of the two groups would then be independent of the effects of body fat.

The major limitation of the cross-sectional study design is that the study outcome and exposure are measured at the same point in time. Therefore, it is impossible to know whether physical activity exposure was actually responsible for the effects observed. That is, cross-sectional study designs do not allow for definitive conclusions about cause and effect because the temporal sequence (timing) of the relationship between the outcome and the exposure is not known.

a p = 0.03, low activity, BMI < 25 vs. BMI 30+
b p = 0.07, BMI 30+, low vs. high activity

Figure 17.3 Body mass index (BMI) versus 2/16 estrogen ratios in active and inactive postmenopausal women.

A final and often overlooked limitation of cross-sectional studies is that a lack of association in a cross-sectional study may not mean that no longitudinal relationship exists between the two factors being examined. For example, on a cross-sectional basis, there can be little or no relationship between dietary fat intake and blood cholesterol levels. In longitudinal studies, however, there is a clear positive relationship between fat intake and cholesterol levels (Jacobs, Anderson, & Blackburn, 1979).

Ecological Designs

Ecological studies use existing data sources for both exposure and disease outcomes to compare and contrast rates of disease by specific characteristics of an entire population. Typical data sources in these studies are census data, vital statistics records (for countries, states, counties), employment records, and national figures for health-related information such as food consumption.

For example, Morris, Heady, Rattle, Roberts, and Parks (1953) reported on a series of studies that used occupational classifications as a surrogate measure of physical activity levels. Perhaps the most famous comparisons were among London busmen working on double-decker buses. Conductors, who had to walk up and down the bus stairs all day, every day, were compared with the bus drivers, who sat and drove the bus all day, every day. Drivers consistently had heart disease mortality rates that were twice as high as those of conductors. Also included in this seminal work was an ecological study of heart disease mortality and occupational physical activity. The investigators tabulated heart disease mortality rates by age and occupational activity level among Englishmen and Welshmen using existing public health statistics for the years 1930 to 1932. The retabulation and classification of these data by occupational types into heavy-, intermediate-, and light-activity categories provided a scientifically crude, but ultimately valid, understanding of the effect of physical activity on heart disease mortality.

Figure 17.4 shows a graded inverse relationship between occupational physical activity and heart disease mortality in each age group. That is, men who were engaged in heavy occupational activity had the lowest death rates, and men in the light-activity occupations had the highest death rates, regardless of age. Men in intermediate-activity categories had death rates that were in between. A clear increase in heart disease mortality with increasing age was also observed.

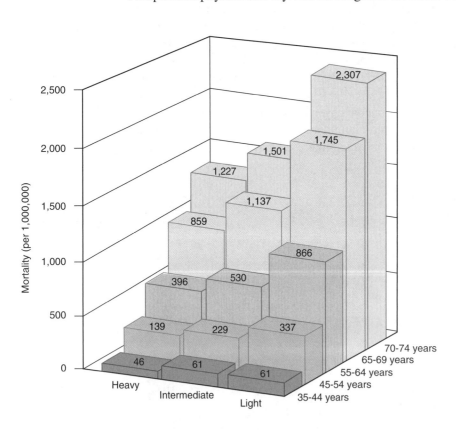

Figure 17.4 Heart disease mortality in heavy-, intermediate-, and light-activity occupational categories by age group among men: England and Wales, 1930–1932.

Reprinted from *The Lancet*, Vol. 262, No.6796, J.N. Morris et al., "Coronary heart-disease and physical activity of work, pp.1111-1120, Copyright 1953, with permission from Elsevier. www.sciencedirect.com/science/article/pii/ S0140673653914950

Like most things that are cheap and easy, ecological studies have limitations that must be considered. First, the level of analysis is a population group rather than an individual. Therefore, there is no way to link individual physical activity levels to a specific heart disease outcome. Second, ecological studies are severely limited in their ability to control for the effects of other factors that could obscure the relationship being examined.

Although ecological analyses can lead to useful and valid estimates of an exposure–disease relationship, as noted in the previous example, they have also been known to produce spurious results. For example, ecological studies first conducted in the 1970s examining the relationship between dietary fat intake and breast cancer mortality revealed strong linear relationships between this exposure and breast cancer. Women in countries whose citizens ate more fat had higher breast cancer mortality rates. Since that time, however, a large body of research using methodologically superior study designs has been unable to demonstrate the strong link between fat intake and breast cancer observed in the early ecological studies (Willett, 1990). Accordingly, ecological studies should be considered only a first line of investigation aimed at the development of testable hypotheses.

In summary, descriptive studies are useful for developing and crudely testing initial hypotheses about exposure–disease relationships. More definitive, scientifically sound results, however, can come only from analytical or experimental study designs.

Analytical Designs

The **analytical designs**, cohort and case-control studies, test hypotheses regarding causal links between exposures and mortality and incidence outcomes using purely observational methods.

To understand analytical designs more thoroughly, we must consider a simplified model of the natural history of disease (see figure 17.5). Because most epidemiological studies that evaluate physical activity as the exposure of main interest are for chronic disease, this model is for chronic disease. Chronic diseases such as heart disease, osteoporosis, and cancer can take 10 to 40 years to develop into a problem sufficient to be diagnosed clinically.

Physical activity epidemiologists are typically interested in how exposure to physical activity early in the disease process alters the course of the natural history of the disease. The disease course could be extended by preventive exposures (an active lifestyle) or hastened by adverse exposures (a sedentary lifestyle).

The induction period of a disease is conceptualized as the period between the point at which the exposure of interest alters the physiology of the disease process and the point at which the disease is clinically diagnosed. The clinical diagnosis always occurs later than the disease onset. After being clinically diagnosed with the disease, a person becomes an incident case. Primary prevention (1°) refers to prevention of the first occurrence of the disease, and secondary prevention (2°) refers to prevention of disease recurrence. The usual goal of analytical studies is to quantify the effect of exposures that occur during the induction period on eventual disease risk. That is, the goal is to identify predictors of the disease outcome during the early portions of the natural history of the disease.

> **analytical designs—** Studies that test hypotheses about causal links between exposures and mortality and incidence outcomes, using only observational research.

Figure 17.5 Natural history of chronic disease.

> **cohort study design—**A study design that involves the examination of a large disease-free population over a period of several years.

Cohort Studies

The terms *follow-up studies, prospective studies,* and *longitudinal studies* have all been used to describe the **cohort study design**. In this design, a large disease-free population is defined and assessment of relevant exposures is obtained. Baseline data are used to categorize the cohort into levels of exposure (i.e., low, medium, high). After this baseline assessment has been made, the follow-up period begins. Because chronic diseases such as colon and breast cancer are relatively rare, the follow-up period can last from as little as 2 years to more than 20 years. At the end of the follow-up period, the number of people within the cohort who died during follow-up and were diagnosed with the disease outcome of interest is tabulated.

The analysis of cohort studies is relatively simple. Because we are interested in how various levels of baseline exposure predict disease occurrence, the basic analysis consists simply of calculating disease rates for the levels of exposure. For example, the mortality rates among people reporting regular exercise would be compared with the mortality rates of nonexercisers. In more refined analyses, mortality rates for three to five levels of physical activity may be calculated, and comparisons between the low activity level and each of the higher activity levels could be made.

Disease rates in cohort studies are often expressed relative to person-years of follow-up. A person-year represents 1 year of observation for one person during the follow-up period. For example, in figure 17.6 five people were observed from time 0 to time 5 (6 full years). A subject's person-year contribution to the cohort begins when he enters follow-up (at time 0 for the top three people, at time 1 for the fourth person, at time 2 for the fifth person). Follow-up ends for each person when the person has an event (X), when the person is lost for follow-up, or when the follow-up period ends. In the example in figure 17.6, each subject's person-year contribution to the cohort is in parentheses, and the summary measure of the incidence rate is 2 events (X) / 22 person-years of observation.

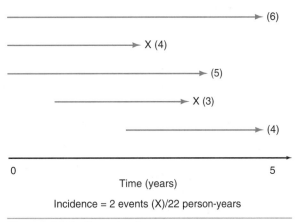

Incidence = 2 events (X)/22 person-years

Figure 17.6 Example of person-years.

relative risk—A measure of the association of the exposure–disease relationship typically employed in cohort studies.

The summary measures of the exposure–disease relationship in cohort studies are typically expressed in absolute or relative terms. Differences in disease rates between exposure groups expressed on an absolute basis can be obtained by simple subtraction of rates. More frequently, estimates of the exposure–disease relationship are expressed as the **relative risk**. Relative risks are calculated as the ratio of the referent category to various exposure levels. The null value, or no effect, of the relative risk is 1.0. Values below 1.0 indicate reduced risk, whereas values above 1.0 indicate increased risk.

A landmark cohort study in the physical activity epidemiology literature is the College Alumni Health Study. Ralph S. Paffenbarger Jr. and colleagues initiated this study by enrolling 16,936 male Harvard alumni aged 35 to 74 years in 1962 and 1966. At baseline, participants completed mailed questionnaires that included questions about their daily physical activities. Using simple questions about walking, stair climbing, and exercise activities, they categorized participants into six levels of physical activity energy expenditure (i.e., <500, 500–999, 1,000–1,999, 2,000–2,999, 3,000–3,999, and ≥4,000 kcal/wk). In a follow-up assessment 6 to 10 years later (1972), it was determined that 572 men had experienced their first heart attack (Paffenbarger, Wing, & Hyde, 1978). Rates of fatal, nonfatal, and all first heart attacks were compared in each of the baseline physical activity exposure categories. These data (presented in figure 17.7) clearly indicated that the rates for each category of heart attack were inversely related to physical activity index levels, up to about 2,000 kcal/week, where the heart attack rates reached a plateau (or increased slightly).

Stratified comparisons of these data by age group, smoking status, blood pressure level, and adiposity revealed that these potentially confounding factors did not account for this relationship between physical activity and first heart attacks. Thus, the results strongly suggested that the effect of physical activity on heart attack risk was independent of many important factors that could potentially confound this relationship. The summary relative risk estimate of 1.64 was obtained by dividing the heart attack rates for less active men (<2,000 kcal/wk, 57.9 attacks per 10,000 person-years) by the rates for highly active men (≥2,000 kcal/wk, 35.3 attacks per 10,000 person-years). That is, less active men were at 1.64 times greater risk of having a heart attack relative to highly active men (Paffenbarger et al., 1978).

Important contributions from this cohort study include the finding that collegiate sports play did little to provide cardiovascular protection later in life (Paffen-

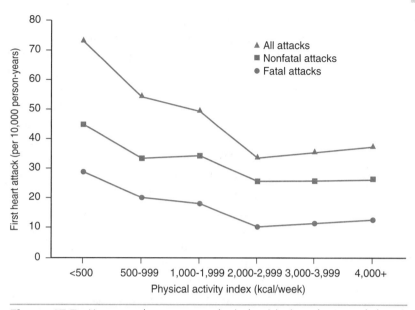

Figure 17.7 Heart attack rates versus physical activity in male Harvard alumni.

From R.S. Paffenbarger, Jr., A.L. Wing, and R.T. Hyde, 1995, "Physical activity as an index of heart attack risk in college alumni," *American Journal of Epidemiology* 142(9): 889-903. Adapted by permission from Oxford University Press.

barger, Hyde, Wing, & Stenimetz, 1984). That is, to reduce cardiovascular risk the most, participants needed to be active in early adulthood (college) through middle age. Finally, this cohort study demonstrated that beginning moderately vigorous physical activity in middle age can extend the life span (Paffenbarger et al., 1993).

Advantages of Cohort Studies

Cohort studies have three primary advantages:

1. The temporal sequence between exposure and outcome is clearly defined. Exposure assessments are obtained before disease onset, so the timing of the exposure in the natural history of the disease is in the correct sequence. This feature is the primary strength of cohort studies.

2. Cohort studies are good for rare exposures. For example, if you wanted to study the effects of a rare exposure such as running, you could specifically recruit a cohort of runners into your study to ensure an adequate number of people in low-, medium-, and high-mileage categories. If you recruited for such a study using a random sample of the adult population of the United States, you would have only 10 to 12% of your study population with your exposure of interest (U.S. Department of Health and Human Services, 1996).

3. Cohort studies are good for understanding the multiple effects of a single exposure. For example, within the Harvard alumni cohort study mentioned earlier, the effects of physical activity on a diverse number of health outcomes were examined, including lung, colon, prostate, and pancreatic cancers, as well as the risk of first heart attack, stroke, hypertension, depression, suicide, Parkinson's disease, and cardiovascular and all-cause mortality.

Disadvantages of Cohort Studies

Cohort studies are difficult and costly to conduct because of the challenges inherent in keeping track of large numbers of people over long periods of time (about 40 years in the case of the College Alumni Health Study). Loss to follow-up can be problematic because large numbers of losses can result in biased estimates of the exposure–disease relationship.

A second limitation of cohort studies is that some disease outcomes are sufficiently rare that even an extremely large cohort may not produce enough cases for meaningful analyses. For example, it took only 6 to 10 years for the Harvard alumni study ($N = 16,936$) to accumulate 572 cases of first heart attack (or, an incidence of 42.2 per 10,000 men/year). By comparison, in a cohort study of the effects of physical activity on the rarer outcome of breast cancer, 14 years of follow-up in 25,624 women were required to accumulate 351 cases of breast cancer (or an incidence of 9.8 per 10,000 women/year).

Case-Control Studies

case-control study—A
study that involves the
identification of causally
related factors for disease
outcome in populations
with and without disease.

odds ratio—A measure
of the exposure–disease
relationship typically
employed in case-control
studies.

The **case-control study**, like the cohort study, aims to identify factors that are causally related to disease outcome. In this design, people with disease (i.e., cases) and without disease (i.e., controls) are recruited into the study over the same period. Frequently, the disease-free control subjects are selected to match cases with respect to potentially important confounding factors, such as age or ethnicity. Both cases and controls are then asked about their exposures to potential causal factors using in-person interviews or self-administered questionnaires. Thus, the objective of this retrospective exposure assessment in case-control studies is to identify factors that influenced the natural history of the disease during its induction period (i.e., before disease onset). The analysis of case-control data contrasts the exposure history of the cases to that of controls and typically expresses this relationship as an **odds ratio (OR)**. The odds ratio is interpreted as an estimate of the relative risk that would have been calculated in the study group if a cohort study had been conducted. Like the relative risk, the null value of the odds ratio is 1.0. Values above 1.0 typically indicate increased risk, and values less than 1.0 indicate reduced risk.

The case-control epidemiological design has been particularly important in studying the relationship between physical activity and relatively rare diseases such as cancer. For example, breast cancer is the leading cancer diagnosis among women in the United States, accounting for 175,000 cases in 1999, or about 30% of all cancers diagnosed in women (Ries et al., 1999). Clearly, it is a major public health concern, yet the relatively low incidence of the disease makes it a challenging outcome to investigate using the cohort design.

An interesting case-control study examining the relationship between physical activity and breast cancer was published by Bernstein, Henderson, Hanisch, Sullivan-Halley, and Ross (1994). The purpose of the investigation was to examine the effect of physical activity exposure accumulated from adolescence through adulthood on the risk of breast cancer in women under 40 years of age. Investigators enrolled 545 cases with breast cancer and 545 controls that were free of the disease. All participants were younger (<40 years) residents of Los Angeles and were recruited between 1983 and 1989. Women in the control group were matched to cases according to age, ethnicity, number of pregnancies, and neighborhood of residence. Upon recruitment, each woman was interviewed about lifetime exercise participation and a number of additional breast cancer risk factors. In the interview, women reported each exercise activity in which they participated on a regular basis during their lifetime, the length of time that they participated in this activity (in years), and the average duration of their exercise (hours per week). Using these data, women were categorized into five levels of average lifetime exercise exposure. The least active women reported no regular exercise (0 hours per week). Active women were categorized into four categories based on the quartile cutoff points for controls reporting some exercise (>0 hours per week).

Table 17.1 was adapted from the Bernstein article and presents univariate (unadjusted) and multivariate (adjusted) odds ratios for this association. Women who exercised were at lower risk of having breast cancer relative to nonexercisers because the value of the odds ratios is below 1.0. The beneficial effect of exercise was particularly evident for the most active women (≥3.8 hours per week). Women who exercised ≥3.8 hours per week were at a 50% reduced risk of breast cancer in comparison with nonexercisers (0 hours per week) in

TABLE 17.1

Odds Ratios for Breast Cancer Associated With Exercise

Lifetime exercise[a]	n cases/n controls	Univariate OR (95% confidence level)	Multivariate OR (95% confidence level)[b]
None	195/154	1.0	1.0
0.1–0.7	103/90	0.92 (0.65–1.32)	0.95 (0.64–1.41)
0.8–1.6	84/102	0.66 (0.46–0.95)	0.65 (0.45–0.96)
1.7–3.7	103/100	0.82 (0.58–1.17)	0.80 (0.54–1.17)
≥3.8	61/99	0.50 (0.34–0.73)	0.42 (0.27–0.64)

[a]Hours per week, from menarche to about 12 months before study entry.

[b]Adjusted for age at menarche, age at first full-term pregnancy, number of full-term pregnancies, months of breast feeding, family history of breast cancer, body mass index at reference date, and oral contraceptive use.

Reprinted from L. Bernstein et al. 1994, "Physical exercise and reduced risk of breast cancer in young women," *Journal of the National Cancer Institute* 86(18): 1371-1372. (www.cancer.gov)

univariate analyses, and at a 58% reduced risk after controlling for important confounding factors.

The findings in this investigation spurred a number of studies of the effect of physical activity on breast cancer risk using both case-control and cohort designs (Friedenreich, Thune, Brinton, & Albanes, 1998). Two recent reports using the cohort study design (Rockhill et al., 1999; Thune, Brenn, Lund, & Gaard, 1997) supported the findings presented by Bernstein and colleagues. That is, high levels of physical activity accumulated during a woman's lifetime appear to reduce the risk of breast cancer.

Advantages of Case-Control Studies

The case-control study design is often the first line of analytical investigation because it can provide valid estimates of exposure–disease relationships in a shorter time and with less monetary expense in comparison to the cohort design. Case-control designs have the following advantages:

1. They are particularly efficient for investigating rare diseases because they obviate the need for extremely large populations or long follow-up periods to accumulate a useful number of cases for analysis.

2. They enable hypothesis testing for multiple exposures for a single disease outcome.

3. They can be used to acquire exposure information in greater detail because more resources can be used to gather this information as a result of the smaller scale of most case-control studies.

Disadvantages of Case-Control Studies

The major disadvantages of case-control studies are that exposure information is obtained after the disease has been diagnosed and that recruiting an appropriate control group can be challenging. The exposure information and the resulting odds ratios may be adversely affected by problems in each of these areas. First, people who have just been diagnosed with a major disease that is often life threatening may recall their exposures differently than control subjects do, simply because of their recent diagnosis. Bias of this type has been called **recall bias**. For example, in a case-control study examining the physical activity–heart disease relationship, participants with disease (cases) may report that they were less active than they truly were because it is well known that exercise is protective against heart disease.

recall bias—Systematic errors introduced by differences in recall accuracy between comparison groups (i.e., between cases and controls).

This recall bias would increase the apparent difference in physical activity levels between the controls who reported their exposure without a recall bias. In this example, recall bias would bias odds ratios away from the null value of 1.0 and suggest that a stronger relationship existed than was actually present.

Second, for the contrast of exposure histories between cases and controls to provide valid odds ratios, control subjects need to be representative of the population from which the cases were obtained. This requirement means that the control group does not have to be generalizable to the general population, but only to the population that produced the case series. If the exposure histories of the control subjects in the study differ importantly from the unknowable exposure distribution of the cohort from which the cases arose, then **selection biases** may cause odds ratios that are not valid estimates of the true exposure–disease relationship.

Case-control methods have been developed and extensively refined in the last 50 years, resulting in a study design that can provide invaluable information about important exposure–disease outcomes in an efficient manner. A full description of the many facets of this study design, however, is well beyond the scope of this chapter. Interested readers are encouraged to consult standard epidemiology texts.

Threats to Validity in Analytical Study Designs

The major threats to obtaining a valid understanding of the exposure–disease relationship in analytical studies are methodological biases and confounding factors. Epidemiological methods, when implemented appropriately, limit the effects of biases and confounding factors. **Bias** in epidemiological research typically refers to systematic deviations of risk estimates (i.e., odds ratios or relative risk) from their true value. Recall and selection biases have previously been discussed in the case-control studies section. Other important biases to consider are observer biases and biases that result from exposure misclassification. Observer bias can be introduced in studies that use interviewers to collect exposure information but do not blind the interviewers to the disease status of the participants. A result of this lack of blinding can be a systematic bias introduced if the interviewer collects exposure data differently from cases and controls.

Biases that result from the misclassification of physical activity exposure levels often arise from measurement errors introduced when the exposures are measured. Because physical activities are often measured rather crudely, or are assessed at only one point in a person's life, the measure of exposure may not adequately reflect the person's true physical activity exposure during the induction period of the disease of interest. Accordingly, the exact relationship between the exposure and the study outcome may not be clearly described by the data collected. If this misclassification of exposure is randomly distributed in the study population, these errors will result in attenuation in the measures of effect. That is, an actual effect on the disease outcome by the exposure could be less pronounced than it actually is, or the effect may be missed completely. This type of bias is usually referred to as bias toward the null (i.e., no effect).

A **confounding factor** in epidemiological studies refers to an additional factor that obscures the observable relationship between an exposure and the outcome. Confounding factors must be associated with both the exposure and the outcome and are identifiable by their effect on the measure of the association. That is, changes in odds ratios or relative risk estimates between unadjusted and adjusted analyses suggest the presence of confounding factors. Confounding factors can be controlled using statistical methods or stratification. For example, if smoking was a confounding factor in the relationship between physical activity and blood cholesterol levels, its confounding effect could be controlled by examining the relationship in smokers and nonsmokers separately.

selection bias—Systematic errors introduced by differences in the characteristics of subjects entering and not entering a study.

bias—The systematic deviation of a calculated (estimated) value from the true value.

confounding factor—A factor that obscures the true relationship between an exposure and outcome of interest.

Experimental Designs With Randomized Trials

Regular physical activity increases cardiorespiratory fitness, promotes weight loss, and if done at regular intervals, reduces the risks for heart attack, colon cancer, and premature mortality (U.S. Department of Health and Human Services, 1996). Experimental designs allow researchers to identify the effects of a specific intervention on a health outcome in a group of people (experimental group) while simultaneously monitoring changes in the same health outcome among people not receiving the intervention (control or comparison group). Randomized trials that are focused on changing health at the individual level are referred to as **clinical trials**. Because denying treatment that is known to cause health benefits is illegal, randomized trials for physical activity often provide a form of physical activity to the comparison group that does not have the same biological effect as the intervention treatment provided to the intervention group, or that compares different types of treatment.

clinical trials—Trials focused on changing health at the individual level.

An example of a randomized trial involving a physical activity intervention is Project Active, which compares the effects of a personal, lifestyle physical activity intervention with a traditional, structured exercise intervention. Subjects were 235 initially sedentary and apparently healthy adults who were randomized into either a lifestyle or a structured intervention group. The hypothesis was that there would be no difference in the physical activity and cardiorespiratory fitness between the two groups at the end of the treatment. Both groups were instructed in programmatically different, but physiologically equivalent, methods to increase their levels of daily physical activity (Kohl, Dunn, Marcus, & Blair, 1998). Results showed increases in physical activity and cardiorespiratory fitness at the end of the 6-month intensive intervention, but these changes were similar between the two intervention styles (Dunn et al., 1998).

Randomized trials that are focused on changing behaviors in communities are referred to as **community trials**. The rationale for community-level interventions are as follows: (a) targeting everyone may prevent more cases of disease than targeting just high-risk people; (b) environmental modifications may be easier to accomplish than large-scale voluntary behavior change; (c) risk-related behaviors are socially influenced; (d) community interventions reach people in their native habitat; and (e) community interventions can be logistically simpler and less costly on a per-person basis.

community trials—Randomized trials focused on changing behaviors in communities.

In community trials, whole communities are randomized to receive multiple treatments in the form of mass media campaigns, school-based programs, point of purchase interventions (e.g., posting signs in grocery stores indicating brands of food that have less saturated fat), and related activities. The comparison communities do not receive the intervention but serve to show the effect of naturally occurring changes in community behaviors over time. Data collection and analysis include monitoring the extent to which an intervention is implemented as intended and the effects of the intervention on community behaviors (process evaluation), individual-level changes in behavior and health outcomes (individual evaluation), and changes in the environment influenced by the intervention (community-level indicator evaluation). Data analysis must account for community-level and individual-level variations in behaviors. An advantage of community trials is that they allow researchers to see the population changes in behaviors as a result of interventions. But this benefit is tempered by the difficulty of changing individual behaviors and the amount of time that may be required (sometimes many years) to see changes in community behaviors.

The simplest design for a community trial involves one intervention and one control community. Examples of these are the North Karelia Project in Finland (Puska, Salonen, & Nissinen, 1983), the Pawtucket Heart Health Program (Carleton et al., 1995), and the Heart-to-Heart Project in South Carolina (Goodman, Wheeler, & Lee, 1995), each of which was designed to reduce the risks for cardiovascular disease. Unfortunately, because these study designs had only one intervention and one control community, investigators were unable to compare community-level variation in the observed behaviors. Thus, it was unknown whether the changes observed were because of the intervention or would have occurred naturally over time without the intervention. Subsequent community trials involving physical activity interventions have included multiple intervention and control cities. These include the Stanford Five-City Project (Winkleby, Taylor, Jatulis, & Fortmann, 1996), which enrolled two interventions and two control communities, and the Minnesota Heart Health Program (Jacobs et al., 1986), which enrolled three interventions and three control communities located in geographically distant locations in Northern California and Minnesota, respectively.

In the Stanford Five-City Project, intervention activities designed to reduce cardiovascular disease risk factors were delivered to two communities from 1979 to 1985. The physical activity interventions were targeted toward increasing moderate and vigorous physical activity in intervention communities. The interventions used electronic and print media, individual and community activities, and school-based functions. Results showed modest but statistically significant changes in physical activity in the intervention communities. Men were more likely to participate in vigorous activities, and women reported spending more time in moderate-intensity activities. Similar changes were not observed in the control communities.

READING AND INTERPRETING A PHYSICAL ACTIVITY EPIDEMIOLOGICAL STUDY

Because epidemiological studies may involve a variety of study designs and large numbers of people, it is a good idea to follow a guideline when reading and interpreting a research paper. The example used in this section is an article titled "Associations of Light, Moderate, and Vigorous Intensity Physical Activity With Longevity: The Harvard Alumni Health Study" (Lee & Paffenbarger, 2000).

Introduction

The introduction should provide a global overview of the rationale for the study, highlighting major research that has shown associations (or lack of associations) between the exposure and outcomes and explaining why this health concern is worthy of study. The rationale should also address the biological mechanism for the association between the physical activity and a health outcome. Research questions should reflect findings and questions arising from earlier studies and should state the purpose of the study.

> *Example:* **The authors recognized the growing acceptance that a healthy lifestyle includes regular physical activity. However, they noted the inconsistent findings about the optimal intensity of physical activity needed to reduce premature mortality. The purpose of the paper was to present associations between bouts of light-, moderate-, and vigorous-intensity activity and mortality among men enrolled in the Harvard Alumni Health Study.**

Materials and Methods

This section should include information about the study design, study population, variables, and analytical methods.

The *study design* should be appropriate to answer the research questions. To describe events in a population (descriptive studies), an ecological or cross-sectional study design is appropriate. To identify causal associations that reflect changes in health status across time, however, prospective cohort studies or experimental study designs are needed. Case-control studies are useful to examine exposure and outcome relationships in small study samples and provide only limited evidence of the existence of a temporal relationship between the exposure and outcome.

The *study population* should be appropriate to answer the research questions asked in the study. Writers should describe their *sample size* to include how many subjects were enrolled in the study, completed the tasks as assigned, and were included in the data analysis. The writers also should provide an explanation of differences in the sample sizes from the start to the end of the study. This is an important issue for avoiding biased studies. The methods used to recruit the subjects and inclusion or exclusion criteria should be listed.

Characteristics of the study population should include subjects' age, sex, race, income, education, or other factors related to the research questions. Is the study population representative of a group used for generalization?

The *study variables* should be described for the exposure (physical activity), outcome (e.g., mortality), and potential confounders. Writers should identify how the exposure was measured and scored and whether data are available that address the validity and reliability of the measures. The *outcome* should be identified by a case definition (e.g., ICD-version 10 codes) and a description of how data were obtained, whether from hospital records, death certificates, physical exams, or self-reports. Disease outcomes should have a case definition that the authors use to guide the researchers in the classification of the outcome status. *Potential confounders* should be identified with the data collection methods for each variable described. Writers should also describe the quality control measures used to ensure accuracy in the data collection process.

Methods used for *data analyses* should reflect the research questions and hypotheses tested in the study.

> *Example: Study population.* Subjects were described as men who matriculated as undergraduates at Harvard University between 1916 and 1950. Inclusion criteria included having returned a survey sent in 1977 ($n = 17,835$). Exclusion criteria were the presence of physician-diagnosed cardiovascular disease, cancer, or chronic obstructive pulmonary disease ($n = 3,706$) or failure to identify information about potential covariate variables on the survey ($n = 644$). The sample used in the analysis was 13,485 men with a mean age of 57.5 ± 8.9 years.

> *Example: Exposure.* Physical activity was measured with a mailed survey that had subjects recall the number of stairs climbed, city blocks walked, and sports and recreational activities performed in the past week. A physical activity index (PAI) was expressed as a composite of these activities in kilojoules per week. The survey has a one-month reliability of $r = .72$ and a correlation with physical activity records of $r = .65$. MET levels associated with intensity levels were as follows: light: <4 METs; moderate: 4–5.5 METs; vigorous: >6 METs.

> *Example: Outcome.* Identification of subjects' deaths was obtained from the National Death Index. The causes of death were determined from death certificates.

> *Example: Potential confounders.* Confounder variables were obtained from self-reported information recorded on the mailed health survey. Variables included age, BMI, cigarette smoking, alcohol consumption, and early (<65 years) parental death.

> *Example: Data analyses.* The association between total energy expenditure in kilojoules per week (kJ/wk) and mortality was examined using proportional hazards models.

Results

The text should present a global overview of the results and refer the reader to the tables and figures that show the data. Tables and figures should be self-explanatory and identify the confounding variables controlled in the analyses in a footnote. Statistical comparisons should present *p*-values or confidence intervals around parameter estimates (e.g., odds ratio, relative risk) and dose responses by presenting comparisons of the data among multiple levels,

Example: Results. **A total of 2,539 men died between 1977 and 1992. Compared with men who expended <4,200 kJ/wk, those who expended more energy through regular physical activity reduced their risk for mortality by 20 to 27%. Trends for associations between type and intensity of activity and mortality were significant at *p* < .05 for distance walked, stairs climbed, and moderate and vigorous activity (see figure 17.8). Light activity was not associated with decreases in mortality rate.**

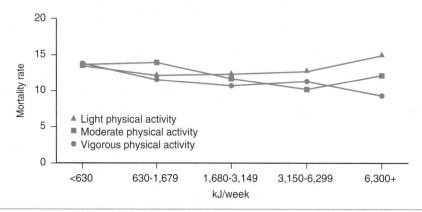

Figure 17.8   Age-adjusted mortality rates (per 1,000) by intensity of physical activity.

From I.M. Lee and R.S. Paffenbarger, Jr., 2000, "Associations of light, moderate, and vigorous intensity physical activity with longevity. The Harvard alumni health study," *American Journal of Epidemiology* 151: 293-299. Used with permission from Oxford University Press.

Discussion and Conclusions

In the discussion, the writers should highlight the main study findings, compare and contrast the results with those of similar studies, and relate the findings to data that highlight the magnitude of the problem. Sources of bias that may have influenced the study results should be addressed as potential weaknesses of the study. The criteria for causality should be discussed as a prelude to determining whether the associations and hypotheses tested support a causal association between the physical activity exposure and a health outcome.

Example: **The authors concluded that greater energy expenditure is associated with increased longevity. Participation in moderate activities showed a trend toward lower mortality rates, and participation in vigorous activities showed a strong association with lower mortality rates. Participation in light activities was unrelated to mortality rates. Biological mechanisms for the association are supported by other studies showing lower blood pressure, glucose, and insulin levels; higher HDL cholesterol; and enhanced cardiac function and hemostatic factors with regular moderate and vigorous physical activity. Strengths of the study include a detailed physical activity assessment and a long follow-up period. Limitations include the possibility of recall bias in using a self-administered physical activity survey and the inability to account for diseases not measured on the survey. The public health impact of the study showed benefits of moderate- and vigorous-intensity exercise in reducing risks for mortality among men.**

SUMMARY

This chapter provided an overview of the research methods used in the study of physical activity epidemiology. Epidemiological studies identify the distribution and determinants of physical activity and health-related states in population groups and apply the research findings to reduce the burden of disease in communities.

Check Your Understanding

1. Describe the differences between observational and experimental research.

2. Briefly describe the strengths and weaknesses of descriptive, analytical, and experimental study designs in examining the relationships between physical activity and health.

3. Compare and contrast cohort and case-control study designs.

4. Write a review of a physical activity research study published in the *American Journal of Epidemiology*, the *Journal of Clinical Epidemiology*, the *International Journal of Epidemiology*, or *Epidemiology*.

5. Compare and contrast the ways epidemiologists study the distribution, determinants, and applications of health issues.

18

Experimental and Quasi-Experimental Research

Never try to teach a pig to sing. It wastes your time and annoys the pig.

—Mark Twain

simple idea underpins science: Trust but verify. To verify, good science must be open to challenge from experimental studies. In modern science we do not verify often enough. For example, in 2012 scientists from a biotech firm could reproduce only 6 of 53 important studies from cancer research; another drug company group could verify only 25% of 67 important papers. Many of these problems are a result of journals publishing mostly papers that find important differences. In fact, negative results (no differences) account for 14% of published papers (Trouble at the lab. (2013, October 19). *The Economist,* p. 2). Although we like to think of scientific research as self-correcting, without the replication of experiments, it may not be. Thus, in this chapter we discuss experimental research and its design, but we certainly advocate for efforts to verify previously found important results.

Experimental research attempts to establish cause-and-effect relationships. That is, an independent variable is manipulated to judge its effect on a dependent variable. But the process of establishing cause and effect is a difficult one. Three criteria must be present to establish cause and effect:

1. *The cause must precede the effect.* For example, the starting gun in a race precedes the runners' beginning the race; the runners' beginning does not cause the starting gun to go off.

2. *The cause and effect must be correlated with each other.* As we have already discussed, just because two variables are correlated does not mean one causes the other; cause and effect, however, cannot exist unless two variables are correlated.

3. *The correlation between cause and effect cannot be explained by another variable.* Recall from chapter 1 that the relationship between the academic performance of elementary school children and shoe size was explained by a third variable, age.

We may think of cause and effect in terms of necessary and sufficient conditions (Krathwohl, 1993). For example, if the condition is necessary and sufficient to produce the effect, then it is the cause. But alternative situations exist as well (see the sidebar Examples of Cause and Effect in Golf):

- Necessary but not sufficient: Some related condition likely produces the effect.
- Sufficient but not necessary: Some alternative condition is likely the cause.
- Neither necessary nor sufficient: Some contributing condition is likely the cause.

Also, remember that statistics do not establish cause and effect. Statistical techniques can only reject the null hypothesis (establish that groups are significantly different) and identify the percentage of variance in the dependent variable accounted for by the independent variable or the effect size. Neither of these procedures establishes cause and effect (they are necessary but not sufficient). Cause and effect can be established only by applying logical thinking to well-designed experiments. This process establishes that no other reasonable explanation exists for the changes in the dependent variable except the manipulation of the independent variable. The application of this logic is made possible by the following:

> Establishing cause and effect requires logical thinking applied to well-designed experiments.

- Selection of a good theoretical framework
- Use of appropriate participants
- Application of an appropriate experimental design
- Proper selection and control of the independent variable (treatment)
- Appropriate selection and measurement of the dependent variable
- Use of the correct statistical model and analysis
- Correct interpretation of the results

In this chapter, we discuss experimental designs by explaining how you can recognize and control sources of invalidity and threats to both internal and external validity. We also explain several types of experimental designs. Before going any further into this chapter, review the following terms (introduced in chapters 1 and 3), which we use throughout the discussion:

- Independent variable
- Dependent variable
- Categorical variable
- Control variable
- Extraneous variable

SOURCES OF INVALIDITY

All the design types we discuss have strengths and weaknesses that pose threats to the validity of the research. The importance of validity was stated well by Campbell and Stanley (1963):

> **Fundamental . . . is a distinction between internal validity and external validity. Internal validity is the basic minimum without which any experiment is uninterpretable: Did in fact the experimental treatments make a difference in this specific experimental instance? External validity asks the question of generalizability: To what populations, settings, or treatment variables can this effect be generalized? (p. 5)**

Both internal validity and external validity are important in experiments, but they are frequently at odds in research planning and design. Gaining internal validity involves controlling all variables so that the researcher can eliminate all rival hypotheses as explanations for the outcomes observed. Yet in controlling and constraining the research setting to gain internal validity, the researcher places the generalization (external validity) of the findings in jeopardy. In studies with strong internal validity, the answer to the question of to whom, what, or where the findings can be generalized may be very uncertain. This is because in ecologically valid settings (perceived by the participants as intended by the researchers), not everything is controlled and operated in the same way as in the controlled laboratory context. Thus, the researcher must decide whether it is more important to be certain that the manipulation of the independent variable caused the observed changes in the dependent variable or to be able to generalize the results to other populations, settings, and so on. We cannot provide an easy answer to that question, which is often debated at scientific meetings and in the literature (e.g., see Martens, 1979, 1987; Siedentop, 1980; Thomas, 1980; Thomas, French, & Humphries, 1986).

Examples of Cause and Effect in Golf

1. No matter how bad your last shot was, the worst is yet to come. This condition does not expire until after the 18th hole.

2. Your best round of golf will be followed almost immediately by your worst round ever. (The probability increases based on the number of people you tell about your best round.)

3. Brand-new golf balls are water magnetic. (The magnetic effect increases with the price of the ball.)

4. The higher your handicap, the more qualified you deem yourself to be an instructor.

5. Every par-three hole has a secret desire to humiliate golfers. (The shorter the hole, the greater its desire.)

6. Electric golf carts always fail at the point farthest from the clubhouse.

7. The person you most hate to lose to will always be the one who beats you.

8. The last three holes of a round will automatically adjust your score to what it really should be.

9. Golf balls never bounce off trees back into play. (If one does, the groundskeeper will remove that tree before your next round.)

10. All vows taken on a golf course are valid only until sundown.

To expect any experiment to meet all research design considerations is unreasonable. A more realistic approach is to identify the specific goals and limitations of the research effort. Is internal validity or external validity the more important issue? After that is decided, the researcher can plan the research with one type of validity as the major focus while maintaining as much of the other type of validity as possible. Another recourse is to plan a series of experiments in which the first experiment has strong internal validity even at the expense of external validity. If the first experiment confirms that changes in the dependent measure are the result of manipulating the independent variable, subsequent experiments can be designed to increase external validity even at the expense of internal validity. This approach allows evaluation of the treatment in settings more like the real world. (Review the related discussion of basic and applied research in chapter 1 and Christina, 1989.)

THREATS TO INTERNAL VALIDITY

Campbell and Stanley (1963) identified eight threats to the internal validity of experiments, and Rosenthal (1966) identified a ninth:

1. **History:** events occurring during the experiment that are not part of the treatment
2. **Maturation:** processes within the participants that operate as a result of time passing (e.g., aging, fatigue, hunger)
3. **Testing:** the effects of one test on subsequent administrations of the same test
4. **Instrumentation:** changes in instrument calibration, including lack of agreement within and between observers
5. **Statistical regression:** the fact that groups selected based on extreme scores are not as extreme on subsequent testing
6. **Selection bias:** choosing comparison groups in a nonrandom manner
7. **Experimental mortality:** loss of participants from comparison groups for nonrandom reasons
8. **Selection–maturation interaction:** the passage of time affecting one group but not the other in nonequivalent group designs
9. **Expectancy:** experimenters' or testers' anticipating that certain participants will perform better

If these threats are uncontrolled, the change in the dependent variable may be difficult to attribute to the manipulation of the independent variable.

History

A history threat to internal validity means that some unintended event occurred during the treatment period. For example, in a study evaluating the effects of a semester of physical education on the physical fitness of fifth-graders, the fact that 60% of the children participated in a recreational soccer program would constitute a history threat to internal validity. The soccer program is likely to produce benefits to physical fitness that would be difficult to separate from the benefits of the physical education program.

Maturation

Maturation as a threat to internal validity is most often associated with aging. This threat occurs frequently in designs in which one group is tested on several occasions over a long period. Elementary school physical education teachers frequently encounter this source of invalidity when they give a physical fitness test in the early fall and again in the late spring. The children nearly always do better in the spring. The teacher would like to claim that the physical education program was the cause. Unfortunately, maturation is a plausible rival hypothesis for the observed increase; that is, the children have grown larger and stronger and thus probably run faster, jump higher, and throw farther.

Testing

A testing threat is the effect that taking a test once has on taking it again. For example, if athletes are administered a multiple-choice test to evaluate their knowledge about steroids today and again two days later, they will do better the second time even though no treatment intervened. Taking the test once helps in taking it again. The same effect is present in physical performance tests, particularly if the participants are not allowed to practice the test a few times. If a class of beginners in tennis attempts to hit 20 forehand shots delivered to them from a ball machine today and again three days later, they will usually do better the second time. They learned something from performing the test the first time.

Instrumentation

Instrumentation is a problem frequently faced in exercise science research. Suppose the researcher uses a spring-loaded device to measure strength. Unless the spring is calibrated regularly, it will decrease in tension with use. Thus, the same amount of applied force produces increased readings of strength compared with earlier readings. Instrumentation problems also occur in research using observers. Unless observers are trained and regularly checked, the same observer's ratings may systematically vary across time or participants (a phenomenon called observer drift), or different observers may not rate the same performance in the same way.

Statistical Regression

Statistical regression can occur when groups are not randomly formed but are selected based on an extreme score on some measure. For example, if someone rates the behavior of a group of children on a playground on an activity scale (very active to very inactive) and two groups are formed—one of very active children and one of very inactive children—statistical regression is likely to occur when the children are next observed on the playground. The children who were very active will be less active (although still active), and the very inactive children will be more active. In other words, both groups will regress (move from the extremes) toward the overall average. This phenomenon reflects only the fact that a participant's score tends to vary about her average performance (estimated true score). Extreme scores may simply reflect the observation of a performance on the high (or low) side of the participant's typical performance. The next performance is usually not as extreme. Thus, when average scores of extreme groups are compared from one time to the next, the high group appears to get worse, whereas the low group appears to get better. Statistical regression is a particular problem in studies that attempt to compare extreme groups selected on some characteristic, such as highly anxious, fit, or skilled participants versus not very anxious, fit, or skilled participants.

Selection Bias

Selection biases occur when groups are formed on some basis other than random assignment. Thus, when treatments are administered, because the groups were different to begin with, always present is the rival hypothesis that any differences found are due to initial selection biases rather than the treatments. Showing that the groups were not different on the dependent variable at the beginning of the study does not overcome this shortcoming. Any number of other unmeasured variables on which the groups differ might explain the treatment effect. Gall, Borg, and Gall (1986) asked important questions that apply to selection (or sampling) bias:

> ***Did the study use volunteers?*** **Use of volunteers is common in research in the study of physical activity. Yet volunteers are often not representative of anyone but other volunteers. They may differ considerably from nonvolunteers in motivation for the experimental task and setting.**

> ***Are participants extremely nonrepresentative of the population?*** **Often we are unable to select participants at random for our studies, but it is useful if we at least believe (and can demonstrate) that they represent some larger group from our culture.**

Recall the discussion of sampling in chapter 6, particularly the concept of a good enough sample.

Experimental Mortality

Experimental mortality refers to the loss of participants. Even when groups are randomly formed, this threat to internal validity can occur. Participants may remain in an experimental group taking part in a fitness program because it is fun, whereas participants in the control group become bored, lose interest, and drop out of the study. Of course, the opposite can occur, too. Participants may drop out of an experimental group because the treatment is too difficult or time-consuming.

Selection–Maturation Interaction

A selection–maturation interaction occurs only in designs in which one group is selected because of some specific characteristic, whereas the other group lacks this characteristic. An example is a study of the differences between 6-year-olds in two school districts; students in one school district form the experimental group that receives a fitness program, and students in the other are a control group. If the school districts have different admission policies so that the 6-year-olds in the experimental group are 5 months older, it would be difficult to determine whether the fitness program or the fitness program combined with the participants' advanced age produced the observed changes.

Expectancy

One additional threat to internal validity not mentioned by Campbell and Stanley (1963) has been identified. **Expectancy** (Rosenthal, 1966) refers to experimenters' or testers' anticipating that certain participants will perform better. This effect, although usually unconscious on the part of the experimenters, occurs where participants or experimental conditions are clearly labeled. For example, testers may rate skilled participants better than unskilled

expectancy—A threat to internal validity in which the researcher anticipates that certain behaviors or results will occur.

You must keep participants' attention when doing interviews for research.

participants, regardless of treatment. This effect is also evident in observational studies in which the observers rate posttest performance better than pretest performance because they expect change. If the experimental and control groups are identified, observers may rate the experimental group better than the control group even before any treatment occurs. The expectancy effect may influence the participants, too. For example, in a youth sport study, coaches may actually cause poorer performance by substitutes (compared with starters) because the substitutes realize that the coach treats them differently (e.g., the coach may show less concern about incorrect practice trials).

Any of these nine threats to internal validity may reduce the researcher's ability to claim that the manipulation of the independent variable produced the changes in the dependent variable. We discuss experimental designs and how they control (or fail to control) the threats to internal validity later in this chapter.

> There are nine threats to internal validity that can undermine the claim of a causal relationship between two variables.

THREATS TO EXTERNAL VALIDITY

Campbell and Stanley (1963) identified four threats to external validity, or the ability to generalize results to other participants, settings, measures, and so on:

1. **Reactive or interactive effects of testing:** The pretest may make the participant more aware of or sensitive to the upcoming treatment. As a result, the treatment is not as effective without the pretest.

2. **Interaction of selection bias and the experimental treatment:** When a group is selected on some characteristic, the treatment may work only on groups possessing that characteristic.

3. **Reactive effects of experimental arrangements:** Treatments that are effective in constrained situations (e.g., laboratories) may not be effective in less constrained settings (more like the real world).

4. **Multiple-treatment interference:** When participants receive more than one treatment, the effects of previous treatments may influence subsequent ones.

Reactive or Interactive Effects of Testing

Reactive or interactive effects of testing may be a problem in any design with a pretest. Suppose a fitness program is the experimental treatment. If a physical fitness test is administered to the sample first, the participants in the experimental group might realize that their levels of fitness are low and be particularly motivated to follow the prescribed program closely. In an untested population, however, the program might not be as effective because the participants might be unaware of their low levels of physical fitness.

Interaction of Selection Bias and Experimental Treatment

The interaction of selection bias and the experimental treatment may prohibit the generalization of the results to participants lacking the particular characteristics (bias) of the sample. For example, a drug education program might be effective in changing the attitudes of college freshmen toward drug use. This same program would probably lack effectiveness for third-year medical students because they would be much more familiar with drugs and their appropriate uses.

Reactive Effects of Experimental Arrangements

Reactive effects of experimental arrangements means that the experimental treatment may not be generalizable to real-world situations. These reactive effects are a persistent problem for laboratory-based research (e.g., in exercise physiology, biomechanics, motor behavior). In such research, is the researcher investigating an effect, process, or outcome that is specific to the laboratory and cannot be generalized to other settings? We have referred to this earlier as a problem of ecological validity. For example, in a study employing high-speed cinematography, the skill to be filmed must be performed in a certain place, and joints must be marked for later analysis. Is the skill performed in the same way during sport participation?

A specific type of reactive behavior labeled the **Hawthorne effect** (Brown, 1954) refers to the fact that participants' performances change when attention is paid to them. This effect may be a threat to both internal and external validity, because it is likely to produce better treatment effects and reduce the ability to generalize the results.

Hawthorne effect—A phenomenon in which participants' performances change when attention is paid to them, which is likely to reduce the ability to generalize the results.

Multiple-Treatment Interference

Multiple-treatment interference is most frequently a problem when the same participants are exposed to more than one level of the treatment. Suppose participants are going to learn to move to the hitting position in volleyball using a lead step or a crossover step. We want to know which step gets the players in a good hitting position most quickly. If the players are taught both types of steps, learning one might interfere with (or enhance) learning the other. Thus, the researcher's ability to generalize the findings may be confounded by the use of multiple treatments. A better design might be to have two separate groups, each of which learns one of the techniques.

The ability to generalize findings from research to other participants or situations is a question of random sampling (or at least good enough sampling) more than any other. Do the participants, treatments, tests, and situations represent any larger populations? Although a few of the experimental designs discussed later control certain threats to external validity, usually researchers control these threats through the way they select the samples, treatments, situations, and tests.

CONTROLLING THREATS TO INTERNAL VALIDITY

Threats to internal and external validity are controlled in different ways by applying specific techniques. In this section, we describe useful approaches to solving threats to internal validity in the design of experiments. Many threats to internal validity are controlled by making the participants in the experimental and control groups as alike as possible. This objective is most often accomplished by randomly assigning participants to groups.

Randomization

As mentioned in chapter 6, randomization allows the assumption that the groups do not differ at the beginning of the experiment. The randomization process controls for history up to the point of the experiment; that is, the researcher can assume that past events are equally distributed among groups. This approach does not control for history effects during the experiment if experimental and control participants are treated at different times or places. The researcher must try to prevent an event (besides the treatment) from occurring in one group but not in the other groups.

Randomization also controls for maturation because the passage of time is equivalent in all groups. Statistical regression, selection biases, and selection–maturation interaction are controlled because they occur only when groups are not randomly formed.

Ways other than random assignment of participants to groups are sometimes used to control threats to internal validity. The matched-pair technique matches pairs of participants who are equal on some characteristic and then randomly assigns each to a different group.

The researcher might want very tight control on previous experience in strength training. Thus, participants would be matched on this characteristic and then randomly assigned to the experimental and control groups.

A matched-group technique may also be used. This method involves nonrandom assignment of participants to experimental and control groups so that the group means are equivalent on some variable. This procedure is generally regarded as unacceptable because the groups may not be equivalent on other unmeasured variables that could affect the outcome of the research.

In within-subjects designs, the participants are used as their own controls. This means that each participant receives both the experimental and the control treatment. In this type of design, the order of treatments should be counterbalanced; that is, half the participants should receive the experimental treatment first and then the control, and the other half should receive the control treatment first and then the experimental treatment. If there are three levels of the independent variable (1 = control, 2 = experimental treatment A, 3 = experimental treatment B), participants should be randomly assigned to one of the six possible treatment orders (1-2-3, 1-3-2, 2-1-3, 2-3-1, 3-1-2, 3-2-1). If the number of treatment levels administered to the same participants is larger than three, experimenters can assign a random order of the treatments to each participant rather than counterbalance treatments.

Placebos, Blind Setups, and Double-Blind Setups

Other ways of controlling threats to internal validity include placebos and blind and double-blind setups. A **placebo** is used to evaluate whether the observed effect is produced by the treatment or is a psychological effect. Frequently, a control condition is used in which participants receive the same attention from and interaction with the experimenter, but the treatment administered does not relate to performance on the dependent variable.

A study in which participants do not know whether they are receiving the experimental or the control treatment is called a **blind setup** (i.e., the participant is blind to the treatment). In a **double-blind setup,** neither the participant nor the tester knows which treatment the participants are receiving. A triple-blind test has also been reported (Day, 1983): The participants do not know what they are getting, the experimenters do not know what they are giving, and the investigators do not know what they are doing. We can only hope the triple-blind technique finds limited use in our field.

All these techniques (except the triple-blind technique) are useful in controlling the Hawthorne, expectancy, and halo effects, as well as what we call the **Avis effect** (a recent version of the John Henry effect), or the fact that participants in the control group may try harder simply because they are in the control group.

A good example of the use of these techniques for controlling psychological effects is the use of steroids to build strength in athletes. A number of studies were done to evaluate the effects of steroids. To combat the fact that athletes may become stronger because they think they should when using steroids, a placebo (a pill that looks just like the steroid) is used. In a blind study, the athletes do not know whether they receive the placebo or the steroid. In a double-blind study, the athlete, the person dispensing the steroids or placebos, and the testers do not know which group received the steroids. Unfortunately, these procedures did not work well in this type of study because taking large quantities of steroids made the athletes' urine smell bad. Thus, the athletes knew whether they were receiving the steroid or the placebo.

Uncontrolled Threats to Internal Validity

Three threats to internal validity remain uncontrolled by the randomization process: reactive or interactive effects, instrumentation, and experimental mortality.

placebo—A method of controlling a threat to internal validity in which a control group receives a false treatment while the experimental group receives the real treatment.

blind setup—A method of controlling a threat to internal validity in which participants do not know whether they are receiving the experimental or control treatment.

double-blind setup—A method of controlling a threat to internal validity in which neither the participant nor the experimenter knows which treatment the participant is receiving.

Avis effect—A threat to internal validity wherein participants in the control group try harder just because they are in the control group.

Reactive or Interactive Effects

Reactive or interactive effects of testing can be controlled only by eliminating the pretest. These effects can be evaluated, however, by two of the designs discussed later in this chapter: pretest–posttest randomized groups and the Solomon four-group design.

Instrumentation

Instrumentation problems cannot be controlled or evaluated by any design. Only the experimenter can control this threat to internal validity. In chapter 11, we went into some detail on techniques for controlling the instrumentation threat to develop valid and reliable tests. Of particular significance is test reliability. Whether the measurement is obtained from a laboratory device (e.g., an oxygen analyzer), a motor performance test (e.g., a standing long jump), an attitude-rating scale (e.g., on feelings about drug use), an observer (e.g., coding the percentage of time a child is active), a knowledge test (e.g., on basketball strategy), or a survey (e.g., on available sport facilities), the answer must be consistent. Controlling this threat frequently involves the assessment of test reliability across situations, between and within testers or observers, and within participants. The validity of the instrument (does it measure what it was intended to measure?) must also be established to control for instrumentation problems. The total process of establishing appropriate instrumentation for research is called psychometrics.

One final point about instrumentation involves the halo effect. As we discussed in chapter 11, this effect occurs in ratings of the same person on several skills. Raters seeing a skilled performance on one task are likely to rate the participant higher on subsequent tasks, regardless of the level of skill displayed. In effect, the skilled behavior has "rubbed off" (created a halo) on later performances. An order effect may also occur in observation. Gymnastics and swimming judges often rate earlier performers lower to save room on the rating scale for better performers. Because gymnastics and swimming coaches know this, they always place better performers later in the event order.

Experimental Mortality

Experimental mortality is not controlled by any type of experimental design. The experimenter can control this threat only by ensuring that participants are not lost (at all, if possible) from groups. Many problems of participant retention can be handled in advance of the research by carefully explaining the research to the participants and emphasizing the need for them to follow through with the project. (During the experiment itself, begging, pleading, and crying sometimes work.)

CONTROLLING THREATS TO EXTERNAL VALIDITY

External validity is generally controlled by selecting the participants, treatments, experimental situation, and tests to represent some larger population. Of course, random selection (or good enough sampling) is the key to controlling most threats to external validity. Remember that elements other than the participants may be randomly selected. For example, the levels of treatment can be randomly selected from the possible levels, experimental situations can be selected from possible situations, and the dependent variable (test) can be randomly selected from a pool of potential dependent variables.

As previously noted, the ability to generalize the situation is called ecological validity. Although the results of a particular treatment can be generalized to a larger group if the sample is representative, this generalization may apply only to the specific situation in the experiment. If the experiment is conducted under controlled laboratory conditions, then the findings may apply only under controlled laboratory conditions. Frequently, the experimenter hopes that the findings can be generalized to real-world exercise, sport, industrial, or instructional settings. Whether the outcomes can be generalized in this way depends largely on how the participants perceive the study, and this influences the way participants respond to study characteristics. The questions of interest are Does the study have enough characteristics of real-world settings that participants respond as if they were in the real world? Is ecological validity present? These questions are not easy to answer, so a number of scholars advocate that more research in kinesiology, physical education, exercise science, and sport science be conducted in field settings (e.g., Costill, 1985; Martens, 1987; Thomas, French, & Humphries, 1986).

The reactive or interactive effects of testing can be evaluated by the Solomon four-group design. Interaction of selection biases and the experimental treatment is controlled by the random selection of participants. The reactive effects of experimental arrangements can be controlled only by the researcher (this is again the issue of ecological validity). Multiple-treatment interference can be partially controlled by counterbalancing or randomly ordering the treatments among participants. But the researcher can control whether the treatments still interfere only through knowledge about the treatment rather than the type of experimental design.

TYPES OF DESIGNS

This section (much of which is taken from Campbell & Stanley, 1963) is divided into three categories: preexperimental designs, true experimental designs, and quasi-experimental designs. We use the following notation:

- Each line indicates a group of participants.
- R signifies the random assignment of participants to groups.
- O signifies an observation or a test.
- T signifies that a treatment is applied; a blank space in a line where a T appears on another line means that the group is a control.
- A dotted line between groups means that the groups are used intact rather than being randomly formed.
- Subscripts indicate either the order of observations and treatments (when they appear on the same line) or observations of different groups or different treatments (when they appear on different lines). For example, when the terms T_1 and T_2 appear on different lines, they refer to different treatments; when they appear on the same line, they mean that the treatment is administered more than once to the same group.

Preexperimental Designs

The three designs discussed in this section are called **preexperimental designs** because they control few of the sources of invalidity. None of the designs has random assignment of participants to groups.

One-Shot Study

In a one-shot study design, a group of participants receives a treatment followed by a test to evaluate the treatment:

$$T \qquad O$$

This design fails all the tests of good research. All that can be said is that at a certain point in time this group of participants performed at a certain level. In no way can the level of performance (O) be attributed to the treatment (T).

One-Group Pretest–Posttest Design

The one-group pretest–posttest design, although very weak, is better than the one-shot design. At least we can observe whether any change in performance has occurred:

$$O_1 \quad T \quad O_2$$

If O_2 is better than O_1, we can say that the participants improved. For example, Bill Biceps (a qualified exercise instructor) conducted an exercise test at a health club. Participants then trained 3 days per week, 40 min per day, at 70% of their estimated $\dot{V}O_2$max, for 12 weeks. After the training period, participants retook the exercise test and significantly improved their scores. Can Mr. Biceps conclude that the exercise program caused the changes that he observed in the exercise test performance? Unfortunately, this design does not allow us to say why the participants improved. Certainly, the improvement could be due to the treatment, but it could also be due to history. Some event other than the treatment (T) may have occurred between the pretest (O_1) and the posttest (O_2); the participants may have exercised at home on the other days. Maturation is a rival hypothesis. The participants may have gotten better (or worse) just as a result of the passage of time. Testing is a rival hypothesis; the increase at O_2 may be the result only of experience with the test at O_1. If the group being tested is selected for some specific reason, then any of the threats involving selection bias can occur. This design is most frequently analyzed by the dependent t test to evaluate whether a significant change occurred between O_1 and O_2.

Static Group Comparison

The static group-comparison design compares two groups, one of which receives the treatment and one of which does not:

$$T \qquad O_1$$
$$\text{-----------------------}$$
$$O_2$$

But we do not know whether the groups were not equivalent when the study began, as indicated by the dotted line between the groups. This means that the groups were selected intact rather than being randomly formed. We are thus unable to determine whether any differences between O_1 and O_2 are because of T or only because the groups differed initially. This design is subject to invalidity because of selection biases and the selection–maturation interaction. A t test for independent groups is used to evaluate whether O_1 and O_2 differ significantly. But even if they do differ, the difference cannot be attributed to T.

The three preexperimental designs are not valid methods of answering research questions (see table 18.1). They do not represent experiments because the change in the dependent variable cannot be attributed to the manipulation of the independent variable. You will not encounter these preexperimental designs in research journals, and we hope that you will not find (or produce) theses and dissertations using these designs. The preexperimental designs represent much wasted effort because little or nothing can be concluded from the findings. If you submit studies using these designs to research journals, you are likely to receive rejection letters similar to one Snoopy (from the *Peanuts* comic strip) received: "Dear Researcher, Thank you for submitting your paper to our research journal. To save time, we are enclosing two rejection letters—one for this paper and one for the next one you send."

true experimental design—Any design used in experimental research in which groups are randomly formed and that controls most sources of invalidity.

True Experimental Designs

The designs discussed in this section are called **true experimental designs** because the groups are randomly formed, allowing the assumption that they were equivalent at the beginning of the research. This controls for past (but not present) history, maturation (which should occur equally in the groups), testing, and all sources of invalidity that are based on nonequivalence

TABLE 18.1

Preexperimental Designs and Their Control of the Threats to Validity

Validity threat	One-shot study	One-group pretest and posttest	Static group
Internal			
History	−	−	+
Maturation	−	−	?
Testing		−	+
Instrumentation		−	
Statistical regression	?	+	
Selection	−	+	−
Experimental mortality	−	−	
Selection × maturation		−	−
Expectancy	?	?	?
External			
Testing × treatment		−	
Selection biases × treatment	−	−	−
Experimental arrangements		?	
Multiple treatments			

Note: + = strength; − = weakness; blank = not relevant; ? = questionable.

of groups (statistical regression, selection biases, and selection–maturation interaction). But only the experimenter can make sure that nothing happens to one group (besides the treatment) and not the other (present history), that scores on the dependent measure do not vary as a result of instrumentation problems, and that the loss of participants is not different between the groups (experimental mortality).

Randomized-Groups Design

Note that the randomized-groups design resembles static group comparison except that groups are randomly formed:

$$R \quad T \quad O_1$$

$$R \quad \quad O_2$$

If the researcher controls the threats to internal validity that are not controlled by randomization (no easy task), has a sound theoretical basis for the study, and meets the necessary-and-sufficient rule, then this design allows the conclusion that significant differences between O_1 and O_2 are due to T. An independent t test is used to analyze the difference between O_1 and O_2. This design as earlier depicted represents two levels of one independent variable. It can be extended to any number of levels of an independent variable:

$$R \quad T_1 \quad O_1$$

$$R \quad T_2 \quad O_2$$

$$R \quad \quad O_3$$

Here, three levels of the independent variable exist, where one is the control and T_1 and T_2 represent two levels of treatment. This design can be analyzed by simple ANOVA, which contrasts the dependent variable as measured in the three groups (O_1, O_2, O_3). For example, T_1 is training at 70% of $\dot{V}O_2$max, T_2 is training at 40% of $\dot{V}O_2$max, and the control group is not training. The variables O_1, O_2, and O_3 are the measures of cardiorespiratory fitness (12 min run) in each group taken at the end of the training.

This design can also be extended into a factorial design; that is, more than one independent variable (IV) could be considered. Example 18.1 shows how this works.

Independent variable 1 (IV_1) has three levels (A_1, A_2, A_3), and independent variable 2 (IV_2) has two levels (B_1, B_2). This results in six cells (A_1B_1, A_1B_2, A_2B_1, A_2B_2, A_3B_1, A_3B_2) to which participants are randomly assigned. At the end of the treatments, each cell is tested on the dependent variable (O_1 through O_6). This design is analyzed by a 3×2 factorial ANOVA that tests the effects of IV_1 (F_A), IV_2 (F_B), and their interaction (F_{AB}).

Example 18.1

		IV$_2$	
		B$_1$	B$_2$
IV$_1$	A$_1$	A$_1$B$_1$	A$_1$B$_2$
	A$_2$	A$_2$B$_1$	A$_2$B$_2$
	A$_3$	A$_3$B$_1$	A$_3$B$_2$
	R	A$_1$B$_1$	O$_1$
	R	A$_1$B$_2$	O$_2$
	R	A$_2$B$_1$	O$_3$
	R	A$_2$B$_2$	O$_4$
	R	A$_3$B$_1$	O$_5$
	R	A$_3$B$_2$	O$_6$

Note: Analyzed in a 3×2 factorial ANOVA. F_A = main effect of A; F_B = main effect of B; F_{AB} = interaction of A and B.

This design can also be extended to address even more independent variables and retain all the controls for internal validity previously discussed. Sometimes this design is used with a categorical independent variable. Look again at example 18.1, and suppose that IV_2 (B_1, B_2) represents two age levels. Clearly, the levels of B could not be randomly formed. The design appears as follows:

	R	A$_1$	O$_1$
B$_1$	R	A$_2$	O$_2$
	R	A$_3$	O$_3$
---	---	---	---
	R	A$_1$	O$_4$
B$_2$	R	A$_2$	O$_5$
	R	A$_3$	O$_6$

The levels of A are randomly formed within B, but the levels of B cannot be randomly formed. This no longer qualifies completely as a true experimental design, but it is frequently used in the study of physical activity. This design is analyzed in a 3×2 ANOVA, but the interpretation of results must be done more conservatively.

Any of the versions of randomized-group design may also have more than one dependent variable. Although the essential design remains the same, the statistical analysis becomes multivariate. Where two or more levels of one independent variable exist and several dependent variables are present, discriminant analysis is the appropriate multivariate statistic. In the factorial versions of this design (two or more independent variables), if multiple dependent variables are used, then MANOVA is typically the most appropriate analysis, although practical concerns (e.g., participant numbers) or theoretical matters may dictate alternative statistical analyses.

Pretest–Posttest Randomized-Groups Design

In the pretest–posttest randomized-groups design, the groups are randomly formed, but both groups are given a pretest as well as a posttest:

$$R \quad O_1 \quad T \quad O_2$$

$$R \quad O_3 \quad \quad O_4$$

The major purpose of this type of design is to determine the amount of change produced by the treatment; that is, does the experimental group change more than the control group? This design threatens the internal validity through testing, but the threat is controlled because the comparison of O_3 to O_4 in the control group as well as the comparison of O_1 to O_2 in the experimental group includes the testing effect. Thus, although the testing effect cannot be evaluated in this design, it is controlled.

This design is used frequently in the study of physical activity, but its analysis is complex. There are at least three common ways to do a statistical analysis of this design. First, a factorial repeated-measures ANOVA can be used. One factor (between subjects) is the treatment versus no treatment, whereas the second factor is pretest versus posttest (within subjects or repeated measures). The interest in this design, however, is usually the interaction: Do the groups change at different rates from pretest to posttest? If you choose a repeated-measures ANOVA for this design (in our opinion, the best choice), pay close attention to our discussion of univariate and multivariate issues in repeated measures in chapter 9. Another choice is simple ANCOVA, using the pretest for each group (O_1 and O_3) to adjust the posttest (O_2 and O_4). Recall that some problems are associated with using the pretest as a covariate (see the discussion of ANCOVA in chapter 9). Finally, the experimenter could subtract each participant's pretest value from the posttest value (called a **difference, or gain, score**) and perform a simple ANOVA (or, with only two groups, an independent t test), using each participant's difference score as the dependent variable. Each of these techniques has strengths and weaknesses, but you will find all three used in the literature.

In this design the important question is Does one group change more than the other group? Although this issue is frequently called the analysis of difference scores, a more appropriate label is the assessment of change. Clearly, in a learning study, the change is a gain. But in an exercise physiology study, the change might be decreased performance caused by fatigue. Regardless, the issues are the same. How can this change be assessed appropriately?

The easiest answer is to obtain a difference score by subtracting the pretest from the posttest. Although this method is intuitively attractive, it has some severe problems. First, these difference scores tend to be unreliable. Second, the level of initial values applies: Participants who begin low in performance can improve more easily than those who begin with high scores. Thus, the initial score is negatively correlated with the difference score. How would you like improvement in your tennis performance evaluated if your initial score was high

> **difference score or gain score**—A score that represents the difference (change) from pretest to posttest.

(e.g., 5 successful forehand drives out of 10 trials) in comparison with that of a friend who began with a low score (e.g., 1 successful hit out of 10)? If you improved to 7 out of 10 on the final test and your friend improved to 5 out of 10 (the level of your initial score), your friend has improved twice as much as you have (a gain of 4 successful hits versus 2). Yet your performance is still considerably better, and it was more difficult for you to improve. For these reasons, the use of difference scores is rarely a good method for measuring change.

The issues involved in the proper measurement of change are complex, and we cannot treat these issues here. Much has been written on this topic. We suggest that you read Schmidt and Lee (1999, chapter 10) about this problem in motor learning and performance. Or, for a classic short book on the topic, see Harris (1963).

This design can also be extended into more complex forms. First, more than two (pretest and posttest) repeated measures can be used. This circumstance is common in the areas of exercise physiology, motor behavior, and exercise psychology. Two randomly formed groups of participants in a motor behavior experiment might be measured 30 or more times as they learn a task. The two groups might differ in the information they are given. The design might use as the statistical analysis a 2 (groups) × 30 (trials) ANOVA with repeated measures on the second factor (trials). Remember from chapter 9 that meeting the assumptions for a repeated-measures ANOVA with many repeated measures is difficult. Thus, in designs like these, trials may be blocked (i.e., several trials averaged, reducing the number of repeated measures) into 3 blocks of 10 trials or 5 blocks of 6 trials.

Sometimes the design is extended in other ways. For example, we could take the design in example 18.1 (a 3 × 2 factorial) and add a third factor of a pretest and a posttest. This design would result in a three-way factorial with repeated measures on the third factor. All the versions of this design are subject to the first threat to external validity: reactive or interactive effects of testing. The pretest may make the participant more sensitive to the treatment and thus reduce the ability to generalize the findings to an unpretested population.

Solomon Four-Group Design

The Solomon four-group design is the only true experimental design to specifically evaluate one of the threats to external validity: reactive or interactive effects of testing. The design is depicted as follows:

$$R \quad O_1 \quad T \quad O_2$$

$$R \quad O_3 \quad \quad O_4$$

$$R \quad \quad T \quad O_5$$

$$R \quad \quad \quad O_6$$

This approach combines the randomized-groups and the pretest–posttest randomized-groups designs. The purpose is explicitly to determine whether the pretest results in increased sensitivity of the participants to the treatment. This design allows a replication of the treatment effect (is $O_2 > O_4$? and is $O_5 > O_6$?), an assessment of the amount of change due to the treatment (is $O_2 - O_1 > O_4 - O_3$?), an evaluation of the testing effect (is $O_4 > O_6$?), and an assessment of whether the pretest interacts with the treatment (is $O_2 > O_5$?). Thus, this experimental design is extremely powerful. Unfortunately, it is also inefficient, because twice as many participants are required. Consequently, this design is in limited use, especially among graduate students doing theses and dissertations. In addition, no good way exists to analyze this design statistically. The best alternative (one that does not use all the data) is a 2 × 2 ANOVA set up as follows:

	No T	T
Pretested	O_4	O_2
Unpretested	O_6	O_5

Thus, IV_1 has two levels (pretested and not pretested), and IV_2 has two levels (treatment and no treatment). In the ANOVA, the F ratio for IV_1 establishes the effects of pretesting, the F for IV_2 establishes the effects of the treatment, and the F for interaction evaluates the external validity threat of interaction of the pretest with the treatment. Table 18.2 summarizes the control of threats to validity for the true experimental designs.

TABLE 18.2

True Experimental Designs and Their Control of the Threats to Validity

Validity threat	Randomized groups	Pretest–posttest randomized groups	Solomon four-group
Internal			
History	+	+	+
Maturation	+	+	+
Testing	+	+	+
Instrumentation			
Statistical regression	+	+	+
Selection	+	+	+
Experimental mortality	+	+	+
Selection × maturation	+	+	+
Expectancy	?	?	?
External			
Testing × treatment	+	–	+
Selection biases × treatment	?	?	?
Experimental arrangements	?	?	?
Multiple treatments			

Note: + = strength; – = weakness; blank = not relevant; ? = questionable.
Data from Campbell and Stanley 1963.

Quasi-Experimental Designs

Not all research in which an independent variable is manipulated fits clearly into one of the true experimental designs. As researchers attempt to increase external and ecological validity, the careful and complete control of the true designs becomes increasingly difficult, if not impossible. The purpose of **quasi-experimental designs** is to fit the design to settings more like the real world while still controlling as many of the threats to internal validity as possible. The use of these types of designs in kinesiology, physical education, exercise science, sport science, and other areas (e.g., education, psychology, and sociology) has increased considerably in recent years. The most authoritative text on quasi-experimental designs is that by Shadish, Cook, and Campbell (2002).

quasi-experimental design—A research design in which the experimenter tries to fit the design to real-world settings while still controlling as many of the threats to internal validity as possible.

In quasi-experimental research, the use of randomization to control threats to internal validity is difficult. It makes sense that random assignment cannot be used in many settings. For example, a pedagogy researcher who wants to investigate the effect of a curriculum intervention could not randomly assign children to classes because schools make these decisions based on other criteria that have educational value; no school district would agree to permit a study if it had to change the students' classes. Randomly assigning classes within a school to different treatments would also be difficult, because teachers are likely to talk to each other (and even trade ideas from a curriculum that they thought was effective) and reduce the strength of the treatment intervention. Similarly, if a researcher were studying the effect of an exercise program on the aged in a community setting, random assignment would not work because people select where to enroll in classes based on various factors in their lives (e.g., convenience, membership, transportation needs), not on whether their choice helps a researcher. Asking people to go to another site or to attend class at another time would probably reduce the number of people who agree to participate and increase participant attrition (or, to say it another way, increase experimental mortality).

Reversal Design

The reversal design is used increasingly in school and other naturalistic settings and is depicted as follows:

$$O_1 \qquad O_2 \qquad T_1 \qquad O_3 \qquad O_4 \qquad T_2 \qquad O_5 \qquad O_6$$

The purpose here (as with the time series) is to determine a baseline measurement (O_1 and O_2), evaluate the treatment (change between O_2 and O_3), evaluate a no-treatment time period (O_3 to O_4), evaluate the treatment again (O_4 to O_5), and evaluate a return to a no-treatment condition (O_5 to O_6). This design is sometimes called A-B-A-B-A (or sometimes just A-B), where A is the baseline condition and B is the treatment condition.

Lines like A, B, and C in figure 18.1 suggest that the treatment is effective, whereas lines like D, E, and F do not support a treatment effect. Statistical analyses for reversal designs also need to be regression tests of the slopes and intercepts of the lines between various observations.

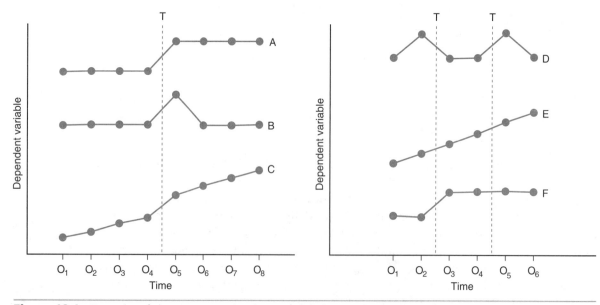

Figure 18.1   Examples of changes across time in reversal designs.

From Campbell and Stanley 1963.

Nonequivalent-Control-Group Design

A design using a nonequivalent control group is frequently used in real-world settings where groups cannot be randomly formed. The design is as follows:

$$O_1 \quad T \quad O_2$$

$$O_3 \qquad O_4$$

You can recognize this as a pretest–posttest design without randomization. Frequently, researchers compare O_1 and O_3 and declare the groups equivalent if this comparison is not significant. Unfortunately, just because the groups do not differ on the pretest does not mean that they are not different on any number of unmeasured characteristics that could affect the outcome of the research. If the groups differ when O_1 and O_3 are compared, ANCOVA is usually employed to adjust O_2 and O_4 for initial differences. Alternatively, a within (the pretest–posttest comparison—a repeated measure) and between (treatment and control group comparison) two-way ANOVA could be used to analyze whether groups changed from pretest to posttest and whether the change was different for those in the treatment and control groups.

Ex Post Facto Design

In its simplest case, ex post facto design is a static group comparison but with the treatment not under the control of the experimenter. For example, we frequently compare the characteristics of athletes versus nonathletes, highly fit versus unfit individuals, female versus male performers, and expert versus novice performers. In effect, we are searching for variables that discriminate between these groups. Our interest usually resides in the question Did these variables influence the way these groups became different? Of course, this design cannot answer this question, but it may provide interesting insights and characteristics for manipulation in other experimental designs. This design is also often called a causal comparative design.

The previously mentioned quasi-experimental designs have been frequently reported in research on the study of physical activity. Several additional designs have considerable potential but have seen less use in our research. We hope that the following presentations about two promising designs will increase interest in and use of these designs.

Switched-Replication Design

The switched-replication design (Shadish, Cook, & Campbell, 2002) can be either a true or quasi experiment depending on whether levels are random or intact groups.

Levels (random or intact groups)	Trials				
	1	2	3	4	5
1	O_1 T	O_2	O_3	O_4	O_5
2	O_6	O_7 T	O_8	O_9	O_{10}
3	O_{11}	O_{12}	O_{13} T	O_{14}	O_{15}
4	O_{16}	O_{17}	O_{18}	O_{19} T	O_{20}

If participants are randomly assigned to levels 1 through 4, the design is a true experiment. If levels 1 through 4 are different intact groups (e.g., tennis players in college, high school, and two age levels of youth leagues), then the design is quasi-experimental. Any number of levels beyond two can be used, but the number of trials must be one greater than the number of levels.

This design has two strong features: The treatment is replicated several times, and long-term treatment effects can be evaluated. There is no standard statistical analysis for this design, but various ANOVAs with repeated measures could be used. Or the design might be analyzed by fitting regression lines to each level and testing how the slopes and intercepts change.

Many opportunities exist to use this design in our field, yet it is seldom used. This design might be particularly useful in research on sport teams in which either different teams or different players within a team could be assigned to the various levels.

Table 18.3 summarizes the threats to validity of quasi-experimental designs discussed so far.

Time-Series Design

The time-series design has only one group but attempts to show that the change that occurs when the treatment is administered differs from the times when it is not. This design may be depicted as follows:

$$O_1 \quad O_2 \quad O_3 \quad O_4 \quad T \quad O_5 \quad O_6 \quad O_7 \quad O_8$$

The basis for claiming that the treatment causes the effect is that a constant rate of change can be established from O_1 to O_4 and from O_5 to O_8, but that this rate of change varies between O_4 and O_5, where T has been administered. For example, in figure 18.2, lines A, B, and C suggest that the treatment (T) results in a visible change between observations, whereas lines D, E, F, and G indicate that the treatment has no reliable effect.

The typical statistical analyses previously discussed do not fit time-series designs very well. For example, a repeated-measures ANOVA with appropriate follow-ups applied to line C in figure 18.2 might indicate that all observations (O_1 to O_8) differ significantly, even though we can visibly see a change in the rate of increase between O_4 and O_5. We do not present the details, but regression techniques can be used to test both the slopes and the intercepts in time-series designs.

This type of design appears to control for a number of the threats to internal validity. For example, maturation is constant between observations. Testing effects can also be evaluated, although they could be difficult to separate from maturation. Selection biases also appear to be controlled because the same participants are used at each observation. Of course, history, instrumentation, and mortality are controlled only to the extent that the researcher controls them. See a humorous example of a time-series design in the sidebar 7 ± 2: Miller Must Have Been an Assistant Professor.

A reminder about using intact groups in quasi experiments: As we noted in chapter 6 in the section on sampling, the use of intact groups and, particularly, the timing of when a treatment is delivered to a group, influences the analysis and the number of groups needed for a study. This issue has been discussed at length elsewhere (Silverman, 2004; Silverman & Solmon, 1998), and we will not duplicate that discussion here. We are compelled to state, however, that when groups receive treatments as a group, the appropriate unit is usually the *group*. This circumstance, then, requires a number of groups to have sufficient power to analyze the data. Those planning quasi experiments need to consider this early so that they are not dealing with the issue after all the data have been collected.

Single-Subject Design

A single-subject design is exactly what it sounds like—a researcher is looking for the effect of an intervention on a single subject. These designs also are sometimes called $N = 1$ designs because they often have only one subject. There are many designs within this family, and we could include them as a type of quasi experiment because a researcher using this design

TABLE 18.3

Quasi-Experimental Designs and Their Control of the Threats to Validity

Validity threat	Time series	Nonequivalent control	Reversal	Ex post facto	Switched replication
Internal					
History	–	–	–	?	?
Maturation	+	+	+	?	+
Testing	+	+	+		+
Instrumentation					
Statistical regression	+	–	+		+
Selection	+	+	+	–	?
Experimental mortality	+	+	+		?
Selection × maturation	+	–	+	–	
Expectancy	?	?	?	?	?
External					
Testing × treatment	–	–	?	–	?
Selection biases × treatment	?	?	?	?	?
Experimental arrangements	?	?	?	?	+
Multiple treatments			?		+

Note: + = strength; – = weakness; blank = not relevant; ? = questionable.

Data from Campbell and Stanley 1963.

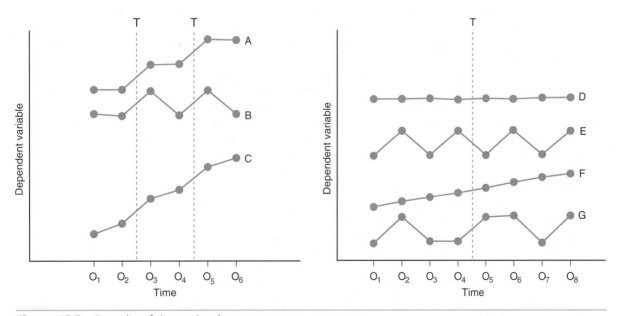

Figure 18.2 Examples of time-series changes.

Data from Campbell and Stanley 1963.

is looking for the effect of a treatment without using randomization. We have not done that here because the focus on individual effects instead of group effects makes these designs different. In addition, those who conduct single-subject research look at the changes on graphs and do not analyze the results with statistics.

7 ± 2: Miller Must Have Been an Assistant Professor

Summary—Miller's magical number for memory span (7 ± 2) is humorously brought to task because of its inability to predict the everyday performance of teenagers, graduate students, and full professors. Explanations are provided for the deficits in memory performance of these subgroups during life span development.

Miller's (1956) classic paper identifying the memory span as 7 ± 2 items must have omitted using full professors as subjects, at least based on an *n* of 1, namely me. I have frequently heard clinical psychologists accused of going into psychology to study their own problems. Maybe that is a more valid explanation than bringing Miller's work to task. For years my graduate students have said that I study memory for movement because (a) I have little or no memory, and (b) I lose my spatial orientation just walking around the block. However, I prefer to think that Miller's results only apply to a subsample of the population—6- to 10-year-old children, college students, and new assistant professors. Six- to 10-year-old children never forget anything you promised to do (or even things you said you might do . . .). However, as soon as they approach the teen years, their memory spans drop to less than 1 unit of information.

Teenagers cannot remember to make their beds as they are getting out of them. Evidently, nothing ever happens to teenagers at school, although I prefer to think that they just cannot recall anything that happened. When these teenagers with little memory capacity go to college, an amazing transition in memory occurs. Between the freshman and senior years, the young adults' memory capacity increases to at least 7 ± 2 units of information. Loosely defined, this means *they know everything;* conversely, parents know nothing.

After working a few years and coming back to graduate school, memory facility is somewhat reduced. Graduate students have a memory capacity of 3 ± 1 units of information: (a) They know that they are graduate students; (b) they remember to pick up their graduate assistantship paycheck; (c) they remember to attend their graduate seminars. The ± 1 refers to the fact that they occasionally remember to do the reading for the seminar (+1) but they sometimes forget to come to class (–1).

As soon as graduate students receive their PhDs, Miller's magical number (7 ± 2) is again a good predictor of memory—new assistant professors know everything. However, movement through the academic ranks gradually reduces capacity to remember until the average capacity of full professors (and parents) is reached, 2 ± 1 units of information. A full professor can remember (a) that he or she is a full professor and (b) his or her paycheck comes regularly. The ± 1 refers to the fact that the full professor sometimes remembers that he or she has graduate students (+1) but occasionally forgets to pick up the regular paycheck (–1).

Based on the failure of this model to meet the assumptions of stage theory (i.e., one should never regress to an earlier stage), we must assume that this uneven transition in memory states (see figure 1) is environmentally induced.

But how can the environment cause these wide variations in memory performance? First, I believe we have to assume that Miller is correct about the structural maximum of memory performance because the deficits are not confined to a single point in the life span (i.e., teenagers, graduate students, full professors). The question of interest then becomes *What causes such a serious depression in memory performance for teenagers, graduate students, and full professors?* Given my extensive study of memory as well as personal experience with all the stages, I can deduce an answer.

For teenagers the drop to about 0 memory with no variance is an interaction of carbonated soft drinks, junk food, and other boys or girls (whichever sex the teenager in question is interested in). This interaction is not a simple one and has an indirect effect; that is, this interaction causes pimples and the need for braces, both of which are very distracting in and of themselves. Taken in combination with Mom

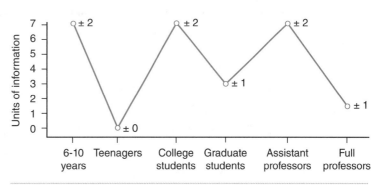

Figure 1 Maximum number of units of information retained in memory across the life span.

saying, "Wash your face" and "Brush your teeth," all of the teenager's memory capacity is occupied.

The deficit that occurs for graduate students is easy to explain. Three factors are involved. First, the graduate student is expected to work full time (either as a research or teaching assistant) for one-third or less of normal pay, a very distracting and disconcerting circumstance. Second, the graduate student is expected to study and do research day and night. One's major professor assumes graduate students do not need sleep. Finally, the graduate student's major professor is continually nagging him or her to read this paper, collect these data, write this paper, and work on his or her dissertation. In combination, these items reduce memory capacity.

But why does the full professor who seems to have everything going for him or her show such poor memory performance? Full professors have a living wage (according to some), cars that run, houses with real furniture, tenure, graduate students to do the work, and time to play golf and tennis. What could possibly explain the deficit in memory performance? Answering that question is the easiest of all:

GRADUATE STUDENTS AND TEENAGERS!
REFERENCE

Miller, G. (1956). The magical number seven, plus or minus two: Some limits on our capacity for processing information. *Psychological Review, 63*, 81–97.

Adapted, by permission, 1987, "7 ± 2: Miller must have been an assistant professor," *NASPSPA Newsletter,* 12(1): 10-11.

In our field, single-subject designs are most often used in clinical settings. Examples include observing physical education instruction, pursuing sport psychology work with athletes, studying an outstanding performer (e.g., an Olympic athlete), or looking at the motor function of a person with a physical impairment (e.g., reaching and grasping by a person with Parkinson's disease). A participant in this type of study is typically measured repeatedly on the task of interest. Many trials are needed for evaluating the influence of the treatment. During some periods a baseline measurement is obtained, and during other periods a treatment is administered. The focus is often on participant variability as well as average values. Quasi-experimental time-series, reversal, and switched-replication designs can work as single-subject designs or group designs. When used with single subjects, these designs are often called A-B or A-B-A-B designs, where A refers to the baseline condition (no treatment) and B refers to when the treatment is administered. Sometimes more than one treatment is administered to the same participant. As in research using a group of participants, counterbalancing the treatment order to separate treatment effects is important. Possible research questions include the following:

- Does the treatment produce the same effect each time?
- Are the effects of the treatment cumulative, or does the participant return to baseline following each treatment period?
- Does the participant's response to the treatment become less variable over multiple treatment periods?
- Is the participant's magnitude of response less sensitive to multiple applications of the treatment?
- Do varying intensities, frequencies, and lengths of treatment produce varying responses?

In figure 18.3 we present a graph with a traditional A-B-A-B design. Note that the A periods are baseline measurements and the B periods are when the intervention occurred. The second A is the reversal, when the treatment was withdrawn.

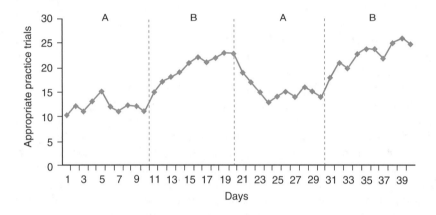

Figure 18.3   Single-subject A-B-A-B design.

The single-subject designs described in this section typically have one subject. Other single-subject designs may have more than one subject undergoing the same intervention. In these studies, the subjects start the intervention at different points in time by extending the baseline measurement for successive subjects. For example, subject 1 has baseline measurements conducted over 5 days, subject 2 over 10 days, and subject 3 over 15 days. This approach permits an examination of the intervention at different points in time. There are many permutations of multiple baseline designs, and reversals also could be added to these designs. In figure 18.4 we present one graphic example.

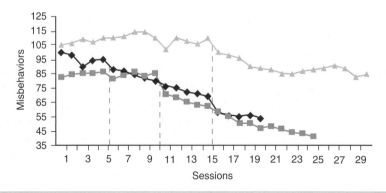

Figure 18.4   Extension of a single-subject design to three subjects over different time spans.

Other quasi-experimental designs exist, but the ones discussed here are the most commonly used. Of course, quasi-experimental designs never quite control internal validity as well as true experimental designs do, but they do allow us to conduct investigations when true experiments cannot be used or when a true experimental design significantly reduces external validity.

SUMMARY

In experimental research, one or more independent variables (the treatment) are manipulated to assess the effects on one or more dependent variables (the response measured). Research studies are concerned with both internal and external validity. Internal validity requires controlling factors so that the results can be attributed to the treatment. Threats to internal validity include history, maturation, testing, instrumentation, statistical regression, selection biases, experimental mortality, selection–maturation interaction, and expectancy.

External validity is the ability to generalize the results to other participants and to other settings. Four threats exist to external validity: reactive or interactive effects of testing, the interaction of selection biases and the experimental treatment, reactive effects of experimental arrangements, and multiple-treatment interference. Having high degrees of both internal and external validity is nearly impossible. The rigid controls needed for internal validity make it difficult to generalize the results to the real world. Conversely, studies with high external validity are usually weak in internal validity. Random selection of participants and random assignment to treatments are the most powerful means of controlling most threats to internal and external validity.

Preexperimental designs are weak because they can control few sources of invalidity. True experimental designs are characterized by random formation of groups, which allows the assumption that the groups were equivalent at the beginning of the study. The randomized-groups, pretest–posttest randomized-groups, and Solomon four-group designs are examples of true experimental designs.

Quasi-experimental designs are often used when it is difficult or impossible to use true experimental designs or when a true experimental design significantly limits external validity. The time-series design, reversal design, nonequivalent-control-group design, and ex post facto design are the most commonly used. The switched-replication design and single-subject design are potentially useful but less frequently used quasi-experimental designs.

Check Your Understanding

Locate two research papers in refereed journals in your area of interest. One paper should have a true experimental design and the other a quasi-experimental design.

1. Describe the type of design. Draw a picture of it using the notation from this chapter.
2. How many independent variables are there? How many levels of each? What are they?
3. How many dependent variables are there? What are they?
4. What type of statistical analysis was used? Explain how it fits the design.
5. Identify the threats to internal validity that are controlled and uncontrolled. Explain each.
6. Identify the threats to external validity that are controlled and uncontrolled. Explain each.

19

Qualitative Research

Although qualitative research in physical education, exercise science, and sport science was once considered relatively new, it is now common in many areas of kinesiology. Researchers in education began adapting ethnographic research design to educational settings in the United States in the 1970s (Goetz & LeCompte, 1984), and a great deal of qualitative research in physical education and sport science has been conducted steadily since the 1980s. In 1989, Locke was invited to write an essay and tutorial on qualitative research that was published in *Research Quarterly for Exercise and Sport,* which promoted further qualitative research in the field.

Qualitative research methods have been employed in anthropology, psychology, and sociology for many years. This general form of research has been called by various names, including ethnographic, naturalistic, interpretive, grounded theory, phenomenological, and participant observational research. Although the approaches are all slightly different, each "bears a strong family resemblance to the others" (Erickson, 1986, p. 119).

Those of you who are unfamiliar with the term *qualitative research* may benefit by reviewing your courses or readings in anthropology. Nearly everyone has heard of classic ethnographic studies (even if not by that name), such as the famous cultural research that Margaret Mead conducted when she lived in Samoa. Mead interviewed the Samoans at length about their society, their traditions, and their beliefs. This research is an example of qualitative research. But you do not need to live in a remote, exotic land to do it. The qualitative research that we refer to in this chapter is done mainly in everyday settings, such as schools, gymnasiums, sport facilities, fitness centers, and hospitals.

It could be argued that the term *qualitative research* may be too restrictive in that it seems to denote the absence of anything quantitative, but this is certainly not the case. Nonetheless, qualitative research seems to be the term used most often in our field. And, as we note in the next chapter, mixed-methods research—using both qualitative and quantitative methods—is being conducted in many fields. Our discussion of qualitative research focuses on the interpretive method as opposed to the so-called thick, rich description that characterized early research in anthropology, psychology, and sociology. This latter approach, involving a long and detailed account of the entity or incident, was espoused by Franz Boas, considered the father of cultural anthropology. In rejecting the armchair speculation that typified late-19th-century work, Boas insisted that researchers not only collect their own data but also report it with as little comment or interpretation as possible (cited in Kirk & Miller, 1986). Now, much of the research produced in kinesiology and related areas involves interpretation. Researchers collect data and make interpretations to address specific questions and to understand what is taking place in the setting(s) or with the person(s) being studied.

We do not intend to present a comprehensive review of qualitative research, because we do not have the space to do so. Moreover, we would need to address the considerable disagreement among qualitative researchers concerning their methodologies and theoretical presuppositions. A comprehensive review would be a major undertaking, so throughout this chapter we will suggest places to find further information. To start, you can gain a solid appreciation of the scope of qualitative research, the approaches to data collection and analysis, and the subtleties of approaches to qualitative research in compact form from Locke, Silverman, and Spirduso (2010) or in a more expansive form from Creswell (2013). Readable, single-source texts on conducting qualitative research include those of Bogdan and Biklen (2007), Creswell (2014), Marshall and Rossman (2011), Morse and Richards (2002), and Patton (2002). You will also learn from reading a qualitative study, such as the one in physical education by McCaughtry and Rovegno (2003) or the one in sport psychology by Bhalla and Weiss (2010).

PROCEDURES IN QUALITATIVE RESEARCH

Qualitative research is done in a variety of ways. Consequently, the procedures outlined in this section should be viewed simply as an attempt to orient readers who are unfamiliar with qualitative research. As with any type of research, the novice will benefit most from reading books that specifically focus on planning research (e.g., Locke, Spirduso, & Silverman, 2014) or qualitative methods, such as those listed in the previous paragraph, and recently completed qualitative research studies that used particular methods.

Defining the Problem

We do not spend much time on defining the problem in this chapter because we have talked about this step in earlier chapters, and it does not differ appreciably from that used with other research methods. We emphasize that researchers have several methods to choose from for any research problem. Different methods can yield different information about the problem. Thus, when researchers decide to use qualitative research as opposed to some other design, the decision is based on what they want to know about the problem.

Formulating Questions and Theoretical Frameworks

In highlighting some differences between quantitative and qualitative research, we stated that qualitative research usually builds hypotheses and theories in an inductive manner—that is, as a result of observations. Quantitative research often begins with research hypotheses that are subsequently tested. Thus, most qualitative studies do not have hypotheses as a part of the introduction. At one time it was common for those doing qualitative research not to state either hypotheses or questions, but in recent years it has become more common for qualitative researchers to state questions that will be the focus of the study. These questions are not as specific as hypotheses in quantitative research. They do, however, provide the direction that is the research focus. For example, the question What is the perspective of children in a competitive youth sport program? focuses the study and provides a great deal of information about the research direction. Generally, the types of qualitative studies that we address in this chapter start with questions that guide the studies. Although researchers may be able to move in other directions and modify questions based on the ongoing research, they are unlikely to totally abandon the original questions.

Qualitative researchers, like other researchers, develop a case for why it is important to do a study. This justification appears in the introduction of a research paper or as part of the rationale or significance of the study sections in a thesis or dissertation. One way this occurs is by providing a theoretical framework—a theory that guides the research. A theoretical framework frames the study and is used throughout to develop questions, design the method, and analyze data. Although many qualitative studies use a theoretical framework, it is possible to provide a rationale for a study by using other published research to show how this research fills a gap in our knowledge (Locke, Silverman, & Spirduso, 2010; Locke, Spirduso, & Silverman, 2014; Shulman, 2003). In either case, in most qualitative research, the questions are situated in previous research and theory and are based on extending what we know about a particular area.

Before moving on to other procedures for qualitative research, we should note that determining the theoretical framework that is appropriate for a study may require a great deal of reading and consultation with others. Because the theoretical framework influences all research aspects, you must thoroughly understand it before using it to frame a study, and so you can write about it in a way that others can easily understand. The process of understanding the theory and then applying it to a study often involves time, discussion, and testing the way it will influence future research. If you are venturing into a new theory or beginning to design a qualitative study that will employ theory, we highly recommend Maxwell (2013) and Rossman and Rallis (2012) as the first resources to consult. Then you may want to read more in-depth resources on theory construction such as Jaccard and Jacoby (2010). Having an advisor or specialist in qualitative research with whom to discuss these readings will help you develop the framework.

Collecting the Data

Several components are involved in collecting data for qualitative research, just as in quantitative research. Training and pilot work are still necessary, and so is selecting participants appropriately. In addition, you need to enter the field setting and be as unobtrusive as possible in your data collection.

Training and Pilot Work

In qualitative research the investigator is the instrument for collecting and analyzing the data. It is imperative, then, that the researcher be adequately prepared. Certainly, coursework, fieldwork reports, and interaction with your advisor are helpful, but ultimately the only way to become competent is through hands-on experience. You should gain this experience with an advisor who has a background in and experience with qualitative research. At many universities students begin this process in a class in which they learn and do qualitative research under the direction of an experienced qualitative researcher.

As always, pilot work is essential, and fieldwork experience in a setting similar to that of the proposed study is recommended. Experience in the field setting (e.g., as an instructor, a coach, or a player) is helpful in most respects. Locke (1989), however, made a good point that familiarity tends to spawn "an almost irresistible flood of personal judgments" (p. 7) that, if not recognized and controlled, could become a significant threat to the integrity of the data in a qualitative study.

Selecting Participants

Basically, selecting participants for a qualitative study is the same as that described for the case study in chapter 16. Qualitative research studies do not attempt to make inferences from their participants to some larger population. Rather, the participants are selected because they have certain characteristics. Obviously, pragmatic concerns arise about the location and the availability of participants because, in nearly every case, numerous other sites and people with similar characteristics exist.

Probability sampling is not used, simply because there is no way to estimate the probability of each person's being selected and no assurance that each person has some chance of being included. Instead, the selection of participants in qualitative research is purposeful, which in essence means that researchers select a sample from which they can learn the most. A researcher, for example, may be looking for people with certain levels of expertise or experience. There are several ways to guide participant selection (e.g., see Marshall & Rossman, 2011). In essence, it involves determining who will provide the data that will help answer the research questions while also deciding where the study will take place and how data will be collected. Further, the researcher must determine whether the research focus is on a particular setting, certain individuals, or both (Creswell, 2013).

> A purposefully chosen sample is one from which the most can be learned.

Entering the Setting

The researcher must have access to the field setting to conduct a qualitative study. Moreover, in most cases, the researcher must be able to observe and interview the participants at the appropriate time and location. These details may seem somewhat trivial, but nothing is more important than site entry. In any type of research, you must have access to the data, whether it is source material in historical research or participants in an experimental or survey study. However, the problem can be of greater magnitude in qualitative research than in most other methods. The quantitative researcher is simply "borrowing" the participants for a short time for some measurements or taking a little class time to administer a questionnaire. In qualitative research, the investigator is often at the site for weeks or months. This outsider is listening, watching, coding, and video-recording, as well as imposing on the time of the teacher (or coach or other person) and the students (or players or other participants) for interviews and observations.

I believe
all of these
participants are
small giraffes.

We are intentionally belaboring this topic because it is so important. Obviously, diplomacy, personality, and artful persuasion are required to gain site entry. Frankly, some people just cannot do this. Even when entry is achieved, some studies have failed because the investigator rubbed people the wrong way and the participants were not motivated to cooperate fully. The negotiation of gaining access to the participants in their natural setting is important and complex. It starts with the first contact by telephone or letter, extends through data collection, and continues after the researcher has left the site (Rossman & Rallis, 2012).

Before we discuss aspects of data collection, we should elaborate a little more on the topic of participant cooperation. Rapport is everything. Unless the participants believe that they can trust you, they will not give you the information you seek. Obviously, formal informed consent must be obtained, and the stipulations embodied in the whole concept of informed consent must be observed.

Qualitative research involves a number of ethical considerations simply because of the intensive personal contact with the participants. Thus, the participants need to know that provisions will be followed to safeguard their privacy and guarantee anonymity. If, for some reason, it is impossible to keep information confidential, this circumstance must be made clear. The researcher must give a great deal of thought to these matters before collecting data, must be able to explain the purpose and significance of the study effectively, and must convey the importance of cooperation in language participants can understand. One of the largest obstacles in the quest for natural behavior and candor on the part of the participants is their suspicion that the researcher will be evaluating them. The researcher must be convincing in this regard. The most successful studies are those in which the participants feel as though they are a part of the project. In other words, a collaborative relationship should be established.

Methods of Collecting Data

The most common sources of data collection in qualitative research are interviews, observations, and documents (Creswell, 2014; Locke, Silverman, & Spirduso, 2010; Marshall & Rossman, 2011). The methodology is planned and pilot-tested before the study. Creswell (2014) placed the data-collecting procedures into four categories: observations, interviews, documents, and audiovisual materials. He provided a concise table of the four methods, the options within each type, and their advantages and limitations.

We noted previously that the researcher typically has some type of framework (subpurposes perhaps) that determines and guides data collection. For example, one phase of the research might pertain to the way expert and nonexpert sport performers perceive aspects of a game. This phase could involve having athletes describe their perceptions of what is taking place in a scenario. A second phase of the study might focus on the interactive thought processes and decisions of the two groups of athletes while they are playing. The data for this phase could be obtained from filming them in action and then interviewing them while they are watching their performances on video. Still another aspect of the study could be directed at the knowledge structure of the participants, which could be determined by a researcher-constructed instrument.

You should not expect qualitative data collection to be quick. It is time intensive. Collecting good data takes time (Locke, Silverman, & Spirduso, 2010), and quick interviews or short observations are unlikely to help you gain more understanding. If you are doing qualitative research, you must plan to be in the environment long enough to collect good data and understand the nuance of what is occurring.

Interviews

The interview is undoubtedly the most common source of data in qualitative studies. The person-to-person format is most prevalent, but occasionally group interviews and focus groups are conducted. Interviews range from the highly structured style, in which questions are determined before the interview, to the open-ended, conversational format. In qualitative research, the highly structured format is used primarily to gather sociodemographic information. For the most part, however, interviews are more open ended and less structured (Merriam, 2001; Rubin & Rubin, 2012). Frequently, the interviewer asks the same questions of all the participants, but the order of the questions, the exact wording, and the type of follow-up questions may vary considerably.

Often, qualitative researchers conduct semistructured interviews in which the main interview questions, developed from theory or the questions guiding the study, are written out along with likely follow-up questions. The major focus of each interview is the same with this format, but follow-up questions can explore other things that come up as the participant answered the main question. This approach allows for great flexibility, while keeping the interview focused.

Being a good interviewer requires skill and experience. We emphasized earlier that the researcher must first establish a rapport with the respondents. If the participants do not trust the researcher, they will not open up and describe their true feelings, thoughts, and intentions. Complete rapport is established over time as people get to know and trust one another. An important skill in interviewing is being able to ask questions in such a way that respondents believes that they can talk freely.

Kirk and Miller (1986) described their field research in Peru, where they tried to learn how much urban, lower-middle-class people knew about coca, the organic source of cocaine. Coca is legal and widely available in Peru. In their initial attempts to get the people to tell them about coca, they received the same culturally approved answers from all the respondents. Only after they changed their style to asking less sensitive questions (e.g., "How did you find out you didn't like coca?") did the Peruvians open up and elaborate on their knowledge of (and sometimes their personal use of) coca. Kirk and Miller made a good point about asking the right questions and the value of using various approaches. Indeed, this is a basic argument for the validity of qualitative research.

Skillful interviewing takes practice! Ways to develop this skill include video-recording yourself conducting an interview, observing experienced interviewers, role-playing, and critiquing peers. It is important that the interviewer appear nonjudgmental. This can be difficult in situations in which the interviewee's views are quite different from those of the interviewer. The interviewer must be alert to both verbal and nonverbal messages and be flexible in rephrasing and pursuing certain lines of questioning. The interviewer must use words that are clear and meaningful to the respondent and must be able to ask questions so that the participant understands what is being asked. Above all, the interviewer has to be a good listener. Developing these skills is not easy. When the participants are very young or come from a cultural background different from that of the interviewer, it is even harder. Targeting the questions to the right level and having the sensitivity to both ask questions and listen well so as to formulate follow-up questions require practice in the technique as well as with the sample being studied.

The use of a digital recorder is undoubtedly the most common method of recording interview data because it has the obvious advantage of preserving the entire verbal part of the interview for later analysis. Although some respondents may be nervous to talk while being recorded, this uneasiness usually disappears in a short time. The main drawback with recording is the possibility of equipment malfunction. This problem is vexing and frustrating when it happens during the interview, but it is devastating when it happens afterward when you are trying to replay and analyze the interview. Certainly, you should have fresh batteries or have recently charged the recorder, and you should make sure it is working properly early in the interview. You should also stop and play back some of the interview to see whether the person is speaking into the microphone loudly and clearly enough and whether you are getting the data. Some participants (especially children) love to hear themselves speak, so playing back the recording for them can also serve as motivation. Remember, however, that machines can malfunction at any time.

Video-recording may seem to be the best method because you preserve not only what people say but also their nonverbal behavior. The drawback to using video, however, is that it can be awkward and intrusive. Therefore, it is used infrequently. Taking notes during the interview is another common method. Occasionally note taking is used in addition to recording, primarily when the interviewer wishes to note certain points of emphasis or make additional notations. Taking notes without recording prevents the interviewer from recording all that is said. Also, because taking notes interferes with the interviewer's thoughts and observations while the respondent is talking, this method is rarely used. In highly structured interviews and when using some types of formal instrument, the interviewer can more easily take notes by checking off items and writing short responses.

In addition to formal interviews, many researchers employ informal interviews to follow up on things they observed or about which they have questions. For example, a researcher who observes a physical education teacher quickly changing to another task during a class can conduct an informal interview to learn why this occurred. These interviews often are very short (e.g., taking place between physical education classes or on a walk from an exercise class to the locker room), and the researcher makes notes afterward. Although taking notes after a formal interview would be disastrous, doing so is common with informal interviews.

Focus Groups

Another type of qualitative research technique employs interviews on a specific topic with a small group of people, called a **focus group**. This technique can be efficient because the researcher can gather information from several people in one session. The group is usually homogeneous, such as a group of students, an athletic team, or a group of teachers.

Patton (2002) argued that focus group interviews might provide quality controls because participants tend to provide checks and balances on one another that can curb false or extreme views. Focus group interviews are usually enjoyable for participants, who may be less fearful of being evaluated by the interviewer because of the group setting. The group members get to hear what others in the group have to say, which may stimulate them to rethink their own views.

In the focus group interview, the researcher is not trying to persuade the group to reach consensus. It is an interview. Taking notes can be difficult, but an audio recorder will solve that problem. Certain group dynamics such as power struggles and a reluctance to state views publicly are limitations of the focus group interview. The number of questions that can be asked in one session is limited. As with most qualitative data collection techniques, the focus group should be used in combination with other data-gathering techniques.

Developing the skills to collect focus group data requires practice and training similar to that required to develop interview skills. The works of Krueger and Casey (2009) and Morgan (1997) are excellent places to begin learning about focus groups. Be sure to practice, as discussed in the preceding section on interviews, to develop the ability to ask follow-up questions and keep the group engaged.

focus group—A small group of people interviewed about a specific topic as a method of qualitative research.

Observation

Observation in qualitative research generally involves spending a prolonged amount of time in the setting. Field notes taken throughout the observations focus on what is seen. Many researchers also record notes to help them determine what the observed events might mean and to help them answer the research questions during subsequent data analysis (Bogdan & Biklen, 2007; Pitney & Parker, 2009). Although some researchers use cameras to record what is occurring at the research site, most use field notes.

Most researchers develop a note-taking technique adapted to the setting. For example, a researcher who is studying participants' perspectives about a group exercise class finds a place to observe where she can see and hear everything and take notes without being obtrusive. She takes notes on what she sees—both what occurred and participants' reactions. She also makes additional comments to help focus future observations or remind herself about how what she is observing ties back to the study's theoretical framework. Additionally, she uses different pages or different colored pens for each type of note to easily identify everything later.

One major drawback to observation is obtrusiveness: a stranger with a notebook and pen, a computer (which is not often used because of participants' reactions), or a camera recording behaviors. The key word here is *stranger*. The task of qualitative researchers is to make sure the participants become accustomed to having them (and, if appropriate, their recording devices) around. They may do this by visiting the site for at least a couple of days before data collection begins.

In an artificial setting, researchers can use one-way mirrors and observation rooms. In a natural setting, the limitations that stem from the presence of an observer can never be ignored. Locke (1989) observed that most naturalistic field studies are reports of what goes on when a visitor is present. The important question is How important and limiting is this? Locke suggested ways to suppress reactivity, such as the visitor's being in the setting long enough that he is no longer a novelty and being as unobtrusive as possible in everything from dress to choice of location in a room.

Other Data-Gathering Methods

Among the many sources of data in qualitative research are self-reports of knowledge and attitude. The researcher can also develop scenarios, in the form of descriptions of situations or pictures, that are acted out for participants to observe. The participants then give their interpretations of what was going on in the scenarios. The participants' responses reveal their perceptions, interpretations, and awareness of the situation and of the interplay of the actors in the scenarios.

Other sources of data include documents used in the setting (e.g., employee handbooks, procedure manuals, handouts and e-mails, and fliers and other material distributed or posted around the research site), photos of the setting, pictures and collages made by participants to represent their perspectives, participants' diaries, Internet chat transcripts, and websites. In most studies, researchers use multiple sources of data to reach conclusions during data analysis.

DATA ANALYSIS

Data analysis in qualitative research is quite different from that in conventional quantitative research. First, analysis is done during and after data collection in qualitative research. During data collection, researchers sort and organize the data and speculate and develop tentative hypotheses to guide them to other sources and types of data. Qualitative research is somewhat similar to multiple-experiment research, in which discoveries made during the study shape each successive phase of the study. Simultaneous data collection and analysis allows researchers to work more effectively. Analysis then becomes more intensive after the data have been collected (Merriam, 2001). Another difference between quantitative and qualitative data analysis is that qualitative data are generally presented in words, descriptions, and images, whereas quantitative data are typically presented in numbers.

Data analysis in a qualitative study can take various forms, depending on the nature of the investigation and the defined purposes. Consequently, we cannot go into great depth in discussing analysis without tying it to a specific study. For this reason, we have sum-

> Subsequent phases of qualitative research are shaped by discoveries made in previous phases.

marized the general phases of analysis synthesized from descriptions in several qualitative research texts. The general phases are sorting and analysis during data collection, analysis and categorization, and interpretation and theory construction. First, however, comes the process of data reduction.

Data Reduction

Prior to doing actual data analysis, all of the data must be reduced so it can be analyzed by the researcher using a computer (Locke, Spirduso, & Silverman, 2014). This involves a number of steps and often is quite labor intensive. Formal interviews and focus group interviews must be transcribed word for word. Often, researchers add comments (e.g., "laughing," "angry," "crying") to denote emotion so that what was said is not confused during analysis. Informal interviews also are transcribed from the researcher's notes. After a formal interview is transcribed, the researcher should read the printout and listen to the interview to make certain it is accurate and to add or adjust comments. Informal interviews also are checked to make sure the notes agree with what was entered into the computer.

Researchers enter field notes into the computer while taking care to clearly delineate what they observed and what they wrote down to inform their analysis or theory. They enter field notes many different ways, and the process is often as idiosyncratic as note taking itself. Documents are also entered into the computer either via scanning (e.g., a flier found on the wall at the research site), importing from other electronic sources (e.g., digital pictures), or cutting and pasting from other documents (e.g., an Internet chat transcript). The accuracy and completeness of the data must be verified before analysis can begin.

Sorting, Analyzing, and Categorizing Data

The simultaneous collection and analysis of data is an important feature of qualitative research. As noted, researchers add comments about their analyses while taking notes. This occurs at other times also: when they are reducing data and then beginning analysis while still other data are collected. This approach enables researchers to focus on certain questions and, in turn, to better direct data collection. Although researchers have specific questions in mind when data collection begins, they may shift their focus as the data unfold.

Researchers need to keep in close touch with the data. Waiting until after the data are collected to analyze them is foolish. Decisions must be made concerning scope and direction, or they may end up with data that are unfocused or repetitious, or an overwhelming volume of material to process (Merriam, 2001). Also, there could be gaps because needed evidence was not collected. Researchers want data that answer their questions. Examining data during data collection will reveal areas in which there is more than enough data (often called data saturation) and areas that are not as complete and need a greater focus.

Researchers typically write many comments to stimulate critical thinking while taking notes or entering data into the computer. They should not be merely human recording machines; rather, they should try out new ideas and consider how certain data relate to large theoretical, methodological, and substantive issues (Bogdan & Biklen, 2007). It is quite common for researchers to reexamine their original questions to make sure their analysis is on track or to decide that the data lead to additional or different questions (Miles, Huberman, & Saldaña, 2014).

Analysis is the process of making sense of the data. Once the data are entered into the computer and the researcher has confirmed that they are accurate, analysis begins. Initial analytical notes, categories from theory, and questions inform the analysis. This is the beginning of the stages of organizing, abstracting, integrating, and synthesizing, which ultimately permit researchers to report what they have seen, heard, and read. They may develop an outline to search for patterns that can be transformed into categories. They should pay attention to how they analyze data, however, or their conclusions may turn out like the quotes from church bulletins in the sidebar.

The qualitative researcher faces a formidable task in sorting the data for content analysis. Obviously, many categories can be devised for any set of data, depending on the problem being studied. For example, a researcher could categorize observations of a physical education class in terms of the teacher's management style, the social interactions among students, sex differences in behavior or treatment, or verbal and nonverbal instructional behaviors. Categories can range in complexity from relatively simple units of behavior types to conceptual typologies or theories (Merriam, 2001; Miles, Huberman, & Saldaña, 2014).

Researchers use a variety of techniques for sorting data. Index cards and file folders were once used but are not common today. Computer software is now used to store, sort, and retrieve data. Most universities at which qualitative researchers are active have one or two of the most popular qualitative software programs (e.g., NVivo or Atlas.ti) available through their computer centers. In addition, many of these universities offer classes (both credit and noncredit) in how to use the software. If you are considering a qualitative study, these classes offer an excellent opportunity to see the characteristics of the software programs

Quotes From Church Bulletins

1. Bertha Belch, a missionary from Africa, will be speaking tonight at Calvary Methodist. Come hear Bertha Belch all the way from Africa.

2. Weight Watchers will meet at 7 p.m. at the First Presbyterian Church. Please use the large double door at the side entrance.

3. Ladies, don't forget the rummage sale. It's a chance to get rid of those things not worth keeping around the house. Don't forget your husbands.

4. Miss Charlene Mason sang "I Will Not Pass This Way Again," giving obvious pleasure to the congregation.

5. The rector will preach his farewell message, after which the choir will sing "Break Forth Into Joy."

6. For those of you who have children and don't know it, we have a nursery downstairs.

7. Next Thursday there will be tryouts for the choir. They need all the help they can get.

8. Barbara remains in the hospital and needs blood donors for more transfusions. She is also having trouble sleeping and requests tapes of Pastor Jack's sermons.

9. Irving Benson and Jessie Carter were married on October 24 in the church. So ends a friendship that began in their school days.

10. At the evening service tonight, the sermon topic will be "What Is Hell?" Come early and listen to our choir practice.

11. Please place your donation in the envelope along with the deceased person you want remembered.

12. The eighth-graders will be presenting Shakespeare's *Hamlet* in the church basement Friday at 7 p.m. The congregation is invited to attend this tragedy.

13. The ladies of the church have cast off clothing of every kind. They may be seen in the basement on Friday afternoon.

14. The pastor would appreciate it if the ladies of the congregation would lend him their electric girdles for the pancake breakfast next Sunday.

15. Low Self-Esteem Support Group will meet Thursday at 7 p.m. Please use the back door.

and evaluate their usefulness. Of course, if you have the opportunity to take a class in which you collect pilot data and then use the software, you can improve your eventual efficiency. And, if you have a faculty member with whom you can consult, you can get help and input throughout the analysis process.

Data categorization is a key facet of true qualitative research. Instead of using mere description, the researcher may use descriptive data as examples of the concepts that are being advanced. The data need to be studied and categorized so that the researcher can retrieve and analyze information across categories as part of the inductive process. Qualitative computer analysis programs permit great flexibility in designing the categories and marking data to examine possible conclusions.

Interpreting the Data

When the data have been organized and sorted, the researcher then attempts to merge them into a holistic portrayal of the phenomenon. An acknowledged goal of qualitative research is to produce a vivid reconstruction of what happened during data collection. This goal is accomplished through the **analytical narrative**, or **analytical memoing** (Miles, Huberman, & Saldaña, 2014). This can consist of a descriptive narrative organized chronologically or topically. In our concept of qualitative research, however, we support the position of Flick (2014, p. 375) that "interpretation is the core activity of qualitative data analysis." As Peshkin (1993) observed, "Pure, straight description is a chimera; accounts that attempt such a standard are sterile and boring" (p. 24). Goetz and LeCompte (1984) and many others (e.g., Creswell, 2013; Rossman & Rallis, 2012) maintained that by leaving readers to their own conclusions, the researcher risks misinterpretation and perhaps trivialization of the data by readers who are unable to make the implied connections. They further suggested that the researcher who can find no implications beyond the data should never have undertaken the study in the first place.

> **analytical narrative**—A short interpretive description of an event or situation used in qualitative research; also called *analytical memoing*.

The analytical narrative or memo is the foundation of qualitative research. Researchers (especially novices) are often reluctant to take the bold action needed to assign meaning to the data. Erickson (1986) suggested that to stimulate analysis early in the process, researchers should force themselves to make an assertion, choose an excerpt from the field notes that substantiates that assertion, and then write a **narrative vignette** that portrays the validity of the assertion. In the process of making decisions concerning the event to report and the descriptive terms to use, the researcher becomes more explicitly aware of the perspectives emerging from the data. This awareness stimulates and facilitates further critical reflection.

> **narrative vignette**—A component of a qualitative research report that gives detailed descriptions of an event, including what people say, do, think, and feel in that setting.

The narrative vignette is one of the fundamental components of qualitative research. As opposed to the typical analysis sections in quantitative research studies (which are about as interesting as watching paint dry), the vignette captures the reader's attention, thereby helping the researcher make his or her point. The vignette gives the reader a sense of being there. A well-written description of a situation can convey a sense of holistic meaning that is definitely advantageous in providing evidence for the researcher's assertions. In characterizing qualitative research, Locke (1989) stated that the researcher can describe the physical education scene so vividly that "you can smell the lockers and hear the thud of running feet" (p. 4). Griffin and Templin (1989, p. 399) provided the following example of a vignette:

> **The second period physical education class at Big City Middle School is playing soccer. The teacher has placed two piles of sweatshirts at each end of a large open field to serve as goals. There are no field markings. Four boys run up and down the field following the ball. Several other students stand silently in their assigned positions until the ball comes near, then they move tentatively toward the ball to kick it away. Three girls stand talking in a tight circle near the far end of the field. They are startled when the ball rolls into their group, and two boys yell at them to get out of the way. They do and then regroup after**

the ball and the boys go to the other side of the field. Two boys, who have not touched the ball during the class, engage in a playful wrestling match near one goal. The teacher stands in the center of the field with a whistle in his mouth. He hasn't said anything since he divided students into teams at the beginning of class. He has blown the whistle twice to call fouls. The students play around him as if he were not there. A bell rings, and all the students drop their pinnies where they are and start toward the school building. Belatedly, the teacher blows his whistle to end the game and begins to move around the field picking up pinnies.

After class, as we walk back to the building, the teacher says, "These kids are wild. If you can just run off some of their energy, they don't get into so much trouble in school. This group especially, not too many smarts (taps his temple), don't get into much game strategy." (He sees a boy and girl from the class standing near the door of the girls' locker room talking.) He yells, "Johnson, get your butt to the shower and stop bothering the ladies." He smiles at me. "You've got to be on them all the time." He looks up and sighs, "Well, two [classes] down, three to go."

We caution you, however, that Locke and other scholars do not believe that richness of detail alone is what makes a narrative vignette valid. Siedentop (1989) warned that whether data are to be trusted should not be based on the narrative skills of the researcher. According to Erickson (1986), a valid account is not simply a description but an analysis: "A story can be an accurate report of a series of events, yet not portray the meaning of the actions from the perspectives taken by the actors in the event. . . . It is the combination of richness and interpretive perspective that makes the account valid" (p. 150). Vignettes should not be left to stand by themselves. The researcher should make interpretive connections between narrative vignettes and other forms of description, such as direct quotations and quantitative materials (Miles, Huberman, & Saldaña, 2014).

Direct quotations from interviews with the participants taken from field notes and audio or video recordings are another form of vignette that enriches the analysis and furnishes documentation for the researcher's point of view. They can demonstrate agreement (or disagreement) about a phenomenon. Direct quotations from the same people on different occasions can provide evidence that certain events are typical or demonstrate a pattern or trend in perceptions over time. For example, the quotes Woods and Lynn (2014) used throughout their results provide some of the original data that helped them arrive at their conclusions. Without these quotes, the material would be more difficult for most readers to grasp and would not represent the vitality of the data collected.

Qualitative researchers should communicate their perspectives clearly to their readers. The function of the narrative is to present the researcher's interpretive point in a clear and meaningful manner.

Theory Construction

theorizing—The cognitive process of discovering abstract categories and the relationships among those categories.

theory—An explanation of some aspect of practice that permits the researcher to draw inferences about future happenings.

Our use of the terms *theory* and *theorizing* should not unduly alarm you or discourage you from undertaking qualitative research. We are not talking about developing a model on the scale of the theory of relativity here. **Theorizing**, according to Goetz and LeCompte (1984), is a cognitive process of discovering abstract categories and the relationships among those categories. A **theory** is an explanation of some aspect of practice that permits the researcher to draw inferences about future events. This process is a fundamental tool for developing or confirming explanations. You process information, compare the findings based on your experience and sets of values, and then make decisions. If the decisions are not correct, you will need to revise the theory or model.

In many qualitative research traditions, but not all, data analysis depends on theorizing. The tasks of theorizing are "perceiving; comparing, contrasting, aggregating, and ordering; establishing linkages and relationships; and speculating" (Goetz & LeCompte, 1984, p. 167). Perceiving involves the consideration of all sources of data and all aspects of the phenomena being studied. Of course, this takes place during as well as after data collection. The perceptual process of determining which factors to analyze guides the collection of data.

Comparing, contrasting, aggregating, and ordering are primary functions in qualitative research. The researcher decides which units are similar or dissimilar and what is important about the differences and similarities. Analytical description cannot occur until the researcher builds the categories of like and unlike properties and carries out a systematic content analysis of the data. Establishing linkages and relationships constitutes a kind of detective work that qualitative researchers do in the theorizing process. They use both inductive and deductive methods of establishing relationships "while developing a theory or hypothesis that is grounded in the data" (Goetz & LeCompte, 1984, p. 172). As Corbin and Strauss (2008) noted, multiple interpretations of the same data are possible, but all interpretations should be grounded in data.

A theory based on and evolving from data is called a **grounded theory** (Glaser & Strauss, 1976). In applied research, grounded theories can be used for explaining observed phenomena, understanding relationships, and drawing inferences about future activities.

grounded theory— A theory based on and evolving from data.

Trustworthiness in Qualitative Research

Qualitative researchers do not attempt to provide numerical evidence that their data are reliable and valid. In fact, the terms *reliability* and *validity* are rarely used in qualitative research. That does not mean, however, that the qualitative researcher is not concerned with getting good data and reaching conclusions in which readers can have faith. Both issues are extremely important, and without attention to them, the quality of the research may be suspect. Just as for quantitative data, those reporting qualitative data make a case for the quality of the data *and* the conclusions that were derived from the data analysis.

A variety of terms are used to describe quality in qualitative research. The terms have changed and still are evolving. Lincoln and Guba (1985) used the term **trustworthiness** to describe the overall quality of the results from the study, and this term is used most often. Rossman and Rallis (2012) broke trustworthiness down into two questions: Is the study competently conducted? and Is it ethically conducted?

trustworthiness—A quality achieved in a study when the data collected generally are applicable, consistent, and neutral.

Conducting an Ethical Qualitative Study

We discuss the issue of ethics first. Ethics often plays a prominent role in judging qualitative research. We can think of ethics in many ways, but it boils down to two major areas. First, the ethical treatment of study participants (Locke et al., 2014) is important in qualitative research. Qualitative researchers often spend a great deal of time with participants, and they should treat them with dignity—from the time they approach them to be a part of the study through the reporting of data. If researchers have promised anonymity, they should maintain it. This idea does not suggest that those doing quantitative research should not treat participants ethically. The relationships and the type of data are different in qualitative research, however, and often require a more sustained commitment to ethics. Second, Rossman and Rallis (2012) provided an in-depth discussion of the ethics of qualitative research. These authors and others (Creswell, 2013; Locke, Spirduso, & Silverman, 2014) suggested confronting ethical issues while designing the study. Without forethought, you can get into an ethical quagmire that can make it impossible to complete or publish the study.

This discussion brings us to the issue of what constitutes a competently conducted qualitative study. Again, we can think of this subject in a variety of ways. Lincoln and Guba (1985) described four concepts that can be used in qualitative research to think about quality that are still used today when designing research and completing data analyses. The first concept

credibility—A quality achieved when the participants and setting of a study are accurately described.

transferability—In qualitative research, a quality achieved when the results have the potential to be transferred to other settings.

is **credibility**. Because the context, participants, and settings are important to interpreting the results of qualitative research, how the researcher understands them and the description presented in a paper are critical to judging other parts of the study. If it is not clear who the participants are and where the study is being completed, readers will have difficulty evaluating the conclusions.

Transferability, the second important concept in evaluating quality, addresses whether the results would be useful to those in other settings or conducting research in similar settings. You may be thinking that this concept is like generalizability, and in some ways it is, but in other ways it is not. All researchers would like their research to help others, but like much of the quantitative research in our area, most qualitative research does not benefit from random selection from a large population. Transferability is a question of argument and perception. The researcher may present reasons a given study might apply in other settings (e.g., many schools operate in a similar fashion, or the participants in an exercise class are similar to those found in many community settings), but ultimately, the readers or users of the research must determine whether the study applies to their work environments or future research. Again, this often is the case in quantitative research—qualitative researchers just start with that concept up front.

The next concept is **dependability**. In many qualitative studies, researchers make changes in the phenomena being studied or in the methods used based on previous data collection. For a relatively straightforward example, a researcher may use a structured interview to ask questions but change how he asks follow-up questions depending on the answers. A second set of interviews might be based on the answers provided during the first set of interviews. If the researcher asked questions in a lockstep manner without adjustments, the quality of the data would suffer. In this example, the researcher could not possibly plan for all possible contingencies. How the researcher deals with the changes determines the dependability of the data.

dependability—A characteristic that addresses the quality of the data in a qualitative study, including how well the researcher deals with change.

confirmability—A characteristic of qualitative research that addresses whether another can place faith in the results.

Finally, **confirmability** deals with the issue of researcher bias. As we discuss later in this section, qualitative research differs from quantitative research in that researchers do not calculate reliability statistics. Rather, they use approaches to confirm their findings so that readers can have faith in the results of their studies.

Providing Evidence of Trustworthiness

Researchers can provide evidence of the trustworthiness of their studies (i.e., whether they were conducted well) in a variety of ways based on the preceding categories. We will review the common techniques that qualitative researchers use to gather good data and reach objective conclusions. These techniques are used during both data collection and data analysis. As noted earlier, analysis often occurs during data collection, so separating these tasks into temporal categories is difficult. Not all researchers use all techniques in every study, but researchers commonly use multiple methods to increase the trustworthiness of their data and conclusions. The following techniques are commonly used:

■ *Prolonged engagement.* The researcher must spend enough time to obtain good data. The collection of qualitative data requires that the researcher spend sufficient time in a setting to develop an in-depth understanding and not reach superficial conclusions.

■ *Audit trail.* The method and focus in qualitative research change during a study. An audit trail describes the changes that occurred and how they influenced the study. These changes are often reported in the methods section, where the researcher explains how the changes improved the study.

■ *Rich, thick description.* As noted earlier, describing the setting and context is important for credibility. A thorough description in the report is needed so that readers can both understand the study and assess whether the setting and results will transfer to their settings or future research.

■ *Clarification of researcher bias.* All researchers come into a study with biases. The management of these biases is particularly important in qualitative research in which the researcher is the instrument of data collection. Evidence that researchers acknowledged their biases and dealt with them is essential. In a move that we believe enhances the quality of research, many qualitative researchers address this directly in the methods section by "coming clean" and presenting their biases and explaining how they worked to control them.

■ *Triangulation.* This technique is often used in qualitative data analysis. It requires the use of three (hence the prefix *tri*) independent sources of data to support a conclusion. For example, a researcher may use student interview data, teacher interview data, and observation to support a result. Note that this is not simple vote counting (i.e., three people said this in an interview), but the use of independent types of data to support a conclusion.

■ *Negative case checking.* This technique is used when researchers look at instances in which what they expected to happen did not. For example, in an exercise class, the researcher notes that as time goes on, many participants seem more upbeat and are integrating the class into their lives. Does this happen for everyone? For those for whom it does not happen, is the phenomenon not as pervasive, did the researcher selectively look at the data to reach a conclusion, or is something different happening for some participants? Negative case checking helps address bias and enables the researcher to investigate the phenomenon further.

■ *Member checking.* A member check occurs when the researcher goes back to the participants to share the conclusions and see whether they agree with them. Sometimes an intermediate step occurs in which the researcher asks the participants to review interview transcripts, clarify statements, and add anything they believe is missing. That by itself is not true member checking. Presenting the participants with the conclusions goes much further and provides a confirmation of the analysis. In some cases in which the participants believe that the researcher may have gotten it wrong, the conclusions may be modified to reflect the participants' views. In other cases, the researcher may believe that bias (e.g., when a report does not reflect positively on a person) influenced a participant's evaluation and would then include in the report an analysis of that person's evaluation.

■ *Peer debriefing.* As in much of what we do, a new set of eyes can bring new light to a data set and our conclusions. A peer debriefer is another person who examines the data and the conclusions and serves as a devil's advocate, questioning the researcher to see whether the findings hold up. Often, the person performing this has a background in the phenomenon being investigated and expertise in qualitative methods and data analysis. This step can improve the conclusions and how they are presented in the research report.

Writing the Report

There is no standard (or correct) format for a qualitative research report, just as there are no formats that are rigorously followed for any other type of research. Here we simply mention some main components of a qualitative research report and their placement in the report. Your department or university may have an order of components that you must follow.

The components of a qualitative study are similar to those of other conventional research reports. The first part introduces the problem and provides background and related literature. A description of method is an integral part of the report. Although this section is not as extensive in a journal article as it is in a thesis, it is usually much more extensive than it is in other forms of research. The reasons are obvious, as we explained. The methodology is integrally related to the analysis and is also important in terms of trustworthiness and credibility.

The results and discussion section forms most of the report. If used, charts, tables, and figures are contained in this section and must be integrated into the narrative. A major contributor to the bulk of this section is the description contained in a qualitative study. As we said earlier, narrative vignettes and direct quotations are basic to this type of research. The qualitative study strives to provide enough detail to show the reader that the author's conclusions make sense (Creswell, 2014; Walcott, 2009). The author is faced with the delicate problem of finding a balance between the rich descriptive materials and the analysis and interpretation.

All academic writing requires rewriting and editing, and this is especially true of qualitative research. Writers should provide the theoretical framework for the study, describe the method, provide data to support the results (which often requires more words than in most quantitative studies), and discuss the contributions of the results to our understanding and to theory. Because this task requires some balancing, writers should take the time to decide which evidence will best elucidate the results and then integrate it with the other parts of the manuscript.

Funny Things That Happened During the Collection of Qualitative Data

1. On two occasions, researcher reported to gymnasium (carrying camera, TV monitor, recorder, and notes) to find door locked. Had to walk all the way around the building—once in the rain.
2. Reported to school to collect data to find classes wouldn't be held because of
    a. a doughnut sale,
    b. teacher being ill (twice), and
    c. holiday music rehearsal.
3. Interview with teacher interrupted by principal due to crisis over a parent and a charity clothing sale. Interview postponed.
4. Audio recorder became inoperable during interview.
5. Audio recorder became inoperable while transcribing. Had to repeat interviews.
6. Group recording session had to be rescheduled because of complete loss of control when a third-grade student belched into the microphone.
7. Had to buy own recorder because department's digital recorder was inoperable.
8. Broke own expensive new recorder. Repair took two weeks.
9. Had to use participating school's TV monitor but couldn't find proper cord. Interviews rescheduled.
10. During the making of backup audio files, the recorder and original recordings were stolen. Police nabbed the culprit and data were recovered, but the researcher was comatose.

It is doubtful that any graduate student is under the false impression that doing qualitative research is a quick-and-easy task. To emphasize this point, we call your attention to the sidebar Funny Things That Happened During the Collection of Qualitative Data, which lists some of the trials and tribulations that befell one doctoral student while she was attempting to collect data.

CONCLUDING REMARKS

At one time, qualitative research was thought to be confined primarily to the area of pedagogy. Getting that impression is easy because of the ever-increasing volume of literature on qualitative research in educational journals and textbooks. Qualitative research, however, is now common in many areas of research, including sport and exercise psychology, sport sociology, athletic training, and exercise prescription and adherence. If you read and use research in your professional work or actively conduct research, you will likely come across many qualitative studies.

This chapter has focused on interpretive qualitative research. Another approach, called **critical theory**, is emerging in the literature. The main difference between the two approaches is found in the research goals (Locke, Silverman, & Spirduso, 2010). Interpretive research is largely free of value judgments, whereas critical theory research is based on value judgments. In other words, in critical theory the aim is to give the research participants the insight necessary to make choices that improve their lives or empower them. Often, critical theory research is grounded in a theory that seeks to improve the human condition (e.g., feminism, neo-Marxism, the empowering pedagogy of Paulo Freire, the writings of Michel Foucault). Such theoretical perspectives challenge the status quo and strive for greater equality. You can begin exploring critical forms of qualitative research by reading Marshall and Rossman (2012); Locke, Silverman, and Spirduso (2010); and Locke, Spirduso, and Silverman (2014).

The sidebar Evaluating the Quality of Qualitative Research has a scale that can be used to evaluate the quality of qualitative research studies. It was originally developed by Linda Bain and presented at the 1992 national convention of the American Alliance for Health, Physical Education, Recreation and Dance. We modified it for this edition of this book. In addition, Locke, Silverman, and Spirduso (2010) provided detailed information for critiquing both qualitative and quantitative research.

> **critical theory**—Qualitative research based on value judgments.

EVALUATING THE QUALITY OF QUALITATIVE RESEARCH

LINDA L. BAIN
California State University, Northridge
AAHPERD, Indianapolis

DEFINITION OF THE PROBLEM
___ Purpose is clearly stated
___ Focuses on a significant issue
___ Seeks to understand meaning of experiences for the participants
___ Provides holistic view of the setting

DATA COLLECTION
___ Researcher has training in methods used
___ Pilot work done in similar setting using similar methods
___ Rationale provided for selection of the sample
___ Researcher has trusting, collaborative relationship with participants
___ Methods for data collection are unobtrusive, where appropriate

(continued)

Evaluating the Quality of Qualitative Research *(continued)*

___ Data collection procedures provide thorough description of events

___ Prolonged engagement in field

DATA ANALYSIS

___ Analysis done during and after data collection

___ Triangulation of data sources and search for convergence

___ Search for negative cases

___ Provides interpretation and theory as well as description of events

___ Provides opportunity for participants to corroborate interpretation (member check)

___ Arranges for peer evaluation of procedures and interpretation

PREPARATION OF REPORT

___ Complete description of setting

___ Complete description of procedures

___ Includes description of researcher's values, assumptions, and bias and how each was addressed

___ Uses vignettes and quotes to support conclusions and interpretation

GENERAL ASSESSMENT

___ Internal validity: How much confidence do you have in the quality of the description and interpretation of events in the particular research setting?

___ External validity: What is your assessment of the extent to which the results of this study apply to a different setting with which you are familiar?

Reprinted, by permission, from L.L. Bain, 1992, *Evaluating the quality of qualitative research*, 350.

SUMMARY

Qualitative research methods include field observations, interviews, and material collection. The researcher gathers data in a natural setting such as a gymnasium, classroom, fitness center, or sport facility.

Qualitative research does not have the preconceived hypotheses that characterize quantitative research. Inductive reasoning is stressed, whereby the researcher seeks to develop hypotheses from observations. The focus is on the essence of the phenomena. The researcher should exhibit sensitivity and perception when collecting and analyzing the data.

We stressed the importance of gaining access to data in the field setting. Establishing rapport and gaining the participants' trust are essential. The most common methods of collecting data are interviews and observations. Data should be analyzed during and after collection. The researcher must input the data into a computer to sort and organize it and develop tentative conclusions from the results that lead to other sources and types of data.

Data analysis involves organizing, abstracting, integrating, and synthesizing. The analytical narrative is the foundation of qualitative research. The narrative vignette gives the reader a sense of being present for the observation; it conveys a sense of holistic meaning to the situation. It is not unusual for a qualitative study to include quantitative analysis.

The qualitative researcher often attempts to construct a theory through the inductive process to explain relationships among categories of data. A theory that evolves from data is called a grounded theory. In the written report, the qualitative researcher must achieve a balance between the rich description and the analysis and interpretation.

Trustworthiness is used to determine whether the study was competent. Ethical issues are important in determining trustworthiness, as are issues related to how the study was conducted (i.e., credibility, transferability, dependability, confirmability). Many techniques are used during data collection and analysis to enhance the quality of the data and the conclusions. Among these are prolonged engagement, keeping an audit trail, providing a thick and rich description, clarifying researcher bias, triangulation, negative case checking, member checking, and peer debriefing. Not all of these can be used in one study, but using most or many is common to enhance the faith that readers can place in the conclusions and the transferability to other settings.

Qualitative research is a viable approach to solving problems in our field. It applies to pedagogy in physical education, to exercise science, and to sport science. Answers to the question What is happening here? can best be obtained in natural settings through the systematic observation and interactional methodology of qualitative research.

Check Your Understanding

1. Locate a qualitative study and write an abstract of approximately 300 words on the methods used in gathering and analyzing the data (observations, interviews, triangulation, member checking, and so on) and in presenting the results (narrative vignettes, quotations, tables, and so on).

2. Locate a qualitative study and write an abstract of approximately 300 words that describes why the study was conducted and what theoretical basis supported the study. Note whether the results answer the questions and add to the theoretical understanding that the authors presented in the introduction.

20

Mixed-Methods Research

> While they were saying among themselves it cannot be done, it was done.
>
> —Helen Keller

Mixed-methods research in kinesiology is relatively new. For a long time, scholars in many areas of research believed that combining quantitative and qualitative methods in a single study or research program was impractical. Qualitative and quantitative researchers bickered over the superiority of their methods (Denzin, 2008; Gelo, Braakmann, & Benetka, 2008; Hatch, 2006; Reichardt & Rallis, 1994; Shavelson & Towne, 2002). Gage (1989) characterized these debates as the paradigm wars and suggested that they would not advance research. Others, however, continued to suggest that one method was superior to another.

The quotation from Helen Keller captures what happened during this debate. Researchers began combining methods—and asking questions that could not be answered using only quantitative or qualitative research. Books were published that discussed mixed-methods research (e.g., see Green & Caracelli, 1997, for one of the first). And researchers found that the use of mixed methods helps answer complementary questions on a similar topic. For example, one of the authors of this book has done research on student attitude in physical education using primarily quantitative methods. While working with a graduate student, it became clear that the quantitative data allow for addressing certain questions (e.g., understanding the underlying structure of attitude and getting student attitude scores for various subgroups of students). It also became apparent that other methods were needed to address questions about what experiences influenced the attitudes of students. This discovery resulted in a series of interrelated studies (Subramaniam & Silverman, 2000, 2002, 2007) that used both quantitative and qualitative methods to approach questions about student attitude. As we discussed earlier in this book, different questions require different methods—even when the questions are closely related.

Teddlie and Tashakkori (2009) noted that mixed-methods research is largely a pragmatic enterprise. People do it because it can help answer questions and advance their research programs. That is not to suggest, however, that mixing methods is easy, because researchers must often work with others who have expertise in another methodological approach or start from scratch and learn new research skills. As with any research enterprise, you must become skilled in the method so that you can plan the study, collect data, analyze the results, and communicate with others about the research (Plano-Clark, Huddleston-Casas, Churchill, Green, & Garrett, 2008).

Besides learning about other research methods, you must learn more about combining methods when engaging in mixed-methods research. This chapter provides an overview of that process and allows you to begin planning a mixed-methods study. Other resources can assist in planning and conducting mixed-methods research. Both Flick (2014) and Robson (2002) provide good, pragmatic introductions to the topic of combining methods. Teddlie and Tashakkori (2009), Creswell and Plano-Clark (2007), and R.M. Thomas (2003) provide more detail in easy-to-read books. If further information is needed, the edited book by Tashakkori and Teddlie (2003) provides an in-depth discussion of mixed-methods research.

COMBINING QUANTITATIVE AND QUALITATIVE METHODS

There is no one way to do mixed-methods research. As we noted earlier, mixed-methods research is pragmatic, and the questions influence the method selected. A study may be primarily quantitative with a part that is qualitative. Conversely, it may be primarily qualitative with a quantitative component. Of course, it may be anything in between (see Morse, 2003, for an example of the many possible weightings of qualitative and quantitative research within one study). Figure 20.1 presents a continuum of mixed-methods research. The balance between quantitative and qualitative components will depend on the study, as will the methods used within each part of the study.

A researcher undertaking a mixed-methods study might collect descriptive quantitative measures and qualitative interview data based on an aspect of the quantitative data. An experimental intervention could be combined with a qualitative study evaluating participants' perceptions of the intervention. Interviews and focus groups could be supplemented by descriptive data. The permutations of mixed-methods designs are infinite. Any combination

Figure 20.1 The continuum of mixed-methods research from highly qualitative to highly quantitative.

of questions that suggest mixed methods can be accommodated as long as the researcher or research team has the skills to conduct all parts of the study.

Although many mixed-method designs are possible (e.g., see Creswell, 2009, or Leech & Onwuegbuzie, 2009, for discussions of other ways to describe designs), the way a study is conducted often is characterized in one of two ways (Creswell, 2014; Tashakkori & Teddlie, 1998; Teddlie & Tashakkori, 2009). The first, and most common, type of mixed-methods design uses a staggered arrangement of quantitative and qualitative components. In this instance, called **sequential mixed methods**, one part comes first and the other part follows. Either the quantitative or the qualitative part of the study can come first. Often, the results of the first part of the study influence what occurs in the second part of the study. For example, a researcher might collect children's fitness test scores to provide a description of the fitness levels in some group. Based on the BMI scores above 30, participants would be recruited for semistructured interviews focusing on how they perceived that their body composition and fitness levels affected their interactions with their peers.

When the quantitative and qualitative components occur at the same time or are independent, the design commonly is labeled **parallel**, or concurrent, **mixed-methods** research. From the beginning of the study, both methodological approaches are used, and the results of one part do not dictate the participants or methods for the second part. For example, a research team may be examining the influence of physical activity and diet on health indicators such as blood pressure, heart rate, fitness, and blood lipids by having participants engage in physical activity of various intensities and diet programs that range from easy to implement to those requiring large changes in eating behavior. The researchers want to know not only the effectiveness of various intervention combinations but also the perceptions of those who were successful and not successful in maintaining the diet and physical activity interventions to which they were assigned. The qualitative portion of the study would use observation and interviews to address this question. The results could indicate that one combination of diet and physical activity was the most effective in improving the health indicators, but that another program was nearly as effective and was perceived by the participants as more enjoyable.

sequential mixed methods—A study design in which either the quantitative or qualitative component comes first and then is followed by the other component.

parallel mixed methods—A study design in which the quantitative and qualitative components occur at the same time or independently; also called *concurrent mixed methods.*

DESIGNING MIXED-METHODS RESEARCH

Everything we noted in the previous chapters applies to planning mixed-methods research. The methods are adapted to the study, and the researcher must plan the study so that all parts of the methods are strong and directly address the questions or hypotheses.

Like all research, a mixed-methods study offers no shortcut to good data. Beginners using mixed-methods research often design a study in which one part is well designed and the other is not. This occurs because researchers are adding a component they are unfamiliar with. Research training and pilot studies are necessary to complete a quality study. Or, as often occurs, a research team works together to plan and conduct a study, and members design and collect data for the parts of the study that draw on their strengths. When the research is cooperative, all members of the team work together so that all parts of the final study are linked.

ISSUES IN MIXED-METHODS RESEARCH

Although the parts of a mixed-methods study are combined to make a whole, a researcher or research team must consider several issues—from planning the study through data analysis and writing the research paper. Some of these decisions are made intuitively. Others require forethought and cooperation among the research team. The following issues should be addressed in a mixed-methods study:

■ *Questions and hypotheses.* If, as the questions and hypotheses for the study are being developed, complementary quantitative and qualitative questions arise, it is important to develop each component and make certain that a strong rationale supports each part of the study. It is not enough to tack on a methodological component because it is simple or because the researcher has strength in that methodological approach. All parts of the study should be interrelated and add to the literature.

■ *Design selection.* After the questions and hypotheses are finalized, a design must be developed. The design should be based on the specific demands of the study, and the researchers should be pragmatic in addressing the qualitative and quantitative portions. Is one part larger than the other? Will data collection occur sequentially or in parallel?

■ *Sampling.* In many mixed-methods studies, the samples of one part of the study are based on the samples of the other part. The study may use a larger sample and a subsample. The quantitative component may have a larger sample based on random selection or on a convenience sample of volunteers, and the qualitative portion may then select from those participants to get a purposeful sample to answer the qualitative questions. Conversely, a qualitative purposeful sample may be supplemented by another sample to obtain descriptive data to answer the quantitative questions. Whether the samples are the same or different, the researcher must plan so that the participants for each part of the study are appropriate.

■ *Data collection.* All parts of data collection should be planned and pilot-tested. If the data are to be collected concurrently, the research team should be ready to start at the same time. Interactions with participants for data collection should be well thought out so that participants' time is well used and they are not turned off by multiple requests that appear to waste time.

■ *Obtaining good data.* As data are collected and analyzed, attention must be paid to whether there is evidence of reliability and validity or trustworthiness and credibility. These decisions must be based on the traditions and demands for each type of data. This step

Sometimes mixing methods is good, sometimes not.

requires researchers to individualize data collection and analysis so that each part relies on good data. All the attention to these issues that separate studies would mandate are required of a mixed-methods study.

■ *Presenting the study.* Many options are available for presenting the study. In a sequential study, the introduction and discussion are typically combined, and the methods and results are separate and based on the order of the parts. In a parallel study, the methods or results may also be separate, or they can be intermixed to suit the presentation. Generally, in a parallel study, more integration is present throughout the discussion section (Creswell, 2014). In some sequential studies in which both the qualitative and quantitative components are themselves major studies, separate papers may be published on each part. If a large study is presented in two papers, the researcher should note this in both papers and inform the editors when submitting them for publication (American Psychological Association, 2010).

EXAMPLES OF MIXED-METHODS RESEARCH

A growing number of mixed-methods research papers are appearing in fields related to physical activity. As researchers see the value in combining methods, more research using this approach is likely. Here we provide two examples that are typical of the types of methods and designs used in mixed-methods research. You will benefit by reading each of the papers as you think about the material in this chapter.

Thøgerson-Ntoumani and Fox (2005) completed a mixed-methods sequential study that addressed well-being typologies in corporate employees. They administered a quantitative questionnaire on the Internet ($N = 312$) and used hierarchical cluster analysis to develop clusters for physical activity and well-being. They employed a variety of statistical techniques to produce evidence that the data provided a valid model for these participants. They followed this with semistructured interviews from participants ($n = 10$) who represented the clusters that were developed in the quantitative phase. This sequential study clearly reported which data were collected in what order. In addition, the authors provided information (e.g., trustworthiness) that showed they had paid attention to getting good data. The results were presented with the quantitative data followed by the qualitative data, and the discussion combines the two parts of the study in examining the results for each cluster.

Harvey and colleagues (2009) provided an example of a parallel, or concurrent, mixed-methods study in which the qualitative portion was a larger part of the study than the quantitative portion. Twelve boys from a purposeful sample, 6 with and 6 without attention-deficit/hyperactivity disorder (ADHD), were tested for motor skill and interviewed using structured and semistructured interview questions to explore their knowledge of and preferences for physical activity. The testing and interviews occurred simultaneously. Comparisons were made between groups on the skills assessment and through thematic analysis of the interviews. The results are presented first for the skill assessment and then for the qualitative results. The results of both types of research are integrated in the discussion, but as would be expected when the qualitative part of the study is much larger, it focuses more on the qualitative results. The authors provided a good overview of the steps they took to obtain good data and were candid about how a pilot study influenced the final method.

SUMMARY

Mixed-methods research is a pragmatic method of addressing interrelated questions that can best be answered by combining quantitative and qualitative approaches. Generally, these methods are used sequentially or concurrently, but as Creswell (2009) noted, the ways mixed-methods designs are described are varied and are likely to change over time. Researchers who do mixed-methods research need to be skilled in both approaches, so two or more researchers often collaborate to address the parts of the study.

Check Your Understanding

1. Locate a mixed-methods study and write an abstract of approximately 300 words about the methods used to gather and analyze the data and present the results. Include information about which methods are qualitative and which are quantitative. Identify the relative emphasis on quantitative and qualitative methods and whether the design is sequential or parallel.

2. Locate a mixed-methods study and write a summary of how the researchers worked to obtain good data. Focus on issues of reliability, validity, trustworthiness, and credibility. Note any differences in the data and discuss whether additional information would have been helpful.

PART IV

Writing the Research Report

We have not succeeded in answering all of your questions. In some ways, we are as confused as you. However, we have succeeded in raising questions so that we are confused at a higher level.

—Click and Clack, National Public Radio's *Car Talk*

Part I discusses the research proposal, its purpose, and the structure of the various parts. Parts II and III provide the details needed for understanding and conducting research, including statistics, measurement, and types of research. This part completes the research process with instructions on preparing the research report. You may also want to refer to chapter 2, which discusses some rules and recommendations for writing the review of literature, because that is an important part of the research report.

Following are the purposes of a thesis or dissertation:

1. Establishing the expertise of the graduate student
2. Warehousing the artifacts that demonstrate scientific competence and chronicling the detailed processes used in research
3. Expanding the knowledge base of the field of kinesiology

Chapter 21 examines all the parts of the research proposal that have already been discussed. In addition, we offer some of our thoughts about the nature of the meeting to review the research proposal. Up until now, we have tried to explain how to understand other research and how to plan your own research. Here we help you organize and write the results and discussion sections (or chapters) about research you have conducted. We also explain how to prepare tables, figures, and illustrations and where to place them in the research report.

Finally, in chapter 22, we suggest ways of using journal and traditional styles to organize and write theses and dissertations. We also present a brief section on writing for scientific journals and a short discourse on preparing and giving oral and poster presentations.

21

Completing the Research Process

Ginger Rogers did everything Fred Astaire did only backwards and in high heels.

—1982 *Frank and Ernest* cartoon

Completing the research process involves writing your proposal, getting it approved, carrying out the research, and writing it up in the results and discussion section of your thesis. In the following sections, we provide guidelines to help you complete this process. After you have read these guidelines, you can find greater detail in Locke, Spirduso, and Silverman (2014).

RESEARCH PROPOSAL

The research proposal contains the definition, scope, and significance of the problem and the methodology that will be used to solve it. If the journal format (advocated in this book and reviewed in detail in chapter 22) is used for the thesis or dissertation, the proposal consists of the introduction and methods sections and appropriate tables, figures, and appendixes (e.g., score sheets, cover letters, questionnaires, informed-consent forms, and pilot-study data). In a four-section thesis or dissertation (introduction, methods, results, and discussion, IMRD), the proposal consists of the first two sections. In studies using a five-section format, in which the introduction and the review of literature are separated into two sections or chapters, the proposal encompasses the first three sections.

One of the goals of this book is to help prepare you to develop a research proposal. We have already discussed the contents of the proposal. Chapters 2, 3, and 4 pertain specifically to the body of the research proposal. Other chapters relate to facets of planning a study: the hypotheses, measurements, designs, and statistical analyses. In this chapter, we attempt to bring the proposal together. We also discuss the proposal meeting and committee actions. Finally, we touch on basic considerations involved in grant proposals—specifically, how they differ from thesis and dissertation proposals.

THESIS AND DISSERTATION PROPOSAL

Most departments require a proposal (sometimes called a prospectus) for a thesis or dissertation prior to completing the research and writing the final copy. However, graduate schools do not typically require the formal filing of a proposal; it is usually a department document. The proposal is formatted according to one of the categories listed previously for the journal style of a thesis or dissertation. It may include previous work completed (published, in press, and submitted), and it presents the plan for the upcoming research. The upcoming research plan is also in journal style including an introduction (with a short literature review) and the method. If you have not completed previous research on this topic (more likely with master's students than with doctoral students), you should include pilot data in the proposal. The purpose of pilot data is to show that the research instrumentation (or materials) and methods actually work under the conditions of the research—that you can administer them and participants can complete them, and that they yield valid and reliable data.

The research proposal should also contain references as well as figures or tables. There may be several appendixes that include a more detailed literature review, the pilot data, other method information, and so on. If the proposal is done correctly, it should serve as the basis for the final research paper(s) and thesis or dissertation.

ADVISOR AND DISSERTATION COMMITTEE

You should work closely with your advisor in developing your thesis or dissertation proposal. The planned research should not come as a surprise to anyone associated with your work. This allows the proposal meeting to be about how to best do the research rather than addressing the question Is this a good idea? Once the proposal is approved by your advisor and the committee, it constitutes an agreement that if the research is completed as planned and written effectively, it meets the requirement for a thesis or dissertation. If adjustments are needed during the research, clear this with your advisor and the dissertation committee.

Special consideration must be given to the role of the dissertation committee when previously published and in-press papers are included in the dissertation. The committee should be involved in the planning and execution of all work. Once work is published, the committee has no avenue for revision; thus, its approval during the process is important. The committee will approve the entire dissertation or thesis, not just the unpublished studies. Therefore, it should be involved as the studies are conducted, completed, and prepared for publication. You should be first author on all studies in your thesis or dissertation; the committee members and your advisor may or may not be co-authors depending on their contributions to the work. It is a good practice to discuss authorship during the process from proposal to submission of articles for review.

THE GOOD SCHOLAR MUST RESEARCH AND WRITE

In today's world, the good scholar must have the knowledge and skill to carry out quality research that answers important questions. However, this is not enough. The good scholar must also publish that research in quality journals. "Thus, the scientist must not only 'do' science but must 'write' science" (Day & Gastel, 2006, p. ix). Of course, changing the format and style of reporting research does not improve the research. Bad science cannot be improved with good formatting or writing, but good science can certainly be negatively affected by bad writing and formatting. Use of the journal style allows good science to be reported in a standard outlet, the scholarly journal, while meeting all requirements of the graduate school for the thesis or dissertation.

SCIENTIFIC WRITING

A writer of scientific papers is not a writer in the literary sense (Day & Gastel, 2006). The writer of science should use simple and straightforward language to report research finding. Do not swallow a thesaurus; keep your words simple and understandable, and use jargon minimally. We reported the following sentence that appeared in an abstract:

> **Amalgamating the decision maker's inputs is a new and unique decision model that can be classified as a fuzzily parametered, multi-staged, forward and backward chained, Displaced Ideal, two-dimensional attribute model of business-level strategic objectives and the functional-level strategies that realize those objectives.**

Although the words are clear, the meaning is not.

Nearly all fields of research have jargon that is used by the scientific writer. However, Day and Gastel (2006, p. 200) explained the use of inappropriate jargon in the following paragraph:

> **The most common type of verbosity that afflicts authors is jargon. This syndrome is characterized, in extreme cases, by the total omission of one-syllable words. Writers with this affliction never *use* anything—they *utilize*. They never *do*—they *perform*. They never *start*—they *initiate*. They never *end*—they *finalize* (or *terminate*). They never *make*—they *fabricate*. They use *initial* for *first*, *ultimate* for *last*, *prior* for *before*, *subsequent to* for *after*, *militate against* for *prohibit*, *sufficient* for *enough*, and *a plethora* for *too much*. An occasional author will slip and use the word *drug*, but most will salivate like Pavlov's dogs in anticipating of using *chemotherapeutic agent*. (We do hope that the name Pavlov rings a bell.) Who would use the three-letter word *now* instead of the elegant expression *at this point in time*?**

Another reason for straightforward scientific writing is that English is generally accepted as the language of science. Yet English is a second language to many scientists around the world. Not only do these scientists need to read and understand papers in English, but also, in most cases, they need to write their scientific papers in English.

FIRST THINGS ARE SOMETIMES BEST DONE LAST

Research writing is often best done by not using the IMRD model in a linear way. For example, the final versions of the paper's title and abstract are typically better done last than first. Greater detail on any of the following topics can be found in Day and Gastel (2006).

Title

Although your thesis and dissertation (and the journal-style papers that are included) will need a working title, often writing the final version of the title is best done last. In today's electronic world, people will locate your paper by the title; thus, it needs to be informative about the research. However, we have often seen titles that consisted of the statement of purpose and the methods. Such a title is likely too long, whereas a three- or four-word title is likely too short. Day and Gastel (2006) defined a good title as "the fewest possible words that adequately describe the contents of the paper" (p. 39).

Here is an excellent title from the June 2012 issue of *Research Quarterly for Exercise and Sport*: "Quiet Eye Duration Is Responsive to Variability of Practice and to the Axis of Target Changes." This title provides the outcome of the research in 16 words. For fun (probably only for a professor), look at the thesis and dissertation titles from a recent graduation program at your institution. It is great entertainment during the boring speeches. Here are a few recent ones we have encountered:

- "Antigravity"—That is a rather broad topic; a little more definition might be useful.

- "Real Analyticity of Hausdorff Dimension of Disconnected Julia Sets of Cubic Parabolic Polynomials"—Although this title may carry meaning to a few others in the specific area, typical well-informed readers would not understand what this research was about.

- "Working Baby Boomers' Knowledge of Retiree Health Benefits and Costs"—This is a good title that conveys the purpose of the study and is understandable by an informed reader.

Always check your titles for waste phrases such as *the effects of, the influence of, a study of, an experiment on,* and *an observation of.* Typically, these phrases can be eliminated from the title without detrimental effects.

Figures and Tables

The journal papers in your thesis or dissertation will typically have tables and figures. Tables are useful for storing data, whereas figures are better for showing changes and trends, particularly as groups or subgroups change differently. Put only important tables and figures in your journal paper(s); relegate the rest to the appendix.

Tables and figures often contain average or mean values. These are useless without estimates of variability. So wherever you give a mean value, provide either the standard deviation or a confidence interval. We personally prefer standard deviations in tables and confidence intervals in figures, but that is just preference.

References

The first rule for references is: Check your references carefully. Make sure every citation in the paper has a reference at the end of the paper. If other references are needed (e.g., for the extended literature review), put them in the appendix. Then make sure all your references are correct. Rates of errors even in journal citations are high—47% for the 1988 and 1999 volumes of *Research Quarterly for Exercise and Sport* (Stull, Christina, & Quinn, 1991) and 50% for the *Journal of the American Medical Association* (Goodrich & Rowland, 1977). Careful scholars check every reference against the text of the paper and against the original source.

DEVELOPING A GOOD INTRODUCTION

Your most important task is to convince the committee (whether the proposal committee, a journal reviewer, or a reviewing committee for a granting agency) that the problem is important and worth investigating. The first section of the proposal should do this, and it should also attract the reader's interest to the problem. The review of literature provides background information and a critique of the previous research done on the topic, pointing out weaknesses, conflicts, and areas needing study. A concise statement of the problem informs the reader of your exact purpose—that is, what you intend to do. Hypotheses or questions are advanced on the basis of previous research and perhaps a theoretical model.

The first part of the introduction defines the constructs that define the study and *briefly* determines the relationships among them. Generally, this section, just a few pages in length, goes from broader to narrower; when the reader gets to the problem or purpose statement, it becomes clear which variables are being studied. For example, in a study of student attitude in physical education, the introduction might first address the valuable health consequences of physical activity. Subsequent paragraphs might discuss how physical education provides and promotes physical activity, how attitude influences whether students participate in physical activity, how elementary school physical education attitudes may influence future attitudes, and how attitudes change with time. The constructs of physical education, attitude, and elementary school children are defined and their relationships are described before the discussion moves on to other sections of the introduction.

In many universities, the introduction includes a section on the significance of or rationale for the study. This is the place to make the case that the study will contribute to the field. This section may address theoretical and practical reasons for the study. Many students find it relatively easy to address practical reasons, but describing how the study contributes to theory is crucial in determining whether a paper from the study can be published. Making a case for theoretical contributions requires going beyond saying that there is not a lot of literature on the topic. You must specifically state how the study will fit with and extend the current literature and theory.

Throughout the proposal, you need to provide operational definitions to explain exactly how you are using certain terms (see the sidebar for an operational definition of cricket). Some departments have the tradition of a section of operational definitions, but it has become more common to define terms operationally as they are used. This approach is similar to the way research papers are written and helps the reader by defining terms when they are encountered. Limitations and possible shortcomings of the study also are acknowledged by the researcher and are generally the result of the delimitations that the researcher imposes. This discussion may be completed in a separate section of the introduction. Generally, a more efficient approach is to include a sentence acknowledging limitations and possible shortcomings when samples and techniques are discussed in the methods section.

Operational Definition of Cricket as Explained to Us by an Australian Friend

You have two sides: one side in the field and one out.

Each player that's on the side that's in goes out, and when out comes in, the next player goes in until out.

When they are all out, the side that's out comes in and the side that has been in goes out and tries to get those coming in out.

Sometimes you get players still in and not out.

When both sides have been in and out including the not out, that ends the game!

Chapter 2 of this book concerns the literature review and includes a discussion on the inductive and deductive reasoning processes used in developing the problem and formulating hypotheses. Chapter 3 covers the other parts typically required in the introduction section or chapter of a proposal.

Innumerable hours are required to create the first section (introduction) of the proposal, especially in preparing the literature review and formulating the significance of or rationale for the problem. You will most likely depend heavily on your advisor and on completed studies for examples of formatting and descriptions. Before we continue, however, we need to mention that you should write your proposal in the future tense: that so many participants will be selected and that certain procedures will be carried out. Of course, pilot studies that have been completed should be in the past tense. Theoretically, if the proposal is carefully planned and well written, you need only change from future to past tense to have the first two sections or chapters of the thesis or dissertation. Realistically, however, you will probably make numerous revisions between the proposal and the final version. To reiterate, the importance of the study and its contribution to the profession are the main focus of the first section of the proposal, and this constitutes the basis for approval or disapproval.

DESCRIBING THE METHODS

The methods section of the proposal frequently draws the most questions from committee members in the proposal meeting. In the methods section or chapter, you must clearly describe how the data will be collected to solve the problem set forth in the first section. You must specify who the participants will be and how they will be chosen, how many participants are planned, whether any special characteristics of importance are present, how the participants' rights and privacy will be protected, and how informed consent will be obtained. Methods of obtaining measurements are detailed, and the validity and reliability of these measures are documented. Next, the procedures are described. If, for example, the study is a survey, discuss the steps in developing the instrument and cover letter, mailing the questionnaires, and following up. If the study is experimental, describe the treatments (or experimental programs) explicitly, along with the control procedures that will be exercised. Finally, explain the experimental design and planned statistical analysis of the data.

We have previously emphasized the importance of conducting pilot studies before gathering data. If pilot work has been done, it should be described, and the results reported. Often, the committee members have major concerns about such questions as whether the treatments can produce meaningful changes, whether the measurements are accurate and can reliably discriminate among participants, and whether the investigator can satisfactorily perform the measurements and administer the treatments. The pilot study should provide answers to these questions.

If the proposed study is qualitative, then it is important to address the method and tailor its presentation to the study. This is similar to what we suggested in the previous paragraph. For example, most qualitative proposal methods sections include the sample, data collection techniques and pilot-testing procedures (e.g., development, pilot studies, and revisions for semistructured interview questions), data analysis, and the ways trustworthiness and credibility (or reliability and validity) are addressed in both data collection and analysis. We present advice for preparing the qualitative proposal later in this chapter.

We recommended in chapter 4 that you use the literature to determine the methodology. Answers to questions about whether certain treatment conditions are sufficiently long, intense, and frequent to produce anticipated changes can be defended with the results of previous studies.

THE PROPOSAL PROCESS

We reiterate the contents of the proposal: the introduction (including the review of literature) and the methods to be used. The proposed purpose—in conjunction with pertinent background information, plausible hypotheses or questions, operational definitions, and delimitations—determines whether the study is worthwhile. Consequently, the first section (or chapter) is instrumental in stimulating interest in the problem and establishing the rationale for and significance of the study. The committee's decision to approve or disapprove rests primarily with the persuasiveness exhibited in the first section.

Actually, the basic decision about the merit of the topic should already have been made before the proposal meeting. You should consult with your advisor and most (if not all) of the committee members to reach a consensus about the worth of the study before the proposal meeting is scheduled. If you cannot convince the majority of the committee that your study is worthwhile, do not convene a formal proposal meeting. We strongly recommend working closely with your major professor and scheduling the proposal meeting only when she or he is certain that the proposal is ready and that the meeting will result in a successful conclusion. You may have a problem if your proposal is returned with a checklist such as the one in the sidebar Interim Thesis/Dissertation Evaluation.

To avoid problems, convince the majority of the committee of your study's worth before convening the proposal meeting.

What to Expect of the Proposal Committee

Let us digress a moment to discuss the composition of the proposal committee. The structure of committees and the number of committee members vary from one institution to another. We can safely say that most thesis committees consist of at least three members, and most dissertation committees have at least five. The major and minor professors are included at institutions where that occurs, although the master's student is often not required to have a minor. Other members should be chosen on the basis of their knowledge about the subject or their expertise in other aspects of the research, such as design and statistical analysis. Sometimes, the institution or department specifies how many members of the committee must be from inside or outside the department. Usually, no limit is placed on the number of committee members.

You should work closely with your advisor in developing your thesis or dissertation proposal. The planned research should not come as a surprise to anyone associated with your work. This allows the proposal meeting to be about how to best do the research rather than addressing the question Is this a good idea? Once the proposal is approved by your advisor and the committee, it constitutes an agreement that if the research is completed as planned and written effectively, it meets the requirement for a thesis or dissertation. If adjustments are needed during the research, clear this with your advisor and the dissertation committee.

Special consideration must be given to the role of the dissertation committee when previously published and 'in press' papers are included in the dissertation. The committee should be involved in the planning and execution of all work. Once work is published the committee has no avenue for revision, thus their approval during the process is important. The committee will approve the entire dissertation or thesis, not just the unpublished studies. Therefore, they should be involved as the studies are conducted, completed, and prepared for publication. The student should be first author on all studies in the thesis or dissertation, the committee members and advisor may or may not be co-authors depending upon their individual contributions to the work. It is a good practice to discuss authorship during the process from proposal to submission of articles for review.

Interim Thesis/Dissertation Evaluation

Dear _____:

Greetings! I regret that my busy schedule prohibits me from rendering a detailed written evaluation of your thesis/dissertation. However, I have checked the appropriate actions or comments that apply to your proposal.

___ If at first you don't succeed, try, try again.

___ If I agreed with you, we would both be wrong.

___ Don't sell your research methods textbook (Thomas, Nelson, and Silverman, of course); you'll need to take the course again.

___ You were not required to write your paper in a foreign language (Burmese, or whatever it was).

___ I couldn't read beyond the third page; one does not have to eat a whole pie to know it is bad.

___ I hear they are hiring at Sam's Diner.

___ You are entitled to your own opinions but not your own facts.

___ Have you paid your tuition and fees yet? If not . . . .

___ Please call my secretary and arrange an appointment with me. Consider taking a tranquilizer before you arrive.

___ I have been serving on thesis/dissertation committees for over 15 years and can now honestly say that I've seen it all.

How to Prepare the Formal Proposal

The formal proposal should be carefully prepared. Typically, the proposal undergoes multiple rounds of editing, review by the major professor, and revisions before it is ready for a proposal meeting. If the proposal contains errors of grammar, spelling, or format, committee members may conclude that you lack the interest, motivation, or competence to do the proposed research. With the availability of computers and word processors, no reason exists to present a poorly prepared proposal. Also, always use a spelling checker. Remember, however, that a spelling checker cannot identify the use of an incorrect word that is spelled correctly, as the following poem by Jerrold H. Zar illustrates:

I have a spelling checker.

It came with my PC.

It plane lee marks four my review

Miss steaks aye can knot see.

When you are ready to distribute your proposal, print it using a good-quality printer and on good paper. Be sure that copies duplicated for distribution are of high quality and that all pages are included (look at each page in each copy—copy machines often do not work perfectly). If you distribute your proposal to committee members via e-mail (which is most common now), be certain that all materials are attached. Also, the subject line of your e-mail should indicate what is included, and the body of your e-mail should contain all pertinent information, such as any expected time lines. This will ensure that committee members recognize what they are receiving, what is attached, and when subsequent events will occur.

Committee members should not ignore errors in proposals with the idea that the student will correct these later. Doing so may lead the student to assume that carelessness is acceptable in data collection or in the final thesis or dissertation. Copies of the proposal should be given to the committee members well before the meeting. The department or university usually specifies the number of days in advance of the meeting that proposals should be distributed.

What Happens at the Proposal Meeting

In the typical proposal meeting, the student is asked to summarize the rationale for the study, its significance, and the methodology (good visual aids enhance this presentation). The remainder of the session consists of questioning by the committee members. Practicing the presentation and answering questions in front of your major professor and other graduate students will improve your presentation and may alert you to areas that require more preparation. If the topic is acceptable, the questions will concern mainly your methods and your competence to conduct the study. Exhibit tactful confidence in presenting the proposal. A common mistake that students make is to be so humble and pliable that they agree to every suggestion made, even those that radically change the study. Your advisor should help ward off these "helpful" suggestions, but you must also be able to defend the scope of your study and your methodology. If adequate planning has gone into the proposal, you should be able to recognize (with the assistance of your major professor) useful suggestions and defend against those that seem to offer minimal aid.

Projected schedules of procedures are sometimes required for thesis or dissertation proposals. Regardless of whether a time frame is formally required, it is definitely something you should address. A common mistake is underestimating how long each phase of the study will take. Students tend to assume, like Pollyanna, that all those involved (participants, helpers, committee members, word processors, and university personnel) will drop everything they are doing to accommodate the study. Moreover, students often assume that all phases of the study will proceed without a hitch. Unfortunately, it just does not work that way. The

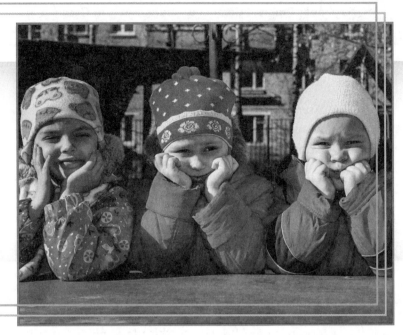

Clearly,
my thesis
committee is
enthralled by
my findings.

© Pavel Losevsky/Fotolia

simple logistics of accomplishing each aspect of the study are usually more complicated and time-consuming than anticipated. Much grief can be avoided if you carefully and realistically project how long the steps in your research process will take (and then add more time).

After the proposal is approved, most institutions treat it as a contract in that the committee expects the study to be done in the manner specified in the proposal. Moreover, you can assume that if your study is conducted and analyzed as planned and is well written, it will be approved. If any unforeseen changes are required during the course of the study, your advisor must approve them. Substantial changes usually must be reviewed by some or all of the committee members.

> Avoid grief and headaches by adding time to careful and realistic projections of the expected duration of each phase.

Questions by Proposal Committees and Answers by Graduate Students

Committee: This myasthenia gravis, does it affect your memory at all?
Student: Yes.
Committee: In what ways does it affect your memory?
Student: I forget.
Committee: Can you give us an example of something that you've forgotten?

Committee: In your qualitative study, can you describe one of the participants?
Student: He was 6 feet tall, had a beard, and trained regularly by distance running.
Committee: Was this a male or female?
Student: ?

Committee: Were you present when this picture of you using the apparatus was taken?
Student: ?

(continued)

Questions by Proposal Committees and Answers by Graduate Students *(continued)*

> Committee: How far apart were the participants at the time of the collision?
> Student: ?
>
> Committee: You say the stairs the participants climbed went down to a basement?
> Student: Yes.
> Committee: And these stairs, did they go up also?
>
> Committee: How many participants were there in your research?
> Student: Forty.
> Committee: Men or women?
> Student: All women.
> Committee: Were there any men?
>
> Committee: Were your participants qualified to give a urine sample?
> Student: ?
>
> Committee: After you sacrificed the mouse, was it still alive?
> Student: No.
> Committee: Did you check for a heartbeat?
> Student: No.
> Committee: So it's possible the mouse was still alive?
> Student: No.
> Committee: How can you be sure?
> Student: His brain was sitting in a jar on my desk.
> Committee: But the mouse could have been alive nonetheless?
> Student: It's possible he could have been alive and serving as a professor somewhere.

PREPARING AND PRESENTING QUALITATIVE RESEARCH PROPOSALS

The content, procedures, and expectations for the proposal that we have discussed so far largely pertain to quantitative theses and dissertations. Although many similarities exist in the kinds of information presented in quantitative and qualitative research proposals, some salient differences should be mentioned.

Preparing the proposal is much easier if all committee members are familiar with qualitative research. Today, many university faculty members have a basic understanding of qualitative research and the methodological differences between qualitative and quantitative research. If proposed members of the committee are not familiar with qualitative research, problems may arise. If you are contemplating doing a qualitative study, we strongly urge you to read Locke, Spirduso, and Silverman's *Proposals That Work* (2014) or Marshall and Rossman's *Designing Qualitative Research* (2011).

One of the problems the qualitative researcher may face concerns the possible shifts in focus and methods that may occur during the study. Earlier, we likened the proposal to a contract, whereby the committee members expect you to carry out the study as specified in the proposal. Unlike quantitative research and much qualitative research, however, some qualitative studies typically undergo changes in the focus of the research question or the sources of data, the methodology, and the analysis of data. This aspect of qualitative research demands an "open contract" (Locke et al., 2014), and the methods for handling changes should be included in the proposal.

Another difference between the quantitative and qualitative proposal may be found in the literature review. In some rare cases in qualitative research, closely related studies may be purposefully omitted when planning the study because the researcher does not want to be influenced by the views and perceptions of others. Still another difference is the need for qualitative researchers to address their own values, biases, and perceptions so that they can more clearly understand the context of the research setting. This discussion often is included as a paragraph in the methods section where data collection and analysis are addressed.

All committee members, whether familiar or unfamiliar with qualitative research, are interested in methodological concerns: Where will the study take place? Who will be the participants? How will the researcher gain access to the site? What data sources will be used? Does the researcher possess the needed research skills? What pilot work has been done? How will ethical concerns be handled? What strategies will be employed in collecting, sorting, categorizing, and analyzing the data?

It should be apparent by now, depending on the study, that there may be differences in the quantitative and qualitative proposal format and in the nature of the proposal process. These differences must be recognized and accepted. As with any type of research, your major professor plays a vital role in preparing you (and, in some cases, the committee members), in showing support, and in helping you get out of tight spots. Most of all, with the support of your major professor, you must be able to assure the committee that you will carry out the study in a competent and scholarly manner.

WRITING PROPOSALS FOR GRANTING AGENCIES

All sources of grants, whether government agencies or private foundations, require research proposals so that they can decide which projects to fund and how much to award. Granting agencies nearly always publish guidelines for applicants to follow in preparing proposals.

A well-written proposal is everything. Researchers rarely get a chance to explain or defend the purpose or procedures. Thus, the decision is based entirely on the written proposal. The basic format for a grant proposal is similar to that of a thesis or dissertation proposal. But some additional information is required, and some procedural deviations may exist.

Researchers must follow the guidelines of the granting agency. Granting agencies tolerate little (if any) deviation from their guidelines. All sections of the application must be addressed, and deadlines must be strictly followed. Frequently, a statement of intent to submit a proposal is required a month or so before the proposal is filed.

Grant proposals include the following:

- Abstract of the proposed project
- Statement of the problem and its relevance to one of the granting agency's priorities
- Methodology to be followed
- Time frame
- Budget
- Curricula vitae of the investigators

A limited review of literature is often required to demonstrate familiarity with previous research. Occasionally, the funding agency imposes some restrictions on design and methods. For example, an agency may prohibit the use of control groups if the treatment is hypothesized to be effective. In other words, the agency may not want anyone to be denied treatment. This limitation can pose problems for the researcher in scientifically evaluating the outcomes of the project

Because every granting agency has stipulations regarding the types of programs it can and cannot fund, a detailed budget is required, as is a justification for each budget item. A time frame is usually expected so that the reviewers can see how the project will be conducted and when the phases will be accomplished. This time frame provides information about the depth and scope of the study, justification for the length of time that support is requested, contributions of the personnel, and when the granting agency can expect progress reports.

The competence of the researcher has to be documented. Each researcher must include a curriculum vitae and often a written statement describing appropriate preparation, experience, and accomplishments. The adequacy of facilities and sources of support must also be addressed. Letters of support are occasionally encouraged. These are included in the appendix. Many agencies now require that the proposal be submitted electronically. This necessitates double-checking that all parts are prepared in accordance with the request for proposals, because after the proposal is uploaded, the funder's site will determine, for example, whether all parts have been addressed and whether the word count exceeds the number allowed. A panel of reviewers evaluates the proposal in accordance with criteria regarding the contribution to knowledge, relevance, significance, and soundness of design and methodology.

The preparation of a grant proposal is a time-consuming and exacting process. Several types of information are required, and time is needed for gathering the information and stating it in the manner prescribed by the guidelines. Applicants are advised to begin preparing the proposal as soon as the guidelines are available.

Finally, it is usually wise to contact the granting agency before preparing and submitting a proposal. Seldom are proposals funded that are submitted without some prior contact with the granting agency. Ascertaining the agency's interests and needs is a time-saving venture. A visit to, or an extended phone conversation with, the research officer is often advisable. Looking at proposals previously funded by the agency or seeking guidance from researchers who have been funded can be helpful.

SUBMITTING INTERNAL PROPOSALS

Many colleges and universities (particularly larger research universities) offer internal funding for graduate student research, although these grants usually are not large. These grants typically require a two- to five-page proposal that has the support of the major professor and often the department chair. The contents usually include an abstract, a budget, and a short narrative focusing on the proposed methodology and why the research is important. Notices that internal funding is available are routinely posted or advertised around the college or university. A good place to find internal funding sources is at the office of your graduate school or vice president for research.

COMPLETING YOUR THESIS OR DISSERTATION

After your proposal meeting, you collect the data to evaluate the hypotheses that you proposed. Of course, you follow the methods that you specified in your proposal carefully and consult with your major professor if problems arise or changes need to be made. After the data are collected, you complete the agreed-upon analysis and discuss the outcomes with your major professor (and possibly some committee members, particularly if a member has statistical expertise). Then you are ready to write the results and discussion to complete your research.

RESULTS AND DISCUSSION

The final sections of a thesis or dissertation are the results and discussion sections. Results are what you have found, and the discussion explains what the results mean. In theses and dissertations, the results and discussion usually are separate sections, although they are sometimes combined (particularly in multiple-experiment papers). We discuss these sections here as separate parts of the research report.

How to Write the Results Section

The results section is the most important part of the research report. The introduction and literature review indicate why you conducted the research, the methods section explains how you did it, and the results section presents your contribution to knowledge; that is, what you found. The results should be concise and effectively organized and should include appropriate tables and figures.

The results section may be organized in any one of a number of ways. The best may be to address each of the tested hypotheses; on other occasions, the results may be organized around the independent or dependent variables of interest. Occasionally, you may want to show first that certain standard and expected effects have been replicated before you go on to discuss other findings. For example, in developmental studies of motor performance tasks, older children typically perform better than younger ones. You may want to report the replication of this effect before discussing the other results. When looking at the effects of training on several dependent variables, you may first want to establish that a standard dependent variable known to respond to training did, in fact, respond. For example, before looking at the effects of cardiorespiratory training as potentially reducing cognitive stress, you need to show that a change occurred in cardiorespiratory response as a result of the training.

Some items should always be reported in the results. The means and standard deviations for all dependent variables under the important conditions should be included. These are basic descriptive data that allow other researchers to evaluate your findings. The main descriptive data should be presented in one table, if possible. Sometimes only the means and standard deviations of important findings are included in the results, but all the remaining means and standard deviations should be included in the appendix.

The results section should also use tables and figures to display appropriate findings. Figures are particularly useful for percentage data, interactions, and summaries of related findings. When using figures representing group effects, always present an estimate of variability with the mean data—either standard deviations or confidence intervals are appropriate. Only the important tables and figures should be included in the results; those remaining should be placed in the appendix, if they are included at all.

Statistical information should be summarized in the text where possible. Statistics from ANOVA and MANOVA should always be summarized in the text, and complete tables should be relegated to the appendix. Make sure, however, that you include the appropriate statistical information in the text. For example, when giving the F ratio, report the degrees of freedom, the probability, and an estimate of the effect size: $F(1,36) = 6.23, p < .02$, ES $= 0.65$. Above all, the statistics reported should be meaningful. Day and Gastel (2006, p. 67) reported a classic case that read "33 1/3% of the mice used in this experiment were cured by the test drug; 33 1/3% of the test population were unaffected by the drug and remained in a moribund condition; the third mouse got away."

Sometimes tables are a better way to present this information. If the perfect scientific paper is ever written, it will read, "Results are in table 1." However, this suggestion does not mean that the results section should consist mostly of tables and figures. Having to thumb through eight tables and figures placed between two pages of text is disconcerting. But even worse is having to turn 50 pages to the appendix to find a necessary table or figure. Read what you have written. Are all the important facts there? Have you provided more information than the reader can absorb? Is some of the information peripheral to the questions or hypotheses guiding the study? If the answer to any of these questions is yes, you should revise the results section.

Do not be redundant and repetitive. A common error is to include a table or figure in the results and then repeat it in the text, or to repeat data from a table in a figure. Describing tables and figures in a general way or pointing out particularly important facts is appropriate, but do not repeat every finding. Also, be sure that you do not call tables figures, and vice versa. As Day and Gastel (2006) reported, some writers are so concerned with reducing verbiage that they lose track of antecedents, particularly for the pronoun *it:*

> **"The left leg became numb at times and she walked it off. . . . On her second day, the knee was better, and on the third day it had completely disappeared." The antecedent for both *its* was presumably "the numbness," but I rather think that the wording in both instances was a result of dumbness. (p. 36)**

Reporting Statistical Data

A consistent dilemma among researchers, statisticians, and journal editors concerns the appropriate reporting of statistical information for published research papers. In recent years, some progress has been made that involves two issues in particular—reporting some estimate of the size and meaningfulness of the finding along with the reliability or significance of the finding. Two organizations of importance to our field—the American Physiological Society (www.the-aps.org/mm/Publications/Info-For-Authors/Composition.aspx) and the American Psychological Association (2010)—have now published guidelines regarding these issues. Following are summaries of general guidelines taken from these two sources:

- Information on how sample size was determined is always important. Indicate the information (e.g., effect sizes) used in the power analysis to estimate sample size. When the study is analyzed, confidence intervals are best used to describe the findings.

- Always report any complications that have occurred in the research, including missing data, attrition, and nonresponse, as well as how these problems were handled in data analysis. Before you compute *any* statistics, *look at your data*. Always screen your data (this is not tampering with data) to be sure the measurements make sense.

- Select minimally sufficient analyses; using complicated methods of quantitative analyses may appropriately fit data and lead to useful conclusions, but many designs fit basic and simpler techniques. When they do, these should be the statistics of choice. Your job is not to impress your reader with your statistical knowledge and expertise but to analyze the research appropriately and present it so that a reasonably well-informed person can understand it.

- Report actual p values; confidence intervals are even better. Always report an estimate of the magnitude of the effect. If the measurement units (e.g., maximal oxygen consumption) have real meaning, then reporting them in an unstandardized way such as mean difference is useful. Otherwise, standardized reporting such as effect size or r^2 is useful. In addition, placing these findings in practical and theoretical context adds much to the report.

- Control multiple comparisons through techniques such as Bonferroni.

- Always report variability using the standard deviation. Standard error characterizes the uncertainty associated with a population and is most useful in determining confidence intervals.

- Report your data at the level (e.g., how many decimal places) that is appropriate for scientific relevance.

Reporting Qualitative Data

Most of our general suggestions for preparing the results section hold for reporting qualitative data. In a qualitative study, you will report the results from the data analysis as themes or conclusions. Each of these themes or subthemes must be supported with data—from interviews, observations, or material collection. Present data to make the case that the results come from multiple sources and are clearly supported. For example, if you said that triangulation among student interviews, teacher interviews, and observations was used during data analysis, all three types of data should be included. Writing a qualitative results section may require multiple revisions so that the results seem plausible and are clear to the committee.

What to Include in the Discussion Section

Although the results are the most important part of the research report, the discussion is the most difficult to write. There are no cute tricks or clear-cut ways to organize the discussion, but the following rules define what to include:

- Discuss your results—not what you wish they were, but what they are.
- Relate your results back to the introduction, previous literature, and hypotheses.
- Explain how your results fit within theory.
- Interpret your findings.
- Recommend or suggest applications of your findings.
- Summarize and state your conclusions with appropriate supporting evidence.

Your discussion should point out both where data support and where they fail to support the hypotheses and important findings. But do not confuse significance with meaningfulness in your discussion. In fact, be especially careful to point out where they may not coincide. In particular, the discussion should point out factual relationships among variables and situations, thus leading to a presentation of the significance of the research. Of course, this is an essential place not to confuse cause and effect with correlation. For example, do not say that a characteristic had an effect or influence on a variable when you mean that they were related to each other.

The discussion should end on a positive note, possibly a summarizing statement of the most important finding and its meaning. Never end your discussion with a variation of the old standby of graduate students: *More research is needed.* Who would have thought otherwise?

The discussion should also point out any methodological problems that occurred in the research. But using a methodological cop-out to explain the results is unacceptable. *If you did not find predicted outcomes and you resort to methodological failure as an explanation, you did not do sufficient pilot work.*

Graduate students sometimes want their results to sound wonderful and to solve all the problems of the world. Thus, in their discussions they often make claims well beyond what their data indicate. Your major professor and committee are likely to know a lot about your topic and therefore are unlikely to be fooled by these claims. They can see the data and read the results. They know what you have found and the claims that can be made. A much better strategy is to make your points effectively in your discussion and not try to generalize these points into grandiose ideas that solve humanity's major problems. Write so that your limited contribution to knowledge is highlighted. If you make broader claims, knowledgeable readers are likely to discount the importance of your legitimate findings. Your discussion should not sound like the *Calvin and Hobbes* cartoon (by Bill Watterson) in which Calvin said, "I used to hate writing assignments, but now I enjoy them. I realized that the purpose of writing is to inflate weak ideas, obscure poor reasoning, and inhibit clarity."

Writing the Discussion

Problem statement: Why did the chicken cross the road?

Method: One chicken observed by several individuals.

Results: Said chicken crossed the road.

Discussion: Following are the explanations given for the chicken crossing the road.

Dr. Seuss—Did the chicken cross the road? Did he cross it with a toad?

Yes, the chicken crossed the road. But why it crossed I've not been told.

Sigmund Freud—I dream of a world where chickens can cross the road without having their motives questioned.

Bill Gates—I have just witnessed *eChicken 7,* which will not only cross roads, but will lay eggs, file your important documents, and balance your checkbook. Internet Explorer is an integral part of *eChicken 7.*

Barack Obama—The chicken crossed the road because it was time for a CHANGE. The chicken wanted CHANGE!

George W. Bush—We don't care why the chicken crossed the road. Is the chicken on our side of the road or not? The chicken is either for us or against us. There is no middle ground here.

Bill Clinton—I did not cross the road with *that* chicken. What is your definition of chicken?

Dick Cheney—Where's my gun?

Pat Buchanan—To steal the job of a decent, hard-working American.

Captain Kirk—To boldly go where no chicken has gone before.

Colonel Sanders—Did I miss one?

Graduate student—Is that regular or extra crispy?

In Other Words

1. As a case in point, other authorities have proposed that slumbering canines are best left in a recumbent position.

2. It has been posited that a high degree of curiosity proved lethal to a feline.

3. There is a large body of experimental evidence which clearly indicates that smaller members of the genus *Mus* tend to engage in recreational activity while the feline is remote from the locale.

4. From time immemorial, it has been known that the ingestion of an "apple" (i.e., the pome fruit of any tree of the genus *Malus,* said fruit being usually round in shape and red, yellow, or greenish in color) on a diurnal basis will with absolute certainty keep a primary member of the health care establishment from one's local environment.

5. Even with the most sophisticated experimental protocol, it is exceedingly unlikely that you can instill in a superannuated canine the capacity to perform novel feats of legerdemain.

6. A sedimentary conglomerate in motion down a declivity gains no addition of mossy material.

7. The resultant experimental data indicate that there is no utility in belaboring a deceased equine.

Another point about writing your discussion is to write so that reasonably informed and intelligent people can understand what you have found. We strongly recommend having others proofread your work and identify any passages with which they had difficulty. Do not use a thesaurus to replace your normal vocabulary with multisyllabic words and complex sentences. Your writing should not look like the examples in the sidebar In Other Words. (By translating, you can probably recognize these sentences as some well-known sayings.)

Your discussion can generally be guided by the following questions taken from the *Publication Manual of the American Psychological Association* (American Psychological Association, 2010):

- What have I contributed here?
- How has my study helped to resolve the original problem?
- What conclusions and theoretical implications can I draw from my study?

The responses to these questions are the core of your contribution, and readers have a right to clear and direct answers. If, after reading your discussion, the reader asks, So what? then you have failed in your research reporting.

HANDLING MULTIPLE EXPERIMENTS IN A SINGLE REPORT

Graduate students are conducting more research that involves multiple experiments. These experiments may ask several related questions about a particular problem or may build on one another, with the outcomes of the first leading to questions for the second. This trend is positive, but it sometimes leads to problems within the traditional (chapter structure) thesis or dissertation format. Chapter 22 discusses the journal and traditional formats for organizing theses and dissertations.

Multiple experiments in journal format typically involve a general introduction and literature review. If the experiments use a common methodology, then a general methods section might follow. This is followed by a presentation of each experiment with its own short introduction and the citation of a few critical studies, methods (specific to this experiment), results, and a discussion (sometimes the results and discussion are combined). The report concludes with a general discussion of the series of experiments and their related findings.

Within the traditional framework, multiple experiments are probably best handled by separate chapters or major sections within the results chapter. The first chapter includes the introduction, theoretical framework, literature review, statement of the research problem, and related definitions and delimitations. Subsequent chapters describe each experiment. Each of these chapters includes a brief introduction; a discussion of the specific problem and hypotheses; and the methods, results, and discussion sections. The final chapter is a general discussion in which the experiments are tied together. It contains the features of the discussion previously presented.

USING TABLES AND FIGURES

Preparing tables and figures is a difficult task. Howard Wainer (1992) wrote one of the best papers on this topic. We begin with a quotation that he used.

> **Drawing graphs, like motor-car driving and love-making, is one of those activities which almost every educator thinks can be done well without instruction. The results are of course usually abominable. (paraphrased, with my [Wainer's] apologies, from Margerison, 1965) (p. 53)**

Wainer suggested that tables and figures should allow the reader to answer questions at three levels:

- Basic: Extraction of data
- Intermediate: Trends in parts of the data
- Advanced: Overall questions involving the deep structure of the data (seeing trends and comparing groupings)

These levels can be thought of as an ordered effect:

1. Variables by themselves (data)
2. One variable in relation to another
3. Overall comparisons and relationships in the data

Preparing Tables

Getting information from a table is like extracting sunlight from a cucumber, to paraphrase Farquhar and Farquhar (1891). Remember, tables are for communicating to the reader, not storing data. The first question is Do you need a table? In settling on an answer, two characteristics are important: Is the material more easily understood in a table? and Does the table interfere with reading the results? After you decide that you need a table (not all numbers require tables), follow these basic rules:

- Like characteristics should read vertically in the table.
- Headings of tables should be clear.
- The reader should be able to understand the table without referring to the text.

Examples of Poor and Good Tables

Table 21.1 is an example of a useless table; the data could be more easily presented in the text. This table can be handled in one sentence: The experimental group ($M = 17.3$, $s = 4.7$) was significantly better than the control group ($M = 12.1$, $s = 3.9$), $t(28) = 3.31$, $p < .05$. Table 21.2 is also unnecessary. Of the 10 comparisons among group means, only 1 was significant. The values in the table are the equivalent of t tests. This table can also be presented in one sentence: The Scheffé test was used to make comparisons among the age-group means, and the only significant difference was between the youngest (7-year-olds) and oldest (15-year-olds) groups, $t = 8.63$, $p < .05$.

We have borrowed an example of a useful table (table 21.3) from Safrit and Wood (1983). As you can see, like characteristics appear vertically. Also, an extensive amount of text would be required to present these results, yet they are easy to understand in this brief table.

TABLE 21.1 *USELESS TABLE 1*

Means, Standard Deviations, and *t* Test for Distance Cartwheeled While Blindfolded

Groups	N	M	s	t
Experimental	15	17.3 m	4.7 m	3.31*
Control	15	12.1 m	3.9 m	

*$p < .05$.

TABLE 21.2   *USELESS TABLE 2*

Scheffé's Test for Difference Among Age Levels in Ability to Wiggle Their Ears

Age	7	9	11	13	15
7	–	1.20	1.08	1.79	8.63*
9		–	1.32	1.42	1.57
11			–	1.58	1.01
13				–	0.61
15					–

*$p < .05$.

TABLE 21.3   *EXAMPLE OF A USEFUL TABLE*

Characteristics of Participants

	Urban	Suburban	Rural	All participants
Gender				
Female	280 (55.8%)	219 (50.7%)	101 (47.2%)	600 (52.3%)
Male	222 (44.2%)	213 (49.3%)	113 (52.8%)	548 (47.7%)
Age				
10	36 (7.2%)	14 (4.0%)	6 (2.8%)	59 (5.1%)
11	103 (20.5%)	96 (22.2%)	42 (19.6%)	241 (21.0%)
12	183 (36.5%)	127 (29.4%)	69 (32.2%)	379 (33.0%)
13	162 (32.3%)	158 (36.6%)	82 (38.3%)	402 (35.0%)
14	18 (3.6%)	34 (7.9%)	15 (7.0%)	67 (5.8%)
BMI				
Under 20	57 (11.4%)	32 (7.4%)	8 (3.7%)	97 (8.4%)
20.0–25.0	193 (38.4%)	162 (37.5%)	117 (54.7%)	472 (41.1%)
25.1–30.0	207 (41.2%)	171 (39.6%)	61 (28.5%)	439 (38.2%)
30.1 and over	45 (9.0%)	67 (15.5%)	28 (13.1%)	149 (12.2%)
Self-report as physically active				
Yes	209 (41.6%)	178 (41.2%)	109 (50.9%)	496 (43.2%)
No	293 (58.4%)	254 (58.8%)	105 (49.1%)	652 (56.8%)

Total N = 1,148.

Improving Tables

An important topic is how to improve tables so that they are more useful, more informative, and easier to interpret. Wainer (1992) offers the following three good rules for developing tables:

1. The columns and rows should be ordered so that they make sense. For example, the row elements are often placed in alphabetical order according to the label for the row (e.g., names, places). This approach is seldom useful. Order the rows naturally, for instance, by time (e.g., from the past to the future) or by size (e.g., put the biggest or smallest value of mean or frequency first).

2.   When values go to multiple decimal places, round them off. Two digits are about the most that people can understand, that can be measured with precision, or that anyone cares about. For example, what does a $\dot{V}O_2$max value mean when it is carried to four decimal places? We do not understand it, we cannot measure it that precisely, and no one cares. Sometimes attempts at precision become humorous, bringing to mind the report that the average American family has 2.4 children. (We thought that children came only in whole units!) A whole child is the smallest (most discrete) unit of measurement available.

3.   Use and pay attention to the summaries of rows and columns. The summary data, often provided as the last row or column, are important because these values (sometimes sums, means, or medians) provide a standard of comparison (or usualness). Often, setting these values apart in some way (e.g., bold type) is valuable.

Let us try these improvements on an actual table. In our research methods class, we often give students the assignment to find a table in *Research Quarterly for Exercise and Sport* and improve it by applying Wainer's suggestions. Our graduate students have not been hesitant about finding tables from our scholarly work to improve. (Professors: It is not only our work that can be improved. Offer your students one of your published figures or tables. They will improve it, too.) A good example from a paper by Thomas, Salazar, and Landers (1991) that appeared in *Research Quarterly for Exercise and Sport* was provided by James D. George when he was a doctoral student at Arizona State University (our thanks to Jim for allowing us to use his work). The example in A Fine Table Made Better—The Original Table shows the data as presented by Thomas, Salazar, and Landers (1991). The one in A Fine Table Made Better—The Better Way is George's rearrangement. Observing the improvements in data presentation and understanding is easy. First, the rows have been reordered by the column ES Info, and all the yes responses are followed by all the no responses (it might be just as well to remove the no responses and list the authors' names at the bottom of the table). Then, they have been reordered from smallest to largest by the next column, *N* (sample size). Finally, an additional label (study's most important effects) was inserted under Primary ES to clarify the meaning of those three columns.

How do these changes relate to Wainer's three levels of questions? The revised table makes clear the sample size and ES information (basic level) for each study (but so did the original table). At the intermediate and advanced levels, however (trends, relationships, and overall structure), the revised table is a considerable improvement. For example, you can more easily observe that sample size and ES are unrelated; that is, neither studies with large samples nor those with small samples are more likely to produce larger or smaller treatment effects (as estimated by ES).

Another example of the mindless use of numbers is often the reporting of statistical values. Just because computer printouts carry the statistics (e.g., *F, r*) and probabilities (*p*) to five or more places beyond the decimal does not mean that the numbers should be reported to that level. Two or (at most) three places are adequate. But applying this guideline can result in rather odd probabilities: $t(22) = 14.73$, $p < .000$. Now, $p < .000$ means no chance of error; this cannot occur because if the chance of error is zero, it cannot be a probability. What happened is that the exact probability was something like $p < .00023$ and the researcher rounded it to $p < .000$. **You must not do this**. As indicated earlier, we believe that it is more appropriate to report the exact probability (e.g., $p = .025$) and whether this probability exceeded the alpha set for the experiment (e.g., $p < .05$). But the last digit in the probability must always be 1 or higher. The previous example, $p < .00023$, if reported to three decimals, should read $p < .001$.

A FINE TABLE MADE BETTER—THE ORIGINAL TABLE

Table 2 Data on Articles in Volume 59, 1988

First author	ES info	N^c	Primary ES*		
Doody	no				
Kamen[b]	yes	9	0.64	0.72	0.14
Alexander	yes	26–48	0.33	0.73*	0.39
Era	yes	5–6	0.50*	0.10	1.42*
Kokhonen	yes	9–12	−1.97*	−2.64*	−1.78*
Farrell	yes	45–368	0.77*	−0.51*	0.37*
Heinert	no				
Kamen	no[a]	10	1.14	0.81	0.90
Ober	no				
Simard	yes	7	−1.59*	0.52	−2.71*
Berger	no				
Stewart	no				
Abernethy	no				
Etnyre	no				
Nelson[b]	yes	13	0.73	1.76	0.85
Wesson	no				
Housh	yes	20	−0.53*	−2.11*	0.25*
Hutcheson	yes	34	−0.06	0.63*	−0.30*

*Comparison of Ms forming ES was significant, $p < .05$.

[a]No significant main effects.

[b]The main effect is significant, but no information is provided regarding the significance of the post hoc comparison.

[c]Per comparison group.

Other numbers are frequently used mindlessly as well. In reviewing for a research journal, one of us encountered a study in which children were given a 12-week treatment. The author reported the mean age and standard deviation of the children before and after the 12-week treatment. Not surprisingly, the children had all aged 12 weeks. The author also calculated a t test between the pre- and the posttreatment means for age that was, of course, significant. That is, the fact that the children had aged 12 weeks during the 12-week period was a reliable finding.

A FINE TABLE MADE BETTER—THE BETTER WAY

Table 2 Data on Articles in Volume 59, 1988

First author	ES info	N^c	Primary ES* (study's most important effects)		
Era	yes	5–6	0.10	0.50*	1.42*
Simard	yes	7	−2.71*	−1.59*	0.52
Kamen[a]	yes	9	0.14	0.64	0.72
Kamen	yes[b]	10	0.81	0.90	1.14
Kokhonen	yes	9–12	−2.64*	−1.97*	−1.78*
Nelson[b]	yes	13	0.73	0.85	1.76

(continued)

A Fine Table Made Better—The Better Way (continued)

First author	ES info	N^c	Primary ES* (study's most important effects)		
Housh	yes	20	–2.11*	–0.53*	0.25*
Hutcheson	yes	34	–0.30*	–0.06	0.63*
Alexander	yes	26–48	0.33	0.39	0.73*
Farrell	yes	45–368	–0.51*	0.37*	0.77*
Abernethy	no				
Berger	no				
Doody	no				
Etnyre	no				
Heinert	no				
Ober	no				
Stewart	no				
Wesson	no				

[a]Per comparison group.

[b]Comparison of Ms forming ES.

[c]The main effect is significant, but no information is provided regarding the significance of the post hoc comparison.

*ES was significant, $p < .05$.

Preparing Figures and Illustrations

Many suggestions about table construction also apply to figures and illustrations. A figure is often another way to present a table. Before using a table or a figure, ask, Does the reader need the actual numbers, or is a picture of the results more useful? A more important question is Do you need either? Can the data be presented more concisely and easily in the text? Figures and tables do not add scientific validity to your research report. In fact, they may only clutter the results. Day and Gastel (2006, p. 94) suggested a reasonable means for deciding whether to use a table or a figure: "If the data show pronounced trends making an interesting picture, use a graph. If the numbers just sit there, with no exciting trend in evidence, a table should be satisfactory."

Several other considerations are important in preparing figures. Selection of the type of figure is somewhat arbitrary, but some distinctions make the choice of one type of figure more appropriate than another (see table 21.4).

To evaluate whether you have used a figure appropriately, make sure that it

- does not duplicate text,
- contains important information,
- does not have visual distractions,
- is easy to read and understand,
- is consistent with other figures in the text, and
- contains a way to evaluate the variability of the data (e.g., standard deviation bars or confidence intervals).

Figures are useful for presenting interactions and data points that change over time (or across multiple trials). The dependent variable is placed on the y-axis, and some independent or categorical variable is placed on the x-axis. If you have more than one independent variable, how do you decide which to put on the x-axis? We have already partially answered that question. If time or multiple trials are used, put them on the x-axis. For example, if a

> A table or figure should be used only if the data is not better presented in the text.

TABLE 21.4

Charts and Diagrams

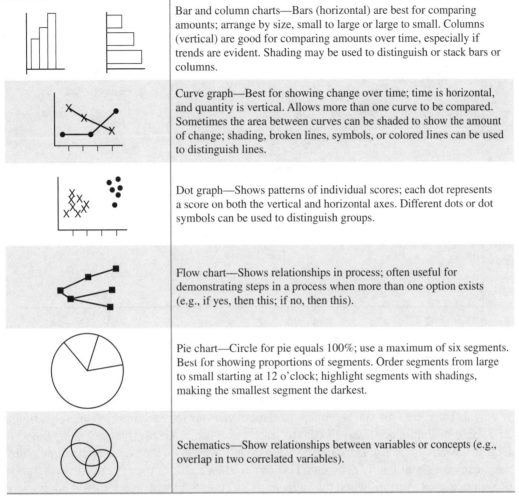	Bar and column charts—Bars (horizontal) are best for comparing amounts; arrange by size, small to large or large to small. Columns (vertical) are good for comparing amounts over time, especially if trends are evident. Shading may be used to distinguish or stack bars or columns.
	Curve graph—Best for showing change over time; time is horizontal, and quantity is vertical. Allows more than one curve to be compared. Sometimes the area between curves can be shaded to show the amount of change; shading, broken lines, symbols, or colored lines can be used to distinguish lines.
	Dot graph—Shows patterns of individual scores; each dot represents a score on both the vertical and horizontal axes. Different dots or dot symbols can be used to distinguish groups.
	Flow chart—Shows relationships in process; often useful for demonstrating steps in a process when more than one option exists (e.g., if yes, then this; if no, then this).
	Pie chart—Circle for pie equals 100%; use a maximum of six segments. Best for showing proportions of segments. Order segments from large to small starting at 12 o'clock; highlight segments with shadings, making the smallest segment the darkest.
	Schematics—Show relationships between variables or concepts (e.g., overlap in two correlated variables).

Developed from J.V. White, 1984, *Using charts and graphs: 1000 ideas for visual presentation* (New Providence, NJ: R.R. Bowker Co.).

study found an interaction between the dependent variable by age level (7-, 9-, 11-, 13-, and 15-year-olds) and the treatment (experimental versus control), age with five levels is usually the more appropriate choice for the *x*-axis. Note that this is a general rule; specific circumstances may dictate otherwise. A good example of the use of a figure to present an interaction is shown in figure 21.1. Note that both age and time are independent variables, so time is placed on the *x*-axis.

Figure 21.2 is a good example of a useless figure. The results can be summarized in two sentences: "During acquisition, all three groups reduced their frequency of errors but did not differ significantly from one another. At retention, experimental group 2 further reduced its number of errors, whereas experimental group 1 and the controls remained at the same level." When results of different groups follow a similar pattern, a figure

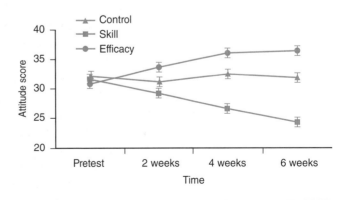

Figure 21.1 Appropriate use of a figure to depict an interaction.

frequently appears cluttered. If figures appear cluttered because standard deviation bars or confidence intervals overlap, try putting the bars from one group going upward on the graph and the bars from a group in close proximity downward.

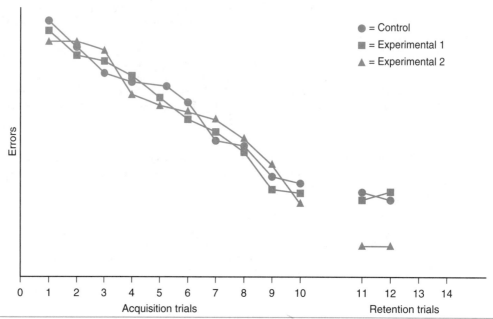

Figure 21.2   Example of a useless figure.

Figures should include error bars representing the variability of each mean data point that is shown. The error bars can be either standard deviations for each mean value (e.g., figure 21.1) or 95% confidence intervals (CI) for each mean value. An advantage of using the 95% CI is that when the 95% CIs for mean values do not overlap on the x-axis at the same point, the means are significantly different. This circumstance allows the effective interpretation of significant interactions when they appear as a figure. Standard errors should not be used as the representation of variability because they represent the variability of sample means about the true population mean, not variability in an individual sample. The most appropriate use of standard error is to calculate the 95% CI.

One final consideration is the construction of the y-axis. In general, use between 8 and 12 intervals to encompass the range of values. To avoid wasting space, do not extend the y-axis outside the range of values. Again, consider whether you need a figure at all. Sometimes, theses and dissertations include examples such as that in figure 21.3. Looking at that figure, you immediately see a strong and significant interaction between knowledge of results (KR) and goal setting. Now look at the y-axis, on which the dependent variable is shown. Note that the scores are given to the nearest hundredth of a second. There is a difference of less than 500 ms among the four groups on a task in which average performance is about 18 s (18,000 ms). In fact, this interaction is not significant and clearly accounts for little variance. The result should merely be reported as nonsignificant with no figure included. The researcher made the interaction appear important by the scale used on the y-axis.

Sometimes multiple figures reporting data across several independent variables are used. For example, a comparison of running speed of Chinese, African, and U.S. boys and girls in five age groups might be reported. A single figure would be cluttered with all this data, so several figures (one for each continent of origin, each age, or each sex) might be used. The y-axis on each figure should have the same 8 to 12 points for running speed. Otherwise, visual comparisons of the data are difficult.

With current computer technology and software, producing figures from data takes a matter of seconds. Try different formats for figures to see which looks the best and displays the data most effectively. Put data in a bar graph and then in a line graph and compare the two. Be creative; you can really improve your presentation of results by using tables and figures effectively.

Illustrations (photographs and line drawings) are also used in research reporting. Most frequently, illustrations are of experimental arrangements and equipment. They should not be used when the equipment is of a standard design or make; a brief description suffices. Any unusual arrangement or novel equipment should be described and either a picture or line drawing included. If specifications and relationships are important to include within the illustration rather than in the text, a line drawing is preferable because it can be more easily labeled.

Remember that tables, figures, and illustrations are generally appropriate for the results chapter but not for the discussion. An exception to this rule is a report of multiple experiments, in which the general discussion section could contain a table or figure to display common findings or a summary of several experiments.

Figure 21.3 A nonsignificant interaction made to appear significant by the scale of the dependent variable.

To review, when determining whether tables, figures, and illustrations should appear in the text or in the appendix, consider the following recommendations:

■ Put important tables, figures, and illustrations in the text and all others in the appendix.

■ Try not to clutter the results with too many tables, figures, and illustrations.

■ Do not put summary tables for ANOVA and MANOVA in the results. Place the important statistics from these tables in the text and put the tables in the appendix.

Remember that all journals have prescribed formats and styles for articles submitted to them (e.g., those of the American Psychological Association or American Physiological Society). The instructions for authors usually include directions for preparing tables and figures. Many of these decisions are arbitrary. Read and look at what you have written, and then use common sense. Select the tables, figures, and illustrations that are needed for reading and understanding the results. Everything else goes in the appendix. The appropriate use of tables, figures, and illustrations can add to readers' interest, understanding, and motivation. We recommend that you get a copy of the *Publication Manual of the American Psychological Association* (APA, 2010) or the style manual used in your department.

SUMMARY

The research proposal describes the definition, scope, and significance of a problem and the methodology that will be used to study it. The proposal is essentially the plan for the study. The introduction provides the background and literature on the problem, the problem statement, hypotheses, definitions, assumptions and limitations, and the significance of the study. The methods section describes the participants, instrumentation, procedures, and design and analysis.

These two sections are presented to the research committee as the plan for inquiry. The committee determines the worth of the study, suggests needed alterations, and ultimately must agree that the study should be done. The proposal must be carefully prepared with appropriate pilot work so that the committee is convinced that you can complete the research plan.

Proposals to granting agencies are similar but are typically required to have specific lengths and formats. Talk with outside agencies and more experienced grant writers before preparing and submitting proposals. Many colleges and universities offer internal grants for which graduate students can apply to support thesis and dissertation research.

The results and discussion sections are written after the data have been collected and analyzed. The results tell what you found; the discussion explains what the results mean. The results are the most important part of the research. They represent the unique findings of your study and your contribution to knowledge. The discussion ties the findings back to the literature review, theory, and empirical findings from other studies. Your findings should be interpreted in the discussion. In journal format, the results and discussion are sections in the body of the paper (more on this in chapter 22). In the traditional format, the results are chapter 3 and the discussion is chapter 4. In multiple-experiment studies, however, each experiment may best be reported in a separate section (or chapter) with its own brief introduction, methods, results, and discussion. A section (or chapter) of general discussion of the multiple experiments often follows.

Tables should be used in the discussion to summarize and present data when they are more effective than text presentations. Figures and illustrations are also used in the results, most frequently to demonstrate more dramatic findings. Careful construction of tables and figures is important in ensuring that information is well communicated.

Check Your Understanding

1. Learn the steps for writing a thesis or dissertation proposal at your school. List them in chronological order (e.g., select a major professor and committee, prepare a proposal, and get the proposal approved). Explain the process at each step.

2. Select a research report of interest from a refereed journal in your area of interest. Read the paper, but concentrate on the results and discussion. Answer the following questions in a brief report.

   a. Results:

      How is the results section organized?

      Compare the order of reported findings with the introduction, literature review, and statement of the problem. Do you see any relationships?

      How else might the results have been organized? Would a different scheme have been better or worse? Why?

   b. Tables, figures, and illustrations:

      Are there any? How many of each?

      Why are they used? Could the data have been reported more easily in the text? When either a table or figure is used, would the other have been as good? Better? Why?

      Can you restructure one of the tables or figures to improve it?

   c. Discussion:

      How is the discussion organized?

      Are all results discussed?

      Is the discussion accurate in terms of the results?

      Are previous findings and theory woven into the discussion?

      Are all conclusions and supporting evidence clearly presented?

      Did the authors use any methodological cop-outs?

22

Ways of Reporting Research

If I agreed with you we'd both be wrong.

—Onelinerz.net

Our favorite authors, Day and Gastel (2006), provide an appropriate introduction to this chapter on ways of reporting research with the following quotation:

A scientific experiment, no matter how spectacular the results, is not completed until the results are published. In fact, the cornerstone of the philosophy of science is based on the fundamental assumption that original research *must* be published; only thus can new scientific knowledge be authenticated and then added to the existing database that we call scientific knowledge. (p. ix)

In chapter 21, we covered the research proposal: how to write the introduction, the literature review, the problem statement, and the methodology for the thesis or dissertation. Then we explained how to organize and write the results and discussion sections. Effectively coordinating all this information into a thesis or a dissertation is the topic of concern in this chapter. We present both the journal style of organization, which we advocate (Thomas, Nelson, & Magill, 1986), and the traditional chapter style. In addition, we include information about writing for publication in research journals, preparing abstracts, and presenting papers orally (including in a poster format).

BASIC WRITING GUIDELINES

A thesis or dissertation demonstrates that you can write in an acceptable scientific style. However, because it is useful to determine the expectations of your advisor, department, and university, you should do the following:

- Collect all the documents that outline university, graduate school, and department policy for theses and dissertations. Then actually read these documents, because someone at some level will eventually check to see whether you have followed them.

- Review the theses and dissertations of past graduate students whose work is well regarded by your institution. Identify common elements in their work, and pattern your work after theirs.

- Allow twice as much time as you think you will need. Remember Murphy's law, "Whatever can go wrong will go wrong," and its special corollary, "When several things can go wrong, the one that will go wrong is the one that will cause the greatest harm."

The thesis and dissertation have basically the same parts found in any scientific paper: an introduction and a literature review and methods, results, and discussion sections. In the traditional format, each of these parts becomes a chapter. Sometimes the introduction and the literature review are separate chapters, leading to five rather than four chapters. However, the model of introduction (and literature review), methods, results, and discussion (IMRD) is pretty standard for scientific theses and dissertations. Qualitative, historical, and philosophy papers often vary considerably from this format. We advocate for the use of journal style in theses and dissertations because this results in a paper ready for journal submission. Nearly all universities allow the use of this style, but you should check with your graduate school office.

A BRIEF WORD ABOUT ACKNOWLEDGMENTS

A section seldom mentioned in articles and books on preparing theses and dissertations is the **acknowledgments**. You should acknowledge the people without whom the research would have been impossible. Acknowledgments prepared by graduate students take some odd turns. We saw one in which a woman acknowledged her ex-husband. She said that if he had not been so difficult to live with, she would never have gone back to the university and done graduate work. She found an area in which she had a tremendous interest, and her life changed significantly. That is about the only positive statement that we have ever heard about an ex-spouse. Some other funny acknowledgments follow:

- My parents, husband, and children provided inspiration and support throughout, but I was able to complete my thesis anyway.
- My major professor, Dr. I.M. Published, coordinated the work and made an occasional contribution.
- Professor B.A. Snobb wants everyone to know he had nothing to do with this thesis.
- Finally, we would like to thank our proofreader, I.D. Best, without whose hell; thes document wld net be pausible.

More seriously, you can acknowledge appropriate people, but keep the list short and to the point and do not be mushy. As in other areas, use correct English. In their acknowledgments, graduate students often use *wish* when they mean *want,* as in *I wish to acknowledge I.B.A. Fink.* Does that mean they might have acknowledged him if his contribution had not been so lousy? The graduate student really wants to acknowledge him. But more appropriately, why not just say it directly: *I acknowledge I.B.A. Fink.*

THESIS AND DISSERTATION FORMAT: TRADITIONAL VERSUS JOURNAL

We devote considerable space here to the journal format for thesis and dissertation writing. Our material is partially reproduced from an article two of us wrote with a colleague (Thomas, Nelson, & Magill, 1986). We acknowledge the contribution of Dr. Richard Magill from New York University, and we appreciate *Quest* (published by Human Kinetics) for allowing us to adapt the article for this book.

Graduate education in the United States, especially at the doctoral level, was modeled after graduate education in Germany. The German university's spirit of the search for knowledge and the concomitant emphasis on productive research were transplanted in large measure in America (Rudy, 1962). Although some changes have occurred over the years in the requirements for the doctoral degree, the basic aims and expectations have in essence remained unaltered. In most cases, the doctoral degree is conferred in recognition of a candidate's scholarship and research accomplishments in a specific field of learning through an original contribution of significant knowledge and ideas (Boyer, 1973). Some universities now have alternatives to the traditional dissertation in which the production of new research is the goal for those pursuing the doctorate for professional practice (e.g., an evaluation study with suggestions to improve practice based on research). These types of dissertations are appropriate for those who will not do research after graduating with their degrees. Our suggestions here, however, are focused on dissertations and theses in which the reporting of research is the focus.

Stated more simply, research is the foundation of the doctoral program, and the dissertation is the most distinctive aspect of the doctoral degree. It has been reported that the dissertation occupies an average of 39% of the time devoted to obtaining a degree in the fields of biochemistry, electrical engineering, psychology, physics, sociology, and zoology (Porter, Chubin, Rossini, Boeckmann, & Connally, 1982).

A Case for the Journal Format

The major purpose of the thesis and the dissertation is to contribute new knowledge with scientific merit (American Psychological Association, 2010; Berelson, 1960; Porter & Wolfe, 1975; Robinson & Dracup, 2008). A purpose typically cited in university bulletins is that the thesis or dissertation provides evidence of competency in planning, conducting, and reporting research. In terms of program objectives, the study is a valuable learning experience in that theses and dissertations are functional exercises in executing the steps in the scientific method of problem solving. Even ABDs ("all but dones," or people who have completed all work for the PhD except the dissertation) acknowledge the contribution that dissertations make to science and the scientific method (Jacks, Chubin, Porter, & Connally, 1983).

The importance of disseminating results as part of the research process is well established. It follows that one of the purposes of the thesis or dissertation is to serve as a vehicle to carry the results of independent investigation undertaken by the graduate student. Thus, the dissertation or thesis becomes part of the dissemination process. A.L. Porter and colleagues (1982) reported that dissertations do make substantial contributions to the knowledge base and that authors' published dissertations are cited more often than are other papers they write.

Despite the potential contribution that the thesis and dissertation can make to a field of study, the fact remains that no more than about one-third to one-half of the dissertations (and even fewer theses), depending on the field of study, become available to the profession through publications (Anwar, 2004; Lee & Kamler, 2008; McPhie, 1960; Porter et al., 1982; Silverman & Manson, 2003). Several reasons may account for why a thesis or dissertation is never published. For one, despite the emphasis placed on research by the institutions, many students do not consider research an important aim, especially those who enter graduate programs focused on professional practice. A.L. Porter and colleagues (1982) reported that 24% of the doctoral recipients surveyed expressed this feeling. Arlin (1977) went so far as to claim that most educators never do another piece of published research after they complete the master's thesis or doctoral dissertation. In addition, job placements fix varying degrees of importance on research and publication. Another inescapable fact is that not all theses and dissertations are worthy of publication.

Conceivably, another contributing factor in the low rate of publication is the traditional style and format often used in the thesis or dissertation. Granted, the highly motivated new PhD recipient will spend the time and effort to rewrite the dissertation into the proper format (IMRD) for journal review, but the less motivated PhD recipient may not. Why should dissertations and theses be written in a format that requires rewriting before publication if a vital part of the research process is publication of the research study?

We contend that the traditional format of dissertations and theses is archaic. Doctoral students (especially those working with productive major professors) may have published in several scholarly refereed publications by the time the PhD is awarded. Should such people, who have shown that their previous work has scientific merit, be required to go through the ritual of writing a dissertation involving separate chapters that spell out every detail of the research process? It seems more logical to have the body of the dissertation prepared in the appropriate format and style for submission to a journal, which is the acceptable model for communicating results of research and scholarly works in many of the arts, sciences, and professions. We suggest restructuring the thesis and dissertation format to one that accomplishes this purpose, and we explain the contents of its parts in this chapter. Professionals in other fields have also been advocating for journals style (e.g., Duke & Beck, April 1999).

Limitations of Chapter Style

Conventional theses and dissertations typically contain four or five chapters. Traditionally, the chapters are intended to reflect the scientific method for solving problems: developing the problem and formulating the hypotheses, gathering the data, and analyzing and interpreting the results. These steps are usually embodied in chapters such as the introduction (which sometimes contains the literature review; at other times the literature review is a separate chapter), methods, results, and discussion chapters.

The thesis or dissertation also has several introductory pages as prescribed by the institution, usually consisting of the title page, acknowledgments, abstract, table of contents, and lists of figures and tables. At the end of the study are the references and one or more appendixes that contain items such as participant consent forms, tabular materials not presented in the text, more detailed descriptions of procedures, instructions to participants, and raw data. A brief biographical sketch (the curriculum vitae) is usually the last entry in the conventional thesis or dissertation.

The chapter format is, of course, steeped in scholastic tradition. In defense of this tradition, developing the discipline required to accomplish the steps involved in the scientific method is usually viewed as an educationally beneficial experience. Moreover, for the master's student, the thesis is normally the first research effort, and there may be merit in formally addressing such steps as operationally defining terms, delimiting the study, stating the basic assumptions, and justifying the significance of the study.

A more serious limitation of the chapter style relates to the dissemination of the results of the study; that is, publishing the manuscript in a research journal. Considerable rewriting is usually required to publish a thesis or dissertation because journal formats differ from chapter format. Granted, the information for the journal article is provided in the thesis and the dissertation, but a considerable amount of deleting, reorganizing, and consolidating is necessary to transform the study into a journal article.

Students usually want to publish this product of months (or years) of time and painstaking effort. But in terms of expediency, the chapter format is decidedly counterproductive. The rewriting required may be made more difficult by the fact that the typical new PhD recipient immediately begins a new job that demands considerable time and energy. Unless the person is motivated, the transformation of the conventional dissertation into journal format may be delayed, sometimes indefinitely. A.L. Porter and colleagues (1982) claimed that new PhD recipients who fail to publish within two years of the awarding of the degree are unlikely to publish later.

The master's thesis is even more unlikely to be published. One reason is that the master's student generally does not consider publishing unless the major professor suggests it. Furthermore, the master's student may not be as well prepared to write for publication as the doctoral student is. Thus, the burden for publication is on the major professor, who understandably is often unwilling to spend the additional time necessary to supervise (or do) the rewriting. Thus, most theses are not submitted for publication. Ironically, the time-honored scholarly chapter style of the thesis and dissertation impedes an integral part of the research process: the dissemination of results.

Structure of the Journal Format

To develop a better model for theses and dissertations, we must overcome the limitations of the chapter format for reporting while maintaining the contents of a complete research report. The format that we suggest has three major parts. Preliminary materials include such items as the title page, table of contents, acknowledgments, and abstract. The body of the thesis and dissertation is a complete manuscript prepared in journal form. Included are the standard parts of a research report, such as the introduction, methods, results, discussion, references, figures, and tables. The appendixes often include a more thorough literature review, additional detail about methods, and additional results not placed in the body of the thesis or dissertation.

Journal style (IMRD) in theses and dissertations is appropriate when the goal is publication as an article in a journal, which should be the goal for all scientific studies. There are exceptions, however; the traditional chapter format may be appropriate for studies of the history or philosophy of sport because the likely publication would be in book form. Also, some modification of the IMRD format may be best for areas such as the sociology of sport.

Following are the parts of the thesis and dissertation for quantitative studies using the journal format:

1.0 Preliminary materials

1.1   Title Page

1.2   Acknowledgments

1.3   Abstract

1.4   Table of contents

1.5   List of tables

1.6   List of figures

2.0 Body of the thesis or dissertation (IMRD)

2.1   Introduction

2.2   Methods

2.3   Results

2.4   Discussion

2.5   References

2.6   Tables

2.7   Figures

3.0 Appendixes

3.1   Extended literature review

3.2   Additional methodology

3.3   Additional results

3.4   Other additional materials

4.0 One-page curriculum vitae

How can this format overcome the limitations of the chapter style? For both master's and doctoral students, a manuscript (body of the thesis or dissertation, IMRD) is developed that is ready for journal submission. All that remains is to add the title page and abstract, and the paper can be sent to a suitable journal.

The advantage of the journal format for doctoral students should be apparent. Because PhD recipients who fail to publish their dissertations within two years are unlikely to publish afterward, a more functional format encourages publication. Especially when we consider that dissertations appear to make important contributions to knowledge, the evaluation and subsequent publication of that knowledge through refereed journals is an important step. Although master's theses are not as likely as dissertations to be published, any format that encourages the publication of quality thesis work is desirable.

We want to make one final point before proceeding to the structure of the journal format: Your graduate school probably requires that the thesis or dissertation follow a standard style manual (or at least the style of a journal). The three most common are the *Publication Manual of the American Psychological Association* (American Psychological Association, 2010), American Physiological Society style (Curran-Everett & Benos, 2004), and *The Chicago Manual of Style* (University of Chicago Press, 2010). The journal format adapts nicely to any of these styles. University regulations usually do not specify a particular style, but frequently, an academic department may adopt one or two styles. If the journal format is to be used, a department might want to allow more than one style. For example, many journals reporting exercise physiology and biomechanical studies use the American Physiological Society style. Journals publishing articles in motor behavior, sport psychology and sociology, and professional preparation frequently use the APA manual. Journals that publish articles on the history and philosophy of sport frequently use *The Chicago Manual of Style*. Graduate students benefit considerably by having the flexibility to choose the style recommended by the journal to which the paper will be submitted.

Preliminary Materials

Most of the information in the preliminary materials is required by the institution and usually appears at the beginning of the thesis or dissertation. The exact nature of this material is often specified by an institution's graduate school, so make sure you have the necessary information. Generally included are the title page, committee approval form, acknowledgments, dedication, abstract, table of contents, list of tables, and list of figures. Many institutions require that the abstract follow the form for *Dissertation Abstracts International,* which sets a 350-word maximum length for the abstract published in ProQuest. We recommend a slightly different length for the abstract. Journals (and the style manuals mentioned previously) typically require abstracts between 100 and 150 words. If you keep the abstract between 100 and 150 words, it can serve both requirements.

Body of the Thesis or Dissertation

The body should be a complete research report using the appropriate style for the journal to which it is to be submitted (or the style required by the department in which the work was done). Included are the introduction and literature review, methods, results, discussion, references, tables, and figures. Keep the length within the bounds set by the journal or the institution. Journals typically set a limit of 20 to 30 pages for single experiments. Most allow additional pages for multiple experiments or unusually complex articles. You and your major professor must be rigorous in keeping the paper within a length that a journal will consider. In our experience, maintaining an acceptable length is the most difficult goal to achieve using this format. If the dissertation has multiple parts and will be published as more than one paper, you would include two or more reports in this section.

You should consult several sources in preparing the body of your paper. One is the journal to which submission is anticipated. Guidelines to authors and instructions regarding the appropriate style manual are typically published in journals (e.g., see each issue of *Research Quarterly for Exercise and Sport*). Read a number of similar papers in the selected journal to see how specific issues are handled (tables, figures, unusual citations, multiple experiments). In addition, you should carefully follow the style manual that has been selected.

Of course, the ultimate goal of this part of the thesis or dissertation is to enable the paper to be submitted to a journal as soon as possible. Improved format and better writing will not aid poor-quality research. On the other hand, high-quality research can easily be hidden by a poor format that makes rewriting difficult, poor reporting that omits important information, or weak and boring writing that makes reading tedious. "Thus the scientist must not only 'do' science; he (or she) must 'write' science. Although good writing does not lead to the publication of bad science, bad writing can and often does prevent or delay the publication of good science" (Day & Gastel, 2006, p. ix). Research is not recognized in any formal sense as having been conducted until it has been shared with and evaluated by the academic community. Additionally, no law says that the thesis or the dissertation must be written in such a way that the reader has difficulty staying awake.

FORMAT FOR TYPICAL SCIENTIFIC STUDIES REPORTED IN JOURNAL STYLE

As we have suggested, good scientific style is often labeled IMRD—introduction (including the literature review), methods, results, and discussion. Here is the format for journal style:

- Introduction and brief literature review—Here is where you catch the reader's attention and say what the study is about, why it is important, what closely related research tells you, and what your hypotheses are.
- Method—What you did and how you did it.
- Results—Your contribution to knowledge and the details of what you found.
- Discussion—What it means and why it is important.
- References—Only those cited in the paper; put the rest in the appendix.
- Tables—Important data; put the rest in the appendix.
- Figures—Important trends and changes; put the rest in the appendix.
- Appendixes—Typically include a more thorough literature review, additional methodological procedures, other tables and figures, and any other material the major professor and committee require.

FORMAT FOR MULTIPLE STUDIES

Many graduate students, particularly doctoral students, have a continuous line of research with their advisor, a desired result in good and productive graduate student–advisor relationships. The dissertation should not be the first research the graduate student has completed but one step along the way of multiple research studies and publications. In fact, the dissertation may include multiple studies and therefore articles for publications.

Some graduate schools have specific formats for multiple studies. Most require that students be first authors on any papers included in the thesis or dissertation. Following are the parts of the thesis or dissertation:

- Standard graduate school materials
- Chapter or section I—introduction to the series of studies
- Multiple chapters or sections—one for each study using the single-study format (IMRD) previously given
- Final chapter or section—overview and summary of the findings from all the studies
- Appendixes—extended literature review, additional methods, additional results, other items required by the major professor or committee

In this case, each chapter or section for a study is a stand-alone paper that includes all the standard parts previously listed. Each paper is ready to submit to a journal.

(continued)

Format for Typical Scientific Studies Reported in Journal Style *(continued)*

SERIES OF RELATED STUDIES

If the paper reports a series of related studies (e.g., experiment 1, experiment 2), they may be included as parts of a single paper that is ready to go to a single journal. Typically, no introduction or final summary is required as a separate chapter or section because this is already part of the paper itself.

SERIES OF STUDIES, SOME PUBLISHED, SOME NOT PUBLISHED

Another situation that may occur is the use of published and in-press papers in the dissertation. This is expected when the graduate student has been doing research and writing papers over several years of work with the advisor, especially when the studies are related. In this instance, the dissertation might include one or more published papers, one or more in-press papers, and one or more studies just completed. Most graduate schools allow this model as long as the student is first author on all the papers. However, a specific format may be required, such as the following:

- Chapter or section I—introduction to the series of papers
- Chapter or section II—previously published work
- Chapter or section III—in-press work
- Chapter or section IV—recently completed studies
- Chapter or section V—summary of studies
- Appendixes—same as previously discussed

Thanks to Dr. Katherine Thomas, University of North Texas, for her contributions to this section.

Appendixes

In the traditional format, the appendixes serve primarily as a depository for nonessential information. To a degree, this notion still characterizes what should be put in the appendixes when the journal format is used, but some additional features give appendixes unique worth. Usually, the number and content of the appendixes are determined by the agreement of the student, the advisor, and the supervisory committee. In the journal format, we suggest four types of appendixes. Other appendixes could be included, but we see these four as essential.

Each appendix should begin with a description of what is in it and how that information relates to the body of the thesis or dissertation. This description enables the reader to get the most use of the information in the appendix.

More Thorough Literature Review

An important and useful appendix is a literature review. The introduction in the body of the thesis or dissertation includes a discussion of related research but presents only minimal information to establish an appropriate background for the one or more studies that follow. One purpose of the thesis or dissertation is to provide an opportunity to demonstrate knowledge of the research literature related to the topic. In the body of the journal format, limited opportunity is available for this demonstration because journals typically have concise introductions. A comprehensive literature review should be included as the first appendix to provide an appropriate mechanism for demonstrating knowledge of the relevant literature. We recommend that this review also be written in journal style.

At least two additional purposes are served by including this literature review as an appendix. First, the information is made available to committee members, some of whom may not be sufficiently familiar with the literature related to the thesis or dissertation. Second, if properly developed, the literature review can go directly to a journal for publication.

The literature review contained in the appendixes can take several forms, the most popular of which is probably the comprehensive narrative that synthesizes and evaluates research. This review links studies and establishes a strong foundation on which to build the research. It also reveals how the research represented by the thesis or dissertation extends the existing body of knowledge related to the research topic.

A second form that the literature review can take is a meta-analysis: a quantitative literature review that synthesizes previous research by analyzing the results of many research studies using specified statistical methods (see chapter 14). An example of a meta-analysis related to the study of physical activity can be seen in an article by Thomas and French (1985).

Additional Methods Information

A second important appendix presents additional methodological information not included in the body of the thesis or dissertation. Journals encourage authors to provide methodology information that is brief yet sufficient to describe essential details related to the participants, apparatuses, and procedures used in the research. The disadvantage of journal articles is that information sufficient to enable someone to replicate the study is often not provided. In the journal format for the thesis or dissertation, this useful additional information becomes an appendix. Information suitable for this section includes more detailed participant characteristics; more comprehensive experimental design information; fuller descriptions (and perhaps photographs) of tests, testing apparatuses, or interview protocols; copies of tests, inventories, or questionnaires; and instructions given to participants.

Additional Results Information

The third appendix should include information that is not essential for the results section of the body of the thesis or dissertation. Journal editors typically want only summary statements about the analyses and a minimal number of figures and tables. Thus, considerable material concerning the results can be placed in the appendix, such as all means and standard deviations not included in the results section, ANOVA tables, multiple-correlation tables, validity and reliability information, and additional tables and figures.

This information serves several purposes. First, it provides evidence to the committee and your advisor that you accurately described the data and properly performed the analyses. Second, the committee members have an opportunity to evaluate the statistical analyses and interpretations from the body of the thesis or dissertation. Third, other researchers have access to more detailed data and statistical information should they want it. For example, you may want to include this thesis or dissertation in a meta-analysis. Because the additional results presented in the appendix may have many future uses, their importance to the total thesis or dissertation cannot be overemphasized.

Other Materials

A fourth appendix includes information that is not appropriate for the body of the thesis or dissertation or the other appendixes, such as the approval form from the human research committee, participants' consent forms, sample data-recording forms, and perhaps the raw data from each participant. Also, detailed descriptions of any pilot work done before the study could be included in this or a separate appendix.

The appendixes in the journal format of a thesis or dissertation provide a mechanism for including information not contained in the journal article that may be significant for you, the committee, and other researchers. They also provide a means for elaborating on some information in the body of the thesis or dissertation. Moreover, the comprehensive literature review provides you with another possible publication. As a result, the appendixes are meaningful components of the total thesis or dissertation.

One-Page Curriculum Vitae

Many colleges and universities request that the last page of the thesis or dissertation be a one-page curriculum vitae about the student. This document should be a professional version of a curriculum vitae and include items such as education and previous work experience.

Examples of Theses and Dissertations Using Journal Format

Numerous theses (e.g., Boorman, 1990; Tinberg, 1993) and dissertations (e.g., French, 1985; T.M. Lee, 1982; McPherson, 1987; Scrabis-Fletcher, 2007; Yan, 1996) use journal format. These theses and dissertations provide excellent examples of the format discussed here.

The information in an unpublished thesis or dissertation remains in the exclusive domain of a few people. We established earlier that a vital part of the research process is the dissemination of knowledge. We maintain that style and format should not impede this process. The proposed journal format accomplishes the traditional objectives of the thesis or dissertation yet facilitates the dissemination of knowledge.

HELPFUL HINTS FOR SUCCESSFUL JOURNAL WRITING

We highly recommend the sixth edition of a work to which we have frequently referred: *How to Write and Publish a Scientific Paper* (Day & Gastel, 2006). In our opinion, this book is the researcher's best resource for preparing a paper for submission to a research journal. The book is short, informative, readable, and funny. Although the small section here cannot replace Day and Gastel's more thorough treatment, we do offer a few suggestions.

First, decide to which journal you will submit your research paper. Carefully read its guidelines (select a recent issue, because the guidelines may have changed from older issues) and follow the recommended procedures. Guidelines usually explain the publication style of the journal; procedures for preparing tables, figures, and illustrations; the process for submitting papers (and the number of copies); acceptable lengths; and sometimes estimated review time. Nearly all journals require that manuscripts not be submitted elsewhere simultaneously. To do so is unethical (see chapter 5).

Journals follow a standard procedure for papers submitted. Most now use electronic submissions. One of us (Thomas) was editor-in-chief of *Research Quarterly for Exercise and Sport* for six years (1983–1989); and another (Silverman), for three years (2002–2005). Here is what happens to a manuscript between submission to the editor and return to the author (the average time of this process for *Research Quarterly for Exercise and Sport* is 75 to 90 days):

1. The editor checks to see whether the paper is within the scope of the journal, is the appropriate length, and uses correct style. If any of these characteristics are inappropriate, it may be returned to the author.

2. The editor looks through the paper to determine that all appropriate materials are included (e.g., abstract, tables, and figures).

3. The editor reads the abstract, looks at the keywords, and evaluates the reference list to identify potential reviewers.

4. Depending on the size of the journal, there may be section editors (*Research Quarterly for Exercise and Sport* has 10 sections: biomechanics, pedagogy, physiology, psychology, and so on) or a board of associate editors. The editor places the paper in the appropriate section or with an appropriate associate editor.

5. The editor (sometimes in consultation with section or associate editors) assigns reviewers (usually two or three).

6. The reviewers and section editor (or associate editor) receive an e-mail informing them that the paper is posted electronically on a particular site. They have a specific date by which they should complete the review.

7. The reviewers submit reviews on the journal's website to the section editor (or associate editor), who evaluates the paper and the reviews and makes a recommendation to the editor.

8. The editor reads the paper, the reviews, and the evaluations and writes the author concerning the status of the paper. Usually, the editor outlines the major reasons for the decision.

Smaller, more narrowly focused journals may not have section or associate editors, so reviews go directly to the editor. Publication decisions usually fall into three or more general categories (e.g., see table 22.1, which shows the categories of numerous kinesiology journals), such as the following:

- Acceptable (sometimes with varying degrees of revision)
- Unacceptable (sometimes called rejected without bias), which means that the editor will have the paper evaluated again if it is revised
- Rejected, which usually means that the journal will no longer consider the paper

TABLE 22.1

Peer Review Procedures in Kinesiology Journals

Journal	Blinding	Criteria*	Decision categories
British Journal of Sports Medicine	Author	2 (e.g., ethics in research and publication)	4
European Journal of Sport Science	Double	6 (e.g., impact, originality, rank)	5
Journal of Applied Biomechanics	Author	9 (e.g., clear question, new insights)	5
Journal of Motor Behavior	Double	6 (e.g., varying levels of analysis)	4
Journal of Sport & Exercise Psychology	Double	8 (e.g., experimental design sound)	4
Journal of Sports Sciences	Double	3 (e.g., quality of data and methods)	5
Journal of Strength & Conditioning Research	Double	4 (e.g., practical applications impact)	4
Journal of Orthopaedic & Sports Physical Therapy	Double	6 (e.g., importance or interest)	6
Journal of Physical Activity and Health	Double	4 (e.g., study design and interpretation)	3
Measurement in Physical Education and Exercise Science	Double	None	4
Medicine & Science in Sports & Exercise	Author	4 (e.g., percentile rank)	4
Perceptual and Motor Skills	Author	2 (e.g., conclusions supported by data)	4
Research Quarterly for Exercise and Sport	Double	4 (e.g., impact to whole field)	4
Sports Biomechanics	Author	10 (e.g., writing and organization)	5

*Note: Because of space limitations, only sample evaluation criteria for each journal are listed.

"Advancing kinesiology through improved peer review," J.R. Thomas, J.K. Nelson, and S.J. Silverman, *Research Quarterly for Exercise and Sport,* Vol. 85, pp. 127-135, 2014, Taylor and Francis, reprinted by permission of the publisher (Taylor & Francis Ltd, www.tandfonline.com).

Review criteria are fairly common and are similar to those presented in chapter 2 for a review of published literature. Ratios of published to rejected papers vary by the quality of the journal and the area of the paper and are frequently available in an issue of the journal. *Research Quarterly for Exercise and Sport* accepts about 20% of the papers submitted (see each June issue for a more detailed description).

If this is your first paper, seek some advice from a more experienced author, such as your major professor or another faculty member. Often, papers are rejected because they fail to provide important information. More experienced authors pick up on this immediately. Journal editors and reviewers do not have time to teach you good scientific reporting. Not everyone is fond of journal editors. One jokester was reported to have said, "Editors have only one good characteristic—if they can understand a scientific paper, anyone can!" Another said, "Editors are, in my opinion, a low form of life—inferior to the viruses and only slightly above academic deans" (Day, 1983, p. 80). Teaching good reporting is the responsibility of your major professor, research methods courses, and books such as this one, but it is your responsibility to acquire the skill.

The peer review process is widely used to judge scientific papers. However, there is disagreement about its value and success. Knudson, Morrow, and Thomas (2014) made several recommendations to improve the peer review process (e.g., publishing clear evaluation standards, establishing collaborative evaluation procedures and editorial team roles, using online submission data to improve reviewer comments, creating author appeal procedures).

REVISING RESEARCH PAPERS

Do not be discouraged if your paper is rejected. Every researcher has had papers rejected. Carefully evaluate the reviews and determine whether the paper can be salvaged. If it can, then rewrite it, taking into account the reviewers' and editors' criticisms, and submit it to another journal. Do not send the paper out to another journal without evaluating the reviews and making appropriate revisions. Often, another journal uses the same reviewers, who will not be happy if you did not take their original advice. If your research and paper cannot be salvaged, learn from your mistakes.

Reviewers and editors will not always agree in what they say about your paper. This is frustrating because sometimes they tell you conflicting things. Be aware that editors select reviewers for different reasons. For example, in a sport psychology study that develops a questionnaire, the editor may decide that one or more content reviewers and a statistical reviewer are needed to evaluate the work. You would not expect the content and statistical reviewers to discuss the same issues. Studies have been done of the peer review system for scholarly journals. For example, Morrow, Bray, Fulton, and Thomas (1992) reported that *Research Quarterly for Exercise and Sport* had an average reviewer agreement of .37 based on 363 manuscripts between 1987 and 1991. They reported that other behavioral journals had reviewer agreements from a high of .70 in *American Psychologist* to a low range of .17 to .40 for a group of medical journals. Whether reviewer agreement is to be expected is debatable. For a good overview of the peer review process, see *Behavioral and Brain Sciences, 14* (1), 1991. Our advice to you as a graduate student is to seek help from your advisor and other faculty members in making revisions when you receive reviews of a paper you have submitted.

If you are fortunate enough to be asked to submit a revision, you should start by reading the reviewers' and editors' comments. Often, these comments seem overwhelming at first, so it may be a good idea to read the comments thoroughly and then put them away for a few days, but no longer. This interval allows your ego to recover from the onslaught of criticism before you prepare a revision. We recommend the following steps for preparing a revision:

1. Reread all the comments from the reviewers and editors. Make notes as you are doing this so that you can prepare the revision and provide the rationale for the changes you made. In some cases the requested changes (e.g., additional data from participants whom you can no longer contact) may be impossible, and you may decide to revise the paper for another journal. This is recommended if the editor makes it clear that something is required that you cannot provide in a revision.

2. Put all the comments from the reviews in a table with three columns: (a) comment number (or original page number if the review was organized that way); (b) the com-

ment from the reviewer or editor; and (c) how you addressed the comment. Have a section for each editor and reviewer, and use the table to make clear which comment is which (i.e., use spaces or lines so that it will be clear to you—and to those reading it with the submitted revision).

3. Look at each comment again and determine whether any themes are running across comments and reviewers. If so, determine whether major changes will address all these comments. For example, reviewers may have commented that they did not understand how the results apply to other settings and that you did not provide enough information about the participants. Providing additional information about the participants and the setting context may address all these comments.

4. Look at individual comments and determine what changes are appropriate for the revised manuscript. You do not need to make all changes, but you should address most of the comments in a revision—particularly those that (a) are straightforward, (b) improve the manuscript, and (c) are identified by the reviewers and editors as important. Our advice is to address as many of the comments as you can if the revisions do not change the paper focus or add issues that may be problematic in a subsequent review. You cannot address every comment, but you should have a research or theoretical reason for not addressing a comment in the revision.

5. As you are making revisions, record how you addressed each comment in the third column of the table you prepared in step 2. Be specific and provide detail (e.g., *We have revised the section on participants and added additional demographic information, as well as adding a description of the study setting. Please see page XX, lines XX–XX*). If the suggestions are grammatical, you may just indicate that the change was made. If you did not make a change, provide a rationale for why you did not (i.e., you thought about making a change but doing so was not appropriate for the following reasons, or you made other changes that eliminated the need to make the first change). This kind of explanation may necessitate a slightly longer response than that required for the changes you did make. If you do not provide a rationale, when the revision is returned to the reviewers and editors, they will think you did not take the time to consider their suggestions and will be less likely to feel positive about your work.

6. After you have made all the changes, including updating the references, proofread the paper to ensure that it is grammatically correct and reads well. If the manuscript now exceeds the journal page limits, consider ways of editing it down to the acceptable limits. Include pertinent information in the responses to the comments.

7. Add any general comments to the file with the comments and revisions table to address major changes. Proofread the table of comments and your responses to ensure that page and line numbers are correct, that everything is clear and well written, and that your comments do not sound negative. Being combative with the reviewers and editors will not help your case that the revision is an improvement and should now be accepted. You may find it helpful to have others read the table of comments to ensure that the tone and writing are appropriate.

8. Submit the revision by the deadline stated in the letter from the editor.

Occasionally, you may feel that you have received an unfair review. If so, write back to the editor, point out the biases, and ask for another assessment by a different reviewer. Editors are generally open to this type of correspondence if it is handled in a professional way. Opening with a statement such as "Look what you and the stupid reviewers said" is unlikely to achieve success (for a humorous example, see the sidebar A Letter We've All Wanted to Write). Recognize that appealing to the editor is less likely to be successful if two or more reviewers have agreed on the important criticisms. Editors cannot be experts in every area,

A Letter We've All Wanted to Write

Dear Sir, Madam, or Other:

Enclosed is our latest version of MS #85-02-22-RRRR; that is, the re-re-re-revised revision of our paper. Choke on it. We have again rewritten the entire manuscript from start to finish. We even changed the #@*! running head! Hopefully we have suffered enough by now to satisfy even you and your bloodthirsty reviewers.

I shall skip the usual point-by-point description of every single change we made in response to the critiques. After all, it is fairly clear that your reviewers are less interested in details of scientific procedure than in working out their personality problems and frustrations by seeking some kind of demented glee in the sadistic and arbitrary exercise of tyrannical power over hapless authors like ourselves who happen to fall into their clutches. We do understand that, in view of the misanthropic psychopaths you have on your editorial board, you need to keep sending them papers, for if they weren't reviewing manuscripts, they'd probably be out mugging old people or clubbing baby seals to death.

Some of the reviewers' comments we couldn't do anything about. For example if (as reviewer C suggested) several of my recent ancestors were indeed drawn from other species, it is too late to change that. Other suggestions were implemented, however, and the paper has improved and benefited. Thus, you suggested that we shorten the manuscript by 5 pages, and we were able to accomplish this very effectively by altering the margins and printing the paper in a different font with a smaller typeface. We agree with you that the paper is much better this way.

Our perplexing problem was dealing with suggestions 13 through 28 by reviewer B. As you may recall (that is, if you even bother reading the reviews before writing your decision letter), that reviewer listed 16 works that he or she felt we should cite in this paper. These were on a variety of different topics, none of which had any relevance to our work that we could see. Indeed, one was an essay on the Spanish-American War from a high school literary magazine. The only common thread was that all 16 were by the same author, presumably someone whom reviewer B greatly admires and feels should be more widely cited. To handle this, we have modified the introduction and added, after the review of relevant literature, a subsection titled Review of Irrelevant Literature that discusses these articles and also duly addresses some of the more asinine suggestions in the other reviews.

We hope that you will be pleased with this revision and will finally recognize how deserving of publication this work is. If not, then you are an unscrupulous, depraved monster with no shred of human decency. You ought to be in a cage. If you do accept it, however, we wish to thank you for your patience and wisdom throughout this process and to express our appreciation of your scholarly insights. To repay you, we would be happy to review some manuscripts for you; please send us the next manuscript that any of these reviewers submits to your journal.

Assuming you accept this paper, we would also like to add a footnote acknowledging your help with this manuscript and pointing out that we liked the paper much better the way we originally wrote it, but that you held the editorial shotgun to our heads and forced us to chop, reshuffle, restate, hedge, expand, shorten, and in general convert a meaty paper into stir-fried vegetables. We couldn't or wouldn't have done it without your input.

Sincerely,

Trying to Publish but Perishing Anyway

Adapted from the Canadian Society for Psychomotor Learning and Sport Psychology, 1991, "A letter we've all wanted to write," *Bulletin of the Canadian Society for Psychomotor Learning and Sport Psychology*. Author unknown.

and they must rely on reviewers. If you find yourself resorting to this tactic often, the problem is likely in your work. You do not want to end up like Snoopy in *Peanuts* (by Charles Schulz), who received this note from the journal to which he had submitted a manuscript: "Dear Contributor, We are returning your manuscript. It does not suit our present needs. P.S. We note that you sent your story by first-class mail. Junk mail may be sent third-class."

As a final point in this part, we have noted the infrequency with which most scientific papers are read (of course, none of ours fall into that category). Some writers have speculated that only two to four people read the average published paper completely.

WRITING ABSTRACTS

Whether you are writing for a journal, using the chapter or journal format for preparing a thesis or dissertation, or readying a conference paper, you need to include an abstract. Abstracts for each purpose require slightly different orientations, but nearly all have constraints on length and form.

Thesis or Dissertation Abstracts

The abstract for your thesis or dissertation probably has several constraints, including length, form, style, and location. First, consult your university or graduate school regulations and then follow these carefully. For dissertations, the graduate school regulations nearly always include the form for submission to *Dissertation Abstracts International*. The exact headings, length, and margins are provided in a handout available from your graduate school office. In writing the abstract, consider who will read it. Computer searches will locate it by the title and keywords. (The importance of these was discussed in chapter 2.) Write the abstract so that anyone reading it can decide whether to look at the total thesis or dissertation. Clearly identify the theoretical framework, the problem, the participants, the measurements, and your findings. Do not use all the space writing about your sophisticated statistical analyses or minor methodological problems. Keep jargon to a minimum. People in related areas read the abstract to see whether your work is relevant to theirs.

Abstracts for Published Papers

When looking for literature related to a topic, a researcher first looks at the title. If that gets his attention, the next thing read is the abstract. Nearly all electronic search engines provide the title and abstract of the paper. Thus, a well-written abstract is essential to the research paper. The abstract should have all the components of the paper—introduction and problem, methods, results, and discussion. Too often graduate students become so enamored with their methods and sophisticated analyses that they spend too much of their abstracts on them. The abstract should tell the purpose and value of the study in the opening sentence or two; then who the participants were and what the main treatments and measurements were, what the outcomes were, and why it was important. Use synonyms for words in the title because this increases the hit rate on e-searches.

There will be two constraints on your abstract's length—by your graduate school (usually following the length of *Dissertation Abstracts International*) and by the journal to which you plan to submit. You must use the length specified by your graduate school at the beginning of the thesis or dissertation. Then have an appropriate length abstract in front of the article(s) you are preparing for the journal(s) (go to the journal sites and see what length is allowed). Be certain that your abstract provides only information given in the article. Finally, the most useless sentence in any abstract is *Results were discussed*. Who would have thought otherwise?

Conference Abstracts

Abstracts for conferences are slightly different from other types of abstracts. Usually, you are allowed a little more space (from 400 to 2,000 words depending on the conference) because the reviewers must be persuaded to accept your paper for presentation. In these abstracts, you should follow these procedures:

1. Write a short introduction to set up the problem statement.
2. State the problem.
3. Describe the methodology briefly, including the following:
   a. Participants
   b. Instrumentation
   c. Procedures
   d. Design and analysis
5. Summarize the results.
6. Explain why the results are important.

Writing an Abstract

Please use the following style for all abstracts.

I.M. Tenured and U.R. Promoted, JR's South Fork School of Hard Knocks, Dallas, TX 00001

How Motor Skill Research Can Get You a Merit Raise

Current research suggests that merit raises are directly related to the number of papers faculty present and publish and inversely related to the quality of the research . . . .

. . . The long-term effect is to increase the number of journals, conferences, and full professors.

If the submission guidelines ask for other formats or information, you must follow those guidelines (for tips on writing conference papers and abstracts, see Jalongo, 2012). If a word limit is in place, use the word count function on your word processor to make certain that you do not exceed the limit. Most conference websites allow only the stated maximum number of words and truncate abstracts when that number is reached.

The sidebar Writing an Abstract is a humorous example of an abstract developed by one of the authors (J.R. Thomas, 1989).

The results and their importance are a critical part of a conference abstract. If you do this in a nondescript way, the reviewer may conclude that you have not completed the study. This judgment generally leads to rejection. The conference planners cannot turn down completed research when it is possible that yours will not be finished.

Finally, most conferences require that the paper be presented before publication. So if you have a paper in review at a journal, submitting it to a conference that is 8 to 12 months away could be hazardous. Also, many conferences require that the paper not have been previously presented. Be aware of this and follow the regulations. Violating them will never enhance your professional status, and other scholars will quickly become aware of it.

MAKING ORAL AND POSTER PRESENTATIONS

After your conference paper is accepted, you are faced with presenting it. The presentation is conducted either orally or in a poster session.

How to Give Oral Presentations

Oral presentations usually cause panic among graduate students and new faculty members. The only way to get over this is to do several oral presentations. However, you can help assuage the apprehension. Usually, the time allowed for oral research reports is 10 to 20 minutes, depending on the conference. You will be notified of the time limit when your paper is accepted. Because you must stay within the time limit and cannot possibly present a complete report within this period, what do you do? We suggest that you present the essential features of the report using the following divisions for a 15-minute presentation. Visual aids, such as slides, computer presentation software, and overheads, are helpful.

> **oral presentation**—A method of presenting a paper in which the author speaks before a group of colleagues at a conference following this format: introduction, statement of the problem, methods, results, discussion, questions.

- Introduction that cites a few important studies: 3 minutes
- Statement of the problem: 1 minute
- Methods: 3 minutes
- Results (present tables and figures; figures are usually better because tables are hard to read): 3 minutes
- Discussion (main points): 2 minutes
- Questions and discussion: 3 minutes
- Total time: 15 minutes

> Prepare different tables and figures for a visual presentation. Paper versions often contain too much information and are too small.

The most frequent errors in oral presentations are spending too much time on methods and presenting results poorly (see the sidebar Do Not Use These Quotations From British Sport Commentators in Your Presentation). Proper use of visual aids is the key to an effective

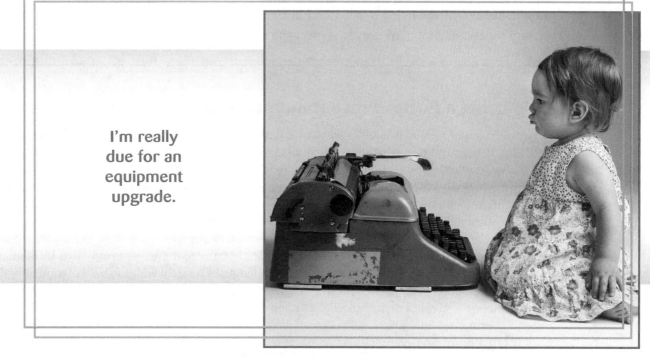

I'm really due for an equipment upgrade.

© Artur Bogacki/Fotolia

presentation, particularly in the results section. Place a brief statement of the problem on a slide and show it while you talk. A slide of the experimental arrangements reduces much of the verbiage in the methods discussion. Always use slides to illustrate the results. Pictures of the results (particularly figures and graphs) are much more effective than either tables or a verbal explanation. Keep the figures and graphs simple and concise. Tables and figures prepared for a written paper are seldom effective when converted for visual presentations. They typically contain too much information and are not large enough. Have a pointer available to indicate significant features. Finally, remember Thomas, Nelson, and Silverman's Four Laws of Oral Presentations as outlined in the sidebar. Practicing your presentation is one solution to these and many other problems. At our universities, we gather graduate students and faculty who are presenting papers at upcoming conferences and conduct practice sessions. Everyone presents her or his paper and has it timed. Then the audience asks questions and offers suggestions to clarify the presentation and visual aids. Practice sessions improve the quality of presentations and strengthen presenters' confidence.

Do Not Use These Quotations From British Sport Commentators in Your Presentation

"Moses Kiptanul—the 19-year-old Kenyan who turned 20 a few weeks ago."

"We now have exactly the same situation we had at the start of the race, only exactly the opposite."

"He's never had major knee surgery on any other part of his body."

"She's not Ben Johnson—but then who is?"

"I owe a lot to my parents, especially my mother and father."

"The Port Elizabeth ground is more of a circle than an oval. It's long and square."

"The racecourse is as level as a billiard ball."

"Watch the time—it gives you an indication of how fast they are running."

"That's inches away from being a millimeter perfect."

"If history repeats itself, I think we can expect the same thing again."

Using a Poster Presentation to Best Advantage

The poster session is another way to present a conference paper. Poster sessions take place in a large room in which presenters place summaries of their research on the wall or on poster stands. The session is scheduled for a specific period, during which presenters stand by their work while those interested walk around, read the material, and discuss items of interest with the presenters.

We prefer this format to oral presentations. The audience can look at the papers in which they are interested and have detailed discussions with the authors. Within 75 minutes, 15 to 40 poster presentations can be made available in a large room, whereas only five 15-minute oral presentations can take place in the same length of time. In addition, in a session of oral presentations the audience must sit through several papers. Audience members may lose interest or create a disturbance by arriving or leaving.

Poster sessions allow audience members to pace the way they look at material. We have often heard oral presentations about which we wanted more information. We may have

Thomas, Nelson, and Silverman's
Four Laws of Oral Presentations

1. Something always goes wrong with the LCD projector. More specifically:
   a. Unless you check it beforehand, it will not work.
   b. Either the electrical cord or the mouse cord (or both) will be too short.
   c. If the projector has worked perfectly for the three previous presenters, the bulb will burn out during your presentation.
2. The screen will be too small for the room.
3. Your paper will be the last one scheduled during the conference. Only the moderator, you, and the previous presenter will be there, the latter of whom will leave on completing his or her paper. Or, conversely, your paper will be scheduled at 7:30 a.m., at which time you will be hung over and everyone else will still be in bed.
4. At your first presentation, the most prestigious scholar in your field will show up, be misquoted by you, and ask you a question.

wished that a table or figure presented had been left up longer. Those concerns are not an issue in poster sessions. Participants are more likely to ask questions or comment on points in a poster session when they have sufficient time to consider the material. Longer and more in-depth conversations that benefit both the presenter and the spectator often ensue.

Good poster design maximizes all the benefits that poster sessions offer. The poster should be arranged to highlight the important points and eliminate unneeded information. Posters are typically mounted on walls or portable stands. Usually, the conference will notify you about how much space you have, but a 4-foot-by-6-foot (1.2 by 1.8 m) space is common. The poster should be arranged in five or six panels that represent the parts of the poster presentation—abstract, introduction and problem statement, methods, results (including figures and tables), discussion, and references. Placing the sections on panels with headings helps readers follow the poster and provides some white space that is pleasing to the eye. (If you print the poster on one large sheet, these suggestions still apply.) In figure 22.1 you can see a logical way to arrange the panels using the vertical flow of information (horizontal flow can also be used, but vertical flow is usually preferred). Often, an effective approach is to place tables and figures at the center of the poster and arrange panels around them. The tables and figures represent what you have found and should be the central feature of the poster.

Posters have text, but this text should have specific features to increase how easily it can be read and understood:

- Short sentences with simple structure
- To-the-point text (no filler information)
- Use of common words with little jargon

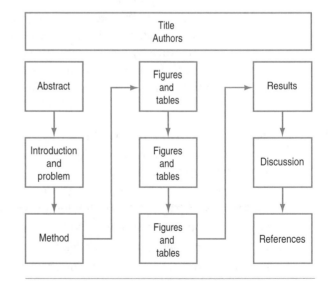

Figure 22.1   Poster with vertical flow.

- Little abstraction
- Use of lists with bullets

Use design elements that improve posters and make them more understandable:

- Select a plain and clear typeface (common ones includes Helvetica and Times).
- Use a font size of at least 20 points for text, 28 to 32 points for headings, and 64 points for titles.
- Most current evidence suggests that using both upper and lower case increases readability.
- Major headings on each panel with section headings inside the panel guide the eye in reading the poster.
- Within sections, use left justification and ragged right rather than full justification.
- If there is a reasonable choice between a figure and a table, use the figure; if you must use a table, put in minimal information, use large numbers, and use labeling (e.g., shading, colors) to guide the reader to important information; *a table or figure prepared for publication is seldom acceptable for a poster (even when enlarged)*.
- Use color wisely; it can help the poster if it sets off points, unifies sections, and guides the reader's eye. If you are having trouble deciding how much color to use, you are probably using too much; tone it down.

If you want your poster to be exceptional and attract interested readers, follow these suggestions, prepare it well in advance, and then have your friends and faculty advisors look at it to make suggestions and comments.

SUMMARY

The journal format for theses and dissertations has the advantage of being the format for journal publication, one of the main ways to disseminate scholarly work. Yet it retains the essential characteristics of the complete reporting that is valued in the thesis or dissertation. This format comprises preliminary materials (e.g., title page, abstract), the body (e.g., introduction, methods, results, discussion), and a set of appendixes (e.g., extended literature review, additional results). The goal of this reporting style is to promote rapid publication of high-quality research.

Abstracts are frequently used as a form for submitting papers to scholarly meetings so that they can be evaluated for possible presentation. Usually, the abstract is limited in length and format by the scholarly group to which it is submitted. If the abstract paper is accepted for presentation, the form may be oral or by poster. Oral presentations usually last 10 to 20 minutes, and poster presentations are typically limited to a display space.

Check Your Understanding

1. Writing: Select a study from a journal and write a 150-word abstract in APA style (or whatever style your department uses).

2. Oral presentations: To emphasize the importance of time limits on oral presentations, we suggest that each student give the following presentations in class:

   a. Select a published research study and present a 2-minute summary.

   b. Select another study but give a 5-minute summary.

3. Poster presentation: Prepare a poster presentation of a research study from a journal. Put your poster on the classroom wall, along with those of the other students in your class. Critique each other's posters (or some percentage of posters for larger classes).

APPENDIX

STATISTICAL TABLES

Table 1 Table of Random Numbers

06 02 34 58 99	35 69 62 28 65	16 94 45 21 60	25 35 57 14 61	49 98 17 05 15
13 68 77 79 46	16 09 56 35 99	11 16 23 35 01	43 98 68 35 42	28 19 82 53 07
85 63 19 10 74	95 26 45 74 38	92 50 27 94 59	73 19 97 40 93	75 74 29 03 31
86 40 22 25 69	39 78 61 81 66	91 78 55 93 04	59 66 95 75 68	93 32 58 13 18
33 55 44 48 16	32 13 64 55 94	38 20 22 68 02	03 58 05 44 23	59 92 92 39 18
36 89 63 60 98	79 52 62 03 91	91 32 73 81 84	79 38 54 96 83	45 65 43 76 99
03 10 77 54 44	04 46 09 63 91	58 46 41 75 98	48 50 36 56 79	21 04 13 67 31
65 87 01 20 35	30 20 04 83 11	08 64 10 77 43	37 33 96 43 97	27 65 85 04 23
59 28 09 39 31	96 95 60 66 64	96 32 92 81 29	30 50 21 97 20	98 25 02 97 57
76 41 40 52 24	29 53 51 46 89	14 81 54 76 27	60 66 06 85 36	18 83 95 67 52
76 39 33 28 47	90 66 20 84 58	44 38 57 99 56	07 20 80 18 23	66 37 45 82 59
37 22 86 47 58	41 96 56 52 19	89 41 49 81 05	07 51 30 94 13	65 49 03 39 70
59 35 93 58 97	02 92 56 08 95	90 64 10 68 37	73 64 69 95 59	45 42 38 14 92
65 38 16 58 51	81 39 78 59 50	38 50 95 42 18	69 57 38 17 59	96 10 15 50 80
57 35 36 63 81	94 99 06 73 17	89 19 07 47 19	40 46 70 11 42	45 24 25 46 90
62 07 15 77 81	35 01 22 18 30	86 60 49 28 61	29 07 39 10 45	25 07 35 09 23
05 64 14 06 93	67 19 68 40 12	94 92 96 37 35	09 48 10 03 02	39 86 29 32 44
30 42 50 92 19	37 99 54 29 63	19 06 36 60 79	44 57 13 79 91	16 61 90 06 42
45 02 03 22 60	94 39 02 36 21	70 67 61 40 88	95 52 56 58 18	69 31 70 81 83
13 99 02 41 98	93 85 06 99 96	08 30 32 92 96	86 46 08 60 96	69 90 51 77 42
21 41 56 49 93	82 13 54 88 36	08 55 84 31 78	17 83 03 82 12	01 06 29 78 51
39 57 21 08 68	67 34 68 57 70	84 03 34 90 59	80 96 92 37 92	73 66 55 31 52
93 85 21 84 22	64 38 99 11 45	53 63 12 64 17	79 81 42 27 33	66 37 45 82 59
83 36 47 05 68	28 32 26 23 02	86 98 20 42 95	29 23 11 63 13	65 49 03 39 70
93 67 46 10 30	67 48 60 93 05	09 04 23 22 90	14 45 02 71 15	45 42 37 14 92
58 53 77 99 08	08 99 77 84 95	16 21 48 57 29	97 56 52 27 99	86 10 15 50 80
74 23 31 88 69	48 16 14 46 86	71 39 26 56 34	55 94 32 89 85	45 24 25 46 90
98 01 34 66 15	51 36 70 52 31	39 80 46 65 52	04 93 08 32 69	25 07 35 09 23
27 53 49 16 41	11 38 17 71 03	03 17 92 80 84	91 83 23 28 09	39 86 29 32 44
65 58 85 04 96	46 10 85 53 12	24 88 86 28 33	53 06 69 27 55	16 61 90 06 43
63 30 03 40 78	71 95 58 60 07	35 78 95 90 22	11 28 46 72 12	69 31 70 81 83
56 68 75 87 76	52 66 01 75 72	91 14 86 28 82	83 88 23 50 71	69 90 51 77 43
76 40 04 44 07	95 89 33 07 42	09 98 15 75 94	39 03 13 12 55	01 06 27 87 51
10 22 98 58 96	93 39 11 86 43	84 17 26 27 40	50 39 33 93 98	73 66 55 31 52
42 34 68 06 23	92 08 60 53 78	05 75 41 49 14	20 23 09 84 99	47 15 52 81 41
26 76 79 35 27	60 03 88 53 78	33 28 06 15 46	75 09 72 12 94	64 44 29 91 67
11 24 10 38 15	93 10 89 78 60	70 22 50 58 41	85 54 33 22 36	85 91 56 13 58
49 07 62 90 13	66 60 74 98 81	98 34 70 06 51	33 97 72 77 94	30 15 35 38 42
88 09 76 81 14	80 50 51 73 91	05 37 40 99 54	11 70 80 40 66	09 23 74 06 43
27 86 83 16 71	71 66 20 59 95	31 27 18 07 96	51 32 92 10 63	40 07 70 53 76
14 85 54 28 61	31 08 99 62 66	50 81 73 33 63	25 11 12 61 99	59 39 25 44 40
01 91 25 42 49	50 05 15 49 03	68 91 59 84 29	42 94 91 55 04	22 46 41 88 78
64 51 37 74 04	17 26 33 50 88	77 08 90 45 65	54 59 42 16 06	99 90 85 86 07
71 33 21 76 22	88 99 34 64 14	59 02 66 33 72	60 49 34 70 59	40 29 39 55 82
99 45 44 19 02	06 16 66 69 10	21 96 88 02 95	65 90 71 61 73	43 61 99 89 73
11 35 41 39 65	51 28 50 37 35	77 03 28 18 61	10 12 56 89 22	07 73 02 82 12
82 62 11 61 43	55 45 28 89 11	08 43 24 79 55	86 02 61 49 30	58 46 67 02 50
56 03 81 96 87	09 21 76 30 52	97 84 63 83 46	51 28 52 92 48	76 47 33 03 66
78 48 50 80 22	56 63 44 66 54	70 61 18 01 16	48 79 52 66 32	24 28 07 05 34
47 34 55 08 07	06 23 04 71 24	71 56 75 85 17	71 31 24 22 19	41 84 01 33 20

```
28 89 32 88 33    49 08 47 10 28    71 40 77 55 11    06 49 53 64 91    57 30 64 92 78
15 65 95 86 72    71 70 79 54 57    71 86 23 29 85    76 59 45 81 37    56 02 63 88 16
07 78 61 49 17    06 81 43 71 31    33 87 07 65 42    57 91 60 64 29    79 39 49 08 23
64 52 43 31 71    19 35 85 56 50    57 62 55 50 96    39 33 64 82 87    94 52 42 18 86
42 67 44 23 71    97 17 09 15 18    88 97 32 84 03    10 57 60 25 35    20 13 89 53 96

59 46 84 64 81    10 02 60 93 36    48 49 15 87 53    44 41 83 45 99    35 88 55 87 14
64 42 12 35 03    62 97 56 17 04    51 44 48 83 35    29 98 10 80 35    73 23 57 70 96
54 35 27 06 41    61 65 62 44 55    59 54 64 40 91    18 93 74 11 26    05 29 58 38 80
17 49 31 47 24    11 14 92 40 18    69 48 89 86 64    20 06 93 33 86    27 81 96 90 39
12 86 02 35 22    91 27 01 38 93    11 18 72 02 46    94 53 22 32 44    54 45 89 53 10

75 70 90 36 23    13 95 56 44 57    08 31 37 55 50    71 38 74 53 62    60 45 73 54 17
12 50 97 40 52    99 24 86 70 31    22 71 56 62 94    37 95 88 19 03    92 81 33 97 65
85 91 34 74 68    12 80 64 54 19    38 36 80 19 52    51 01 78 61 49    73 37 15 84 46
65 71 09 50 29    26 88 61 96 04    80 53 54 55 04    79 39 84 10 16    65 22 07 46 82
56 35 26 14 46    11 78 24 93 16    46 75 09 15 88    89 66 34 10 93    29 48 42 96 67

63 17 70 21 30    64 24 07 59 29    72 88 24 58 27    71 80 14 05 61    25 11 58 73 42
15 98 06 69 34    51 76 27 29 38    32 66 37 11 24    25 05 96 11 16    39 95 31 94 91
65 60 02 84 19    42 17 27 91 45    40 59 61 02 80    06 43 48 77 08    75 84 20 51 09
40 31 78 80 66    09 64 15 24 79    73 04 28 25 48    08 08 30 06 88    75 12 73 35 53
08 91 49 85 63    78 66 52 91 25    04 14 09 41 38    20 93 04 16 52    58 92 95 05 77

67 79 50 87 99    47 67 53 05 97    60 62 86 13 90    05 10 69 94 37    87 88 72 23 65
20 13 48 51 91    46 83 15 54 72    71 78 30 88 65    41 84 59 14 69    67 65 78 38 01
41 96 55 12 36    37 28 35 93 07    21 91 85 61 66    51 46 71 52 19    48 14 09 34 50
44 98 38 30 52    38 65 19 66 30    81 91 21 36 87    60 54 06 82 41    03 92 48 75 08
06 90 97 72 33    35 01 98 97 99    47 87 52 61 25    98 02 41 37 16    08 88 72 23 56

15 25 10 88 95    75 69 81 52 43    13 28 63 48 96    92 48 24 77 35    60 99 37 39 83
66 73 40 29 75    23 07 03 20 86    41 52 58 51 13    50 17 13 74 46    87 28 96 13 75
90 87 72 40 43    71 23 03 88 96    38 99 63 42 82    74 55 32 87 24    79 04 59 70 96
56 60 89 70 13    61 04 39 93 14    65 67 31 70 22    42 61 65 34 24    73 16 29 82 94
23 97 04 75 51    37 09 52 95 39    78 89 55 14 96    84 24 11 55 96    50 29 49 12 13

37 25 77 53 16    98 01 08 17 36    33 25 66 04 54    33 72 49 32 52    69 59 25 09 89
22 30 77 51 44    81 66 73 27 06    15 19 54 65 07    88 32 10 06 30    97 76 74 82 60
08 62 61 27 28    87 96 97 76 60    04 46 99 39 48    86 61 69 84 59    64 53 76 42 61
09 46 39 63 16    82 32 44 55 33    25 63 78 46 99    64 49 37 74 33    13 98 45 47 54
59 33 30 32 12    15 36 34 74 25    77 35 05 89 82    52 33 65 92 49    59 11 80 87 53

73 62 27 47 57    30 59 95 55 36    85 68 33 66 45    84 54 64 85 60    26 34 81 47 85
79 01 85 30 93    09 55 08 27 21    14 23 73 44 33    43 66 56 78 67    60 89 59 08 09
23 79 05 67 04    18 30 50 09 05    09 58 94 80 79    08 54 87 37 50    02 30 38 09 97
62 50 69 64 89    18 63 24 34 43    37 11 13 06 85    46 79 31 10 80    15 05 67 06 11
75 39 76 85 82    21 58 31 29 77    34 39 07 22 55    20 41 22 56 45    46 78 63 85 30

18 47 61 10 46    61 94 44 74 79    56 45 46 59 92    22 27 18 45 33    32 87 42 96 06
76 37 69 64 32    56 53 39 57 70    52 41 32 49 79    93 89 02 61 65    44 71 19 94 92
75 43 79 86 70    68 05 42 80 45    95 13 48 83 69    67 65 78 38 01    44 53 96 02 92
31 84 98 21 55    56 35 10 02 87    72 58 47 89 81    48 14 09 34 50    08 71 98 15 93
66 24 11 10 47    87 91 06 34 92    48 24 74 18 95    03 92 48 75 08    81 10 78 35 19

67 36 74 15 39    44 76 35 29 52    57 11 06 38 15    18 52 99 22 85    32 14 55 67 12
12 83 99 69 55    87 45 17 74 96    46 94 89 54 19    05 08 80 53 01    04 03 11 83 92
74 73 70 16 03    91 46 84 90 97    13 43 81 03 02    73 52 75 51 69    11 07 23 32 55
73 03 30 66 39    75 06 90 05 11    11 08 10 74 44    38 36 86 34 52    44 13 67 75 08
38 04 28 20 64    93 70 13 19 57    96 37 40 76 22    27 19 81 78 80    31 44 09 39 82
```

Table 2 The Standard Normal Curve

	One-tailed		Two-tailed			One-tailed		Two-tailed	
z	π beyond	π remainder	π beyond	π remainder	z	π beyond	π remainder	π beyond	π remainder
1.75	0.040059	0.959941	0.080118	0.919882	2.88	0.001988	0.998012	0.003977	0.996023
1.76	0.039204	0.960796	0.078408	0.921592	2.89	0.001926	0.998074	0.003852	0.996148
1.77	0.038364	0.961636	0.076727	0.923273	2.90	0.001866	0.998134	0.003732	0.996268
1.78	0.037538	0.962462	0.075076	0.924924	2.91	0.001807	0.998193	0.003614	0.996386
1.79	0.036727	0.963273	0.073454	0.926546	2.92	0.001750	0.998250	0.003500	0.996500
1.80	0.035930	0.964070	0.071861	0.928139	2.93	0.001695	0.998305	0.003390	0.996610
1.81	0.035148	0.964852	0.070296	0.929704	2.94	0.001641	0.998359	0.003282	0.996718
1.82	0.034380	0.965620	0.068759	0.931241	2.95	0.001589	0.998411	0.003178	0.996822
1.83	0.033624	0.966376	0.067250	0.932750	2.96	0.001538	0.998462	0.003076	0.996924
1.84	0.032884	0.967116	0.065768	0.934232	2.97	0.001489	0.998511	0.002978	0.997022
1.85	0.032157	0.967843	0.064314	0.935686	2.98	0.001441	0.998559	0.002882	0.997118
1.86	0.031443	0.968557	0.062886	0.937114	2.99	0.001395	0.998605	0.002790	0.997210
1.87	0.030742	0.969258	0.061484	0.938516	3.00	0.001350	0.998650	0.002700	0.997300
1.88	0.030054	0.969946	0.060108	0.939892	3.01	0.001306	0.998694	0.002612	0.997388
1.89	0.029379	0.970621	0.057858	0.942142	3.02	0.001264	0.998736	0.002528	0.997472
1.90	0.028717	0.971283	0.057433	0.942567	3.03	0.001223	0.998777	0.002446	0.997554
1.91	0.028067	0.971933	0.056133	0.943867	3.04	0.001183	0.998817	0.002366	0.997634
1.92	0.027429	0.972571	0.054858	0.945142	3.05	0.001144	0.998856	0.002288	0.997712
1.93	0.026803	0.973197	0.053607	0.946393	3.06	0.001107	0.998893	0.002213	0.997787
1.94	0.02619	0.973810	0.062380	0.937620	3.07	0.001070	0.998930	0.002141	0.997859
1.95	0.025588	0.974412	0.051176	0.948824	3.08	0.001035	0.998965	0.002070	0.997930
1.96	0.024998	0.975002	0.049996	0.950004	3.09	0.001001	0.998999	0.002002	0.997998
1.97	0.024419	0.975581	0.048838	0.951162	3.1	0.009680	0.990320	0.001935	0.998065
1.98	0.023852	0.976148	0.047704	0.952296	3.11	0.009350	0.990650	0.001871	0.998129
1.99	0.023295	0.976705	0.046591	0.953409	3.12	0.000904	0.999096	0.001809	0.998191
2.00	0.022750	0.977250	0.045500	0.954500	3.13	0.000874	0.999126	0.001748	0.998252
2.01	0.022216	0.977784	0.044431	0.955569	3.14	0.000845	0.999155	0.001689	0.998311
2.02	0.021692	0.978308	0.043383	0.956617	3.15	0.000816	0.999184	0.001633	0.998367
2.03	0.021178	0.978822	0.042357	0.957643	3.16	0.000789	0.999211	0.001578	0.998422
2.04	0.020675	0.979325	0.041350	0.958650	3.17	0.000762	0.999238	0.001524	0.998476
2.05	0.020182	0.979818	0.040364	0.959636	3.18	0.000736	0.999264	0.001473	0.998527
2.06	0.019699	0.980301	0.039399	0.960601	3.19	0.000711	0.999289	0.001423	0.998577
2.07	0.019226	0.980774	0.038452	0.961548	3.2	0.000687	0.999313	0.001374	0.998626
2.08	0.018763	0.981237	0.037526	0.962474	3.21	0.000664	0.999336	0.001327	0.998673
2.09	0.018309	0.981691	0.036618	0.963382	3.22	0.000641	0.999359	0.001282	0.998718
2.10	0.017864	0.982136	0.035729	0.964271	3.23	0.000619	0.999381	0.001238	0.998762
2.11	0.017429	0.982571	0.034858	0.965142	3.24	0.000598	0.999402	0.001195	0.998805
2.12	0.017003	0.982997	0.034006	0.965994	3.25	0.000577	0.999423	0.001154	0.998846
2.13	0.016586	0.983414	0.033172	0.966828	3.26	0.000557	0.999443	0.001114	0.998886
2.14	0.016177	0.983823	0.032355	0.967645	3.27	0.000538	0.999462	0.001075	0.998925

	One-tailed		Two-tailed			One-tailed		Two-tailed	
z	π beyond	π remainder	π beyond	π remainder	z	π beyond	π remainder	π beyond	π remainder
2.15	0.015778	0.984222	0.031555	0.968445	3.28	0.005190	0.994810	0.001038	0.998962
2.16	0.015386	0.984614	0.030773	0.969227	3.29	0.000501	0.999499	0.001002	0.998998
2.17	0.015003	0.984997	0.030007	0.969993	3.30	0.000483	0.999517	0.000967	0.999033
2.18	0.014629	0.985371	0.029257	0.970743	3.31	0.000466	0.999534	0.000933	0.999067
2.19	0.014262	0.985738	0.028524	0.971476	3.32	0.000450	0.999550	0.000900	0.999100
2.20	0.013903	0.986097	0.0278707	0.9721293	3.33	0.000434	0.999566	0.000868	0.999132
2.21	0.013553	0.986447	0.027105	0.972895	3.34	0.000419	0.999581	0.000838	0.999162
2.22	0.013209	0.986791	0.026419	0.973581	3.35	0.000404	0.999596	0.000808	0.999192
2.23	0.012874	0.987126	0.025747	0.974253	3.36	0.000390	0.999610	0.000779	0.999221
2.24	0.012545	0.987455	0.025091	0.974909	3.37	0.000376	0.999624	0.000752	0.999248
2.25	0.012224	0.987776	0.024449	0.975551	3.38	0.000362	0.999638	0.000725	0.999275
2.26	0.011911	0.988089	0.023821	0.976179	3.39	0.000349	0.999651	0.000699	0.999301
2.27	0.011604	0.988396	0.023208	0.976792	3.4	0.000337	0.999663	0.000674	0.999326
2.28	0.011304	0.988696	0.022608	0.977392	3.41	0.000325	0.999675	0.000650	0.999350
2.29	0.011011	0.988989	0.022021	0.977979	3.42	0.000313	0.999687	0.000626	0.999374
2.30	0.010724	0.989276	0.021448	0.978552	3.43	0.000302	0.999698	0.000604	0.999396
2.31	0.010444	0.989556	0.020888	0.979112	3.44	0.000091	0.999909	0.000582	0.999418
2.32	0.010170	0.989830	0.020341	0.979659	3.45	0.000280	0.999720	0.000561	0.999439
2.33	0.009903	0.990097	0.019806	0.980194	3.46	0.000270	0.999730	0.000540	0.999460
2.34	0.000964	0.999036	0.019284	0.980716	3.47	0.000260	0.999740	0.000520	0.999480
2.35	0.009387	0.990613	0.018773	0.981227	3.48	0.000251	0.999749	0.000501	0.999499
2.36	0.000914	0.999086	0.018275	0.981725	3.49	0.000242	0.999758	0.000483	0.999517
2.37	0.008894	0.991106	0.017788	0.982212	3.50	0.000233	0.999767	0.000465	0.999535
2.38	0.008656	0.991344	0.017313	0.982687	3.51	0.000224	0.999776	0.000448	0.999552
2.39	0.008424	0.991576	0.016848	0.983152	3.52	0.000216	0.999784	0.000432	0.999568
2.40	0.008198	0.991802	0.016395	0.983605	3.53	0.002080	0.997920	0.000416	0.999584
2.41	0.007976	0.992024	0.015953	0.984047	3.54	0.000200	0.999800	0.000400	0.999600
2.42	0.007760	0.992240	0.015521	0.984479	3.55	0.000193	0.999807	0.000385	0.999615
2.43	0.007549	0.992451	0.015099	0.984901	3.56	0.000185	0.999815	0.000371	0.999629
2.44	0.007344	0.992656	0.014687	0.985313	3.57	0.000178	0.999822	0.000357	0.999643
2.45	0.007143	0.992857	0.014286	0.985714	3.58	0.000172	0.999828	0.000344	0.999656
2.46	0.006947	0.993053	0.013894	0.986106	3.59	0.000165	0.999835	0.000331	0.999669
2.47	0.006756	0.993244	0.013511	0.986489	3.60	0.000159	0.999841	0.000318	0.999682
2.48	0.000657	0.999343	0.013138	0.986862	3.61	0.000153	0.999847	0.000306	0.999694
2.49	0.006387	0.993613	0.012774	0.987226	3.62	0.000147	0.999853	0.000295	0.999705
2.50	0.006210	0.993790	0.012419	0.987581	3.63	0.000142	0.999858	0.000283	0.999717
2.51	0.006037	0.993963	0.012073	0.987927	3.64	0.000136	0.999864	0.000273	0.999727
2.52	0.005868	0.994132	0.011735	0.988265	3.65	0.000131	0.999869	0.000262	0.999738
2.53	0.005703	0.994297	0.011406	0.988594	3.66	0.000126	0.999874	0.000252	0.999748
2.54	0.005543	0.994457	0.011085	0.988915	3.67	0.000121	0.999879	0.000243	0.999757

(continued)

Table 2 (*continued*)

z	One-tailed π beyond	One-tailed π remainder	Two-tailed π beyond	Two-tailed π remainder	z	One-tailed π beyond	One-tailed π remainder	Two-tailed π beyond	Two-tailed π remainder
2.55	0.005386	0.994614	0.010772	0.989228	3.68	0.000117	0.999883	0.000233	0.999767
2.56	0.005234	0.994766	0.010467	0.989533	3.69	0.000112	0.999888	0.000224	0.999776
2.57	0.005085	0.994915	0.010170	0.989830	3.70	0.000108	0.999892	0.000216	0.999784
2.58	0.004940	0.995060	0.009880	0.990120	3.71	0.000104	0.999896	0.000207	0.999793
2.59	0.004799	0.995201	0.009598	0.990402	3.72	0.000100	0.999900	0.000199	0.999801
2.60	0.004661	0.995339	0.009322	0.990678	3.73	0.000096	0.999904	0.000191	0.999809
2.61	0.004527	0.995473	0.009054	0.990946	3.74	0.000092	0.999908	0.000184	0.999816
2.62	0.004396	0.995604	0.008793	0.991207	3.75	0.000088	0.999912	0.000177	0.999823
2.63	0.004269	0.995731	0.008538	0.991462	3.76	0.000085	0.999915	0.000170	0.999830
2.64	0.004145	0.995855	0.008291	0.991709	3.77	0.000082	0.999918	0.000163	0.999837
2.65	0.004025	0.995975	0.008049	0.991951	3.78	0.000078	0.999922	0.000157	0.999843
2.66	0.003907	0.996093	0.007814	0.992186	3.79	0.000075	0.999925	0.000151	0.999849
2.67	0.003793	0.996207	0.007585	0.992415	3.80	0.000072	0.999928	0.000145	0.999855
2.68	0.003681	0.996319	0.007362	0.992638	3.81	0.000069	0.999931	0.000139	0.999861
2.69	0.003573	0.996427	0.007145	0.992855	3.82	0.000670	0.999330	0.000133	0.999867
2.70	0.003467	0.996533	0.006934	0.993066	3.83	0.000064	0.999936	0.000128	0.999872
2.71	0.003364	0.996636	0.006728	0.993272	3.84	0.000062	0.999938	0.000123	0.999877
2.72	0.003264	0.996736	0.006528	0.993472	3.85	0.000059	0.999941	0.000118	0.999882
2.73	0.003167	0.996833	0.006333	0.993667	3.86	0.000057	0.999943	0.000113	0.999887
2.74	0.003072	0.996928	0.006144	0.993856	3.87	0.000054	0.999946	0.000109	0.999891
2.75	0.002980	0.997020	0.005960	0.994040	3.88	0.000052	0.999948	0.000104	0.999896
2.76	0.002890	0.997110	0.005780	0.994220	3.89	0.000050	0.999950	0.000100	0.999900
2.77	0.002803	0.997197	0.005606	0.994394	3.9	0.000048	0.999952	0.000097	0.999904
2.78	0.002718	0.997282	0.005436	0.994564	3.91	0.000046	0.999954	0.000093	0.999908
2.79	0.002635	0.997365	0.005271	0.994729	3.92	0.000044	0.999956	0.000089	0.999911
2.80	0.002555	0.997445	0.00511	0.994890	3.93	0.000042	0.999958	0.000085	0.999915
2.81	0.002477	0.997523	0.004954	0.995046	3.94	0.000041	0.999959	0.000081	0.999919
2.82	0.002401	0.997599	0.004802	0.995198	3.95	0.000039	0.999961	0.000078	0.999922
2.83	0.002327	0.997673	0.004655	0.995345	3.96	0.000037	0.999963	0.000075	0.999925
2.84	0.002256	0.997744	0.004511	0.995489	3.97	0.000036	0.999964	0.000072	0.999928
2.85	0.002186	0.997814	0.004372	0.995628	3.98	0.000034	0.999966	0.000069	0.999931
2.86	0.002118	0.997882	0.004236	0.995764	3.99	0.000033	0.999967	0.000066	0.999934
2.87	0.002052	0.997948	0.004105	0.995895	4.00	0.000032	0.999968	0.000063	0.999937

Table 3 Critical Values of Correlation Coefficients

	One-tailed								One-tailed				
	.25	.10	.05	.025	.01	.005		.25	.10	.05	.025	.01	.005
	Two-tailed								Two-tailed				
df	.50	.20	.10	.05	.02	.01	df	.50	.20	.10	.05	.02	.01
1	.7071	.9511	.9877	.9969	.9995	.9999	51	.0947	.1789	.2284	.2706	.3188	.3509
2	.5000	.8000	.9000	.9500	.9800	.9900	52	.0938	.1772	.2262	.2681	.3158	.3477
3	.4040	.6870	.8054	.8783	.9343	.9587	53	.0929	.01755	.2241	.2656	.3129	.3445
4	.3473	.6084	.7293	.8114	.8822	.9172	54	.0912	.1723	.2201	.2609	.3074	.3385
5	.3091	.5509	.6694	.7545	.8329	.8745	55	.0912	.1723	.2201	.2609	.3074	.3385
6	.2811	.5067	.6215	.7067	.7887	.8343	56	.0904	.1708	.2181	.2586	.3048	.3357
7	.2596	.4716	.5822	.6664	.7498	.7977	57	.0896	.1693	.2162	.2564	.3022	.3328
8	.2423	.4428	.5494	.6319	.7155	.7646	58	.0888	.1678	.2144	.2542	.2997	.3301
9	.2281	.4187	.5214	.6021	.6851	.7348	59	.0880	.1664	.2126	.2521	.2972	.3274
10	.2161	.3981	.4973	.5760	.6581	.7079	60	.0873	.1650	.2108	.2500	.2948	.3248
11	.2058	.3802	.4762	.5529	.6339	.6835	61	.0866	.1636	.2097	.2480	.2925	.3223
12	.1968	.3646	.4575	.5324	.6120	.6614	62	.0858	.1623	.2075	.2461	.2902	.3198
13	.1890	.3507	.4409	.5140	.5923	.6411	63	.0852	.1610	.2058	.2441	.2880	.3173
14	.1820	.3383	.4259	.4973	.5742	.6226	64	.0845	.1598	.2042	.2423	.2858	.3150
15	.1757	.3271	.4124	.4821	.5577	.6055	65	.0828	.1586	.2027	.2606	.2837	.3126
16	.1700	.3170	.4000	.4683	.5425	.5897	66	.0832	.1574	.2012	.2387	.2816	.3104
17	.1649	.3077	.3887	.4555	.5285	.5751	67	.0826	.1562	.1997	.2367	.2796	.3081
18	.1602	.2992	.3783	.4438	.5155	.5614	68	.0820	.1550	.1982	.2352	.2776	.3060
19	.1558	.2914	.3687	.4329	.5034	.5487	69	.0814	.1539	.1968	.2335	.2756	.3038
20	.1518	.2841	.3598	.4227	.4921	.5368	70	.0808	.1528	.1954	.2319	.2737	.3017
21	.1481	.2774	.3515	.4132	.4815	.5256	71	.0802	.1517	.1940	.2303	.2718	.2997
22	.1447	.2711	.3438	.4044	.4716	.5151	72	.0796	.1507	.1927	.2287	.2700	.2977
23	.1415	.2653	.3365	.3961	.4622	.5052	73	.0791	.1497	.1914	.2272	.2682	.2957
24	.1384	.2598	.3297	.3882	.4534	.4869	74	.0786	.1486	.1901	.2257	.2664	.2938
25	.1356	.2546	.3233	.3809	.4451	.4869	75	.0780	.177	.1888	.2242	.2647	.2919
26	.1330	.2497	.3172	.3739	.4372	.4785	76	.0775	.1467	.1876	.2227	.2630	.2900
27	.1305	.2451	.3115	.3673	.4297	.4705	77	.0770	.1457	.1864	.2213	.2613	.2882
28	.1280	.2407	.3061	.3610	.4226	.4629	78	.0765	.1448	.1852	.2199	.2597	.2864
29	.1258	.2366	.3009	.3550	.4158	.4556	79	.0760	.1439	.1841	.2186	.2581	.2847
30	.1237	.2327	.2960	.3494	.4093	.4487	80	.0755	.1430	.1829	.2172	.2565	.2830
31	.1217	.2289	.2914	.3440	.4032	.4421	81	.0751	.1421	.1818	.2159	.2550	.2813
32	.1197	.2254	.2869	.3388	.3916	.4296	82	.0746	.1412	.1807	.2146	.2535	.2796
33	.1179	.2220	.2826	.3338	.3916	.4296	83	.0742	.1404	.1796	.2133	.2520	.2780
34	.1161	.2187	.2785	.3291	.3862	.4238	84	.0737	.1396	.1786	.2120	.2505	.2764
35	.1144	.2156	.2746	.3246	.3810	.4182	85	.0733	.1387	.1775	.2108	.2491	.2748
36	.1128	.2126	.2709	.3202	.3760	.4128	86	.0728	.1379	.1756	.2096	.2477	.2732
37	.1113	.2097	.2673	.3160	.3712	.4076	87	.0724	.1371	.1755	.2084	.2463	.2717
38	.1098	.2070	.2638	.3120	.3665	.4026	88	.0720	.1364	.1745	.2072	.2449	.2790
39	.1084	.2043	.2605	.3081	.3621	.3978	89	.0716	.1356	.1735	.2061	.2435	.2687
40	.1070	.2018	.2573	.3044	.3578	.3932	90	.0712	.1348	.1726	.2050	.2422	.2673
41	.1057	.1993	.2542	.3088	.3536	.3887	91	.0708	.1341	.1716	.2039	.2409	.2659
42	.1044	.1970	.2512	.2973	.3496	.3843	92	.0704	.1334	.1707	.2028	.2396	.2645
43	.1032	.1947	.2483	.2940	.3457	.3801	93	.0700	.1327	.1698	.2017	.2384	.2631
44	.1020	.1925	.2455	.2907	.3420	.3761	94	.0697	.1320	.1689	.2006	.2371	.2617
45	.1008	.1903	.2429	.2876	.3374	.3721	95	.0693	.1313	.1680	.1996	.2359	.2604
46	.0997	.1883	.2403	.2845	.3348	.3683	96	.0689	.1306	.1671	.1986	.2347	.2591
47	.0987	.1863	.2377	.2816	.3314	.3646	97	.0686	.1299	.1663	.1975	.2335	.2578
48	.0976	.1843	.2343	.2787	.2381	.3610	98	.0682	.1292	.1654	.1966	.2324	.2565
49	.0966	.1825	.2329	.2759	.3249	.3575	99	.0679	.1286	.1646	.1956	.2312	.2552
50	.0956	.1806	.2306	.2732	.3218	.3542	100	.0675	.1279	.1638	.1946	.2301	.2540

Table 4 Transformation of *r* to *Zr*

r	z_r	*r*	z_r	*r*	z_r
0.00	0.0000	0.33	0.3428	0.67	0.8107
0.01	0.0100	0.34	0.3541	0.68	0.8291
0.02	0.0200	0.35	0.3654	0.69	0.8480
0.03	0.0300	0.36	0.3769	0.70	0.8673
0.04	0.0400	0.37	0.3884	0.71	0.8872
0.05	0.0500	0.38	0.4001	0.72	0.9076
0.06	0.0601	0.39	0.4118	0.73	0.9287
0.07	0.0701	0.40	0.4236	0.74	0.9505
0.08	0.0802	0.41	0.4356	0.75	0.9730
0.09	0.0902	0.42	0.4477	0.76	0.9962
0.10	0.1003	0.43	0.4599	0.77	1.0203
0.11	0.1104	0.44	0.4722	0.78	1.0454
0.12	0.1206	0.45	0.4847	0.79	1.0714
0.13	0.1307	0.46	0.4973	0.80	1.0986
0.14	0.1409	0.47	0.5101	0.81	1.1270
0.15	0.1511	0.48	0.5230	0.82	1.1568
0.16	0.1614	0.49	0.5361	0.83	1.1881
0.17	0.1717	0.50	0.5493	0.84	1.2212
0.18	0.1820	0.51	0.5627	0.85	1.2562
0.19	0.1923	0.52	0.5763	0.86	1.2933
0.20	0.2027	0.53	0.5901	0.87	1.3331
0.21	0.2132	0.54	0.6042	0.88	1.3758
0.22	0.2237	0.55	0.6184	0.89	1.4219
0.23	0.2342	0.56	0.6328	0.90	1.4722
0.24	0.2448	0.57	0.6475	0.91	1.5275
0.25	0.2554	0.58	0.6625	0.92	1.5890
0.26	0.2661	0.59	0.6777	0.93	1.6584
0.27	0.2769	0.60	0.6931	0.94	1.7380
0.28	0.2877	0.61	0.7089	0.95	1.8318
0.29	0.2986	0.62	0.7250	0.96	1.9459
0.30	0.3095	0.63	0.7414	0.97	2.0923
0.31	0.3205	0.64	0.7582	0.98	2.2976
0.32	0.3316	0.65	0.7753		

Table 5 Critical Values of *t*

	One-tailed					
	.25	.10	.05	.025	.01	.005
	Two-tailed					
df	.50	.20	.10	.05	.02	.01
df = 1	1.000	3.078	6.314	12.706	31.821	63.657
2	0.816	1.886	2.920	4.303	6.965	9.925
3	0.765	1.638	2.353	3.182	4.541	5.841
4	0.741	1.533	2.132	2.776	3.747	4.604
5	0.727	1.476	2.015	2.571	3.365	4.032
6	0.718	1.440	1.943	2.447	3.143	3.707
7	0.711	1.415	1.895	2.365	2.998	3.499
8	0.706	1.397	1.860	2.306	2.896	3.355
9	0.703	1.383	1.833	2.262	2.821	3.250
10	0.700	1.372	1.812	2.228	2.764	3.169
11	0.697	1.363	1.796	2.201	2.718	3.106
12	0.695	1.356	1.782	2.179	2.681	3.055
13	0.694	1.350	1.771	2.160	2.650	3.012
14	0.692	1.345	1.761	2.145	2.624	2.977
15	0.691	1.341	1.753	2.131	2.602	2.947
16	0.690	1.337	1.746	2.120	2.583	2.921
17	0.689	1.333	1.740	2.110	2.567	2.898
18	0.688	1.330	1.734	2.101	2.552	2.878
19	0.688	1.328	1.729	2.093	2.539	2.861
20	0.687	1.325	1.725	2.086	2.528	2.845
21	0.686	1.323	1.721	2.080	2.518	2.831
22	0.686	1.321	1.717	2.074	2.508	2.819
23	0.685	1.319	1.714	2.069	2.500	2.807
24	0.685	1.318	1.711	2.064	2.492	2.797
25	0.684	1.316	1.708	2.060	2.485	2.787
26	0.684	1.315	1.706	2.056	2.479	2.779
27	0.684	1.314	1.703	2.052	2.473	2.771
28	0.683	1.313	1.701	2.048	2.467	2.763
29	0.683	1.311	1.699	2.045	2.462	2.756
30	0.683	1.310	1.697	2.042	2.457	2.750
40	0.681	1.303	1.684	2.021	2.423	2.704
50	0.679	1.299	1.676	2.009	2.403	2.678
60	0.679	1.296	1.671	2.000	2.390	2.660
70	0.678	1.294	1.667	1.994	2.381	2.648
80	0.678	1.292	1.664	1.990	2.374	2.639
90	0.677	1.291	1.662	1.987	2.368	2.632
100	0.677	1.290	1.660	1.984	2.364	2.626
z	0.674	1.282	1.645	1.960	2.326	2.576

Table 6 Critical Values of F

n_1 degrees of freedom (for numerator of F)

df	1	2	3	4	5	6	7	8	9	10	15	20	30	40	60	120	∞
1	161.4476	199.5000	215.7073	224.5832	230.1619	233.9860	236.7684	238.8827	240.5433	241.8817	245.9499	248.0131	250.0951	251.1432	252.1957	253.2529	254.3144
	4,052.1807	4,999.5000	5,403.3520	5,624.5833	5,763.6496	5,858.9861	5,928.3557	5,981.0703	6,022.4732	6,055.8467	6,157.2846	6,208.7302	6,260.6486	6,286.7821	6,313.0301	6,339.1275	6,365.5460
2	18.5128	19.0000	19.1643	19.2468	19.2964	19.3295	19.3532	19.371	19.3848	19.3959	19.4291	19.4458	19.4624	19.4707	19.4791	19.4874	19.4957
	98.5025	99.0000	99.1662	99.2494	99.2993	99.3326	99.3564	99.3742	99.3881	99.3992	99.4325	99.4492	99.4658	99.4742	99.4825	99.4908	99.4991
3	10.1280	9.5521	9.2766	9.1172	9.0135	8.9406	8.8867	8.8452	8.8123	8.7855	8.7029	8.6602	8.6166	8.5944	8.5720	8.5494	8.5264
	34.1162	30.8165	29.4567	28.7099	28.2371	27.9107	27.6717	27.4892	27.3452	27.2287	26.8722	26.6880	26.5045	26.4108	26.3165	26.2211	26.1263
4	7.7086	6.9443	6.5914	6.3882	6.2561	6.1631	6.0942	6.0410	5.9988	5.9644	5.8578	5.8025	5.7459	5.7170	5.6877	5.6581	5.6281
	21.1977	18.0000	16.6944	15.9770	15.5219	15.2069	14.9758	14.7989	14.6591	14.5459	14.1982	14.0196	13.8377	13.7454	13.6522	13.5581	13.4642
5	6.6079	5.7861	5.4095	5.1922	5.0503	4.9503	4.8759	4.8183	4.7725	4.7351	4.6188	4.5581	4.4957	4.4638	4.4314	4.3985	4.3650
	16.2582	13.2739	12.0600	11.3919	10.9670	10.6723	10.4555	10.2893	10.1578	10.0510	9.7222	9.5526	9.3793	9.2912	9.2.20	9.1118	9.0215
6	5.9874	5.1433	4.7571	4.5337	4.3874	4.2839	4.2067	4.1468	4.0990	4.0600	3.9381	3.8742	3.8082	3.7743	3.7398	3.7047	3.6689
	13.7450	10.9248	9.7795	9.1483	8.7459	8.4661	8.2600	8.1017	7.9761	7.8741	7.5590	7.3958	7.2285	7.1432	7.0567	6.9690	6.8811
7	5.5914	4.7374	4.3468	4.1203	3.9715	3.8660	3.7870	3.7257	3.6767	3.6365	3.5107	3.4445	3.3758	3.3404	3.3043	3.2674	3.2298
	12.2464	9.5466	8.4513	7.8466	7.4504	7.1914	6.9928	6.8400	6.7188	6.6201	6.3143	6.1554	5.9920	5.9084	5.8236	5.7373	5.6506
8	5.3177	4.4590	4.0662	3.8379	3.6875	3.5806	3.5005	3.4381	3.3881	3.3472	3.2184	3.1503	3.0794	3.0428	3.0053	2.9669	2.9276
	11.2586	8.6491	7.5910	7.0061	6.632	6.3707	6.1776	6.0289	5.9106	5.8143	5.5151	5.3591	5.1981	5.1156	5.03162	4.9461	4.8599
9	5.1174	4.2565	3.8625	3.6331	3.4817	3.3738	3.2927	3.2296	3.1789	3.1373	3.0061	2.9365	2.8637	2.8259	2.7872	2.7475	2.7067
	10.5614	8.0215	6.9919	6.4221	6.0569	5.8018	5.6129	5.4671	5.3511	5.2565	4.9621	4.8080	4.6486	4.5666	4.4831	4.3978	4.3116
10	4.9646	4.1028	3.7083	3.4780	3.3258	3.2172	3.1355	3.0717	3.0204	2.9782	2.8450	2.7740	2.6996	2.6609	2.6211	2.5801	2.5379
	10.0443	7.5594	6.5523	5.9943	5.6363	5.3858	5.2001	5.0567	4.9424	4.8491	4.5581	4.4054	4.2469	4.1653	4.0819	3.9965	3.9100
11	4.8443	3.9823	3.5874	3.3567	3.2039	3.0946	3.0123	2.948	2.8962	2.8536	2.7186	2.6464	2.5705	2.5309	2.4901	2.4480	2.4045
	9.6460	7.2057	6.2167	5.6683	5.3160	5.0692	4.8861	4.7445	4.6315	4.5393	4.2509	4.0990	3.9311	3.8596	3.7761	3.6904	3.6035
12	4.7472	3.8853	3.4903	3.2592	3.1059	2.9961	2.9134	2.8486	2.7964	2.7534	2.6169	2.5436	2.4663	2.4259	2.3842	2.3410	2.2962
	9.3302	6.9266	5.9525	5.4120	5.0643	4.8206	4.6395	4.4994	4.3875	4.2961	4.0096	3.8584	3.7008	3.6192	3.5355	3.4494	3.3619
13	4.6672	3.8056	3.4105	3.1791	3.0254	2.9153	2.8321	2.7669	2.7144	2.6710	2.5331	2.4589	2.3803	2.3392	2.2966	2.2524	2.2064
	9.0738	6.7010	5.7394	5.2053	4.8616	4.6204	4.4410	4.3021	4.1911	4.1003	3.8154	3.6646	3.5070	3.4253	3.3413	3.2548	3.1665
14	4.6001	3.7389	3.3439	3.1122	2.9582	2.8477	2.7642	2.6987	2.6458	2.6022	2.4630	2.3879	2.3082	2.2664	2.2229	2.1778	2.1307
	8.8616	6.5149	5.5639	5.0354	4.6950	4.4558	4.2779	4.1399	4.0297	3.9394	3.6557	3.5052	3.3476	3.2656	3.1813	3.0942	3.0051
15	4.5431	3.6823	3.2874	3.0556	2.9013	2.7905	2.7066	2.6408	2.5876	2.5437	2.4034	2.3275	2.2468	2.2043	2.1601	2.1141	2.0658
	8.6831	6.3589	5.4170	4.8932	4.5556	4.3183	4.1415	4.0045	3.8948	3.8049	3.5222	3.3719	3.2141	3.1319	3.0471	2.9595	2.8695
16	4.4940	3.6337	3.2389	3.0069	2.8524	2.7413	2.6572	2.5911	2.5377	2.4935	2.3522	2.2756	2.1938	2.1507	2.1058	2.0589	2.0096
	8.5310	6.2262	5.2922	4.7726	4.4374	4.2016	4.0259	3.8896	3.7804	3.6909	3.4089	3.2587	3.1007	3.0182	2.9330	2.8447	2.7540
17	4.4513	3.5915	3.1968	2.9647	2.8100	2.6987	2.6143	2.5480	2.4943	2.4499	2.3077	2.2304	2.1477	2.1040	2.0584	2.0107	1.9604
	8.3997	6.1121	5.1850	4.6670	4.3360	4.1015	3.9267	3.7910	3.6822	3.5931	3.3117	3.1615	3.0032	2.9205	2.8348	2.7459	2.6432

Degrees of freedom for denominator of F

Degrees of freedom for denominator of F — F distribution critical values (upper value: p = .05 level; lower value: p = .01 level)

df																	
18	4.4139	3.5546	3.1599	2.9277	2.7729	2.6613	2.5767	2.5102	2.4563	2.4117	2.2686	2.1906	2.1071	2.0629	2.0166	1.9681	1.9168
	8.2854	6.0129	5.0919	4.5790	4.2479	4.0146	3.8406	3.7054	3.5971	3.5082	3.2273	3.0771	2.9185	2.8354	2.7493	2.6597	2.5671
19	4.3807	3.5219	3.1274	2.8951	2.7401	2.6283	2.5435	2.4768	2.4227	2.3779	2.2341	2.1555	2.0712	2.0264	1.9795	1.9302	1.8780
	8.1849	5.9259	5.0103	4.5003	4.1708	3.9286	3.7653	3.6305	3.5225	3.4338	3.1533	3.0031	2.8442	2.7608	2.6742	2.5839	2.4904
20	4.3512	3.4928	3.0984	2.8661	2.7109	2.5990	2.5140	2.4471	2.3928	2.3479	2.2033	2.1242	2.0391	1.9938	1.9464	1.8963	1.8432
	8.0960	5.8489	4.9382	4.4307	4.1027	3.8714	3.6987	3.5644	3.4567	3.3682	3.0880	2.9377	2.7785	2.6847	2.6077	2.5168	2.4224
21	4.3248	3.4668	3.0725	2.8401	2.6848	2.5727	2.4876	2.4205	2.3660	2.3210	2.1757	2.0960	2.0102	1.9645	1.9165	1.8657	1.8117
	8.0166	5.7804	4.8740	4.3688	4.0421	3.8117	3.6396	3.5056	3.3981	3.3098	3.0300	2.8796	2.7200	2.6359	2.5484	2.4568	2.3615
22	4.3009	3.4434	3.0491	2.8167	2.6613	2.5491	2.4638	2.3965	2.3419	2.2967	2.1508	2.0707	1.9842	1.9380	1.8894	1.8380	1.7831
	7.9454	5.7190	4.8166	4.3134	3.9880	3.7583	3.5867	3.4530	3.3458	3.2576	2.9779	2.8274	2.6675	2.5831	2.4951	2.4029	2.3067
23	4.2793	3.4221	3.0280	2.7955	2.6400	2.5277	2.4422	2.3748	2.3201	2.2747	2.1282	2.0476	1.9605	1.9139	1.8648	1.8128	1.7570
	7.8811	5.6637	4.7649	4.2636	3.9392	3.7102	3.5390	3.4057	3.2986	3.2106	2.9311	2.7805	2.6202	2.5355	2.4471	2.3542	2.2571
24	4.2597	3.4028	3.0088	2.7763	2.6207	2.5082	2.4226	2.3551	2.3002	2.2547	2.1077	2.0267	1.9390	1.8920	1.8424	1.7896	1.7330
	7.8229	5.6136	4.7181	4.2184	3.8951	3.6667	3.4959	3.363	3.2560	3.1681	2.8887	2.7380	2.5773	2.4923	2.4035	2.3100	2.2119
25	4.2417	3.3852	2.9912	2.7587	2.6030	2.4904	2.4047	2.3371	2.2821	2.2365	2.0889	2.0075	1.9192	1.8718	1.8217	1.7684	1.7110
	7.7698	5.5680	4.6755	4.1774	3.8550	3.6272	3.4568	3.3239	3.2172	3.1294	2.8502	2.6993	2.5383	2.4530	2.3637	2.2696	2.1706
26	4.2252	3.3690	2.9752	2.7426	2.5868	2.4741	2.3883	2.3205	2.2655	2.2197	2.0716	1.9898	1.9010	1.8533	1.8027	1.7488	1.6906
	7.7213	5.5263	4.6366	4.1400	3.8183	3.5911	3.4210	3.2884	3.1818	3.0941	2.8150	2.6640	2.5026	2.4170	2.3273	2.2325	2.1327
27	4.2100	3.3541	2.9604	2.7278	2.5719	2.4591	2.3732	2.3053	2.2501	2.2043	2.0558	1.9736	1.8842	1.8361	1.7851	1.7306	1.6717
	7.6767	5.4881	4.6009	4.1056	3.7848	3.5580	3.3882	3.2558	3.1494	3.0618	2.7827	2.6316	2.4699	2.3840	2.2938	2.1985	2.0978
28	4.1960	3.3404	2.9467	2.7141	2.5581	2.4453	2.3593	2.2913	2.2360	2.1900	2.0411	1.9586	1.8687	1.8203	1.7689	1.7138	1.6541
	7.6356	5.4529	4.5681	4.0740	3.7539	3.5276	3.3581	3.2259	3.1196	3.0320	2.7530	2.6017	2.4397	2.3535	2.2629	2.1670	2.0655
29	4.1830	3.3277	2.9340	2.7014	2.5454	2.4324	2.3463	2.2783	2.2229	2.1768	2.0275	1.9446	1.8543	1.8055	1.7537	1.6981	1.6376
	7.5977	5.4204	4.5378	4.0449	3.7254	3.4995	3.3303	3.1982	3.0920	3.0045	2.7256	2.574	2.4118	2.3253	2.2344	2.1379	2.0355
30	4.1709	3.3158	2.9223	2.6896	2.5336	2.4205	2.3343	2.2662	2.2107	2.1646	2.0148	1.9317	1.8409	1.7918	1.7396	1.6835	1.6223
	7.5625	5.3903	4.5097	4.0179	3.6990	3.4735	3.3045	3.1726	3.0665	2.9791	2.7002	2.5487	2.3860	2.2992	2.2079	2.1108	2.0075
40	4.0847	3.2317	2.8387	2.6060	2.4495	2.3359	2.2490	2.1802	2.1240	2.0772	1.9245	1.8389	1.7444	1.6928	1.6373	1.5766	1.5089
	7.3141	5.1785	4.3126	3.8283	3.5138	3.2910	3.1238	2.9930	2.8876	2.8005	2.5216	2.3689	2.2034	2.1142	2.0194	1.9172	1.8061
60	4.0012	3.1504	2.7581	2.5252	2.3683	2.2541	2.1665	2.0970	2.0401	1.9926	1.8364	1.7480	1.6491	1.5943	1.5343	1.4673	1.3893
	7.0771	4.9774	4.1239	3.6490	3.3388	3.1187	2.9530	2.8233	2.7185	2.6318	2.3523	2.1978	2.0285	1.9360	1.8363	1.7263	1.6023
120	3.9201	3.0718	2.6802	2.4472	2.2899	2.1750	2.0868	2.0164	1.9588	1.9105	1.7505	1.6587	1.5543	1.4952	1.4290	1.3519	1.2539
	6.8509	4.7865	3.9491	3.4795	3.1735	2.9559	2.7918	2.6629	2.5586	2.4721	2.1915	2.0346	1.8600	1.7628	1.6557	1.5330	1.3827
∞	3.8415	2.9957	2.6049	2.3719	2.2141	2.0986	2.0096	1.9384	1.8799	1.8307	1.6664	1.5705	1.4591	1.3940	1.3180	1.2214	1.0000
	6.637	4.6073	3.7836	3.3210	3.0191	2.8038	2.6411	2.5130	2.4091	2.3227	2.0403	1.8801	1.6983	1.5943	1.4752	1.3273	1.0476

Degrees of freedom for denominator of F

p = .05 level; p = .01 level.

455

Table 7 Critical Values of Chi Square

Probability under H_0 that $\chi^2 \geq$ chi square

df	.99	.95	.90	.85	.80	.75	.50	.25	.15	.10	.05	.02	.01	.001
1	0.00016	0.00393	0.0158	0.0358	0.06418	0.1015	0.4549	1.3233	2.0722	2.7055	3.841	5.412	6.6349	10.8276
2	0.02010	0.10259	0.2107	0.3250	.4463	0.5754	1.386	2.7726	3.7942	4.6052	5.9915	7.8240	9.2103	13.8155
3	0.1148	0.3518	0.5844	0.7978	1.0052	1.2125	2.3660	4.1083	5.3170	6.2514	7.8147	9.8274	11.345	16.266
4	0.2971	0.7107	1.0636	1.3665	1.6488	1.9226	3.3567	5.3853	6.7449	7.7794	9.4877	11.6678	13.2767	18.4668
5	0.5543	1.1455	1.61031	1.9938	2.3425	2.6746	4.3514	6.6257	8.1152	9.2364	11.0705	13.3882	15.08627	20.5150
6	0.8721	1.6354	2.2041	2.6613	3.0701	3.4546	5.3481	7.8408	9.4461	10.6446	12.5916	15.0332	16.8119	22.4577
7	1.2390	2.1673	2.8331	3.3583	3.8223	4.2549	6.3458	9.0371	10.7479	12.0170	14.0671	16.6224	18.4753	24.3219
8	1.6465	2.7326	3.4895	4.0782	4.5936	5.0706	7.3441	10.2189	12.0271	13.3616	15.5073	18.1682	20.0902	26.1245
9	2.0880	3.3251	4.1682	4.8165	5.3801	5.8988	8.3428	11.3888	13.2880	14.6837	16.9190	19.6790	21.6660	27.8772
10	2.5582	3.9403	4.8652	5.5701	6.1791	6.7372	9.3418	12.5489	14.5339	15.9872	18.3070	21.6077	23.2093	29.5883
11	3.0535	4.5748	5.5778	6.3364	6.9887	7.5841	10.3410	13.7007	15.7671	17.2750	19.6751	22.6180	24.7250	31.2641
12	3.5706	5.2260	6.3038	7.1138	7.8073	8.4384	11.3403	14.8454	16.9893	18.5493	21.0261	24.0540	26.2170	32.9095
13	4.1069	5.8919	7.0415	7.9008	8.6339	9.2991	12.3398	15.9839	18.2020	19.8119	22.3620	25.4715	27.6882	34.5282
14	4.6604	6.5706	7.7895	8.6963	9.4673	10.1653	13.3393	17.1169	19.4062	21.0641	23.6848	26.7828	29.1412	36.1233
15	5.2293	7.2610	8.5468	9.4993	10.3070	11.0365	14.3389	18.2451	20.6030	22.3071	24.9958	28.2595	30.5779	37.6973
16	5.8122	7.9616	9.3122	10.3090	11.1521	11.9122	15.3385	19.3589	21.7931	23.5418	26.2962	29.6332	31.9999	39.2524
17	6.4078	8.6718	10.0852	11.1249	12.0023	12.7919	16.3382	20.4887	22.9770	24.7690	27.5871	30.9950	33.4087	40.7902
18	7.0149	9.3905	10.8649	11.946	12.8570	13.6753	17.3379	21.6049	24.1555	25.9894	28.8693	32.3462	34.8053	42.3124
19	7.6327	10.1170	11.6509	12.7727	13.7158	14.5620	18.3377	22.7178	25.3289	27.2036	30.1435	33.6874	36.1909	43.8202
20	8.2604	10.8508	12.4426	13.6039	14.5784	15.4518	19.3374	23.8277	26.4976	28.4120	31.4104	35.0196	37.5662	45.3147
21	8.8972	11.5913	13.2396	14.4393	15.4446	16.3444	20.3372	24.9348	27.6620	29.6151	32.6706	36.3434	38.9322	46.7970
22	9.5425	12.3380	14.0415	15.2788	16.3140	17.2396	21.3370	26.0393	28.8225	30.8133	33.9244	37.6595	40.2894	48.2679
23	10.1957	13.0905	14.8480	16.1219	17.1865	18.1373	22.3369	27.1413	29.9792	32.0069	35.1725	38.9683	41.6384	49.7282
24	10.8564	13.8484	15.6587	16.9686	18.0618	19.0373	23.3367	28.2412	31.1325	33.1962	36.4150	40.2703	42.9798	51.1786
25	11.5240	14.6114	16.4734	17.8184	18.9398	19.9393	24.3366	29.3389	32.2825	34.3816	37.6525	41.5661	44.3141	52.6197
26	12.1981	15.3792	17.2919	18.6714	19.8202	20.8434	25.3365	30.4346	33.4295	35.5632	38.8851	42.8558	45.6417	54.0520
27	12.8785	16.1514	18.1139	19.5272	20.7030	21.7494	26.3363	31.5284	34.5736	36.7412	40.1133	44.1400	46.9629	55.4760
28	13.5647	16.9279	18.9392	20.3857	21.5880	22.6572	27.3362	32.6205	35.7150	37.9160	41.3371	45.4188	48.2782	56.8923
29	14.2565	17.7084	19.7677	21.2468	22.4751	23.5666	28.3361	33.7109	36.8538	39.0875	42.5570	46.6927	49.5879	58.3012
30	14.9525	18.4927	20.5992	22.1103	23.3641	24.4776	29.3360	34.7997	37.9903	40.2560	43.7730	47.9618	50.8921	59.7031
40	22.1643	26.5093	29.0505	30.8563	32.3450	33.6603	39.3353	45.6160	49.2439	51.8051	55.7585	60.4361	63.6907	73.4020
50	29.7067	34.7643	37.6886	39.7539	41.4492	42.9421	49.3349	56.3336	60.3460	63.1671	67.5048	72.6133	76.1539	86.6608
60	37.4849	43.1880	46.4589	48.7587	50.6406	52.2938	59.3347	66.9815	71.3411	74.3970	79.0819	84.5799	88.3794	99.6072
70	45.4417	51.7393	55.3289	57.8443	59.8978	61.6983	69.3345	77.5767	82.2554	85.5270	90.5312	96.3875	100.4252	112.3169
80	53.5401	60.3915	64.2778	66.9938	69.2069	71.1445	79.3343	88.1303	93.1058	96.5782	101.8794	108.0693	112.3288	124.8392
90	61.7541	69.1260	73.2911	76.1954	78.5584	80.6247	89.3342	98.6499	103.9041	107.5650	113.1453	119.5485	124.1163	137.2084
100	70.6489	77.9295	82.3581	85.4406	87.9453	90.1332	99.3341	109.1412	114.6588	118.4980	124.3421	131.1417	135.8067	149.4493

REFERENCES

Adams, J.A. (1971). A closed-loop theory of motor learning. *Journal of Motor Behavior, 3*, 111–150.

Adelman, M. (1986). *A sporting time: New York City and the rise of modern athletics, 1820–70*. Urbana, IL: University of Illinois Press.

Adelman, M.L. (1983). Academicians and American athletics: A decade of progress. *Journal of Sport History, 10*, 80–106.

Ainsworth, B.E., Haskell, W.L., Herrmann, S.D., Meckes, N., Bassett, D.R., Jr., Tudor-Locke, C., Greer, J.L., Vezina, J., Whitt-Glover, M.C., & Leon, A.S. (2011). 2011 compendium of physical activities: A second update of codes and MET values. *Medicine & Science in Sports & Exercise 43* (8), 1575–1581.

American Alliance for Health, Physical Education, Recreation and Dance. (1980). *AAHPERD health related physical fitness test manual*. Reston, VA: Author.

American Association for Health, Physical Education and Recreation. (1958). *AAHPER youth fitness test manual*. Washington, DC: Author.

American College of Sports Medicine. (1994). ACSM's guidelines for exercise testing and prescription. Baltimore: Williams & Wilkins.

American Psychological Association. (2010). *Publication manual of the American Psychological Association* (6th ed.). Washington, DC: American Psychological Association.

Anderson, D. (2001). Recovering humanity: Movement, sport, and nature. *Journal of the Philosophy of Sport 28, 2*, 140–150.

Anshel, M.H., & Marisi, D.Q. (1978). Effects of music and rhythm on physical performance. *Research Quarterly, 49*, 109–115.

Anwar, M.A. (2004). From doctoral dissertation to publication: A study of 1995 American graduates in library and information sciences. *Journal of Librarianship and Information Science, 36*, 151–157.

APA Statement on Authorship of Research Papers. (1983, September 14). *Chronicle of Higher Education, 27*, 7.

Arlin, M. (1977). One-study publishing typifies educational inquiry. *Educational Researcher, 6* (9), 11–15.

Atkinson, R.F. (1978). *Knowledge and explanation in history*. Ithaca, NY: Cornell University Press.

Bain, L.L. (1989). Interpretive and critical research in sport and physical education. *Research Quarterly for Exercise and Sport, 60*, 21–24.

Baker, W.J. (1983). The state of British sport history. *Journal of Sport History, 10*, 53–66.

Baker, W.J. (1994). Press games: Sportswriters and the making of American football. *Reviews in American History, 22*, 530–537.

Barnett, V., & Lewis, T. (1978). *Outliers in statistical data*. New York: Wiley.

Bass, A. (2002). *Not the triumph but the struggle: The 1968 Olympics and the making of the black athlete*. Minneapolis: University of Minnesota Press.

Baumgartner, T.A. (1989). Norm-referenced measurement: Reliability. In M.J. Safrit & T.M. Wood (Eds.), *Measurement concepts in physical education and exercise science* (pp. 45–72). Champaign, IL: Human Kinetics.

Baumgartner, T.A., & Jackson, A.S. (1991). *Measurement for evaluation in physical education* (4th ed.). Dubuque, IA: Brown.

Bentham, J. (1970). *Introduction to the principles of morals and legislation*. London: Athalone Press.

Berelson, B. (1960). *Graduate education in the United States*. New York: McGraw-Hill.

Berg, B.L. (2001). *Qualitative research methods for the social sciences* (4th ed.). Boston: Allyn & Bacon.

Bernstein, L., Henderson, B.E., Hanisch, R., Sullivan-Halley, J., & Ross, R.K. (1994). Physical exercise and reduced risk of breast cancer in young women. *Journal of the National Cancer Institute, 86*, 1403–1408.

Berryman, J.W. (1973). Sport history as social history? *Quest, 20*, 65–73.

Berryman, J.W. (1982). Sport, health, and the rural-urban conflict: Baltimore and John Stuart Skinner's American farmer 1819-1820. *Conspectus of History, 1*, 43–61.

Berryman, J.W., & Park, R.J. (Eds.). (1992). *Sport and exercise science: Essays in the history of sports medicine*. Urbana: University of Illinois Press.

Betts, J.R. (1951). *Organized sport in industrial America*. Doctoral presentation, Columbia University, New York.

Betts, J.R. (1974). *America's sporting heritage, 1850-1950*. Reading, MA: Addison Wesley.

Betz, N.E. (1987). Use of discriminant analysis in counseling psychology research. *Journal of Counseling Psychology, 34*, 393–403.

Bhalla, J.A., & Weiss, M.R. (2010). A cross-cultural perspective of parental influence on female adolescents' achievement beliefs and behaviors in sport and school domains. *Research Quarterly for Exercise and Sport, 81*, 494–505.

Biddle, S.J.H., Markland, D., Gilbourne, D., Chatzisarantis, N.L.D., & Sparkes, A.C. (2001). Research methods in sport and exercise psychology: Quantitative and qualitative issues. *Journal of Sport Sciences, 19*, 777–809.

Blair, S.N. (1993). Physical activity, physical fitness, and health (1993 C.H. McCloy Research Lecture). *Research Quarterly for Exercise and Sport, 64*, 365–376.

Blair, S.N., Kohl, H.W., III, Paffenbarger, R.S., Clark, D.G., Cooper, K.H., & Gibbons, L.W. (1989). Physical fitness and all-cause mortality: A prospective study of healthy men and women. *JAMA, 262*, 2395–2401.

Bloom, J. (2000). *To show what an Indian can do: Sports at Native American boarding schools*. Minneapolis: University of Minnesota Press.

Bogdan, R.C., & Biklen, S.K. (2007). *Qualitative research for education: An introduction to theories and method* (5th ed.). Boston: Allyn & Bacon.

Boorman, M.A. (1990). Effect of age and menopausal status on cardiorespiratory fitness in masters women endurance athletes. Master's thesis, Arizona State University, Tempe.

Booth, D. (1997). Sports history: What can be done? *Sport, Education and Society, 2*, 191–204.

Booth, D. (1999). Sport history. In D. Booth & A. Jutel (Eds.), *Sporting Traditions, 16*, 5–15.

Booth, D. (2001). From ritual to record: Allen Guttmann's insights into modernization and modernity. *Sport History Review, 32*, 19–27.

Booth, D. (2003). Theory: Distorting or enriching sport history? *Sport History Review, 34*, 1–32.

Booth, D. (2005). *The field: Truth and fiction in sport history*. New York: Routledge.

Booth, D., & Loy, J. (1999). Sport, status, and style. *Sport History Review, 30*, 1–26.

Borg, G.A. (1962). *Physical performance and perceived exertion*. Lund, Sweden: Gleerup.

Borg, W.R., & Gall, M.D. (1989). *Educational research* (5th ed.). New York: Longman.

Borish, L.J. (1996). Do not neglect exercise nor recreation: Rural New Englanders, sport and health concerns. *Colby Quarterly, 32*, 25–35.

Bouchard, C., Shepard, R.J., & Stephens, T. (Eds.). (1994). *Physical activity, fitness, and health*. Champaign, IL: Human Kinetics.

Bourque, L.B., & Fielder, E.P. (2003). *How to conduct in-person interviews for surveys* (2nd ed.). Thousand Oaks, CA: Sage.

Boyer, C.J. (1973). *The doctoral dissertation as an informational source: A study of scientific information flow*. Metuchen, NJ: Scarecrow Press.

Briesach, E. (1994). *Historiography: Ancient, medieval, & modern* (2nd ed.). Chicago: University of Chicago Press.

Brown, J.A.C. (1954). *The social psychology of industry*. Middlesex, England: Penguin.

Brownell, S. (2001). The problems with ritual and modernization theory, and why we need Marx: A commentary on From Ritual to Record. *Sport History Review, 32,* 28–41.

Buchowski, M.S., Darud, J.L., Chen, K.Y., & Sun, M. (1998). Work efficiency during step aerobic exercise in female instructors and noninstructors. *Research Quarterly for Exercise and Sport, 69,* 82–88.

Burke, P. (1992). *History and social theory*. Ithaca, NY: Cornell University Press.

Cahn, S. (1994). *Coming on strong: Gender and sexuality in twentieth-century women's sport*. New York: The Free Press.

Campbell, D.T., & Stanley, J.C. (1963). *Experimental and quasi-experimental designs for research* (pp. 5–6). Chicago: Rand McNally.

Cardinal, B.J., & Thomas, J.R. (2005). The 75th anniversary of *Research Quarterly for Exercise and Sport*: An analysis of status and contributions. *Research Quarterly for Exercise and Sport, 76* (Suppl. 2), S122–S134.

Carlberg, C.C., Johnson, D.W., Johnson, R., Maruyama, G., Kavale, K., Kulik, C., Kulik, J.A., Lysakowski, R.S., Pflaum, S.W., & Walberg, H. (1984). Meta-analysis in education: A reply to Slavin. *Educational Researcher, 13* (4), 16–23.

Carleton, R.A., Lasater, T.M., Assaf, A.R., Feldman, H.A., McKinaly, S., & the Pawtucket Heart Health Program Writing Group. (1995). The Pawtucket Heart Health Program: Community changes in cardiovascular risk factors and projected disease risk. *American Journal of Public Health, 85,* 777–785.

Carron, A.V., Hausenblas, H.A., & Mack, D. (1996). Social influence and exercise: A meta-analysis. *Journal of Sport and Exercise Psychology, 18,* 1–16.

Carron, A.V., Widmeyer, W.N., & Brawley, L.R. (1985). The development of an instrument to assess cohesion in sport teams: The group environment questionnaire. *Journal of Sport Psychology, 7,* 244–266.

Caspersen, C.J. (1989). Physical activity epidemiology: Concepts, methods, and applications to exercise science. *Exercise and Sport Sciences Reviews, 17,* 423–474.

Cavello, D. (1981). *Muscles and morals: Organized playgrounds and urban reform*. Philadelphia: University of Pennsylvania Press.

Cheffers, J.T.F. (1973). The validation of an instrument design to expand the Flanders' system of interaction analysis to describe nonverbal interaction, different varieties of teacher behavior and pupil responses (Doctoral dissertation, Temple University, Philadelphia, 1972). *Dissertation Abstracts International, 34,* 1674A.

Chien, I. (1981). Appendix: An introduction to sampling. In L.H. Kidder (Ed.), *Selltiz, Wrightsman, and Cook's research methods in social relations* (4th ed.). New York: Holt, Rinehart & Winston.

Christina, R.W. (1989). Whatever happened to applied research in motor learning? In J.S. Skinner et al. (Eds.), *Future directions in exercise and sport science research* (pp. 411–422). Champaign, IL: Human Kinetics.

Clarke, H.H. (Ed.). (1968, December). *Physical Fitness Newsletter, 14* (4).

Clarke, H.H., & Clarke, D.H. (1970). *Research processes in physical education, recreation, and health*. Englewood Cliffs, NJ: Prentice Hall.

Cohen, D.J., & Rosenzweig, R. (2006). *Digital history: A guide to gathering, preserving, and presenting the past on the web*. Philadelphia: University of Pennsylvania Press.

Cohen, J. (1988). *Statistical power analysis for the behavioral sciences* (2nd ed.). New York: Academic Press.

Cohen, J. (1990). Things I have learned (so far). *American Psychologist, 45,* 1304–1312.

Cohen, J. (1994). The earth is round ($p < .05$). *American Psychologist, 49,* 997–1003.

Cohen, J., & Cohen, P. (1983). *Applied multiple regression in behavioral research*. New York: Holt, Rinehart & Winston.

Cooper, H., & Hedges, L.V. (Eds.). (1994). *The handbook of research synthesis*. New York: Sage Foundation.

Coorough, C., & Nelson, J.K. (1997). The dissertation in education from 1950 to 1990. *Educational Research Quarterly, 20* (4), 3–14.

Corbin, J., & Strauss, A. (2008). *Basics of qualitative research* (3rd ed.). Los Angeles: Sage.

Costill, D.L. (1985). Practical problems in exercise physiology research. *Research Quarterly for Exercise and Sport, 56,* 378–384.

Cox, R.W., & Salter, M.A. (1998). The IT Revolution and the practice of sport history: An overview and reflection on Internet research and teaching resources. *Journal of Sport History, 25,* 283–302.

Cozens, F., & Stumpf, F. (1953). *Sports in American life*. Chicago: University of Chicago Press.

Crase, D., & Rosato, F.D. (1992). Single versus multiple authorship in professional journals. *Journal of Physical Education, Recreation and Dance, 63* (7), 28–31.

Creswell, J.W. (2009). Mapping the field of mixed methods research. *Journal of Mixed Methods Research, 3,* 95–108.

Creswell, J.W. (2013). *Qualitative inquiry: Choosing among five approaches* (3rd ed.). Thousand Oaks, CA: Sage.

Creswell, J.W. (2014). *Research design: Qualitative, quantitative and mixed methods approaches* (4th ed.). Los Angeles: Sage.

Creswell, J.W., & Plano-Clark, V.L. (2006). *Designing and conducting mixed methods research*. Thousand Oaks, CA: Sage.

Cronbach, L. (1951). Coefficient alpha and the internal structure of tests. *Psychometrika, 16,* 297–334.

Crouter, S.E., Schneider, P.I., Karabulut, M., & Bassett, D.R., Jr. (2003). Validity of 10 electronic pedometers for measuring steps, distance, and energy expenditure. *Medicine & Science in Sports & Exercise 35* (8), 1455–1460.

Curran-Everett, D., & Benos, D.J. (2004). Guidelines for reporting statistics in journals published by the American Physiological Society. *Journal of Applied Physiology, 97,* 457–459.

Davidson, M.L. (1972). Univariate versus multivariate tests in repeated-measures experiments. *Psychological Bulletin, 77,* 446–452.

Day, R.A. (1983). *How to write and publish a scientific paper* (2nd ed.). Philadelphia: Institute for Scientific Information.

Day, R.A., & Gastel, B. (2006). *How to write and publish a scientific paper* (6th ed.). Westport, CT: Greenwood Press.

Dennett, D. (1991). *Consciousness explained*. Boston: Little, Brown.

Denton, J.J., & Tsai, C. (1991). Two investigations into the influence of incentives and subject characteristics on mail survey responses in teacher education. *Journal of Experimental Education, 59,* 352–366.

Denzin, N.K. (2008). The new paradigm dialog and qualitative inquiry. *International Journal of Qualitative Studies in Education, 21,* 315–325.

Drowatzky, J.N. (1993). Ethics, codes, and behavior. *Quest 45,* 22–31.

Drowatzky, J.N. (1996). *Ethical decision making in physical activity research*. Champaign, IL: Human Kinetics.

Duke, N.K. & Beck, S.W. (April 1999). Education should consider alternative formats for the dissertation. *Educational Researcher, 28(3),* 31–36.

Dulles, F.R. (1965). *A history of recreation: America learns to play*. New York: Appleton-Century-Crofts.

Dunn, A.L., Garcia, M.E., Marcus, B.H., Kampert, J.B., Kohl, H.W., III, & Blair, S.N. (1998). Six-month physical activity and fitness changes in Project Active, a randomized trial. *Medicine & Science in Sports & Exercise, 30,* 1076–1083.

Dyreson, M. (1998). *Making the American team: Sport, culture and the Olympic experience*. Urbana: University of Illinois Press.

Edwards, A.L. (1957). *Techniques of attitude and scale construction*. New York: Appleton-Century-Crofts.

Eisen, G., & Wiggins, D.K. (1994). *Ethnicity and sport in North American history and culture*. Westport, CT: Greenwood Press.

Erickson, F. (1986). Qualitative methods in research on teaching. In M.C. Wittrock (Ed.), *Handbook of research on teaching* (3rd ed., pp. 119–161). New York: Macmillan.

Evans, L., & Rees, S. (2012). An interpretation of digital humanities. In D.M. Barry (Ed.), *Understanding digital humanities*. London: Palgrave MacMillan.

Eye, A., & Mun, E. (2012). *Log-linear modeling: Concepts, interpretation, and application*. Hoboken, NJ: John Wiley & Sons.

Eyler, M.H. (1974). Objectivity and selectivity in historical inquiry. *Journal of Sport History, 1,* 63–76.

Fahlberg, L.L., & Fahlberg, L.A. (1994). A human science for the study of movement: An integration of multiple ways of knowing. *Research Quarterly for Exercise and Sport, 65,* 100–109.

Fair, J.D. (1999). *Muscletown USA: Bob Hoffman and the manly culture of York barbell*. University Park: Pennsylvania State University Press.

Farquhar, A.B., & Farquhar, H. (1891). *Economic and industrial delusions: A discourse of the case for protection*. New York: Putnam.

Feltz, D.L., & Landers, D.M. (1983). The effects of mental practice on motor skill learning and performance: A meta-analysis. *Journal of Sport Psychology, 5,* 25–57.

Festle, M.J. (1996). *Playing nice: Politics and apologies in women's sports*. New York: Columbia University Press.

Fine, M.A., & Kurdek, L.A. (1993). Reflections on determining authorship credit and authorship order on faculty–student collaborations. *American Psychologist, 48,* 1141–1147.

Fink, A. (2003). *How to sample in surveys* (2nd ed.). Thousand Oaks, CA: Sage.

Fink, A. (2009). *How to sample in surveys: A step-by-step guide* (3rd ed.). Thousand Oaks, CA: Sage.

Fischman, M.G., Christina, R.W., & Anson, J.G. (2008). Memory drum theory's C movement: Revelations from Franklin Henry. *Research Quarterly for Exercise and Sport, 79,* 312–318.

Flanders, N.A. (1970). *Analyzing teaching behavior*. Reading, MA: Addison-Wesley.

Flick, U. (2014). *An introduction to qualitative research* (5th ed.). Los Angeles: Sage.

Fowler, F.J., Jr. (2009). *Survey research methods* (4th ed.). Thousand Oaks, CA: Sage.

Franks, B.D., & Huck, S.W. (1986). Why does everyone use the .05 significance level? *Research Quarterly for Exercise and Sport, 57,* 245–249.

French, K.E. (1985). *The relation of knowledge development to children's basketball performance*. Unpublished doctoral dissertation, Louisiana State University, Baton Rouge.

French, K.E., & Thomas, J.R. (1987). The relation of knowledge development to children's basketball performance. *Journal of Sport Psychology, 9,* 15–32.

Friedenreich, C.M., Thune, I., Brinton, L.A., & Albanes, D. (1998). Epidemiologic issues related to the association between physical activity and breast cancer. *Cancer, 83,* 600–610.

Gage, N.L. (1989). The paradigm wars and their aftermath: A "historical" sketch of research on teaching since 1989. *Educational Researcher, 18* (7), 4–10.

Gall, M.D., Borg, W.R., & Gall, J.P. (1996). *Educational research: An introduction* (6th ed.). White Plains, NY: Longman Publishing.

Gall, M.D., Gall, J.P., & Borg, W.R. (2006). *Educational research: An introduction* (8th ed.). Boston: Allyn & Bacon.

Gelo, O., Braakmann, D., & Benetka, G. (2008). Quantitative and qualitative research: Beyond the debate. *Integrative Psychological and Behavioral Science, 42,* 266–290.

Gems, G.R. (1997). *Windy City wars: Labor, leisure, and sport in the making of Chicago*. Lanham, MD: Scarecrow Press.

Gill, D.L., & Deeter, T.E. (1988). Development of the Sport Orientation Questionnaire. *Research Quarterly for Exercise and Sport, 59,* 191–202.

Glaser, B.G., & Strauss, A.L. (1976). *The discovery of grounded theory*. Chicago: Aldine.

Glass, G.V. (1976). Primary, secondary, and meta-analysis. *Educational Researcher, 5,* 3–8.

Glass, G.V. (1977). Integrating findings: The meta-analysis of research. *Review of Research in Education, 5,* 351–379.

Glass, G.V., McGaw, B., & Smith, M. (1981). *Meta-analysis in social research*. Beverly Hills, CA: Sage.

Glass, G.V., & Smith, M.L. (1979). Meta-analysis of research on the relationship of class-size and achievement. *Evaluation and Policy Analysis, 1,* 2–16.

Glassford, R.G. (1987). Methodological reconsideration: The shifting paradigms. *Quest, 39* (3), 295–312.

Goetz, J.P., & LeCompte, M.D. (1984). *Ethnography and qualitative design in educational research*. Orlando, FL: Academic Press.

Goldstein, W. (1989). *Playing for keeps: A history of early baseball*. Ithaca, NY: Cornell University Press.

Goodman, R.M., Wheeler, F.C., & Lee, P.R. (1995). Evaluation of the Heart to Heart Project: Lessons from a community-based chronic disease prevention project. *American Journal of Health Promotion, 9,* 443–445.

Goodrich, J.E., & Roland, C.G. (1977). Accuracy of published medical reference citations. *Journal of Technical Writing and Communication, 7,* 15–19.

Gordon, E.S., Tucker, P., Burke, S.M., & Carron, A.V. (2013). Effectiveness of physical activity interventions for preschoolers: A meta-analysis. *Research Quarterly for Exercise and Sport, 84,* 287–294.

Gorn, E.J. (1986). *The manly art: Bare-knuckle prize fighting in America*. New York: Cornell University Press.

Gorn, E.J., & Oriard, M. (1995, March 24). Taking sports seriously. *Chronicle of Higher Education*, p. A52.

Gottschalk, L. (1969). *Understanding history: A primer of historical method*. New York: Knopf.

Green, H. (1986). *Fit for America: Health, fitness and sport in American society*. Baltimore: Johns Hopkins University Press.

Green, J.C., & Caracelli, V.J. (Eds.). (1997). *New directions for evaluation, number 74: Advances in mixed-method evaluation, the challenges and benefits of integrating diverse paradigms*. San Francisco: Jossey-Bass.

Green, K.E. (1991). Reluctant respondents: Differences between early, late, and nonresponders to a mail survey. *Journal of Experimental Education, 59,* 268–276.

Green, S.B. (1991). How many subjects does it take to do a regression analysis? *Multivariate Behavioral Research, 26,* 499–510.

Greenockle, K.M., Lee, A.M., & Lomax, R. (1990). The relation between selected student characteristics and activity patterns in required high school physical education class. *Research Quarterly for Exercise and Sport, 61,* 59–69.

Griffin, P., & Templin, T.J. (1989). An overview of qualitative research. In P.W. Darst, D.B. Zakrajsek, & V.H. Mancini (Eds.), *Analyzing physical education and sport instruction* (2nd ed., pp. 399–410). Champaign, IL: Human Kinetics.

Gruneau, R. (1983). *Class, sports, and social development*. Amherst: University of Massachusetts Press.

Guba, E.G., & Lincoln, Y.S. (1981). *Effective evaluation*. San Francisco: Jossey-Bass.

Guttmann, A. (1978). *From ritual to record: The nature of modern sport*. New York: Columbia University Press.

Guttmann, A. (1983). Recent work in European sport history, *Journal of Sport History, 10,* 35–52.

Guttmann, A. (1984). *The games must go on: Avery Brundage and the Olympic movement*. New York: Columbia University Press.

Guttmann, A. (1991). *Women's sport: A history*. New York: Columbia University Press.

Guttmann, A. (1992). *The Olympics: A history of the modern games*. Urbana: University of Illinois Press.

Guttmann, A. (2001). *From Ritual to Record: A retrospective critique. Sport History Review, 32,* 2–11.

Haase, R.F., & Ellis, M.V. (1987). Multivariate analysis of variance. *Journal of Counseling Psychology, 34,* 404–413.

Hagen, R.L. (1997). In praise of the null hypothesis statistical test. *American Psychologist, 52,* 15–24.

Halverson, L.E., Roberton, M.A., & Langendorfer, S. (1982). Development of the overarm throw: Movement and ball velocity changes by seventh grade. *Research Quarterly for Exercise and Sport, 53,* 198–205.

Hamlyn, D.W. (1990). *In and out of the black box.* Oxford, England: Basil Blackwell.

Hardy, S. (1981). The city and the rise of American sport, 1820-1920. *Exercise and Sports Sciences Reviews, 9,* 183–229.

Hardy, S. (1982). *How Boston played: Sport, recreation and community, 1865-1915.* Boston: Northeastern University Press.

Hardy, S. (1990). Adopted by all the leading clubs: Sporting goods and the shaping of leisure. In R. Butsch (Ed.), *For fun and profit: The transformation of leisure into consumption* (pp. 71–92). Philadelphia: Temple University Press.

Harris, C. (1963). *Problems in measuring change.* Madison: University of Wisconsin Press.

Harris, R.J. (1985). *A primer of multivariate statistics* (2nd ed.). Orlando, FL: Academic Press.

Hartmann, D. (2003). *Race, culture, and the revolt of the black athlete: The 1968 Olympic protests and their aftermath.* Chicago: University of Chicago Press.

Harvey, W.J., Reid, G., Bloom, G.A., Staples, K., Grizenko, N., Mbebou, V., Ter-Stepanian, M., & Joober, R. (2009). Physical activity experiences of boys with and without ADHA. *Adapted Physical Activity Quarterly, 26,* 131–150.

Harwell, M.R. (1990). A general approach to hypothesis testing for nonparametric tests. *Journal of Experimental Education, 58,* 143–156.

Haskell, W.L., Lee, I.-M., Pate, R.R., Powell, K.E., Blair, S.N., Franklin, B.A., Macera, C.A., Heath, G.W., Thompson, P.D., & Bauman, A. (2007). Physical activity and public health: Updated recommendation for adults from the American College of Sports Medicine and the American Heart Association. *Medicine & Science in Sports & Exercise, 39* (8), 1423–1434.

Hatch, A. (2006). Research: Musings of a former *QSE* editor. *International Journal of Qualitative Studies in Education, 19,* 403–409.

Hedges, L.V. (1981). Distribution theory for Glass's estimator of effect size and related estimators. *Journal of Educational Statistics, 6,* 107–128.

Hedges, L.V. (1982a). Estimation of effect size from a series of independent experiments. *Psychological Bulletin, 92,* 490–499.

Hedges, L.V. (1982b). Fitting categorical models to effect sizes from a series of experiments. *Journal of Educational Statistics, 7,* 119–137.

Hedges, L.V. (1984). Estimation of effect size under nonrandom sampling: The effects of censoring studies yielding statistically insignificant mean differences. *Journal of Educational Statistics, 9,* 61–85.

Hedges, L.V., & Olkin, I. (1980). Vote counting methods in research synthesis. *Psychological Bulletin, 88,* 359–369.

Hedges, L.V., & Olkin, I. (1983). Regression models in research synthesis. *American Statistician, 37,* 137–140.

Hedges, L.V., & Olkin, I. (1985). *Statistical methods for meta-analysis.* New York: Academic Press.

Helmstadter, G.C. (1970). *Research concepts in human behavior.* New York: Appleton-Century-Crofts.

Henderson, K.A., Bialeschki, M.D., Shaw, S.M., & Freysinger, V.J. (1996). *Both gains and gaps: Feminist perspectives on women's leisure.* State College, PA: Venture.

Henry, F.M., & Rogers, D.E. (1960). Increased response latency for complicated movements and a "memory drum" theory of neuromotor reaction. *Research Quarterly, 31,* 448–458.

Herkowitz, J. (1984). Developmentally engineered equipment and playgrounds. In J.R. Thomas (Ed.), *Motor development during childhood and adolescence.* Minneapolis, MN: Burgess.

Hill, J. (1996). British sports history: A post-modern future. *Journal of Sport History, 23,* 1–19.

Hoberman, J. (1997). *Darwin's athletes: How sport has damaged black America and preserved the myth of race.* Boston: Houghton Mifflin.

Holliman, J. (1931). *American sports (1785-1835).* Durham, NC: Seeman Press.

Hollingsworth, M.A., & Fassinger, R.E. (2002). The role of faculty mentors in the research training of counseling psychology doctoral students. *Journal of Counseling Psychology, 49* (3), 324–330.

Holt, R. (1998). Sport and history: British and European traditions. In L. Allison (Ed.), *Taking sport seriously* (pp. 7–30). Aachen, Germany: Meyer and Meyer.

Howell, C. (1998). On Metcalfe, Marx, and materialism: Reflections on the writing of sport history in the postmodern age. *Sport History Review, 29,* 96–102.

Howell, C.D. (2001). Of remembering and forgetting: *From Ritual to Record* and beyond. *Sport History Review, 32,* 12–18.

Hunter, J.E., & Schmidt, F.L. (2004). *Methods of meta-analysis: Correcting error and bias in research findings* (2nd ed.). Thousand Oaks, CA: Sage.

Hyde, J.S. (1981). How large are cognitive gender differences? A meta-analysis using v^2 and d. *American Psychologist, 36* (8), 892–901.

Iacovetta, F. & Mitchinson, W. (Eds.). (1998). *On the case: Explorations in social history.* Toronto: University of Toronto Press.

Iacovetta, F., & Mitchinson, W. (1998). Social history and case files research. In F. Iacovetta and W. Mitchinson (Eds.), *On the case: Explorations in social history.* Toronto: University of Toronto Press.

ICSSPE Bulletin, 27, Fall 1999.

Iggers, G.G. (1997). *Historiography in the twentieth century: From scientific objectivity to postmodern challenge.* Middletown, CT: Wesleyan University Press.

Jaccard, J, & Jacoby, J. (2012). *Theory construction and model-building skills.* New York: Guilford.

Jacks, P., Chubin, D.E., Porter, A.L., & Connally, T. (1983). The ABCs of ABDs: An interview study of incomplete doctorates. *Improving College and University Teaching, 31,* 74–81.

Jackson, A.W. (1978). *The twelve-minute swim as a test for aerobic endurance in swimming.* Unpublished doctoral dissertation, University of Houston.

Jacobs, D., Anderson, J., & Blackburn, H. (1979). Diet and serum cholesterol: Do zero correlations negate the relationship? *American Journal of Epidemiology, 110,* 77–87.

Jacobs, D.R., Jr., Luepker, R.V., Mittelmark, M.B., Folsom, A.R., Pirie, P.L., Mascioli, S.R., Hannan, P.J., Pechacek, T.F., Bracht, N.F., Carlaw, R.W., et al. (1986). Community-wide prevention strategies: Evaluation design of the Minnesota Heart Health Program. *Journal of Chronic Disease, 39,* 775–778.

Jalongo, M.R. (2012). Getting on the conference program and writing a practical article: Templates for success. *Early Childhood Education Journal.* doi: 10.1007/s10643-012-0533-x

Jewczyn, N. (2013). Theory guidance in social science and finance application. *International Journal of Social Science, 1,* 72–82.

Johnson, B.L., & Nelson, J.K. (1986). *Practical measurements for evaluation in physical education* (4th ed.). Minneapolis: Burgess.

Johnson, R.L. (1979). *The effects of various levels of fatigue on the speed and accuracy of visual recognition.* Unpublished doctoral dissertation, Louisiana State University, Baton Rouge.

Joint Committee on Quantitative Assessment of Research. (2008). *Citation statistics.* Berlin, Germany: International Mathematical Union.

Jones, E.R. (1988, Winter). Philosophical tension in a scientific discipline: So what else is new? *NASP–SPA Newsletter, 14* (1), 10–16.

Keating, X.D., & Silverman, S. (2004). Physical education teacher attitudes toward fitness test scale: Development and validation. *Journal of Teaching in Physical Education, 23,* 143–161.

Kendall, M.G. (1959). Hiawatha designs an experiment. *American Statistician, 13,* 23–24.

Kennedy, M.M. (1979). Generalizing from single case studies. *Evaluation Quarterly, 3,* 661–679.

Kirk, J., & Miller, M.L. (1986). *Reliability and validity in qualitative research.* Newbury Park, CA: Sage.

Kirk, R.E. (1982). *Experimental design: Procedures for the behavioral sciences* (2nd ed.). Belmont, CA: Brooks/Cole.

Kirk, R.E. (1985). *Experimental design: Procedures for the behavioral sciences* (3rd ed.). Pacific Grove, CA: Brooks.

Kirsch, G.B. (1989). *The creation of American team sports: Baseball and cricket, 1838-72*. Urbana: University of Illinois Press.

Knudson, D.V., Morrow, J.R., & Thomas, J.R. (2014). Advancing kinesiology through improved peer review. *Research Quarterly for Exercise and Sport, 85* (2), 127–135.

Kohl, H.W., III, Dunn, A.L, Marcus, B.H., & Blair, S.N. (1998). A randomized trial of physical activity interventions: Design and baseline data from Project Active. *Medicine & Science in Sports & Exercise, 30*, 275–283.

Kokkonen, J., Nelson, A.G., & Cornwell, A. (1998). Acute muscle stretching inhibits maximal strength performance. *Research Quarterly for Exercise and Sport, 69*, 411–415.

Krathwohl, D.R. (1993). *Methods of educational and social science research.* New York: Longman.

Kraus, H., & Hirschland, R.P. (1954). Minimum muscular fitness tests in school children. *Research Quarterly, 25*, 177–188.

Kretchmar, R.S. (1997). Philosophy of sport. In J.D. Massengale & R.A. Swanson (Eds.), *The history of exercise and sport science* (pp. 181–202). Champaign, IL: Human Kinetics.

Kretchmar, R.S. (2005). *Practical philosophy of sport and physical activity* (2nd ed.). Champaign, IL: Human Kinetics.

Kretchmar, R.S. (November, 2007). What to do with meaning? A research conundrum. *Quest, 59*, 373–383.

Kretchmar, R.S. (2015). Pluralistic internalism. *Journal of the Philosophy of Sport 42(1)*, 83–100.

Kretchmar, S. (2005). Jigsaw puzzles and riverbanks: Two ways of picturing our future. *Quest, 57*, 1, 171–177.

Kretchmar, S. (2008). The utility of silos and bunkers in the evolution of kinesiology. *Quest, 60*, 1, 3–12.

Kretchmar, S., & Elcombe, T. (2007). In defense of competition and winning. In W. Morgan (Ed.), *Ethics in sport* (2nd ed., pp. 181–194). Champaign, IL: Human Kinetics.

Kroll, W.P. (1971). *Perspectives in physical education*. New York: Academic Press.

Krout, J.A. (1929). *Annals of American sport*. New Haven, CT: Yale University Press.

Krueger, R.A., & Casey, M.A. (2009). *Focus groups: A practical guide for applied research* (4th ed.). Los Angeles: Sage.

Kruger, A. (1990). Puzzle solving: German sport historiography of the eighties. *Journal of Sport History, 17*, 261–277.

Kruskal, W., & Mosteller, F. (1979). Representative sampling, III: The current statistical literature. *International Statistical Review, 47*, 245–265.

Kuhn, T.S. (1970). *The structure of scientific revolutions*. Chicago: University of Chicago Press.

Kulinna, P.H., Scrabis-Fletcher, K., Kodish, S. Phillips, S., & Silverman, S. (2009). A decade of research literature in physical education pedagogy. *Journal of Teaching in Physical Education, 28*, 119–140.

Kulinna, P.H., & Silverman, S. (1999). The development and validation of scores on a measure of teachers' attitudes toward teaching physical activity and fitness. *Educational and Psychological Measurement, 59*, 507–517.

Lane, K.R. (1983). *Comparison of skinfold profiles of black and white boys and girls ages 11–13*. Unpublished master's thesis, Louisiana State University, Baton Rouge.

Lariviere, V. & Gingras, Y. (2009). The impact factor's Matthew Effect: A natural experiment in bibliometrics. *Journal of the Association for Information Science and Technology, 61*, 424–427.

Last, J.M. (1988). *A dictionary of epidemiology*. New York: Oxford University Press.

Lawrence, P.A. (2003). The politics of publication. *Nature, 422*, 259–261.

Lawson, H.A. (1993). After the regulated life. *Quest, 45* (4), 523–545.

Lee, A., & Kamler, B. (2008). Bringing pedagogy to doctoral publishing. *Teaching in Higher Education, 13*, 511–523.

Lee, I.M., & Paffenbarger, R.S., Jr. (2000). Associations of light, moderate, and vigorous intensity physical activity with longevity: The Harvard Alumni Health Study. *American Journal of Epidemiology, 151*, 293–299.

Lee, T.D. (1982). *On the locus of contextual interference in motor skill acquisition.* Unpublished doctoral dissertation, Louisiana State University, Baton Rouge.

Leech, N.L., & Onwuegbuzie, A.J. (2009). *Quality and Quantity, 43*, 265–275.

Leon, A.S., Connett, J., Jacobs, D.R., Jr., & Rauramaa, R. (1987). Leisure-time physical activity levels and risk of coronary heart disease and death: The Multiple Risk Factor Intervention Trial. *JAMA, 258*, 2388–2395.

Levine, P. (1992). *Ellis Island to Ebbetts Field: Sport and the American Jewish experience*. New York: Oxford University Press.

Levine, R.V. (1990, September/October). The pace of life. *American Scientist, 78*, 450–459.

Light, R.J., Singer, J.D., & Willett, J.B. (1994). The visual presentation and interpretation of meta-analysis. In H. Cooper and L.V. Hedges (Eds.), *The handbook of research synthesis* (pp. 439–453). New York: Russell Sage Foundation.

Lincoln, Y.S., & Guba, E.G. (1985). *Naturalistic inquiry*. Newbury Park, CA: Sage.

Lippe, G. von der. (2001). Sportification processes: Whose logic? Whose rationality? *Sport History Review, 32*, 42–55.

Locke, L.F. (1989). Qualitative research as a form of scientific inquiry in sport and physical education. *Research Quarterly for Exercise and Sport, 60*, 1–20.

Locke, L.F., Silverman, S.J., & Spirduso, W.W. (2010). *Reading and understanding research* (3rd ed.). Thousand Oaks, CA: Sage.

Locke, L.F., Spirduso, W.W., & Silverman, S.J. (2014). *Proposals that work: A guide for planning dissertations and grant proposals* (6th ed.). Los Angeles: Sage.

Looney, M.A., Feltz, C.S., & VanVleet, C.N. (1994). The reporting and analysis of research findings for within subjects designs: Methodological issues for meta-analysis. *Research Quarterly for Exercise and Sport, 65*, 363–366.

Lord, F.M. (1969). Statistical adjustments when comparing preexisting groups. *Psychological Bulletin, 72*, 336–337.

Mabley, J. (1963, January 22). Mabley's report. *Chicago American*, p. 62.

MacAloon, J. (1981). *This great symbol: Pierre de Coubertin and the origins of the modern Olympic Games*. Chicago: University of Chicago Press.

Manchester, H. (1931). *Four centuries of sport in America, 1490-1890*. New York: The Derrydale Press.

Mangan, J.A., & Park, R. (Eds.). (1987). *From "fair sex" to feminism: Sport and the socialization of women in the Industrial and Post-Industrial Eras*. London: Frank Cass.

Margerison, T. (1965, January 3). Review of *Writing Technical Reports* (by B.M. Copper). *Sunday Times*. In R.L. Weber (Compiler) & E. Mendoza (Ed.), *A random walk in science* (p. 49). New York: Crane, Russak.

Marsh, H.W., Marco, I.T., & Abçý, F.H. (2002). Cross-cultural validity of the physical self-description questionnaire. Comparison of factor structures in Australia, Spain, and Turkey. *Research Quarterly for Exercise and Sport, 73*, 257–270.

Marshall, C., & Rossman, G.B. (2011). *Designing qualitative research* (5th ed.). Los Angeles: Sage.

Martens, R. (1973, June). People errors in people experiments. *Quest, 20*, 16–20.

Martens, R. (1977). *Sport competition anxiety test*. Champaign, IL: Human Kinetics.

Martens, R. (1979). About smocks and jocks. *Journal of Sport Psychology, 1*, 94–99.

Martens, R. (1987). Science, knowledge, and sport psychology. *Sport Psychologist, 1*, 29–55.

Matt, K.S. (1993). Ethical issues in animal research. *Quest, 45* (1), 45–51.

Matthews, P.R. (1979). The frequency with which the mentally retarded participate in recreation activities. *Research Quarterly, 50,* 71–79.

Mattson, D.E. (1981). *Statistics: Difficult concepts, understandable explanations.* St. Louis, MO: Mosby.

Maxwell, J.A. (2013). *Qualitative research design* (3rd ed.). Los Angeles: Sage.

McCaughtry, N., & Rovegno, I. (2003). Development of pedagogical content knowledge: Moving from blaming students to predicting skillfulness, recognizing motor development, and understanding emotion. *Journal of Teaching in Physical Education, 22,* 355–368.

McCloy, C.H. (1930). Professional progress through research. *Research Quarterly, 1,* 63–73.

McCullagh, P., & Meyer, K.N. (1997). Learning versus correct models: Influence of model type on the learning of a free-weight squat lift. *Research Quarterly for Exercise and Sport, 68,* 56–61.

McDonald, R.P. (1999). *Test theory: A unified treatment.* Mahwah, NJ: Lawrence Erlbaum.

McLeroy, K.R., Bibeau, D., Steckler, A., & Glanz, K. (1988). An ecological perspective on health promotion programs. *Health Education Quarterly, 15,* 351–377.

McNamara, J.F. (1994). *Surveys and experiments in education research.* Lancaster, PA: Nechnomic.

McNeil, D. (2001). The spectacle of protest and punishment: Newspaper coverage of the Melksham weavers' riot of 1738. *Media History, 7,* 71–86.

McPherson, S.L. (1987). *The development of children's expertise in tennis: Knowledge structure and sport performance.* Unpublished doctoral dissertation, Louisiana State University, Baton Rouge.

McPherson, S.L. (1999). Expert–novice differences in performance skills and problem representations of youth and adults during tennis competition. *Research Quarterly for Exercise and Sport, 70,* 233–251.

McPherson, S.L., & Thomas, J.R. (1989). Relation of knowledge and performance in boys' tennis: Age and expertise. *Journal of Experimental Child Psychology, 48,* 190–211.

McPhie, W.E. (1960). Factors affecting the value of dissertations. *Social Education, 24,* 375–377, 385.

Merriam, S.B. (1988). *Case study research in education.* San Francisco: Jossey-Bass.

Merriam, S.B. (2001). *Qualitative research and case study approaches in education.* San Francisco: Jossey-Bass.

Micceri, T. (1989). The unicorn, the normal curve, and other improbable creatures. *Psychological Bulletin, 105,* 156–166.

Miles, M.B., Huberman, A.M., & Saldaña, J. (2014). *Qualitative data analysis: A methods sourcebook* (3rd ed.). Los Angeles: Sage.

Montoye, H., Kemper, H., Saris, W., & Washburn, R. (1996). *Measuring physical activity and energy expenditure.* Champaign, IL: Human Kinetics.

Mood, D.P. (1989). Measurement methodology for knowledge tests. In M.J. Safrit & T.M. Morgan, D.L. (1996). *Focus groups as qualitative research* (2nd ed.). Thousand Oaks, CA: Sage.

Morgan, W. (Ed.). (2007). *Ethics in sport* (2nd ed.). Champaign, IL: Human Kinetics.

Morgenstern, N.L. (1983). Cogito ergo sum: Murphy's refutation of Descartes. In G.H. Scherr (Ed.), *The best of* The Journal of Irreproducible Results (p. 112). New York: Workman Press.

Morland, R.B. (1958). A philosophical interpretation of the educational views held by leaders in American physical education. *Health, Physical Education and Recreation Microform Publications, 1,* (October 1949–March 1965), PE394.

Morris, J.N., Heady, J.A., Rattle, P.A.B., Roberts, C.G., & Parks, J.W. (1953). Coronary heart disease and physical activity of work. *Lancet, 2,* 1053–1057.

Morrow, D. (1983). Canadian sport history: A critical essay. *Journal of Sport History, 10,* 67–79.

Morrow, J.R., Jr., Bray, M.S., Fulton, J.E., & Thomas, J.R. (1992). Inter-rater reliability of 1987–1991 *Research Quarterly for Exercise and Sport* reviews. *Research Quarterly for Exercise and Sport, 63,* 200–204.

Morrow, J.R., & Frankiewicz, R.G. (1979). Strategies for the analysis of repeated and multiple measure designs. *Research Quarterly, 50,* 297–304.

Morse, J.M. (2003). Principles of mixed methods and multimethod research design. In A. Tashakkori & C. Teddlie (Eds.), *Handbook of mixed methods in social and behavioral research* (pp. 189–208). Thousand Oaks, CA: Sage.

Morse, J.M., & Richards, L. (2002). *Read me first for a user's guide to qualitative methods.* Thousand Oaks, CA: Sage.

Murphy, S.L., Xu, J., & Kochanek, K.D. (2013). Deaths: Final data for 2010. *National Vital Statistics Reports, 62* (24).

Nathan, D. (2003). *Saying it's so: A cultural history of the Black Sox scandal.* Champaign, IL: University of Illinois Press.

National Children and Youth Fitness Study. (1985). *Journal of Physical Education, Recreation and Dance, 56* (1), 44–90.

National Children and Youth Fitness Study II. (1987). *Journal of Physical Education, Recreation and Dance, 56* (9), 147–167.

Nauright, J. (1999). The end of sports history? From sports history to sport studies. In D. Booth & A. Jutel (Eds.), *Sporting Traditions, 16,* 5–15.

Nelson, J.K. (1988, March). Some thoughts on research, measurement, and other obscure topics. LAHPERD Scholar Lecture presented at the LAHPERD Convention, New Orleans, LA.

Nelson, J.K. (1989). Measurement methodology for affective tests. In M.J. Safrit & T.M. Wood (Eds.), *Measurement concepts in physical education and exercise science* (pp. 229–248). Champaign, IL: Human Kinetics.

Nelson, J.K., Thomas, J.R., Nelson, K.R., & Abraham, P.C. (1986). Gender differences in children's throwing performance: Biology and environment. *Research Quarterly for Exercise and Sport, 57,* 280–287.

Nelson, J.K., Yoon, S.H., & Nelson, K.R. (1991). A field test for upper body strength and endurance. *Research Quarterly for Exercise and Sport, 62,* 436–447.

Newell, K.M. (1987). On masters and apprentices in physical education. *Quest, 39* (2), 88–96.

Newell, K.M., & Hancock, P.A. (1984). Forgotten moments: A note on skewness and kurtosis as influential factors in inferences extrapolated from response distributions. *Journal of Motor Behavior, 16,* 320–335.

Nieman, D.C. (1994). Exercise, upper respiratory tract infection, and the immune system. *Medicine & Science in Sports & Exercise, 26B,* 128–139.

Nunnaly, J.C. (1978). *Psychometric theory* (2nd ed.). New York: McGraw-Hill.

Oishi, S.M. (2003). *How to conduct in-person interviews for surveys* (2nd ed.). Thousand Oaks, CA: Sage.

O'Malley, M., & Rosenzweig, R. (1997). Brand new world or blind alley? American history on the World Wide Web. *Journal of American History, 84,* 132–155.

Oriard, M. (1993). *Reading football: How the popular press created an American spectacle.* Chapel Hill: University of North Carolina Press.

Oriard, M. (2001). *King Football: Sport & spectacle in the golden age of radio & newsreels, movies & magazines, the weekly & the daily press.* Chapel Hill: The University of North Carolina Press.

Oriard, M. (2007a). *Brand NFL: Making and selling America's favorite sport.* Chapel Hill: The University of North Carolina Press.

Oriard, M. (2007b). Bowled over: Big-time college football from the sixties to the BCS era. Chapel Hill: The University of North Carolina Press.

Paffenbarger, R.S., Hyde, R., Wing, A., Lee, I., Jung, D., & Kampert, J. (1993). The association of changes in physical activity level and other lifestyle characteristics with mortality among men. *New England Journal of Medicine, 328,* 538–545.

Paffenbarger, R.S., Hyde, R.T., Wing, A.L., & Steinmetz, C.H. (1984). A natural history of athleticism and cardiovascular health. *JAMA, 252,* 491–495.

Paffenbarger, R.S., Jr., Wing, A.L., & Hyde, R.T. (1978). Physical activity as an index of heart attack risk in college alumni. *American Journal of Epidemiology, 108,* 161–175.

Park, R. (1983). Research and scholarship in the history of physical education and sport: The current state of affairs. *Research Quarterly for Exercise and Sport, 54,* 93–103.

Parratt, C.M. (1998). About turns: Reflecting on sport history in the 1990s. *Sport History Review, 29,* 4–17.

Pate, R.R., Pratt, M., Blair, S.N., Haskell, W.L., Macera, C.A., Bouchard, C.,…Wilmore, J.H. (1995). Physical activity and public health: A recommendation from the Centers for Disease Control and Prevention and the American College of Sports Medicine. *Journal of the American Medical Association, 273,* 402–407.

Patton, M.Q. (1990). *Quality evaluation and research methods.* UK: Sage Publications.

Patton, M.Q. (2002). *Qualitative research and evaluation methods* (3rd ed.). Thousand Oaks, CA: Sage.

Payne, V.G., & Morrow, J.R., Jr. (1993). Exercise and V̇O₂max in children: A meta-analysis. *Research Quarterly for Exercise and Sport, 64,* 305–313.

Paxson, F.L. (1917). The rise of sport. *Mississippi Valley Historical Review, 4,* 143–168.

Pedhazur, E.J. (1982). *Multiple regression in behavioral research: Explanation and prediction* (2nd ed.). New York: Holt, Rinehart & Winston.

Peshkin, A. (1993). The goodness of qualitative research. *Educational Researcher, 22* (2), 23–29.

Peters, E.M., & Bateman, E.D. (1983). Ultramarathon running and upper respiratory tract infections: An epidemiological survey. *South African Medical Journal, 64,* 582–584.

Phillips, M. (1999). Navigating uncharted waters: The death of sports history? In D. Booth & A. Jutel (Eds.), *Sporting Traditions, 16,* 5–15.

Phillips, M. (2006). Deconstructing sport history: A postmodern analysis. Albany: State University of New York Press.

Phillips, M.G. (2001). *Deconstructing sport history: The postmodern challenge. Journal of Sport History, 28,* 327–343.

Phillips, M.G. (2005). *Deconstructing sport history: A postmodern analysis.* Albany: State University of New York Press.

Pitney, W.A., & Parker, J. (2009). *Qualitative research in physical activity and the health professions.* Champaign, IL: Human Kinetics.

Plano-Clark, V.L., Huddleson-Casas, C.A., Churchill, S., Green, D.O., & Garrett, A.L. (2008). Mixed methods approaches in family science research. *Journal of Family Studies, 29,* 1543–1566.

Polanyi, M. (1958). *Person knowledge: Toward a post-critical philosophy.* Chicago: University of Chicago Press.

Pope, S.W. (1997a). *Patriotic games: Sporting traditions in the American imagination, 1876-1926.* London: Oxford University Press.

Pope, S.W. (1997b). *The new American sport history: Recent approaches and perspectives.* Urbana: University of Illinois Press.

Pope, S.W. (1998). Sport history: Into the 21st century. *Journal of Sport History, 25,* i–x.

Porter, A.C., & Raudenbush, S.W. (1987). Analysis of covariance: Its model and use in psychological research. *Journal of Counseling Psychology, 34,* 383–392.

Porter, A.L., Chubin, D.E., Rossini, F.A., Boeckmann, M.E., & Connally, T. (1982, September/October). The role of the dissertation in scientific careers. *American Scientist,* 475–481.

Porter, A.L., & Wolfe, D. (1975). Utility of the doctoral dissertation. *American Psychologist, 30,* 1054–1061.

Porter, D.H. (1981). *The emergence of the past: A theory of historical explanation.* Chicago: University of Chicago Press.

Punch, K.F. (2003). *Survey research: The basics.* Thousand Oaks, CA: Sage.

Punch, M. (1986). *The politics and ethics of fieldwork.* Beverly Hills, CA: Sage.

Puri, M.L., & Sen, P.K. (1969). A class of rank order tests for a general linear hypothesis. *Annals of Mathematical Statistics, 40,* 1325–1343.

Puri, M.L., & Sen, P.K. (1985). *Nonparametric methods in general linear models.* New York: Wiley.

Puska, P., Salonen, J.T., & Nissinen, A. (1983). Change in risk factors for coronary heart disease during 10 years of community intervention programme (North Karelia Project). *British Medical Journal, 287,* 1840–1844.

Rail, G. (Ed.). (1998). *Sport and postmodern times.* Albany: State University of New York Press.

Rawdon, T. Sharp, R.L., Shelley, M., & Thomas, J.R. (2012). Meta-analysis of the placebo effect in nutritional supplement studies of muscular performance. *Kinesiology Reviews, 1,* 137–148.

Reah, D. (1998). *The language of newspapers.* London: Routledge.

Regalado, S. (1998). *Viva baseball! Latin major leaguers and their special hunger.* Urbana: University of Illinois Press.

Reichardt, C.S., & Rallis, S.F. (Eds.). (1994). *The qualitative-quantitative debate: New perspectives.* San Francisco: Jossey-Bass.

Ries, L.A.G., Kosary, C.L., Hankey, B.F., Miller, B.A., Clegg, L., & Edwards, B.K. (Eds.). (1999). SEER Cancer Statistics Review, 1973–1996. Bethesda, MD: National Cancer Institute.

Riess, S.A. (1980). *Touching base: Professional baseball and American culture in the progressive era.* Westport, CT: Greenwood Press.

Riess, S.A. (1989). *City games: The evolution of American urban society and the rise of sports.* Urbana: University of Illinois Press.

Riess, S.A. (Ed.). (1998). *Sports and the American Jew.* Syracuse, NY: Syracuse University Press.

Rink, J., & Werner, P. (1989). Qualitative measures of teaching performance scale. In P. Darst, D. Zakrajsak, & P. Mancini (Eds.), *Analyzing physical education and sports instruction* (2nd ed., pp. 269–276). Champaign, IL: Human Kinetics.

Roberts, G.C. (1993). Ethics in professional advising and academic counseling of graduate students. *Quest, 45,* 78–87.

Roberts, R. (Ed.). (2000). *Pittsburgh sports: Stories from the Steel City.* Pittsburgh: University of Pittsburgh Press.

Robertson, M.A., & Konczak, J. (2001). Predicting children's overarm throw ball velocities from their developmental levels in throwing. *Research Quarterly for Exercise and Sport, 72,* 91–103.

Robinson, S., & Dracup, K. (2008). Innovative options for the doctoral dissertation in nursing. *Nursing Outlook, 56,* 174–178.

Robson, C. (2002). *Real world research: A resource for social scientists and practitioner-researchers* (2nd ed.). Malden, MA: Wiley-Blackwell.

Rockhill, B., Willett, W.C., Hunter, D.J., Manson, J.E., Hankinson, S.E., & Colditz, G.A. (1999). A prospective study of recreational physical activity and breast cancer risk. *Archives of Internal Medicine, 159,* 2290–2296.

Rosenthal, R. (1966). Sport, art, and particularity: The best equivocation. *Journal of the Philosophy of Sport, 13,* 49–63.

Rosenthal, R. (1991). Cumulating psychology: An appreciation of Donald T. Campbell. *Psychological Science, 2,* 213, 217–221.

Rosenthal, R. (1994). Parametric measures of effect size. In H. Cooper & L.V. Hedges (Eds.), *The handbook of research synthesis* (pp. 231–244). New York: Russell Sage Foundation.

Rosenzweig, R. (1983). *Eight hours for what we will: Workers and leisure in an industrial city, 1870-1920.* Cambridge, England: Cambridge University Press.

Rosenzweig, R. (2003). Scarcity of abundance? Preserving the past in a digital era. *The American Historical Review, 108,* 735–762.

Rosenzweig, R., & Blackmar, E. (1992). *The park and the people: A history of Central Park.* Ithaca, NY: Cornell University Press.

Rosnow, R.L., & Rosenthal, R. (1989). Statistical procedures and the justification of knowledge in psychological science. *American Psychologist, 44,* 1276–1284.

Rossman, G.B., & Rallis, S.F. (2012). *Learning in the field: An introduction to qualitative research* (3rd ed.). Los Angeles: Sage.

Rothman, K.J. (1986). *Modern epidemiology.* Boston: Little Brown.

Rubin, H.J., & Rubin, I.S. (2012). *Qualitative interviewing: The art of hearing data* (3rd ed.). Los Angeles: Sage.

Rudy, W. (1962). Higher education in the United States, 1862–1962. In W.W. Brickman & S. Lehrer, (Eds.), *A century of higher education: Classical citadel to collegiate colossus* (pp. 20–21). New York: Society for the Advancement of Education.

Ryan, E.D. (1970). The cathartic effect of vigorous motor activity on aggressive behavior. *Research Quarterly, 41*, 542–551.

Sachtleben, T.R., Berg, K.E., Cheatham, J.P., Felix, G.L., & Hofschire, P.J. (1997). Serum lipoprotein patterns in long-term anabolic steroid users. *Research Quarterly for Exercise and Sport, 68*, 110–115.

Safrit, M.J. (Ed.). (1976). *Reliability theory*. Washington, DC: American Alliance for Health, Physical Education and Recreation.

Safrit, M.J., Cohen, A.S., & Costa, M.G. (1989). Item response theory and the measurement of motor behavior. *Research Quarterly for Exercise and Sport, 60* (4), 325–335.

Safrit, M.J., & Wood, T.M. (1983). The health-related fitness test opinionnaire: A pilot survey. *Research Quarterly for Exercise and Sport, 54*, 204–207.

Sage, G., Dyreson, M., & Kretchmar, S. (June, 2005). Sociology, history, and philosophy in the *Research Quarterly*. *Research Quarterly Supplement, 76* (2), 88–107.

Salzinger, K. (2001, February 16). Scientists should look for basic causes, not just effects. *Chronicle of Higher Education, 157* (23), B14.

Sammarco, P.W. (2008). Journal visibility, self-citation, and reference limits: Influences in impact factor and author performance review. *Ethics in Science and Environmental Politics, 8*, 121–125.

Sammons, J.T. (1988). *Beyond the ring: The role of boxing in American society*. Urbana: University of Illinois Press.

Scheffé, H. (1953). A method for judging all contrasts in analysis of variance. *Biometrika, 40*, 87–104.

Schein, E.H. (1987). *The clinical perspective in fieldwork*. Newbury Park, CA: Sage.

Scherr, G.H. (Ed.). (1983). *The best of* The Journal of Irreproducible Results (p. 152). New York: Workman Press.

Schieffer, T.M., & Thomas, K.T. (2012). Fifteen years of promise in school-based physical activity interventions: A meta-analysis. *Kinesiology Reviews, 1*, 155–169.

Schiller, J.S., Lucas, J.W., & Peregoy, J.A. (2012). Summary health statistics for U.S. adults: National Health Interview Survey, 2011. National Center for Health Statistics. *Vital Health Stat 10* (256).

Schmidt, F.L., & Hunter, J.E. (2014). *Methods of meta-analysis: Correcting error and bias in research findings* (3rd ed.). Los Angeles: Sage.

Schmidt, R.A. (1975). A schema theory of discrete motor skill learning. *Psychological Review, 82*, 225–260.

Schmidt, R.A., & Lee, T.D. (1999). *Motor control and learning* (3rd ed.). Champaign, IL: Human Kinetics.

Schumacker, R.E., & Lomax, R.G. (2004). *A beginner's guide to structural equation modeling* (2nd ed.). Mahwah, NJ: Erlbaum.

Schutz, R.W. (1989). Qualitative research: Comments and controversies. *Research Quarterly for Exercise and Sport, 60*, 30–35.

Schutz, R.W., & Gessaroli, M.E. (1987). The analysis of repeated measures designs involving multiple dependent variables. *Research Quarterly for Exercise and Sport, 58*, 132–149.

Scrabis-Fletcher, K. (2007). *Student attitude, perception of competence, and practice in middle school physical education*. Available from ProQuest Dissertations and Theses database. (UMI No. 3285162)

Seidman, I. (2006). *Interviewing as qualitative research: A guide for researchers in education and the social sciences* (3rd ed.). New York: Teachers College Press.

Senn, A.E. (1999). *Power, politics, and the Olympic Games: A history of the power brokers, events, and controversies that shaped the Games*. Champaign, IL: Human Kinetics.

Serlin, R.C. (1987). Hypothesis testing, theory building, and the philosophy of science. *Journal of Counseling Psychology, 34*, 365–371.

Shadish, W.R., Cook, T.D., & Campbell, D.T. (2002). *Experimental and quasi-experimental designs for generalized causal inference*. Boston: Houghton Mifflin.

Shafer, R. (1980). *A guide to historical method* (3rd ed.). Homewood, IL: Dorsey Press.

Shavelson, R.J., & Towne, L. (Eds.). (2002). *Scientific research in education*. Washington, DC: National Academy Press.

Sheets-Johnstone, M. (1999). *The primacy of movement*. Amsterdam and Philadelphia: John Benjamins.

Shore, E.G. (1991, February). Analysis of a multi-institutional series of completed cases. Paper presented at Scientific Integrity Symposium. Harvard Medical School, Boston.

Shropshire, K.L. (1996). *In black and white: Race and sports in America*. New York: New York University Press.

Shulman, L.S. (2003, April). Educational research and a scholarship of education. Charles DeGarmo Lecture presented at the annual meeting of the American Educational Research Association, Chicago.

Siedentop, D. (1980). Two cheers for Rainer. *Journal of Sport Psychology, 2*, 2–4.

Siedentop, D. (1989). Do the lockers really smell? *Research Quarterly for Exercise and Sport, 60*, 36–41.

Siedentop, D., Birdwell, D., & Metzler, M. (1979, March). A process approach to measuring teaching effectiveness in physical education. Paper presented at the American Alliance for Health, Physical Education, Recreation and Dance national convention, New Orleans, LA.

Siedentop, D., Trousignant, M., & Parker, M. (1982). *Academic learning time—Physical education: 1982 coding manual*. Columbus: Ohio State University, School of Health, Physical Education, and Recreation.

Silverman, S. (2004). Analyzing data from field research: The unit of analysis issue. *Research Quarterly for Exercise and Sport, 74*, iii–iv.

Silverman, S. (Ed.). (2005). *Research Quarterly for Exercise and Sport: 75th Anniversary Issue, 76* (supplement).

Silverman, S., & Keating, X.D. (2002). A descriptive analysis of research methods classes in departments of kinesiology and physical education in the United States. *Research Quarterly for Exercise and Sport, 73*, 1–9.

Silverman, S., Kulinna, P.H., & Phillips, S.R. (2014). Physical education pedagogy faculty perceptions of journal quality. *Journal of Teaching in Physical Education, 33*, 134–154.

Silverman, S., & Manson, M. (2003). Research on teaching in doctoral programs: A detailed investigation of focus, method, and analysis. *Journal of Teaching in Physical Education, 22*, 280–297.

Silverman, S., & Solmon, M. (1998). The unit of analysis in field research: Issues and approaches to design and data analysis. *Journal of Teaching in Physical Education, 17*, 270–284.

Silverman, S., & Subramaniam, P.R. (1999). Student attitude toward physical education and physical activity: A review of measurement issues and outcomes. *Journal of Teaching in Physical Education, 19*, 97–125.

Singer, R.N., Hausenblas, H.A., & Janelle, C.M. (Eds.). (2001). *Handbook of sport psychology* (2nd ed.). New York: Wiley.

Slavin, R.E. (1984a). Meta-analysis in education: How it has been used. *Educational Researcher, 13* (4), 6–15.

Slavin, R.E. (1984b). A rejoinder to Carlberg et al. *Educational Researcher, 13* (4), 24–27.

Smith, M.L. (1980). Sex bias in counseling and psychotherapy. *Psychological Bulletin, 87*, 392–407.

Snyder, C.W., Jr., & Abernethy, B. (Eds.). (1992). *The creative side of experimentation*. Champaign, IL: Human Kinetics.

Somers, D.A. (1972). *The rise of sports in New Orleans, 1850-1900*. Baton Rouge: Louisiana State University Press.

Sparling, P.B. (1980). A meta-analysis of studies comparing maximal oxygen uptake in men and women. *Research Quarterly for Exercise and Sport, 51*, 542–552.

Spray, J.A. (1987). Recent developments in measurement and possible applications to the measurement of psychomotor behavior. *Research Quarterly for Exercise and Sport, 58*, 203–209.

Spray, J.A. (1989). New approaches to solving measurement problems. In M.J. Safrit & T.M. Wood (Eds.), *Measurement concepts in physical education and exercise science* (pp. 229–248). Champaign, IL: Human Kinetics.

Stamm, C.L., & Safrit, M.J. (1975). Comparison of significance tests for repeated measures ANOVA design. *Research Quarterly, 46,* 403–409.

Starkes, J.L., & Allard, F. (Eds.). (1993). *Cognitive issues in motor expertise.* Amsterdam: North Holland.

Steinhauser, G. (2009). The nature of navel fluff. *Medical Hypotheses, 10,* 1016.

Stock, W.A. (1994). Systematic coding for research synthesis. In H. Cooper & L.V. Hedges (Eds.), *The handbook of research synthesis.* New York: Sage Foundation.

Stodden, D., Langendorfer, S., & Roberton, M.A. (2009). The association between motor skill competences and physical fitness in young adults. *Research Quarterly for Exercise and Sport, 80,* 223–229.

Struna, N.L. (1985). In glorious disarray: The literature of American sport history. *Research Quarterly for Exercise and Sport, 56,* 151–160.

Struna, N.L. (1997). Sport history. In J.D. Massengale & R.A. Swanson (Eds.), *The history of exercise and sport science* (pp. 143–179). Champaign, IL: Human Kinetics.

Struna, N.L. (2000). Social history in sport. In J. Coakley & E. Dunning (Eds.), *Handbook of sport studies* (pp. 187–203). London: Sage.

Stull, G.A., Christina, R.W., & Quinn, S.A. (1991). Accuracy of references in the *Research Quarterly for Exercise and Sport. Research Quarterly for Exercise and Sport, 62,* 245–248.

Subramaniam, P.R., & Silverman, S. (2000). The development and validation of an instrument to assess student attitude toward physical education. *Measurement in Physical Education and Exercise Science, 4,* 29–43.

Subramaniam, P.R., & Silverman, S. (2002). Using complimentary data: An investigation of student attitude in physical education. *Journal of Sport Pedagogy, 8,* 74–91.

Subramaniam, P.R., & Silverman, S. (2007). Middle school students' attitudes toward physical education. *Teaching and Teacher Education, 22,* 602–611.

Sydnor, S. (1998). A history of synchronized swimming. *Journal of Sport History, 25,* 252–267.

Tamburrini, C.M. (2000). What's wrong with doping? In T. Tansjo & C. Tamburrini (Eds.), *Values in sport: Elitism, nationalism, and gender equality and the scientific manufacture of winners* (pp. 200–216). London: Spon Press.

Tarnas, R. (1991). *The passion of the Western mind: Understanding the ideas that have shaped our world view.* New York: Ballantine Books.

Tashakkori, A., & Teddlie, C. (1998). *Mixed methodology: Combining qualitative and quantitative approaches.* Thousand Oaks, CA: Sage.

Tashakkori, A., & Teddlie, C. (Eds.) (2003). *Handbook of mixed methods in social and behavioral research.* Thousand Oaks, CA: Sage.

Teddlie, C., & Tashakkori, A. (2009). *Foundations of mixed methods research: Integrating quantitative and qualitative approaches in the social and behavioral sciences.* Thousand Oaks, CA: Sage.

Teramoto, M., & Golding, L.A. (2009). Regular exercise and plasma lipid levels associated with the risk of coronary health disease: A 20-year longitudinal study. *Research Quarterly for Exercise and Sport, 80,* 138–145.

Tew, J., & Wood, M. (1980). *Proposed model for predicting probable success in football players.* Houston, TX: Rice University Press.

Thøgersen-Ntoumani, C., & Fox, K.R. (2005). Physical activity and mental well-being typologies in corporate employees: A mixed methods approach. *Work & Stress, 19,* 50–67.

Thomas, J.R. (1977). A note concerning analysis of error scores from motor-memory research. *Journal of Motor Behavior, 9,* 251–253.

Thomas, J.R. (1980). Half a cheer for Rainer and Daryl. *Journal of Sport Psychology, 2,* 266–267.

Thomas, J.R. (Ed.). (1983). Publication guidelines. *Research Quarterly for Exercise and Sport, 54,* 219–221.

Thomas, J.R. (Ed.). (1984). *Motor development during childhood and adolescence.* Minneapolis, MN: Burgess.

Thomas, J.R. (Ed.). (1986). Editor's viewpoint: Research notes. *Research Quarterly for Exercise and Sport, 57,* iv–v.

Thomas, J.R. (1989). An abstract for all seasons. *NASPSPA Newsletter, 14* (2), 4–5.

Thomas, J.R. (2014) Commentary—Improved data reporting in *RQES*: From volumes 45, 59, to 84. *Research Quarterly for Exercise and Sport, 85,* 446–448.

Thomas, J.R., & French, K.E. (1985). Gender differences across age in motor performance: A meta-analysis. *Psychological Bulletin, 98,* 260–282.

Thomas, J.R., & French, K.E. (1986). The use of meta-analysis in exercise and sport: A tutorial. *Research Quarterly for Exercise and Sport, 57,* 196–204.

Thomas, J.R., French, K.E., & Humphries, C.A. (1986). Knowledge development and sport skill performance: Directions for motor behavior research. *Journal of Sport Psychology, 8,* 259–272.

Thomas, J.R., & Gill, D.L. (Eds.). (1993). The academy papers: Ethics in the study of physical activity [special issue]. *Quest, 45* (1).

Thomas, J.R., Lochbaum, M.R., Landers, D.M., & He, C. (1997). Planning significant and meaningful research in exercise science: Estimating sample size. *Research Quarterly for Exercise and Sport, 68,* 33–43.

Thomas, J.R., Nelson, J.K., & Magill, R.A. (1986, Spring). A case for an alternative format for the thesis/dissertation. *Quest, 38,* 116–124.

Thomas, J.R., Nelson, J.K., & Thomas, K.T. (1999). A generalized rank-order method for non-parametric analysis of data from exercise science: A tutorial. *Research Quarterly for Exercise and Sport, 70,* 11–23.

Thomas, J.R., Salazar, W., & Landers, D.M. (1991). What is missing in $p <$ .05? Effect size. *Research Quarterly for Exercise and Sport, 62,* 344–348.

Thomas, J.R., Thomas, K.T., Lee, A.M., Testerman, E., & Ashy, M. (1983). Age differences in use of strategy for recall of movement in a large scale environment. *Research Quarterly for Exercise and Sport, 54,* 264–272.

Thomas, K.T., Keller, C.S., & Holbert, K. (1997). Ethnic and age trends for body composition in women residing in the U.S. Southwest: II. Total fat. *Medicine & Science in Sport & Exercise, 29,* 90–98.

Thomas, K.T., & Thomas, J.R. (1999). What squirrels in the trees predict about expert athletes. *International Journal of Sport Psychology, 30,* 221–234.

Thomas, R.M. (2003). *Blending qualitative and quantitative research methods in theses and dissertations.* Thousand Oaks, CA: Corwin Press.

Thomson Reuters. (2014). *Journal Citation Reports.* Retrieved from http://admin-apps.webofknowledge.com/JCR/JCR?wsid=2FIwwEcIWwo5G6DWD9I&ssid=&SID=2FIwwEcIWwo5G6DWD9I

Thune, I., Brenn, T., Lund, E., & Gaard, M. (1997). Physical activity and the risk of breast cancer. *New England Journal of Medicine, 336,* 1269–1275.

Tinberg, C.M. (1993). The relation of practice time to coaches' objectives, players' improvement, and level of expertise. Master's thesis, Arizona State University, Tempe.

Tinsley, H.E., & Tinsley, D.J. (1987). Uses of factor analysis in counseling psychology research. *Journal of Counseling Psychology, 34,* 414–424.

Todd, J. (1998). *Physical culture and the body beautiful: Purposive exercise in the lives of American women, 1800-1870.* Macon, GA: Mercer University Press.

Tolson, H. (1980). An adjustment to statistical significance: ω^2. *Research Quarterly for Exercise and Sport, 51,* 580–584.

Toothaker, L.E. (1991). *Multiple comparisons for researchers.* Newbury Park, CA: Sage.

Tuckman, B.W. (1978). *Conducting educational research* (2nd ed.). New York: Harcourt Brace Jovanovich.

Tudor-Locke, C., Williams, J.E., Reis, J.P., & Pluto, D. (2004). Utility of pedometers for assessing physical activity construct validity. *Sports Medicine, 34* (5), 281–291.

Turner, A.P., & Martinek, T.J. (1999). An investigation into teaching games for understanding effects on skill, knowledge, and game play. *Research Quarterly for Exercise and Sport, 70,* 286–296.

Uminowicz, G. (1984). Sport in a middle-class utopia: Asbury Park, New Jersey, 1871-1895. *Journal of Sport History, 11,* 51–73.

University of Chicago Press. (2010). *The Chicago manual of style* (16th ed.). Chicago: Author.

U.S. Department of Health and Human Services. *Physical activity and health: A report of the Surgeon General* (1996). Washington, DC: U.S. Department of Health and Human Services.

U.S. Department of Health and Human Services. Office of Disease Prevention and Health Promotion. (2013). *Healthy people 2020*. Washington, DC. Retrieved from www.healthypeople.gov/2020/default.aspx

U.S. Department of Health and Human Services. Physical Activity Guidelines Advisory Committee. (2008). *Physical Activity Guidelines Advisory Committee report, 2008*. Washington, DC: U.S. Department of Health and Human Services.

Vealey, R.S. (1986). The conceptualization of sport-confidence and competitive orientation: Preliminary investigation and instrument development. *Journal of Sport Psychology, 8,* 221–246.

Verbrugge, M.H. (1988). *Able-bodied womanhood: Personal health and social change in nineteenth-century Boston*. New York: Oxford University Press.

Verducci, F.M. (1980). *Measurement concepts in physical education*. St. Louis, MO: Mosby.

Vermilio, R.E., Thomas, K., Thomas, J.R., & Morrow, J.R. (2013). Good to great: Consistent PGA tour money winners—A 6-year longitudinal study—2005-2010.

Vertinsky, P. (1990). *The eternally wounded woman: Women, exercise, and doctors in the late nineteenth century*. Manchester, England: Manchester University Press.

Vertinsky, P. & McKay, S. (2004). *Disciplining bodies in the gymnasium: Memory, movement, modernism*. London: Routledge.

Vockell, E.L. (1983). *Educational research*. New York: Macmillan.

Wainer, H. (1992). Understanding graphs and tables. *Educational Researcher, 21* (1), 14–23.

Walcott, H.F. (2009). *Writing up qualitative research* (3rd ed.). Los Angeles: Sage.

Wallace, B.A. (2000). *The taboo of subjectivity: Toward a new science of consciousness*. New York: Oxford University Press.

Walvin, J. (1984). Sport, social history and the historian. *The British Journal of Sports History, 1,* 5–13.

Ward, P. (2013). Last man picked: Do mainstream historians need to play with sport historians? *International Journal of the History of Sport, 30,* 6–13.

Webb, E.J., Campbell, D.T., Schwartz, R.D., & Sechrest, L. (1966). *Unobtrusive measures: Nonreactive research in the social sciences*. Chicago: Rand McNally.

Weiss, M.R., Bredemeier, B.J., & Shewchuk, R.M. (1985). An intrinsic/extrinsic motivation scale for the youth sport setting: A confirmatory factor analysis. *Journal of Sport Psychology, 7,* 75–91.

Weiss, M.R., McCullagh, P., Smith, A.L., & Berlant, A.R. (1998). Observational learning and the fearful child: Influence of peer models on swimming skill performance and psychological responses. *Research Quarterly for Exercise and Sport, 69,* 380–394.

White, P.A. (1990). Ideas about causation in philosophy and psychology. *Psychological Bulletin, 108,* 3–18.

Whorton, J.C. (1982). *Crusaders for fitness: The history of American health reformers*. Princeton, NJ: Princeton University Press.

Wiggins, D.K. (1983a). Sport and popular pastimes: Shadow of the slavequarter. *Canadian Journal of the History of Sport and Physical Education, X,* 61–88.

Wiggins, D.K. (1983b). The play of slave children in the plantation communities of the old south, 1820-1860. *Journal of Sport History, 7,* 21–39.

Wiggins, D.K. (1986). From plantation to playing field: Historical writing on the black athlete in American sport. *Research Quarterly for Exercise and Sport, 57,* 101–116.

Wiggins, D.K. (1995). Victory for Allah: Muhammad Ali, the Nation of Islam, and American society. In E.J. Gorn (Ed.), *Muhammad Ali, the people's champ* (pp. 88–116). Urbana: University of Illinois Press.

Wiggins, D.K. (1997). *Glory bound: Black athletes in a White America*. Syracuse, NY: Syracuse University Press.

Wiggins, D.K. (2000). The African American athlete experience. In A.E. Strickland & R.E. Weems, Jr., *The African American experience: An historiographical and bibliographical guide* (pp. 255–277). Westport, CT: Greenwood Press.

Wiggins, D.K., & Miller, P.B. (2003). *The unlevel playing field: A documentary history of the African American experience in sport*. Urbana: University of Illinois Press.

Willett, W. (1990). *Nutritional epidemiology*. New York: Oxford University Press.

Winkleby, M.A., Taylor, C.B., Jatulis, D., & Fortmann, S.P. (1996). The long-term effects of a cardiovascular disease prevention trial: The Stanford Five-City Project. *American Journal of Public Health, 86,* 1773–1779.

Woods, A.M., & Lynn, S.K. (2014). One physical educator's career cycle: Strong start, great run, approaching finish. *Research Quarterly for Exercise and Sport, 85,* 68–80.

Yan, J.H. (1996). *Development of motor programs across the lifespan: Arm movement control*. Doctoral dissertation, Arizona State University, Tempe.

Yan, Z., & Cardinal, B.J. (2013). Perception of physical activity participation of Chinese female graduate students: A case study. *Research Quarterly for Exercise and Sport, 84,* 384–396.

Yin, R.K. (2003). *Case study research: Design and methods* (3rd ed.). Thousand Oaks, CA: Sage.

Young, D.C. (1996). *The modern Olympics: A struggle for revival*. Baltimore: The Johns Hopkins University Press.

Zelaznik, H.N. (1993) Ethical issues in conducting and reporting research: A reaction to Kroll, Matt, and Safrit. *Quest 45,* 62–68.

Ziv, A. (1988). Teaching and learning with humor: Experiment and replication. *Journal of Experimental Education, 57,* 5–15.

AUTHOR INDEX

SUBJECT INDEX

Note: The italicized *f* and *t* following page numbers refer to figures and tables, respectively.

Jerry R. Thomas, EdD, is professor and dean of the College of Education at the University of North Texas. Besides writing the previous editions of this book, Thomas has authored more than 200 publications, 120 of which are refereed publications with numerous contributions in research methods. In 1999 he was named the C.H. McCloy Lecturer for his production of research throughout his career. Thomas has been editor in chief of *Research Quarterly for Exercise and Sport* and a reviewer for most major research journals in kinesiology and numerous journals in psychology. He has also served as president of the National Academy of Kinesiology, American Kinesiology Association, AAHPERD Research Consortium, and North American Society for Psychology of Sport and Physical Activity. In 1990 he was named an AAHPERD Alliance Scholar and in 2003 was named a NASPSPA Distinguished Scholar based on lifetime achievement in research. Thomas received an honorary doctorate of science from his undergraduate institution, Furman University, in the spring of 2015.

Courtesy of Jerry Thomas

Jack K. Nelson, EdD, is professor emeritus in the department of kinesiology at Louisiana State University. Nelson conducted and published research and taught research methods for 35 years. He has been an adviser on more than 50 doctoral dissertations and more than 50 master's theses. He has more than 80 publications and has served as editor of research publications. A fellow in the Research Consortium, he has been a member of AAHPERD, the American Educational Research Association, and the American College of Sports Medicine. He has also served as president of the Association for Research, Administration, Professional Councils and Societies (now AAALF) and vice president of AAHPERD.

Courtesy of Jack Nelson

Stephen Silverman, EdD, has taught and written about research methods for more than 30 years. He is a professor of education at Teachers College at Columbia University and has conducted research on teaching in physical education focusing on how children learn motor skill and develop attitudes. He has published more than 75 research articles in addition to many books and book chapters. Silverman is a fellow and past president of the National Academy of Kinesiology and the American Alliance for Health, Physical Education, Recreation and Dance (AAHPERD) Research Consortium and a fellow of the American Educational Research Association (AERA). A former coeditor of the *Journal of Teaching in Physical Education* and editor in chief of the *Research Quarterly for Exercise and Sport*, Silverman was an AERA Physical Education scholar lecturer, a Research Consortium scholar lecturer, and Weiss lecturer for AAHPERD.

Courtesy of Stephen Silverman